D1084283

Green Chemistry for Dyes Removal from Wastewater

Scrivener Publishing
100 Cummings Center, Suite 541J
Beverly, MA 01915-6106

Publishers at Scrivener
Martin Scrivener(martin@scrivenerpublishing.com)
Phillip Carmical (pcarmical@scrivenerpublishing.com)

Green Chemistry for Dyes Removal from Wastewater

Research Trends and Applications

Edited by

Sanjay K. Sharma, FRSC

Scrivener
Publishing

WILEY

Co-published by John Wiley & Sons, Inc. Hoboken, New Jersey, and Scrivener Publishing LLC, Salem, Massachusetts.
Published simultaneously in Canada.

For general information on our other products and services or for technical support, please contact our Customer Care Department within the United States at (800) 762-2974, outside the United States at (317) 572-3993 or fax (317) 572-4002.

Wiley also publishes its books in a variety of electronic formats. Some content that appears in print may not be available in electronic formats. For more information about Wiley products, visit our web site at www.wiley.com.

For more information about Scrivener products please visit www.scrivenerpublishing.com.

Cover design by Russell Richardson

Library of Congress Cataloging-in-Publication Data:

ISBN 978-1-118-72099-8

Printed in the United States of America

10 9 8 7 6 5 4 3 2 1

This book is for Kunal – Kritika....my twin angels on their 15th Birthday, with love.

Contents

Preface

Writing a preface for a book always has been a challenge as things are to be looked upon not only from the eyes of an editor, but also from a reader's perception and expectations; all the while keeping in mind not to do any injustice to the zeal of a contributor who has worked so hard to pen the text.

"Green Chemistry" two decade's old philosophy, has been attracting the attention of scientists worldwide. Academicians as well as industrialists are equally interested in this *new* stream of chemical science. Researchers, all over the world, are conducting active research in different fields of engineering, science and technology by adopting green chemistry principles and methodologies to devise new processes with a view towards helping, protecting, and ultimately saving the environment of our planet from further anthropogenic interruptions and damage. Achieving sustainability and renewability of resources is the basic spirit of green chemistry; it inspires us to try alternative "green" approaches in place of traditional "gray" practices in everyday industrial and scientific activities.

Water pollution is a matter of great concern. It's quality and potability is equally important for both domestic purposes and industrial needs. But, at the same time, industrial effluents pollute the available water resources. Dyes, as one of the pollutants, cause various serious health hazards and socioeconomic problems. It spoils the "productivity" of soil; which in turn may be the reason for other related issues, especially in developing countries. Removal of dyes from water or wastewater is therefore an important task. But, removing dyes at a cost to the environment should be avoided when considering which technique to use. So, the far important challenge is to make a removal technique sufficiently "green."

Water pollution is often discussed with respect to various pollutants and their treatments, but water pollution due to the presence of synthetic dyes has not been discussed sufficiently in the literature. So, the treatment of wastewater produced from industries using dyes (directly or indirectly)

xiii

has tremendous scope worldwide. That is why dye removal is an important issue which needs to be addressed seriously.

The chapters in this book are the outcome of the scholarly writing of researchers of international repute with stellar credentials, who have tried to present an overview of the problem and its solution from different angles. These problems and solutions are presented in a genuinely holistic way using valuable research-based text from world-renowned researchers. Discussed herein are various promising techniques to remove dyes, including the use of nanotechnology, ultrasound, microwave, catalysts, biosorption, enzymatic treatments, advanced oxidation processes, etc., all of which are "green." The book contains eleven chapters, all of which focus on the theme of green chemistry and discuss tools and techniques which are eco-friendly, non-hazardous and, moreover, low waste generating.

The textile industry produces a large amount of dye effluents which are highly toxic as they contain a large number of metal complex dyes. The use of synthetic chemical dyes in various industrial processes, including paper and pulp manufacturing, plastics, dyeing of cloth, leather treatment and printing, has increased considerably over the last few years, resulting in the release of dye-containing industrial effluents into the soil and aquatic ecosystems. The textile industry generates highly polluting wastewaters and their treatment is a very serious problem due to high total dissolved solids (TDS), presence of toxic heavy metals, and the non-biodegradable nature of the dyestuffs present in the effluent. There are many processes available for the removal of dyes by conventional treatment technologies including biological and chemical oxidation, coagulation and adsorption, but they cannot be effectively used individually. Different types of dyes, their working and methodologies and various physical, chemical and biological treatment methods employed so far are comprehensively discussed in Chapter 1.

Adsorption is widely acknowledged as the most promising and efficient method because of its low capital investment, simplicity of design, ease of operation, insensitivity to toxic substances and ability to remove pollutants even from diluted solutions. In recent years, nanotechnology has introduced a myriad of novel nanomaterials that can have promising outcomes in environmental cleanup and remediation. Particularly, carbon-based nanomaterials such as carbon nanotubes and graphene are being intensively studied as new types of adsorbents for removal of toxic pollutants from aquatic systems. This extraordinary interest stems from their unique morphology, nanosized scale and novel physicochemical properties. Thus, Chapter 2 focuses on the use of nanotechnology in the treatment of dye removal.

Textile dyeing industries expend large volumes of water, which is ultimately discharged with intense color, chemical oxygen demand (COD), suspended/dissolved solids and recalcitrant material as unfixed dye residuals and spent auxiliaries. A typical reactive dyebath effluent contains 20–30% of the input dye mass (1500–2200 mgL^{-1}) and traces of heavy metals (i.e., cobalt, chromium and copper) that arise from the use of metal-complex azo dyes. The challenge to destroy dye residuals in biotreated wastewater effluents seems to be resolved by the introduction of advanced oxidation processes (AOP), whereby highly reactive hydroxyl radicals are generated chemically, photochemically and/or by radiolytic/sonolytic means. Hence, AOPs not only offer complete decolorization of aqueous solutions without the production of huge volumes of sludge, but also promise a considerable degree of mineralization and detoxification of the dyes and their oxidation/hydrolysis byproducts. The potential of ultrasound as an AOP is based on cavitation phenomenon, i.e., the formation, growth and implosive collapse of acoustic cavity bubbles in water and the generation of local hot spots with very extreme temperatures and pressures. Application of AOPs in dye removal is comprehensively discussed in Chapters 3 and 7.

The heterogeneous photocatalysis process has shown huge potential for water and wastewater treatment over the last few decades. Chapter 4 summarizes the photocatalytic oxidation process for dye degradation under both UV and visible light, application of solar light and solar photoreactor in dye degradation, and then finally discusses the dependence of different parameters (pH, photocatalyst loading, initial dye concentration, electron scavenger, light intensity) on dye degradation.

Several technologies have been developed to treat dye-containing effluents (DCEFs) such as coagulation-flocculation, filtration, sedimentation, precipitation-flocculation, electrocoagulation-electroflotation, biodegradation, photocatalysis, oxidation, electrochemical treatment, membrane separation, ion-exchange, incineration, irradiation, advanced oxidation, bacterial decolorization, electrokinetic coagulation and adsorption on activated carbon. From an industrial viewpoint, no single process provides adequate treatment, being that significant reduction of expenses and enhancement of dye removal can be achieved by the combination of different methods in hybrid treatments. "Biosorption" can be employed to treat DCEFs because it combines the advantages of adsorption with the use of natural, low-cost, eco-friendly and renewable biosorbents. Biosorption of organic dyes and related research opportunities and challenges are beautifully discussed in length in Chapters 5 and 8.

Laccase + Peroxidases

The enzymatic process using ligninolytic enzymes, such as laccases and peroxidases, is a relatively new emerging technology for the degradation of xenobiotics, including synthetic dyes in textile wastewater. This unique process employs a hybrid of chemical and biological oxidation using a combination of crude or purified enzymes from plant materials or fungal cultures as a biocatalyst and dissolved molecular oxygen or hydrogen peroxide as a chemical oxidant. This enzymatic process has a number of advantages over conventional physical, chemical and biological processes. Chapter 6 provides a comprehensive literature review on the enzymatic treatment of various synthetic dyes and discusses the recent progress and challenges associated with this technology. In addition, the fungal treatment of synthetic dyes and contaminated effluents, as well as the enzymology of the key ligninolytic enzymes, are covered in this chapter to explore the important roles of fungal enzymes in synthetic dye decolorization.

Adsorption is one of the best treatment methods due to its flexibility, simplicity of design, and insensitivity to toxic pollutants. Recently, clay and its modified forms have been used as adsorbents, and there has been an upsurge of interest in the interactions between dyes and clay particles. Clay may serve as an ideal adsorbent because of its low cost. It has relatively large specific surface area, excellent physical and chemical stability, and other advantageous structural and surface properties. Use of clay (especially three-layer clays) as adsorbent has been elaborately presented by Tolga Depci and Mehmet S. Çelik in Chapter 9.

Chapter 10 is about non-conventional adsorbents including clays, siliceous materials, zeolites, agricultural solid wastes, industrial byproducts, peat, chitin and chitosan, biomass, starch-based derivatives and miscellaneous adsorbents. Their effectiveness as an alternative green approach for the removal of dyes from wastewater and industrial effluents is discussed.

Hen feather is an abundantly available waste material found at poultry houses. It possesses marvelous and proficient structures, which are flexible as well as strong. Hen feather is composed of keratin and is biochemically similar to the substance responsible for creating the fur of mammals, scales of reptiles, horns of animals and fingernails of humans.

It is now well established that hen feather can be used as a potential adsorbent for the removal of hazardous pollutants. Before the year 2006, the use of hen feather as adsorbent was limited to the removal of metal ions only. However, in an innovative initiative first made by Alok Mittal and Jyoti Mittal, it was found that hen feather can also be exploited as a dye scavenger for wastewater. Chapter 11 summarizes the results of the removal of dye contaminants from water using hen feather as an adsorbent. The chapter provides comparable consequences of the effects of various parameters influencing the adsorption, various adsorption isotherms, kinetics, etc., of the developed dye removal processes.

The main outcome of reading this book will be that the reader is going to have a holistic view of the immense potential and ongoing research in dye removal by green chemistry, and its close connection with modern research and engineering applications. Furthermore, this book can be used as an important platform to inspire researchers in any related fields to develop greener processes for important techniques for use in several fields.

I gratefully acknowledge all the contributors of this book, without whom these valuable chapters could not have been completed. I express my highest gratitude and thankfulness to all of them.

Sanjay K. Sharma, FRSC
Jaipur, India
1st January 2015

Acknowledgements

When you complete a task and take time to rewind your journey and relive it through memories, you find some smiling and encouraging faces that have motivated you to complete the task with untiring efforts to your full ability. Such smiling faces remove the pain of stress which we occasionally face during any journey and encourage us to "Go ahead." They deserve a special mention and gratitude, love and affection.

It is time for me to express my feelings about my friends, colleagues, supporters and well- wishers and to let them know that I was so fortunate to have them and their valuable cooperation during the writing of this book, *Green Chemistry for Dye Removal from Wastewater: Research Trends and Applications.*

First of all I want to express my special thanks to all esteemed contributors of this book, who deserve special mention for contributing their writings, without which this book would not have been possible.

I deeply acknowledge my parents, Dr. M.P. Sharma and Mrs. Parmeshwari Devi, for their never-ending encouragement, moral support and blessings.

My wife Dr. Pratima Sharma deserves the highest appreciation for being beside me all the way and encouraging me in every hour of crisis. I appreciate her patience over the course of this book.

I also wish to thank Mr. Amit Agarwal and Mr. Arpit Agarwal (Vice Chairpersons, JECRC University, Jaipur) for their never ending support and encouragement, Prof. Victor Gambhir (President, JECRC University, Jaipur), Prof. J.K. Sharma (Pro-President, JECRC University, Jaipur), Prof. R.N. Prasad (Dean, School of Sciences) and Prof. D.P. Mishra (Registrar, JECRC University, Jaipur) for their appreciation and guidance.

My kids Kunal and Kritika always deserve special mention as they are my best companions, who energize me to work with a refreshed mood and renewed motivation.

Special thanks go to Martin and his team behind this publication, without whose painstaking efforts this work could not have been completed in a timely manner.

I am also thankful to many others whose names I have not been able to mention but whose association and support has not been less in any way.

About the Editor

 **Prof. (Dr.) Sanjay K. Sharma** is a very well-known author and editor of many books, research journals and hundreds of articles over the last twenty years.

Presently Prof. Sharma is working as Professor and Head of the Department of Chemistry, JECRC University, Jaipur, India, where he is teaching Engineering Chemistry and Environmental Chemistry to B. Tech Students; Green Chemistry, Spectroscopy and Organic Chemistry to undergraduate and post-graduate students; and pursuing his research interest in the domain of Green Chemistry with special reference to Water Pollution, Corrosion Inhibition and Biopolymers.

Dr. Sharma has had 16 books published on Chemistry by national-international publishers and over 61 research papers of national and international repute to his credit.

He has also been appointed as a Series Editor by Springer, UK, for their prestigious book series "Green Chemistry for Sustainability," where he has been involved in editing 14 different titles by various international contributors so far. Dr. Sharma is also serving as Editor-in-Chief for the *RASAYAN Journal of Chemistry*

He is a Fellow of the Royal Society of Chemistry (UK), member of the American Chemical Society (USA), and International Society for Environmental Information Sciences (ISEIS, Canada) and is also a lifetime member of various international professional societies including the

International Society of Analytical Scientists, Indian Council of Chemists, International Congress of Chemistry and Environment, Indian Chemical Society, etc.

drsanjay1973@gmail.com
sk.sharmaa@outlook.com

Removal of Organic Dyes from Industrial Effluents: An Overview of Physical and Biotechnological Applications

Mehtap Ejder-Korucu[1], Ahmet Gürses[*,2], Çetin Doğar[3], Sanjay K. Sharma[4] and Metin Açıkyıldız[5]

[1]*Kafkas University, Faculty of Science and Arts, Department of Chemistry, Kars, Turkey*
[2]*Ataturk University, K.K. Education Faculty, Department of Chemistry, Erzurum, Turkey*
[3]*Erzincan University, Education Faculty, Department of Science Education, Erzincan, Turkey*
[4]*Green Chemistry & Sustainability Research Group, Department of Chemistry, JECRC University, Jaipur, India*
[5]*Kilis 7 Aralık University, Faculty of Science and Arts, Department of Chemistry, Kilis, Turkey*

Abstract

The textile industry produces a large amount of dye effluents, which are highly toxic as they contain a large number of metal complex dyes. The use of synthetic chemical dyes in various industrial processes, including paper and pulp manufacturing, plastics, dyeing of cloth, leather treatment and printing has increased considerably over the last few years, resulting in the release of dye-containing industrial effluents into the soil and aquatic ecosystems. The textile industry generates highly polluted wastewater and its treatment is a very serious problem due to high total dissolved solids (TDS), the presence of toxic heavy metals and the non-biodegradable nature of the dyestuffs present in the effluents. There are many processes available for the removal of dyes by conventional treatment technologies including biological and chemical oxidation, coagulation and adsorption, but they cannot be effectively used individually.

Many approaches, including physical, chemical and/or biological processes have been used in the treatment of industrial wastewater containing dye, but such

---

*Corresponding author: ahmetgu@yahoo.com

Sanjay K. Sharma (ed.) Green Chemistry for Dyes Removal from Wastewater, (1–34)

methods are often very costly and not environmentally safe. Furthermore, the large amount of sludge generated and the low efficiency of treatment with respect to some dyes have limited their use.

Keywords: Natural dyes, acid dyes, disperse dyes, cationic dyes, adsorption, membrane filtration, ion exchange, irradiation, electrokinetic coagulation, aerobic and anaerobic degradation

1.1   Introduction

Water, which is one of the abundant compounds found in nature, covers approximately three-fourths of the surface of the earth. Over 97% of the total quantity of water is in the oceans and other saline bodies of water and is not readily available for our use. Over 2% is tied up in polar ice caps and glaciers and in atmosphere and as soil moisture. As an essential element for domestic, industrial and agricultural activities, only 0.62% of water found in fresh water lakes, rivers and groundwater supplies, which is irregularly and non-uniformly distributed over the vast area of the globe, is accessible [1].

A reevaluation of the issue of environmental pollution made at the end of the last century has shown that wastes such as medicines, disinfectants, contrast media, laundry detergents, surfactants, pesticides, dyes, paints, preservatives, food additives, and personal care products which have been released by chemical and pharmaceutical industries, are a severe threat to the environment and human health on a global scale [2]. The progressive accumulation of more and more organic compounds in natural waters is mostly a result of the development of chemical technologies towards organic synthesis and processing. The population explosion and expansion of urban areas have had an increased adverse impact on water resources, particularly in regions in which natural resources are still limited. Currently, water use or reuse is a major concern which needs a solution. Population growth leads to a significant increase in default volumes of wastewater, which makes it an urgent imperative to develop effective and low-cost technologies for wastewater treatment [3].

Especially in the textile industry, effluents contain large amounts of dye chemicals which may cause severe water pollution. Also, organic dyes are commonly used in a wide range of industrial applications. Therefore, it is very important to reduce the dye concentration of wastewater before discharging it into the environment. Discharging large amounts of dyes into water resources, organics, bleaches, and salts, can affect the physical and chemical properties of fresh water. Dyes in wastewater that can obstruct light penetration and are highly visible, are stable to light irradiation and heat and also toxic to microorganisms. The removal of dyes is a very complex process due to their structure and synthetic origins [4].

Dyes that interfere directly or indirectly in the growth of aquatic organisms are considered hazardous in terms of the environment. Nowadays a growing awareness has emerged on the impact of these contaminants on ground water, rivers, and lakes [5–8].

The utilization of wastewater for irrigation is an effective way to dispose of wastewater [9]. Although various wastewater treatment methods including physical, chemical, and physicochemical have been studied, in recent years a wide range of studies have focused on biological methods with some microorganisms such as fungi, bacteria and algae [10]. The application of microorganisms for dye wastewater removal offers considerable advantages which are the relatively low cost of the process, its environmental friendliness, the production of less secondary sludge and completely mineralized end products which are not toxic [11]. Numerous researches on dye wastewater removal have been conducted which have proven the potential of microorganisms such as *Cunninghamella elegans* [12], *Aspergillus nigerus* [13], *Bacillus cereus* [14], *Chlorella sp.* [15] and also *Citrobacter sp.* [16,17].

1.1.1 Dyes

A dye or a dyestuff is usually a colored organic compound or mixture that may be used for imparting color to a substrate such as cloth, paper, plastic or leather in a reasonably permanent fashion. The dye that is generally described as a colored substance should have an affinity for the substrate or should fix itself on the substrate to give it a permanent colored appearance, but all the colored substances are not the dye [18,19]. Unlike many organic compounds, the dyes which contain at least one chromophore group and also a conjugated system and absorb light in the visible spectrum (400–700 nm) and exhibit the resonance of electrons, possess special colors [20].

The relationships between wavelength of visible and color absorbed/ observed [21] are given on Table 1.1.

In general, a small amount of dyes in aqueous solution can produce a vivid color because they have high molar extinction coefficients. Color can be quantified by spectrophotometry (visible spectra), chromatography (usually high performance liquid, HPLC) and high performance capillary electrophoresis [19].

Table 1.1 Wavelengths of light absorption versus the color of organic dyes.

Wavelength range absorbed (nm)	Color absorbed	Color observed
400–435	Violet	Yellow-Green
435–480	Blue	Yellow
480–490	Green-Blue	Orange
490–500	Blue-Green	Red
500–560	Green	Purple
560–580	Yellow-Green	Violet
580–595	Yellow	Blue
595–605	Orange	Green-Blue
605–700	Red	Blue-Green

With regard to their solubility, organic colorants fall into two classes, dyes and pigments. The key distinction is that dyes are soluble in water and/or an organic solvent, while pigments are insoluble in both types of liquid media. Dyes are used to color substrates to which they have a specific affinity, whereas pigments can be used to color any polymeric substrate by a mechanism quite different than that of dyes [22,21].

1.1.2   Historical Development of Dyes

Humans discovered that certain roots, leaves, or bark could be manipulated, usually into a liquid form, and then used to dye textiles. They used these techniques to decorate clothing, utensils, and even the body, as a religious and functional practice. Records and cloth fragments dating back over 5000 years ago indicate intricate dyeing practices. Certain hues have historical importance and denote social standing [23]. The dye made from the secretions of shellfish, which is a clear fluid that oxidizes when exposed to the air, was used to produce a red to bluish purple. This dye was difficult to create and used only on the finest garments; hence it became associated with aristocrats and royalty [23]. Until the middle of the last century most of the dyes were derived from plants or animal sources by long and elaborate processes. Ancient Egyptian hieroglyphs contain a thorough description of the extraction of natural dyes and their application in dyeing [18]. In the past, only organic matter was available for use in making dyes. Today, there are numerous options and methods for the colorization of textiles. While today's methods capitalize on efficiency, there is question as to whether the use of chemicals is harmful to the environment.

In 1856, Sir William Perkin discovered a dye for the color mauve, which was the first synthetic dye. The method related to the dyeing of this color using coal and tar led to many scientific advances and the development of synthetic dyes [24,25].

Initially the dye industry was based on the discovery of the principal that dye chromogens associated with a basic arrangement of atoms were responsible for the color of a dye. Essentially, apart from one or two notable exceptions, all the dye types used today were discovered in the 1800s. The discovery of reactive dyes in 1954 and their commercial launch in 1956 heralded a major breakthrough in the dyeing of cotton; intensive research into reactive dyes followed over the next two decades and, indeed, is still continuing today. The oil crisis in the early 1970s, which resulted in a steep increase in the prices of dyestuff, created a driving force for more low-cost dyes, both by improving the efficiency of the manufacturing processes and by replacing tinctorially weak chromogens, such as anthraquinone, with tinctorially stronger chromogens, such as (heterocyclic) azo and benzodifuranone [26,27].

1.1.3   Natural Dyes

Natural dyes which are obtained from plants, insects/animals and minerals are renewable and sustainable bioresource products with minimum environmental impact. They have been known since antiquity for their use in coloring of textiles, food substrate, natural protein fibers like wool, silk and cotton, and leather as well as food ingredients and cosmetics [28–32].

Also, natural dyes are known for their use in dye-sensitized solar cells [33], histological staining [34], as a pH indicator [35] and for several other disciplines [36,37].

Over the last few decades, there has been increasing attention on various aspects of natural dye applications, and extensive research and development activities in this area are underway worldwide [29].

1.2   Classification of Dyes

Dyes may be classified according to their chemical structures and their usage or application methods. Dyes have different chemical structures derived from aromatic and hetero-aromatic compounds, and their chromophor and auxochrom groups mainly differ [18].

The most appropriate system for the classification of dyes is by chemical structure, which has many advantages. First, it readily identifies dyes as belonging to a group that has characteristic properties, for example, azo dyes (strong, good all-round properties, low-cost) and anthraquinone dyes (weak, expensive). Second, there are a number of manageable chemical groups. Most importantly, it is the classification used most widely by both the synthetic dye chemist and technologist. Thus, both chemists and technologists can readily identify with phrases such as an azo yellow, an anthraquinone red, and a phthalocyanine blue [27].

The application classification of dyes arranged according to the C. I. (Color Index) is given in Table 1.2, which includes the principal substrates, the methods of application, and the representative chemical types for each application class [27]. Although not shown in Table 1.2, dyes are also used in high-tech applications, such as in medical, electronics, and especially the nonimpact printing technologies [27,38].

Acid Dyes, which are water-soluble anionic dyes, are applied to nylon, wool, silk and modified acrylics. They are also used to some extent for paper, leather, inkjet printing, food, and cosmetics.

Direct Dyes are water-soluble anionic dyes. When dyed from aqueous solution in the presence of electrolytes, they are substantive to, i.e., have high affinities for cellulosic fibers. Their principal use is in the dyeing of cotton and regenerated cellulose, paper, leather, and, to a lesser extent, nylon. Most of the dyes in this class are polyazo compounds, along with some stilbenes, phthalocyanines, and oxazines. Treatments applied to the dyed material to improve wash fastness properties include chelation with salts of metals, which are usually copper or chromium, and treatment with formaldehyde or a cationic dye-complexing resin.

Azoic Dyes are applied via combining two soluble components impregnated in the fiber to form an insoluble color molecule. These dye components, which are sold as paste-type dispersions and powders, are chiefly used for cellulosic fibers, especially cotton. Dye bath temperatures of 16–27°C (60–80°F) are generally used to make the shade [39].

Disperse Dyes, which are substantially water-insoluble nonionic dyes for application to hydrophobic fibers from aqueous dispersion, are used predominantly on polyester and to a lesser extent on nylon, cellulose, cellulose acetate, and acrylic fibers. Thermal transfer printing and dye diffusion thermal transfer (D_2T_2) processes for electronic photography represent rich markets for selected members of this class.

Sulfur Dyes are used primarily for cotton and rayon. The application of sulfur dyes requires carefully planned transformations between the water-soluble reduced state of the dye and the insoluble oxidized form. Sulfur dyes, which generally have a poor resistance to chlorine, and are not applicable to wool or

Table 1.2 Classification of dyes according to their usage and chemical types.

Class	Principal substrates	Method of application	Chemical types
Acid	nylon, wool, silk, paper, inks, and leather	usually from neutral to acidic dyebaths	azo (including premetallized), anthraquinone, triphenylmethane, azine, xanthene, nitro and nitroso
Azoic components and compositions	cotton, rayon, cellulose acetate and polyester	fiber impregnated with coupling component and treated with a solution of stabilized diazonium salt	Azo
Basic	paper, polyacrylonitrile, modified nylon, polyester and inks	applied from acidic dyebaths	cyanine, hemicyanine, diazahemicyanine, diphenylmethane, triarylmethane, azo, azine, xanthene, acridine, oxazine, and anthraquinone
Direct	cotton, rayon, paper, leather and nylon	applied from neutral or slightly alkaline baths containing additional electrolyte	azo, phthalocyanine, stilbene, and oxazine

(Continued)

Table 1.2 (Cont.)

Class	Principal substrates	Method of application	Chemical types
Disperse	polyester, polyamide, acetate, acrylic and plastics	fine aqueous dispersions often applied by high temperature/pressure or lower temperature carrier methods; dye may be padded on cloth and baked on or thermofixed	azo, anthraquinone, styryl, nitro, and benzodifuranone
Fluorescent brighteners	soaps and detergents, all fibers, oils, paints, and plastics	from solution, dispersion or suspension in a mass	stilbene, pyrazoles, coumarin, and naphthalimides
Food, drug, and cosmetic	foods, drugs, and cosmetics		azo, anthraquinone, carotenoid and triarylmethane
Mordant	wool, leather, and anodized aluminum	applied in conjunction with Cr salts	azo and anthraquinone
Oxidation bases	hair, fur, and cotton	aromatic amines and phenols oxidized on the substrate	aniline black and indeterminate structures
Reactive	cotton, wool, silk, and nylon	reactive site on dye reacts with functional group on fiber to bind dye covalently under influence of heat and pH (alkaline)	azo, anthraquinone, phthalocyanine, formazan, oxazine, and basic

Solvent	plastics, gasoline, varnishes, lacquers, stains, inks, fats, oils, and waxes	dissolution in the substrate	azo, triphenylmethane, anthraquinone, and phthalocyanine
Sulfur	cotton and rayon	aromatic substrate vatted with sodium sulfide and re-oxidized to insoluble sulfur-containing products on fiber	indeterminate structures
Vat	cotton, rayon, and wool	water-insoluble dyes solubilized by reducing with sodium hydrogen sulfite, then exhausted on fiber and re-oxidized	anthraquinone (including polycyclic quinones) and indigoids

silk dyeing, can be applied in both batch and continuous processes; continuous applications are preferred because of the lower volume of dye required. In general, sulfur blacks are the most commercially important colors and are used where good color fastness is more important than shade brightness [39].

Vat Dyes, which are water-insoluble dyes, are mainly applied to cellulosic fibers as soluble leuco salts after reduction in an alkaline bath, usually with sodium hydrogensulfite. Following exhaustion onto the fiber, the leuco forms are re-oxidized to the insoluble keto forms and after treated, usually by soaping, to redevelop the crystal structure. The principal chemical classes of vat dyes are known as anthraquinone and indigoid.

Cationic (Basic) Dyes, which are water-soluble and present as colored cations in solution, and thus frequently referred to as cationic dyes, are applied to paper, polyacrylonitrile (e.g., Dralon), modified nylons, and modified polyesters. Their original use was for silk, wool, and tannin-mordanted cotton when brightness of shade was more important than fastness to light and washing. The principal chemical classes are diazahemicyanine, triarylmethane, cyanine, hemicyanine, thiazine, oxazine, and acridine. Some basic dyes show biological activity and are used in medicine as antiseptics.

Solvent Dyes, which are water-insoluble but solvent-soluble, are devoid of polar solubilizing groups such as sulfonic acid, carboxylic acid, or quaternary ammonium. The dyes, which are used for colored plastics, gasoline, oils, and waxes, are predominantly azo and anthraquinone, and also phthalocyanine and triarylmethane dyes.

Reactive Dyes form a covalent bond with the fiber, usually cotton, although they are used to a small extent on wool and nylon. This class of dyes, first introduced commercially in 1956 by Imperial Chemical Industries (ICI), made it possible to achieve extremely high wash-fastness properties by relatively simple dyeing methods. A marked advantage of reactive dyes over direct dyes is that their chemical structures are much simpler, their absorption spectra show narrower absorption bands, and the dyeings are brighter. The principal chemical classes of reactive dyes are azo (including metallized azo), triphendioxazine, phthalocyanine, formazan, and anthraquinone. High-purity reactive dyes are used in the inkjet printing of textiles, especially cotton.

1.3   Technologies for Color Removal

Most probably, the progressive accumulation of many more organic compounds in natural waters has resulted in the development and growth of chemical technologies toward organic synthesis and processing. The dye industry corresponds to a relatively small part of the overall chemical industry. Dyes and pigments are highly visible materials and so even the minimum amount released into the environment may cause the appearance of color in open waters [3,40].

Colored dye wastewater is created as a direct result of the production of the dye and also as a consequence of its use in the textile and related industries. There are more than 100,000 commercially available dyes with over 700,000 tons produced annually. It is estimated that 2% of the dyes are discharged as effluent from manufacturing operations, while 10% is discharged from textile and associated industries. Among industries, textile factories consume large volumes of water and chemicals for processing of textiles. Wastewater stream from the textile dyeing operation contains unutilized dyes (about 8–20% of the total pollution load due to incomplete exhaustion of the dye) and auxiliary chemicals along with a large amount of water. The rate of loss is approximated to be 1–10% for pigments, paper and leather dyes. Effluent treatment processes for dyes are currently able to eliminate only half of the dyes lost in wastewater streams. Therefore, hundreds of tons daily find their way into the environment, primarily dissolved or suspended in water [41,42]. Dyes are synthetic aromatic compounds which are embodied with various functional groups. Some dyes are reported to cause allergy, dermatitis, skin irritation, cancer, and mutations in humans [43]. Beyond aesthetic considerations, the most important environmental problems related to dyes is their absorption and reflection of sunlight entering the water, which interferes with the growth of bacteria, limiting it to levels insufficient to biologically degrade impurities in the water. It is evident that the decolorization of aqueous effluents is of environmental, technical, and commercial importance worldwide in terms of meeting environmental requirements and water reuse [44]. Textile wastewaters exhibit a considerable resistance to biodegradation, due to the presence of the dyes, which have a complex chemical structure and are resistant to light, heat and oxidation agents. Hence the removal of dyes in an economic and effectual manner by the textile industry appears to be a most imperative problem [45,46].

The dyestuffs in wastewaters cannot be efficiently decolorized by conventional methods. There also are the high cost and disposal problems for treating dye wastewater on a large scale in the textile and paper industries [47]. The technologies for color removal can be divided into three categories as chemical, biological and physical [48].

1.3.1 Chemical Methods

Chemical methods consist of many techniques, such as coagulation or flocculation combined with flotation and filtration, precipitation-flocculation with Fe, Al and Ca hydroxides, electroflotation, electrokinetic coagulation, conventional oxidation methods by oxidizing agents, irradiation or electrochemical processes. These chemical techniques are often high cost, and the accumulation of concentrated sludge, along with the decolorization leads to a disposal problem. These techniques may also cause a secondary pollution problem based on excessive chemical use. Recently, there have been other emerging techniques known as advanced oxidation process and ozonation. The advanced oxidation process (AOP), which is based on the generation of very powerful oxidizing agents such as hydroxyl radicals, has been applied with success for pollutant degradation [49,50]. Oxidation by ozone (ozonation) is capable of degrading chlorinated hydrocarbons, phenols, pesticides and aromatic hydrocarbons [51,52]. The dosage applied for the dye-containing effluent is dependent on the total color level and residual COD. Ozonation shows a preference for double-bonded dye molecules, which leaves the colorless and low-COD effluent suitable for discharge into environment [50, 52–55]. A major advantage of ozonation is that ozone is applied in its gaseous state and therefore does not increase the volume of wastewater and sludge. Lin and Liu [56] used a combination of ozonation and coagulation for treatment of textile wastewater.

Although these methods are efficient for the treatment of waters contaminated with pollutants, they are very costly and commercially unattractive. The high electrical energy demand and the consumption of chemical reagents are common problems.

1.3.2 Physical Methods

Physical methods, which are widely used in industry because of their high dye removal potentials and low operating costs, such as adsorption, ion-exchange and irradiation, filtration and membrane-filtration processes (nanofiltration, reverse osmosis, electrodialysis), are the most applicable methods for treatment of textile wastewater in plants [57,58]. Some adsorbents such as activated carbon [59] and coal [60], fly ash [61,62], silica, wood, clay material [63,64], agriculture wastes and cotton waste are used in dye effluent treatment processes. The irradiation process is more suitable for decolorization at low volumes within a wide range, but degradation of dye in textile effluents requires very high dissolved oxygen. Ion exchange has huge limitations for removal of dyes in textile effluents, and is very specific for dyes and other impurities present in wastewater, which reduces its effectiveness [48,65].

The membrane processes have major disadvantages, such as a limited lifetime and the high cost of periodic replacement. Liquid-phase adsorption is one of the most popular methods for the removal of pollutants from wastewater and also an attractive alternative for the treatment of contaminated waters, especially if the sorbent is inexpensive and does not require an additional pretreatment step before its application.

Adsorption, which is a well-known equilibrium separation process, has been found to be superior to other techniques for water reuse in terms of initial cost, flexibility and simplicity of design, ease of operation and insensitivity to toxic pollutants. Decolorization, which is influenced by many physicochemical factors such as dye/sorbent ratio, sorbent surface area, particle size, temperature, pH, and contact time, is mainly a result of two mechanisms: adsorption and ion exchange [66,57,50]. Also, adsorption generally does not result in the formation of harmful substances.

1.3.2.1   Adsorption

The use of adsorption method for wastewater treatment has become more popular in recent years owing to its efficiency in the removal of pollutants too stable for biological methods. Adsorption is an economically feasible process that can produce high quality water [67].

Because synthetic dyes cannot be efficiently removed from the wastewaters by conventional methods, the adsorption of synthetic dyes on inexpensive and efficient solid supports is considered as a simple and economical method for their removal from wastewaters. The adsorption characteristics of a wide variety of inorganic and organic supports have been measured and their capacity to remove synthetic dyes has been evaluated [11].

Physical adsorption occurs reversibly via weak interactions, such as van der Waals interactions, hydrogen bonding and dipole-dipole interaction, between the adsorbate and adsorbent. Chemical adsorption, chemisorption, occurs irreversibly via strong interactions, such as covalent and ionic bond formation, between adsorbate and adsorbent [68]. A summary of commonly used adsorbents follows.

Activated carbon is the most commonly used adsorbent for dye removal by adsorption and is very effective for the adsorption of cationic dye, mordant, and acid dyes and to a slightly lesser extent, dispersed, direct, vat, pigment and reactive dyes [49, 69–73]. The performance is dependent on the type of carbon used and the characteristics of the wastewater. Due to its highly porous nature, activated carbon has a much larger surface area, and hence has a higher capacity in terms of the adsorption.

Peat has a cellular structure that makes it an ideal choice as an adsorbent for the adsorption of transition metals and polar organic compounds from dye-containing effluents. Peat requires no activation, unlike activated carbon, and also costs much less [74].

Wood chips exhibit a high adsorption capacity toward acid dyes, although due to their hardness they are not as good as other available sorbents and longer contact times are required [75,76].

Fly ash and coal mixture is used as an adsorbent for dye adsorption from colored wastewaters. A high fly ash ratio increases the adsorption capacity of the mixture due to its increased surface area available for adsorption. This combination may be substituted for activated carbon, with a ratio of fly ash:coal, 1:1 [77].

Silica gel is an effective material for removing basic dyes, although side interactions such as air binding and fouling with particulate matter prevent its commercial use.

Natural clays as well as substrates such as corn cobs and rice hulls are commonly used for dye removal. Their main advantages are widespread availability and cheapness. These substrates are more attractive economically for dye removal, compared to the other ones [69,75,78].

Several adsorbents have been studied to determine their ability of adsorption toward dyes from aqueous effluents and are given in Table 1.3. Several studies focused on the economic removal of dyes using different adsorbents such as sawdust [79], banana and orange peels [80], wheat straw, corncobs, barley husks [81], tree fern [82], eucalyptus barks [83,84], wood [85], peat [86], rice husk [87], chitin [88], algal biomass, metal hydroxide sludge [90], soil [91], clays [92,93] and fly ash [94], and coal [95].

A number of low-cost adsorbents were studied to determine their ability to adsorb dyes from aqueous effluents. However, the most widely used and the most easily reached adsorbent for dyes is activated carbon as granule or powder [145].

1.3.2.2   *Membrane Filtration*

This process has the ability to clarify, concentrate and, most importantly, to separate dyes continuously from effluent [52,146, 147]. It has some special

Table 1.3 Some examples of adsorbent-adsorbate pairs used in the adsorption of dyes.

Dyes	Adsorbents	Reference
Acid Dyes	Polyaniline/γ-alumina nanocomposite	[96]
	Ferromagnetic ordered mesoporous carbon	[97]
	Magnetic graphene/chitosan nanocomposite	[98]
	Functionalized mesoporous organosilicas	[99]
	Functionalized graphene oxide-based hydrogels	[100]
	Modified montmorillonite	[101]
	Modified zeolite	[102]
	Zinc aluminum hydroxide	[103]
	Calcium aluminate hydrates	[104]
	Modified wheat straw	[105]
	Maize stem tissue	[106]
	Framboidal vaterite	[107]
	Reduced graphene oxide	[108]
	Nanoporous carbon	[109]
	Modified wheat straw	[110]
Cationic Dyes	Granular composite hydrogel	[111]
	Hydrotalcite-iron oxide magnetic organocomposite	[112]
	Ferromagnetic ordered mesoporous carbon	[97]
	Reduced graphene oxide	[108, 113]
	Kaolinite and montmorillonite	[114]
	Chemically modified biomass	[115]
	Bentonite-based composite	[116]
	Polydopamine microspheres	[117]
	Clay	[118,92]
	Clinoptilolite and amberlite	[119]
	Zeolite	[120]
	Silk worm pupa	[121]
	Modified titanium dioxide nanotube	[122]

(*Continued*)

Table 1.3 (*Cont.*)

Dyes	Adsorbents	Reference
Direct Dyes	Cotton fiber	[123]
	Polyaniline/γ-alumina nanocomposite	[96]
	Zinc aluminum hydroxide	[124]
	Core–shell magnetic nanoparticles	[103]
	Corn Stalk	[125]
	Garlic Peel	[126]
	Cellulose	[127]
	Modified activated carbon	[128]
	Multi-walled carbon nanotubes and activated carbon	[129]
Reactive dyes	Activated carbon	[130–132]
	Modified activated carbon	[133]
	Sugar beet pulp	[134]
	Multiwalled carbon nanotubes	[135]
	Polyaniline/γ-Alumina Nanocomposite	[96]
	Activated red mud	[136]
	Sugarcane bagasse	[137]
	Iron oxide nanospheres.	[138]
	Modified walnut shell	[139]
	Gold nanoparticle-activated carbon	[140]
Disperse dyes	Metal oxide aerogels	[141]
	Carboxymethyl cellulose-acrylic acid	[142]
	Ultrafine silk fibroin powder	[143]
	Biomass char	[144]

features unrivaled by other methods; resistance to temperature, adverse chemical environment and microbial attack.

Wu *et al.* [148] used a combination of membrane filtration with ozonation process for treatment of reactive-dye wastewater. Ciardelli *et al.* [149] combined activated sludge oxidation and ultrafiltration. Zheng and Liu [150] worked on a dyeing and printing wastewater treatment using a membrane bioreactor with gravity drain. A laboratory-scale membrane bioreactor (MBR) with a gravity drain was designed for the treatment of dyeing and printing wastewaters and was tested on wastewater from a wool mill. The MBR was used continuously as a gravity-controlled system without chemical cleaning for 135 days. The average removal ratios of BOD_5, COD, turbidity and color were found to be 80.3%, 95%, 99.3% and 58.7% respectively.

Hai *et al.* [151] developed a submerged membrane fungi reactor for textile wastewater treatment. A submerged microfiltration membrane bioreactor implementing the white rod fungi, *Coriolus versicolor*, was developed for the treatment of textile dye wastewater with different fouling prevention techniques. It was found that the color removal ratio from synthetic wastewater by the reactor is about 99%.

1.3.2.3 Ion Exchange

In this method, wastewater is passed over the ion exchange resin until the available exchange sites are saturated. Both cationic and anionic dyes can be removed from dye-containing effluent this way. Advantages of this method include no loss of adsorbent on regeneration, reclamation of solvent after use and the removal of soluble dyes. Ion exchange is not widely used for the treatment of dye-containing effluents. This is because the ion exchange resins are not effective for a range of a wide variety of dyes [50]. A major disadvantage is cost. Organic solvents are expensive, and the ion exchange method is not very effective for disperse dyes [146].

Commercial anionic exchange resins were applied to the water contaminated with a broad range of reactive dyes by Karcher *et al.* [152,153], and they reported that anionic exchangers possess excellent adsorption capacity (200–1200 µmol/g) as well as efficient regeneration property for their removal and recovery.

1.3.2.4 Irradiation

Sufficient quantities of dissolved oxygen are required for organic substances to be broken down effectively by radiation. The dissolved oxygen is consumed very rapidly, and therefore for photocatalyzed oxidation a constant and adequate supply of oxygen is required. Dye-containing effluent may be treated in a dual-tube bubbling rector. An application of this method showed that some dyes and phenolic molecules can only be oxidized effectively at a laboratory scale [154].

1.3.2.5   *Electrokinetic Coagulation*

Electrokinetic coagulation is an economically feasible method for dye removal. It involves the addition of ferrous sulphate and ferric chloride, allowing excellent removal of direct dyes from wastewaters. Unfortunately, the poor results with acid dyes; as well as the high cost of the ferrous sulphate and ferric chloride, mean that it is not a widely used method.

The optimum coagulant concentration is dependent on the static charge of the dye in the solution, and the difficulty in removing the sludge formed as part of the coagulation is a problem. Production of large amounts of sludge occurs, and this causes high disposal costs [146,53].

1.3.3   Biological Methods

In recent years, an enormous amount of attention has emerged on biological methods with some microorganisms such as fungi, bacteria and algae, which are highly capable of biodegrading and adsorbing dyes from wastewater [155]. The application of biological processes for dye wastewater removal offers many considerable advantages such as being relatively low cost, environmentally friendly, and producing less secondary sludge and nontoxic end products of complete mineralization [11]. Much of the research conducted on the use of microorganisms for dye wastewater removal has proven the potential of microorganisms such as *Cunninghamella elegans* [12], *Aspergillus niger* [13], *Bacillus cereus* [14], *Chlorella sp.* [15] and also *Citrobacter sp.* [16]. The adaptability and the activity of each microorganism are the most significant factors that influence the effectiveness of microbial decolorization [156]. Hence, to develop a practical bioprocess for dye wastewater treatment, it is need to continuously examine the microorganisms that are capable of degrading azo dyes [17].

However, the application of microorganisms is often restricted because of technical constraints. Bhattacharyya and Sharma [157] have suggested that biological treatment requires a large land area, and is constrained by sensitivity towards diurnal variation, as well as the toxicity of some chemicals, and less flexibility in design and operation. Further biological treatment is not able to provide satisfactory color elimination using current conventional biodegradation processes [48]. Moreover, although many organic molecules are degraded, many others are not degradable due to their complex chemical structure and synthetic origin [158].

The biological processes, which can be implemented for both municipal and industrial wastewaters, are classified as aerobic and anaerobic.

The main idea of all biological methods of wastewater treatment is to provide contact with bacteria (cells), which feed on the organic materials in the wastewater, and thereby reduce its biological oxygen demand (BOD). In other words, the purpose of biological treatment is BOD reduction. The natural process of microbiological metabolism in aquatic environment is capitalized on in the biological treatment of wastewater. Under proper environmental conditions, the soluble organic substances of the wastewater are completely destroyed by biological oxidation. A part of it is oxidized while the rest is converted into biological mass in the biological reactors. The biological treatment system usually consists of biological reactors and a settling tank to remove the produced biomass or sludge [159].

1.3.3.1   Aerobic and Anaerobic Degradation

Aerobic means in the presence of air (oxygen); while anaerobic means in the absence of air (oxygen). These two terms are directly related to the type of bacteria or microorganisms that are involved in the degradation of organic impurities in a given wastewater and the operating conditions of the bioreactor. Therefore, aerobic treatment processes take place in the presence of air and utilize those microorganisms (also called aerobes), which use molecular/free oxygen for the assimilation of organic impurities, which are converted into carbon dioxide, water and biomass. On the other hand, the anaerobic treatment processes occur in the absence of air (i.e., molecular/free oxygen) by those microorganisms (also called anaerobes) which do not require air for the assimilation of organic impurities. The final products of organic assimilation in anaerobic treatment are methane and carbon dioxide and biomass. The simplified principles of the two processes are illustrated in Figure 1.1 [160].

Anaerobic and aerobic treatments have been used together or separately for the treatment of textile effluents. Hence aerobic treatment is not

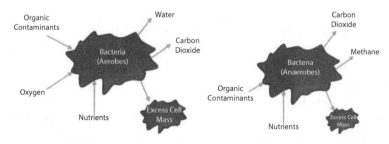

Figure 1.1 A simple illustration showing the principles of aerobic and anaerobic degradation [160].

effective in color removal from textile wastewater containing azo dyes. Conventional biological processes are not effective for treating dyestuff wastewater because many commercial dyestuffs are toxic to the organism being used and result in problems of sludge bulking, rising sludge and pin flock. Because of the low biodegradability of many textile chemicals and dyes, biological treatment is not always effective for textile industry wastewater [161,162].

Even though dyes are difficult to degrade biologically, many researchers are interested in dye-containing wastewaters. Bell et al. [163] treated dye wastewaters using an anaerobic baffled reactor, and obtained a significant reduction in the COD (chemical oxygen demand), as well as the lower color level; (about 55%, and 95%, respectively).

Under anaerobic conditions, it was reported that many bacteria reduce azo dyes by the activity of unspecific, soluble, cytoplasmic reductases, also known as azoreductases. Initially, the bacteria bring about the reductive cleavage of the azo linkage, which results in dye decolorization and the production of colorless aromatic amines. It has been expressed that such compounds threat human health and the environment [164].

Sponza and Işık [165] investigated color and high organic impurities (COD) removal efficiencies using anaerobic-aerobic sequential processing for treatment of 100 mg/L of di-azo dye with glucose as the carbon source and found a high decolorization efficiency of 96%. Işık and Sponza [166] reported 92.3 and 95.3% color and COD removal efficiencies, respectively, when using an upflow anaerobic sludge blanket–aerobic stirred tank reactor sequential system to treat Congo red dye.

Zaoyan et al. [167] obtained 65% color and 74% COD removal efficiencies in textile wastewater contaminated with azo dyes using an anaerobic-aerobic rotating biodisc system.

Supaka et al. [168] obtained 78.2% color removal and 90% COD removal in a sequential anaerobic-aerobic system that was used to treat Remozal Black B dye.

Kapdan and Öztekin [169] investigated Remozal Red dye and reported over 90% color removal and about 85% COD removal.

Khehra et al. [170] reported 98% color removal and 95% COD removal efficiency in an anoxicaerobic sequential bioreactor system that was used to treat Acid Red 88 azo dye.

In a study by Hu [171], three Gram-negative bacteria (*Aeromonas sp.*, *Pseudomonas luteola* and *Escherichia coli*), two Gram-positive bacteria (*Bacillus subtilis* and *Staphylococcus aureus*) and activated sludge (consisting of both Gram-negative and Gram-positive bacteria) were used as biosorbents for the removal of reactive dyes such as Reactive Blue, Reactive Red, Reactive Violet and Reactive Yellow. Dead cells of test genera showed a higher uptake than living cells due to an increased surface area and/or metabolic resistance and Gram-negative bacteria had a higher adsorption capacity than Gram-positive bacteria due to higher lipid contents in the cell wall portion.

A screening of bacteria with the ability of degrading several structurally different dyes such as Poly R-478, Methyl Orange, Lissamine Green B and Reactive Black 5 was carried out by Deive *et al.* [172].

Both aerobic and anaerobic strains were detected, but they have observed that a facultative anaerobic strain was the one leading to the most promising results.

The decolorization ability of *Anoxybacillus flavithermus* in an aqueous effluent containing two representative textile finishing dyes (Reactive Black 5 and Acid Black 48) was investigated. It has been observed that the decolorization efficiency for a mixture of both dyes reached almost 60% in less time than 12 h, which points out the suitability of the selected microorganism [173].

An effective decolorizing bacterial strain, *Bacillus sp.*, for Reactive Black 5 (RB-5) was isolated by Wang *et al.* [174]. This bacterial strain showed great capability to decolorize various reactive textile dyes, including azo dyes. Optimum conditions for the decolorizing of RB-5 were determined to be pH 7.0 and 40°C. *Bacillus sp.*, which grew well in medium containing high concentration of dye (100 mg/L), provides approximately 95% decolorization in 120 h and has a practical application potential in the biotransformation of various dye effluents.

On the other hand, the ionic forms of the dye in solution and the surface electrical charge of the biomass depend on solution pH. Therefore, solution pH generally influences both the fungal biomass surface dye binding sites and the dye chemistry in the medium [10,175–178].

Most textile and other dye effluents are produced at relatively high temperatures and hence temperature will be an important factor in real application of biosorption in the future. Arica and Bayramoglu [177] found that heating the biomass, *Lentinus sajor-caju*, at 100°C for 10 min significantly enhanced the biosorption capacity, while base-treatment with 0.1 M NaOH lowered the biosorption capacity of the fungi to remove Reactive Red 120 and also the biosorption of dye increased with increasing temperature. Aksu and Cagatay [179] reported that for *R. arrhizus* the biosorption of dye increased with increasing temperature.

Hu [171] has investigated the ability of bacterial cells isolated from activated sludge to adsorb reactive dyes, including Reactive Blue, Reactive

Red, Reactive Violet, Reactive Yellow and Procion Red G. Ozer *et al.* [180] have revealed that the potential of decolorization of the algae, *S. rhizopus*, for synthetic wastewaters containing an initial concentration of the dye Acid Red 274 from 25 to 1000 mg/L is high. Almost complete removal of Acid Red 274 dye from the synthetic wastewaters was achieved by using *S. rhizopus* with mechanisms of biocoagulation and biosorption.

Santos *et al.* [181,182] have widely reported the necessity of introducing redox mediators to achieve high decolorization efficiencies of several azo dyes by using anaerobic thermophiles.

Mahadwad *et al.* [183] studied the photocatalytic degradation of Reactive Black 5 dye by using TiO_2-zeolite adsorbent as a semiconductor catalyst system at a batch reactor. The composition of the synthesized photocatalyst is composed of zeolite (ZSM-5) and TiO_2, as adsorbent and as photoactive component. The optimum formulation of the supported catalyst was found to be (TiO_2/ZSM-5 = 0.15/1), which received the highest efficiency with 98% degradation in 50 mg/L solution of Reactive Black 5 within 90 min. The TiO_2 together with the zeolite was found to be stable for repeated use.

Erdal and Taskin [184] investigated the decolorization of Reactive Black 5 dye by *Penicillium chrysogenum MT-6,* and they determined that the dye uptake is strongly dependent on mycelial morphology. The small uniform pellets (2 mm) and poor nutrient medium were found to be better for dye uptake. The optimal conditions for the dye uptake were determined to be an initial pH of 5.0, a shaking rate of 150 rpm, temperature of 28°C, spore concentration of 10^7 /mL, 10 g/L sucrose, and 1 g/L ammonium chloride. The maximum dye removal ratio of the fungus was found to be 89% with biomass production of 3.83 g/L at 0.3 g/L initial dye concentration for 100 h. The fungus was understood to be a good alternative system for the decolorization of a medium containing Reactive Black 5 dye.

References

1. T. Srinivas., *Environmental Biotechnology*, New Age International Ltd. 2008.
2. M. Grassi, G. Kaykioglu, V. Belgiorno, and G. Lofrano, Removal of emerging contaminants from water and wastewater by adsorption process. In: *Emerging Compounds Removal from Wastewater: Natural and Solar Based Treatments,* G. Lofrano, ed., Springer Briefs in Green Chemistry for Sustainability, 2012.
3. A. E. Segneanu, C. Orbeci, C. Lazau, P. Sfirloaga, P. Vlazan, C. Bandas, and I. Grozescu, 2013. Waste water treatment methods. In: *Water Treatment.* W. Elshorbagy and R. K. Chowdhury, eds., InTech Publishing, 53–80.

4. S. N. B. S. Jaafar, 2006. *Dye removal using clay. Adsorption study*, Faculty of Chemical Engineering and Natural Resources. University College of Engineering & Technology Malaysia degree of Bachelor of Chemical Engineering.

5. C. Fernandez, M. S. Larrechi, M. P. Callao, 2010. An analytical overview of processes for removing organic dyes from wastewater effluents. *Trends in Analytical Chemistry* 29(10), 1202–1211.

6. U. G. Akpan, B. H. Hameed, 2009. Parameters affecting the photocatalytic degradation of dyes using TiO_2-based photocatalysts: A review. *Journal of Hazardous Materials* 170, 520–529.

7. F. P. van der Zee, S. Villaverde, 2005. Combined anaerobic–aerobic treatment of azo dyes – A short review of bioreactor studies. *Water Research* 39, 1425–1440.

8. K. Kümmerer, 2009. The presence of pharmaceuticals in the environment due to human use: Present knowledge and future challenges. *J Environ Management* 90, 2354–2366.

9. K. Gomes, *Wastewater Management*. Global Media, 2009.

10. A. Srinivasan, and T. Viraraghavan, 2001. Decolorization of dye wastewaters by biosorbents: A review. *Journal of Environmental Management* 91, 1915–1929.

11. E. Forgacs, T. Cserhati, G. Oros, 2004. Removal of synthetic dyes from wastewaters: A review. *Environment International* 30, 953–971.

12. T. A. Sandra, C. V. J. Jose, A. A. Carlos, O. Kaoru, E. N. Aline, L. L. Ricardo and M. C. Galba, 2012. A biosorption isotherm model for the removal of reactive azo dyes by inactivated mycelia of *Cunninghamella elegans* UCP542. *Molecules* 17, 452–462.

13. K. Karthikeyan, K. Nanthakumar, K. Shanthi, and P. Lakshmanaperumalsamy, 2010. Response surface methodology for optimization of culture conditions for dye decolorization by a fungus, *Aspergillus niger* HM11 isolated from dye affected soil. *Iranian Journal of Microbiology* 2(4), 213–222.

14. A. Lamia, K. Chaieb, A. Ceheref, and A. Bakhrouf, 2010. Biodegradation and decolorization of triphrnylmethane dyes by *Staphylococcus epidermidis*. *Desalination* 260, 137–146.

15. N. Daneshvar, M. Ayazloo, A. R. Khatee, and M. Pourhassan, 2007. Biological decolorization of dye solution containing malachite green by microalgae *Cosmarium sp. Bioresources Technology* 98, 1176–1182.

16. W. Hui, Q. S. Jian, W. Z. Xiao, T. Yun, J. X. Xiao, and L. Z. Tian, 2010. Bacterial decolorization and degradation of the reactive dye Reactive Red 180 by *Citrobacter sp.* CK3. *International Biodeterioration & Biodegradation* 63, 395–399.

17. M. S. Zuraida, C. R. Nurhaslina, K. H. K. Hamid, 2013. Removal of synthetic dyes from wastewater by using bacteria *Lactobacillus delbruckii*. *International Refereed Journal of Engineering and Science* 2(5), 1–7.

18. G. R. Chatwal, *Synthetic Dyes*. Himalaya Publishing House Pvt. Ltd. 2009.

19. L. Pereira, and M. Alves, 2012. Dyes: Environmental impact and remediation. In: *Environmental Protection Strategies for Sustainable Development Strategies for Sustainability*. A. Malik, E. Grohmann, eds., Springer, 111–162.

20. E. N. Abrahart, 1977. *Dyes and Their Intermediates*. New York: Chemical Publishing. 1–12.

21. IARC Monographs on the Evaluation of Carcinogenic Risks to Humans Vol. 99 *Some Aromatic Amines, Organic Dyes, and Related Exposures*, Lyon, France 2010.

22. R. L. M. Allen, 1971. *Colour Chemistry*. London: Thomas Nelson and Sons Ltd. 11–13.

23. N. Belfer, 1973. *Designing in Batik and Tie Dye*. Worcester, MA: Davis Publications.

24. C. Goetz, 2008. Textile dyes: Techniques and their effects on the environment with a recommendation for dyers concerning the green effect. Senior Honors Thesis. Honors Program of Liberty University.

25. S. Garfield, 2001. *Mauve: How One Man Invented a Color that Changed the World*. New York: W. W. Norton & Company.

26. P. F. Gordon, P. Gregory, *Organic Chemistry in Colour*. Springer-Verlag, Berlin, 1983.

27. K. Hunger, *Industrial Dyes: Chemistry, Properties, Applications*. Wiley-VCH. Verlag GmbH & Co. KGaA, Weinheim 2007.

28. A. K. Samanta and P. Agarwal, 2009. Application of natural dyes on textiles. *Indian Journal of Fibre & Textile Research* 34, 384–399.

29. M. Shahid, S. Islam, F. Mohammad, 2013. Recent advancements in natural dye applications: A review. *Journal of Cleaner Production* 53, 310–331.

30. S. J. Kadolph, 2008. Natural dyes: A traditional craft experiencing new attention. *Delta Kappa Gamma Bull.* 75(1), 14–17.

31. A. C. Dweck, 2002. Natural ingredients for colouring and styling. *International Journal of Cosmetic Science* 24, 287–302.

32. D. Frick, 2003. The coloration of food. *Review of Progress in Coloration and Related Topics* 33(1), 15–32.

33. S. Hao, J. Wu, Y. Huang, J. Lin, 2006. Natural dyes as photosensitizers for dyesensitized solar cell. *Solar Energy* 80, 209–214.

34. E. Tousson, B. Al-Behbehani, 2011. Black mulberries (Morus nigra) as a natural dye for animal tissues staining. *Anim. Biol.* 61, 49–58.

35. P. K. Mishra, P. Singh, K. K. Gupta, H. Tiwari, P. Srivastava, 2012. Extraction of natural dye from Dahlia variabilis using ultrasound. *Indian J. Fibre Text. Res.* 37, 83–86.

36. B. Kuswandi, Jayus, T. S. Larasati, A. Abdullah, L. Y. Heng, 2012. Real-time monitoring of shrimp spoilage using on-package sticker sensor based on natural dye of curcumin. *Food Analytical Method* 5, 881–889.

37. A. Zyoud, N. Zaatar, I. Saadeddin, M. H. Helal, G. Campet, M. Hakim, D. H. Park, H. S. Hilal, 2011. Alternative natural dyes in water purification: Anthocyanin as TiO_2-sensitizer in methyl orange photo-degradation. *Solid State Sci.* 13, 1268–1275.

38. P. Gregory, *High Technology Applications of Organic Colorants*, Plenum, New York, 1991.
39. S. V. Kulkarni, C. D. Blackwell, A. L. Blackard, C. W. Stackhouse, and M. W. Alexander, *Textile Dyes and Dyeing Equipment: Classification, Properties, and Environmental Aspects*. EPA 1985.
40. M. Hasan, Dye: Technologies for Colour Removal. Home News Articles, 2014 UniMAP School of Environmental Engineering, Universiti Malaysa Perlis, Kompleks Pusat Pengajian Jejawi 3, 02600 Arau, Perlis, Malaysia.
41. A. K. Mukherjee, B. Gupta, S. M. S. Chowdhury, 1999. Separation of dyes from cotton dyeing effluent using cationic polyelectrolytes. *American Dyestuff Reporter* 88, 25–28.
42. S. J. Allen, G. Mckay, J. F. Porter, 2004. Adsorption isotherm models for basic dye adsorption by peat in single and binary component systems. *Journal of Colloid and Interface Science* 280(2), 322–333.
43. A. Bhatnagar, A. K. Jain, 2005. A comparative adsorption study with different industrial wastes as adsorbents for the removal of cationic dyes from water. *Journal of Colloid and Interface Science* 281(1), 49–55.
44. G. McKay, B. Al Duri, 1987. Simplified model for the equilibrium adsorption of dyes from mixtures using activated carbon. *Chemical Engineering and Processing: Process Intensification* 22(3), 145–156.
45. K. Ravikumar, S. Ramalingam, S. Krishnan, K. Balu, 2006. Application of response surface methodology to optimize the process variables for reactive red and acid brown dye removal using a novel adsorbent. *Dyes and Pigments* 70(1): 18–26.
46. B. Noroozi, G. A. Sorial, 2013. Applicable models for multi-component adsorption of dyes: A review. *Journal of Environmental Sciences* 25(3), 419–429.
47. M. Ghoreishi, and R. Haghighi, 2003. Chemical catalytic reaction and biological oxidation for treatment of non-biodegradable textile effluent. *Chemical Engineering Journal* 95, 163–169.
48. T. Robinson, G. McMullan, R. Marchant, P. Nigam, 2001. Remediation of dyes in textiles effluent: A critical review on current treatment technologies with a proposed alternative. *Bioresource Technology* 77, 247–255.
49. C. Raghavacharya, 1997. Colour removal from industrial effluents: A comparative review of available technologies. *Chem. Eng. World* 32, 53–54.
50. Y. M. Slokar and A. M. Le Marechal, 1998. Methods of decoloration of textile wastewaters. *Dyes and Pigments* 37, 335–356.
51. S. H. Lin, and C. M. Lin, 1993. Treatment of textile waste effluents by ozonation and chemical coagulation. *Water Res.* 27, 1743–1748.
52. Y. Xu, and R. E. Lebrun, 1999. Treatment of textile dye plant effluent by nanofiltration membrane. *Separ. Sci. Technol.* 34, 2501–2519.
53. F. Gahr, F. Hermanutz, W. Opperman, 1994. Ozonation: An important technique to comply with new German law for textile wastewater treatment. *Water Science and Technology* 30, 255–263.
54. N. H. Ince, D. T. Gonenc, 1997. Treatability of a textile azo dye by UV/H_2O_2. *Environ. Technol.* 18, 179–185.

55. A. L. Ahmad, W. A. Harris, Syafiie, and O. B. Seng, 2002. Removal of dye from wastewater of textile industry using membrane technology. *Jurnal Teknologi* 36(F), 31–44.

56. S. H. Lin, and W. Y. Liu, 1994. Continuous treatment of textile water by ozonation and coagulation. *Journal of Environmental Engineering* 120(2), 437–446.

57. A. Dabrowski, 2001. Adsorption: From theory to practice. *Advances in Colloid and Interface Science* 93, 135–224.

58. M. J. Iqbal, M. N. Ashiq, 2007. Adsorption of dyes from aqueous solutions on activated charcoal, *Journal of Hazardous Materials* 139 (1–2), 57–66.

59. M. F. R. Pereira, S. F. Soares, J. J. M. Orfao, J. L. Figueiredo, 2003. A desorption of dyes on activated carbons: Influence of surface chemical groups. *Carbon* 41(4), 811–821.

60. G. McKay, G. Ramprasad, P. P. Mowli, 1986. Equilibrium studies for the adsorption of dyestuffs from aqueous solutions by low-cost materials. *Water Air and Soil Pollution* 29(3), 273–283.

61. S. K. Khare, K. K. Panday, R. M. Srivastava, V. N. Singh, 1987. Removal of victoria blue from aqueous solution by fly ash. *Journal of Chemical Technology and Biotechnology* 38(2), 99–104.

62. B. K. Singh, and N. S. Rawat, 1994. Comparative sorption equilibrium studies of toxic phenols on fly ash and impregnated fly ash. *Journal of Chemical Technology and Biotechnology* 61(4), 307–317.

63. B. K. G. Theng, N. Wells, 1995. Assessing the capacity of some New Zealand clays for decolorizing vegetable oil and butter. *Applied Clay Science* 9, 321–326.

64. R. S. Juang, F. C. Wu, R. L. Tseng, 1997. The ability of activated clay for the adsorption of dyes from aqueous solutions. *Environment Technology* 18(5), 525–531.

65. K. Dajka, E. Takacs, D. Solpan, L. Wojnarovits, O. Güven, 2003. High-energy irradiation treatment of aqueous solutions of C. I. Reactive Black 5 azo dye: Pulse radiolysis experiments. *Radiation Physics and Chemistry* 67(3), 535–538.

66. M. N. V. Kumar, T. R. Sridhari, K. D. Bhavani, P. K. Dutta, 1998a. Trends in color removal from textile mill effluents. *Bioresource Technology* 77, 25–34.

67. S. J. Allen, B. Koumanova, 2005. Decolourisation of water/wastewater using adsorption. *Journal of the University of Chemical Technology and Metallurgy* 40(3), 175–192.

68. D. M. Ruthven, *Principles of Adsorption and Adsorption Processes*, Wiley-Interscience. 1984.

69. N. M. Nasser, and M. El-Geundi, 1991. Comparative cost of color removal from textile effluents using natural adsorbents. *Journal of Chemical Technology and Biotechnology* 50, 257–264.

70. I. S. Thakur, *Environmental Biotechnology: Basic Concepts and Applications*. I. K. International Pvt. Ltd., New Delhi, 2006.

71. S. Karaca, A. Gürses, M. Açıkyıldız, M. Ejder (Korucu), 2008. Adsorption of cationic dye from aqueous solutions by activated carbon. *Microporous and Mesoporous Materials* 115(3), 376–382.

72. A. Rodriguez, J. Garcia, G. Ovejero, M. Mestanza, 2009. Adsorption of anionic and cationic dyes on activated carbon from aqueous solutions: Equilibrium and kinetics. *Journal of Hazardous Materials* 172(2–3), 1311–1320.

73. Y. Bao, and G. Zhang, 2012. Study of adsorption characteristics of methylene blue onto activated carbon made by *Salix Psammophila. Energy Procedia* 16, 1141–1146.

74. V. J. P. Poots, and J. J. McKay, 1976a. The removal of acid dye from effluent using natural adsorbents – I Peat. *Water Res.* 10, 1061–1066.

75. P. Nigam, G. Armour, I. M. Banat, D. Singh, R. Marchant, 2000. Physical removal of textile dyes and solid state fermentation of dye-adsorbed agricultural residues. *Bioresour. Technol.* 72, 219–226.

76. V. J. P. Poots, and J. J. McKay, 1976b. The removal of acid dye from effluent using natural adsorbents – II Wood. *Water Res.* 10, 1067–1070.

77. G. S. Gupta, G. Prasad, V. H. Singh, 1990. Removal of chrome dye from aqueous solutions by mixed adsorbents: Fly ash and coal. *Water Res.* 24, 45–50.

78. S. S. Nawar, and H. S. Doma, 1989. Removal of dyes from effluents using low cost agricultural by-products. *Sci. Tot. Environ.* 79, 271–279.

79. V. K. Garg, R. Gupta, A. B. Yadav, R. Kumar, 2003. Dye removal from aqueous solution by adsorption on treated sawdust. *Bioresource Technology* 89(2), 121–124.

80. G. Annadurai, R.-S. Juang, D.-J. Lee, 2002. Use of cellulose-based wastes for adsorption of dyes from aqueous solutions. *Journal of Hazardous Materials* 92(3), 263–274.

81. T. Robinson, B. Chandran, and P. Nigam, 2002. Effect of pretreatments of three waste residues, wheat straw, corncobs and barley husks on dye adsorption. *Bioresource Technology* 85(2), 119–124.

82. Y. S. Ho, T. H. Chiang, Y. M. Hsueh, 2005. Removal of basic dye from aqueous solution using tree fern as a biosorbent. *Process Biochemistry* 40(1), 119–124.

83. L. C. Morais, O. M. Freitas, E. P. Goncalves, L. T. Vasconcelos, C. G. Gonzalez Beca, 1999. Reactive dyes removal from wastewaters by adsorption on eucalyptus bark: Variables that define the process. *Water Research* 33(4), 979–988.

84. B. Balci, O. Keskinkan, M. Avci, 2011. Use of BDST and an ANN model for prediction of dye adsorption efficiency of *Eucalyptus camaldulensis* barks in fixed-bed system. *Expert Systems with Applications* 38(1), 949–956.

85. Y. S. Ho, and G. McKay, 1998a. Kinetic models for the sorption of dye from aqueous solution by wood. *Process Safety and Environmental Protection* 76(2), 183–191.

86. Y. S. Ho, and G. McKay, 1998b. Sorption of dye from aqueous solution by peat. *Chemical Engineering Journal* 70 (2), 115–124.

87. R. Han, D. Ding, Y. Xu, W. Zou, Y. Wang, Y. Li, L. Zou, 2008. Use of rice husk for the adsorption of Congo red from aqueous solution in column mode. *Bioresource Technology* 99(8) 2938–2946.

88. S. A. Figueiredo, J. M. Loureiro, R. A. Boaventura, 2005. Natural waste materials containing chitin as adsorbents for textile dyestuffs: Batch and continuous studies, *Water Research* 39(17), 4142–4152.

89. Ç. Doğar, A. Gürses, M. Açıkyıldız, and E. Özkan, 2010. Thermodynamics and kinetic studies of biosorption of a basic dye from aqueous solution using green algae Ulothrix sp. *Colloids and Surfaces B: Biointerfaces* 76(1), 279–285.

90. S. Netpradit, P. Thiravetyan, and S. Towprayoon, 2003. Application of 'waste' metal hydroxide sludge for adsorption of azo reactive dyes. *Water Research* 37(4), 763–772.

91. H. Ketelsen, S. Meyer-Windel, 1999. Adsorption of brilliant blue FCF by soils. *Geoderma* 90(1–2), 131–145.

92. A. Gürses, Ç. Doğar, M. Yalçın, M. Açıkyıldız, R. Bayrak, S. Karaca, 2006. The adsorption kinetics of the cationic dye, methylene blue, onto clay. *Journal of Hazardous Materials* 131(1–3), 217–228.

93. M. Doğan, M. H. Karaoğlu, M. Alkan, 2009. Adsorption kinetics of maxilon yellow 4GL and maxilon red GRL dyes on kaolinite. *Journal of Hazardous Materials* 165(1–3), 1142–1151.

94. D. Mohan, K. P. Singh, G. Singh, K. Kumar, 2002. Removal of dyes from wastewater using flyash, a low-cost adsorbent. *Industrial Engineering Chemistry Research* 41(15), 3688–3695.

95. T. A. Khan, V. V. Singh, and D. Kumar, 2004. Removal of some basic dyes from artifical textile wastewater by adsorption on Akash Kinari coal. *Journal of Scientific & Industrial Research* 63, 355–364.

96. H. Javadian, M. T. Angaji, M. Naushad, 2014. Synthesis and characterization of polyaniline/γ-alumina nanocomposite: A comparative study for the adsorption of three different anionic dyes. *Journal of Industrial and Engineering Chemistry* 20(5), 3890–3900.

97. X. Peng, D. Huang, T. Odoom-Wubah, D. Fu, J. Huang, Q. Qin, 2014. Adsorption of anionic and cationic dyes on ferromagnetic ordered mesoporous carbon from aqueous solution: Equilibrium, thermodynamic and kinetics. *Journal of Colloid and Interface Science* 430, 272–282.

98. S. Sheshmani, A. Ashori, S. Hasanzadeh, 2014. Removal of Acid Orange 7 from aqueous solution using magnetic graphene/chitosan: A promising nano-adsorbent. *International Journal of Biological Macromolecules* 68, 218–224.

99. J. R. Deka, C.-L. Liu, T.-H. Wang, W.-C. Chang, H.-M. Kao, 2014. Synthesis of highly phosphonic acid functionalized benzene-bridged periodic mesoporous organosilicas for use as efficient dye adsorbents. *Journal of Hazardous Materials* 278, 539–550.

100. X. Wang, Z. Liu, X. Ye, K. Hu, H. Zhong, J. Yu, M. Jin, Z. Guo, 2014. A facile one-step approach to functionalized graphene oxide-based hydrogels used as effective adsorbents toward anionic dyes. *Applied Surface Science* 308, 82–90.

101. Z. Qin, P. Yuan, S. Yang, D. Liu, H. He, J. Zhu, 2014. Silylation of Al13-intercalated montmorillonite with trimethylchlorosilane and their adsorption for Orange II. *Applied Clay Science*. (In Press).
102. S. Liu, Y. Ding, P. Li, K. Diao, X. Tan, F. Lei, Y. Zhan, Q. Li, B. Huang, Z. Huang, 2014. Adsorption of the anionic dye Congo red from aqueous solution onto natural zeolites modified with N, N-dimethyl dehydroabietyl-amine oxide. *Chemical Engineering Journal* 248, 135–144.
103. N. M. Mahmoodi, 2014. Synthesis of core–shell magnetic adsorbent nanoparticle and selectivity analysis for binary system dye removal. *Journal of Industrial and Engineering Chemistry* 20(4), 2050–2058.
104. P. Zhang, T. Wang, G. Qian, D. Wu, R. L. Frost, 2014. Removal of methyl orange from aqueous solutions through adsorption by calcium aluminate hydrates. *Journal of Colloid and Interface Science* 426, 44–47.
105. R. Zhang, J. Zhang, X. Zhang, C. Dou, R. Han, 2014. Adsorption of Congo red from aqueous solutions using cationic surfactant modified wheat straw in batch mode: Kinetic and equilibrium study. *Journal of the Taiwan Institute of Chemical Engineers*. (In Press).
106. V. M. Vucurovic, R. N. Razmovski, U. D. Miljic, V. S. Puskas, 2014. Removal of cationic and anionic azo dyes from aqueous solutions by adsorption on maize stem tissue. *Journal of the Taiwan Institute of Chemical Engineers* 45(4), 1700–1708.
107. J. Saikia, G. Das, 2014. Framboidal vaterite for selective adsorption of anionic dyes. *Journal of Environmental Chemical Engineering* 2(2), 1165–1173.
108. H. Kim, S.-O. Kang, S. Park, H. S. Park, 2014. Adsorption isotherms and kinetics of cationic and anionic dyes on three-dimensional reduced graphene oxide macrostructure. *Journal of Industrial and Engineering Chemistry*. (In Press).
109. F. Güzel, H. Saygili, G. A. Saygili, F. Koyuncu, 2014. Elimination of anionic dye by using nanoporous carbon prepared from an industrial biowaste. *Journal of Molecular Liquids* 194, 130–140.
110. B. Zhao, Y. Shang, W. Xiao, C. Dou, R. Han, 2014. Adsorption of Congo red from solution using cationic surfactant modified wheat straw in column model. *Journal of Environmental Chemical Engineering* 2(1), 40–45.
111. Y. Zheng, Y. Zhu, A. Wang, 2014. Highly efficient and selective adsorption of malachite green onto granular composite hydrogel. *Chemical Engineering Journal* 257, 66–73.
112. L. D. L. Miranda, C. R. Bellato, M. P. F. Fontes, M. F. de Almeida, J. L. Milagres, L. A. Minim, 2014. Preparation and evaluation of hydrotalcite-iron oxide magnetic organocomposite intercalated with surfactants for cationic methylene blue dye removal. *Chemical Engineering Journal* 254, 88–97.
113. P. Sharma, B. K. Saikia, M. R. Das, 2014. Removal of methyl green dye molecule from aqueous system using reduced graphene oxide as an efficient adsorbent: Kinetics, isotherm and thermodynamic parameters. *Colloids and Surfaces A: Physicochemical and Engineering Aspects* 457, 125–133.

114. K. G. Bhattacharyya, S. S. Gupta, G. K. Sarma, 2014. Interactions of the dye, Rhodamine B with kaolinite and montmorillonite in water. *Applied Clay Science* 99, 7–17.

115. Z. Ding, X. Hu, A. R. Zimmerman, B. Gao, 2014. Sorption and cosorption of lead (II) and methylene blue on chemically modified biomass. *Bioresource Technology* 167, 569–573.

116. M. S. Randelovic, M. M. Purenovic, B. Z. Matovic, A. R. Zarubica, M. Z. Momcilovic, J. M. Purenovic, 2014. Structural, textural and adsorption characteristics of bentonite-based composite. *Microporous and Mesoporous Materials* 195, 67–74.

117. J. Fu, Z. Chen, M. Wang, S. Liu, J. Zhang, J. Zhang, R. Han, Q. Xu, 2014. Adsorption of methylene blue by a high-efficiency adsorbent (polydopamine microspheres): Kinetics, isotherm, thermodynamics and mechanism analysis. *Chemical Engineering Journal.* (In Press).

118. A. Gürses, S. Karaca, C. Doğar, R. Bayrak, M. Açıkyıldız, M. Yalçın, 2004. Determination of adsorptive properties of clay/water system: Methylene blue sorption. *Journal of Colloid and Interface Science* 269, 310–314.

119. J. Yener, T. Kopac, G. Dogu, T. Dogu, 2006. Adsorption of Basic Yellow 28 from aqueous solutions with clinoptilolite and amberlite. *Journal of Colloid and Interface Science* 294(2), 255–264.

120. S. B. Wang, and E. Ariyanto, 2007. Competitive adsorption of malachite green and Pb ions on natural zeolite. *Journal of Colloid and Interface Science* 314(1), 25–31.

121. B. Noroozi, G. A. Sorial, H. Bahrami, M. Arami, 2007. Equilibrium and kinetic adsorption study of a cationic dye by a natural adsorbent-silkworm pupa. *Journal of Hazardous Materials* 139(1), 167–174.

122. T. S. Natarajan, H. C. Bajaj, R. J. Tayade, 2014. Preferential adsorption behaviour of methylene blue dye onto surface hydroxyl group enriched TiO2 nanotube and its photocatalytic regeneration. *Journal of Colloid and Interface Science.* (In Press).

123. L. F. M. Ismail, H. B. Sallam, S. A. Abo Farha, A. M. Gamal, G. E. A. Mahmoud, 2014. Adsorption behaviour of direct yellow 50 onto cotton fiber: Equilibrium, kinetic and thermodynamic profile. *Spectrochimica Acta Part A: Molecular and Biomolecular Spectroscopy* 131, 657–666.

124. N. M. Mahmoodi, O. Masrouri, A. M. Arabi, 2014. Synthesis of porous adsorbent using microwave assisted combustion method and dye removal. *Journal of Alloys and Compounds* 602, 210–220.

125. M. R. Fathi, A. Asfaram, A. Farhangi, 2014. Removal of Direct Red 23 from aqueous solution using corn stalks: Isotherms, kinetics and thermo-dynamic studies. *Spectrochimica Acta Part A: Molecular and Biomolecular Spectroscopy.* (In Press).

126. A. Asfaram, M. R. Fathi, S. Khodadoust, M. Naraki, 2014. Removal of Direct Red 12B by garlic peel as a cheap adsorbent: Kinetics, thermodynamic and equilibrium isotherms study of removal. *Spectrochimica Acta Part A: Molecular and Biomolecular Spectroscopy* 127, 415–421.

127. N. B. Douissa, S. Dridi-Dhaouadi, M. F. Mhenni, 2014. Study of antago-
     nistic effect in the simultaneous removal of two textile dyes onto cellulose
     extracted from Posidonia oceanica using derivative spectrophotometric
     method. *Journal of Water Process Engineering* 2, 1–9.
128. N. Gopal, M. Asaithambi, P. Sivakumar, V. Sivakumar, 2014. Adsorption
     studies of a direct dye using polyaniline coated activated carbon prepared
     from Prosopis juliflora. *Journal of Water Process Engineering* 2, 87–95.
129. L. D. T. Prola, F. M. Machado, C. P. Bergmann, F. E. de Souza, C. R. Gally,
     E. C. Lima, M. A. Adebayo, S. L. P. Dias, T. Calvete, 2013. Adsorption of
     Direct Blue 53 dye from aqueous solutions by multi-walled carbon nanotubes
     and activated carbon. *Journal of Environmental Management* 130, 166–175.
130. Y. S. Al-Degs, M. I. El-Barghouthi, A. H. El-Sheikh, G. M. Walker, 2008.
     Effect of solution pH, ionic strength, and temperature on adsorption behav-
     ior of reactive dyes on activated carbon. *Dyes and Pigments* 77, 16–23.
131. B. Heibati, S. Rodriguez-Couto, A. Amrane, M. Rafatullah, A. Hawari, M. A.
     Al-Ghouti, 2014. Uptake of Reactive Black 5 by pumice and walnut activated
     carbon: Chemistry and adsorption mechanisms. *Journal of Industrial and
     Engineering Chemistry* 20(5), 2939–2947.
132. M. A. Ahmad, N. A. A. Puad, O. S. Bello, 2014. Kinetic, equilibrium and
     thermodynamic studies of synthetic dye removal using pomegranate peel
     activated carbon prepared by microwave-induced KOH activation. *Water
     Resources and Industry.* (In Press).
133. G. Z. Kyzas, E. A. Deliyanni, N. K. Lazaridis, 2014. Magnetic modification
     of microporous carbon for dye adsorption. *Journal of Colloid and Interface
     Science* 430, 166–173.
134. Z. Aksu, I. A. Isoglu, 2007. Use of dried sugar beet pulp for binary biosorption
     of Gemazol Turquoise Blue-G reactive dye and copper (II) ions: Equilibrium
     modeling. *Chemical Engineering Journal* 127(1–3), 177–188.
135. M. A. Baghapour, S. Pourfadakari, A. H. Mahvi, 2014. Investigation of
     Reactive Red Dye 198 removal using multiwall carbon nanotubes in
     aqueous solution. *Journal of Industrial and Engineering Chemistry* 20(5),
     2921–2926.
136. M. Shirzad-Siboni, S. J. Jafari, O. Giahi, I. Kim, S.-M. Lee, J.-K. Yang, 2014.
     Removal of acid blue 113 and reactive black 5 dye from aqueous solutions
     by activated red mud. *Journal of Industrial and Engineering Chemistry* 20(4),
     1432–1437.
137. S. Noreen, H. N. Bhatti, 2014. Fitting of equilibrium and kinetic data for
     the removal of Novacron Orange P-2R by sugarcane bagasse. *Journal of
     Industrial and Engineering Chemistry* 20(4), 1684–1692.
138. M. Khosravi, and S. Azizian, 2014. Adsorption of anionic dyes from aque-
     ous solution by iron oxide nanospheres. *Journal of Industrial and Engineering
     Chemistry* 20(4), 2561–2567.
139. J.-S. Cao, J.-X. Lin, F. Fang, M.-T. Zhang, Z.-R. Hu, 2014. A new absorbent by
     modifying walnut shell for the removal of anionic dye: Kinetic and thermo-
     dynamic studies. *Bioresource Technology* 163, 199–205.

140. R. H. Nia, M. Ghaedi, A. M. Ghaedi, 2014. Modeling of reactive orange 12 (RO 12) adsorption onto gold nanoparticle-activated carbon using artificial neural network optimization based on an imperialist competitive algorithm. *Journal of Molecular Liquids* 195, 219–229.

141. J. M. Rankin, S. Baker, K. J. Klabunde, 2014. Mesoporous aerogel titanium oxide–silicon oxide combinations as adsorbents for an azo-dye. *Microporous and Mesoporous Materials* 190, 105–108.

142. G. Zhang, L. Yi, H. Deng, P. Sun, 2014. Dyes adsorption using a synthetic carboxymethyl cellulose-acrylic acid adsorbent. *Journal of Environmental Sciences* 26(5), 1203–1211.

143. S. Xiao, Z. Wang, H. Ma, H. Yang, W. Xu, 2014. Effective removal of dyes from aqueous solution using ultrafine silk fibroin powder. *Advanced Powder Technology* 25(2), 574–581.

144. F. Ates, and U. T. Un, 2013. Production of char from hornbeam sawdust and its performance evaluation in the dye removal. *Journal of Analytical and Applied Pyrolysis* 103, 159–166.

145. R. Y. L. Yeh, and A. Thomas, 1995. Color removal from dye wastewaters by adsorption using powdered activated carbon: Mass transfer studies. *Journal of Chemical Technology and Biotechnology* 63(1), 48–54.

146. G. Mishra, and M. Tripathy, 1993. A critical review of the treatments for decolourization of textile effluent. *Colourage* 40, 35–38.

147. D. Mantzavinos, R. Hellenbrand, A. G. Livingston, and I. S. Metcalfe, 2000. Beneficial combination of wet oxidation, membrane separation and bio-degradation process for treatment of polymer processing wastewater. *The Canadian Journal of Chemical Engineering* 78, 418–422.

148. J. Wu, M. A. Eiteman, and S. E. Law, 1998. Evaluation of membrane filtration and ozonation process for treatment of reactive-dye wastewater. *Journal of Environmental Engineering* 124(3), 272–277.

149. G. Ciardelli, L. Corsi, and C. Marcucci, 2000. Membrane separation for wastewater reuse in the textile industry. *Resource, Conservation and Recycling* 31, 189–197.

150. X. Zheng, and J. Liu, 2006. Dyeing and printing wastewater treatment using membrane bioreactor with gravity drain. *Bioprocess. Eng.* 23, 205–301.

151. F. I. Hai, K. Yamamoto, K. Fukushi, 2006. Development of a submerged membrane fungi reactor for textile wastewater treatment. *Desalination* 192, 315–322.

152. S. Karcher, A. Kornmuller, M. Jekel, 2001. Cucurbituril for water treatment. Part I: Solubility of cucurbituril and sorption of reactive dyes. *Water Res.* 35(14), 3309–3316.

153. S. Karcher, A. Kornmuller, M. Jekel, 2002. Anion exchange resins for removal of reactive dyes from textile wastewaters. *Water Res.* 36, 4717–4724.

154. M. Hosono, H. Arai, M. Aizawa, I. Yamamoto, K. Shimizu, M. Sugiyama, 1993. Decoloration and degradation of azo dye in aqueous solution of super-saturated with oxygen by irradiation of high-energy electron beams. *Appl. Rad. Iso.* 44, 1199–1203.

155. F. Yuzhu, and T. Viraraghavan, 2001. Fungal decolorization of dye wastewaters: A review. *Bioresource Technology* 79, 251–262.
156. K.-C. Chen, J.-Y. Wu, D.-J. Liou, S.-C. J. Hwang, 2003. Decolorization of the textiles dyes by newly isolated bacterial strains. *Journal of Biotechnology* 101, 57–68.
157. K. G. Bhattacharyya and A. Sharma, 2003. Adsorption characteristics of the dye Brilliant Green. *Dyes and Pigments* 57, 211–222.
158. V. Kumar, L. Wati, P. Nigam, I. M. Banat, B. S. Yadav, D. Singh, and R. Marchant, 1998b. Decolorization and biodegradation of anaerobically digested sugarcane molasses spent wash effluent from biomethanation plants by white-rot fungi. *Process Biochemistry* 33, 83–88.
159. M. N. Hedaoo, A. G. Bhole, N. W. Ingole, and Y.-T. Hung, Biological wastewater treatment. In: *Handbook of Environment and Waste Management: Air and Water Pollution Control*. Y.-T. Hung, L. K. Vang, N. K. Shammas, eds., World Scientific, 2012.
160. A. Mittal, 2011. Biological Wastewater Treatment. *Water Today* 1, 32–44.
161. J. P. A/L A. Dhas, 2008. Removal of COD and colour from textile wastewater using limestone and activated carbon. Thesis Master of Science, University Sains, Malaysia.
162. R. C. Senan, and T. E. Abraham, 2004. Bioremediation of textile azo dyes by aerobic bacterial consortium. *Biodegradation* 15, 275–280.
163. J. Bell, J. J. Plumb, C. A. Buckley, and D. V. Stuckey, 2000. Treatment and decolorization of dyes in an anaerobic baffled reactor. *Journal of Environmental Engineering* 126(11), 1026–1032.
164. G. McMullan, C. Meehan, A. Conneely, N. Kirby, T. Robinson, P. Nigam, I. M. Banat, R. Marchant, W. F. Smyth, 2001. Microbial decolorisation and degradation of textile dyes. *Appl. Microbiol. Biotechnol.* 56, 81–87.
165. D. T. Sponza, and M. Işık, 2002. Decolorization and azo dye degradation by anaerobic/aerobic sequential process. *Enzyme Microb. Techn.* 31(1), 102–110.
166. M. Isık, D. T. Sponza, 2004. Aneorobic/aeorobic treatment of simulated textile wastewater. *Ecology* 14(1), 1–8.
167. Y. Zaoyan, S. Ke, S. Guangliang, Y. Fan, D. Jinshan, M. Huanian,1992. Anaerobic–aerobic treatment of a dye wastewater by combination of RBC with activated sludge. *Water Sci. Technol.* 26, 2093–2096.
168. N. Supaka, K. Juntongjin, S. Damronglerd, M. Delia, P. Strehaiano, 2004. Microbial decolorization of reactive azo dyes in a sequential anaerobic–aerobic system. *Chem. Engin. J.* 99(1), 169–176.
169. IK Kapdan, R. Oztekin, 2006. The effect of hydraulic residence time and initial COD concentration on color and COD removal performance the anaerobic–aerobic SBR system. *J. Hazard. Mater.* 136(8), 896–901.
170. M. S. Khehra, H. S. Saini, D. K. Sharma, B. S. Chadha, S. S. Chimni, 2006. Biodegradation of azo dye C. I. Acid Red 88 by anoxic–aerobic sequential bioreactor. *Dyes Pigments* 70(1), 1–7.
171. T. L. Hu, 1996. Removal of reactive dyes from aqueous solution by different bacterial genera. *Water Science and Technology* 34, 89–95.

172. F. J. Deive, A. Dominguez, T. Barrio, F. Moscoso, P. Moran, M. A. Longo, M. A. Sanroman, 2010. Decolorization of dye Reactive Black 5 by newly isolated thermophilic microorganisms from geothermal sites in Galicia (Spain). *Journal of Hazardous Materials* 182, 735–742.

173. M. S. Alvarez, F. Moscoso, A. Rodriguez, A. Sanroman, F. J. Deive. 2013. Novel physico-biological treatment for the remediation of textile dyes-containing industrial effluents. *Bioresource Technology* 146, 689–695.

174. Z. W. Wang, J. S. Liang, Y. Liang, 2013. Decolorization of Reactive Black 5 by a newly isolated bacterium Bacillus sp. *YZU1. International Biodeterioration & Biodegradation* 76, 41–48.

175. Y. Fu, T. Viraraghavan, 2000. Removal of a dye from an aqueous solution by fungus *Aspergillus niger. Water Quality Research Journal of Canada* 35(1), 95–111.

176. Y. Fu, and T. Viraraghavan, 2001. Removal of Acid Blue 29 from an aqueous solution by *Aspergillus niger. American Association of Textile Chemists and Colorists Review* 1(1), 36–40.

177. M. Arica, and G. Bayramoglu, 2007. Biosorption of Reactive Red 120 dye from aqueous solution by native and modified fungus biomass preparations of *Lentinus sajor-caju. Journal of Hazardous Materials* 149, 499–507.

178. T. O'Mahony, E. Guibal, J. M. Tobin, 2002. Reactive dye biosorption by *Rhizopus arrhizus* biomass. *Enzyme Microb. Technol.* 31, 456–463.

179. Z. Aksu, and S. S. Cagatay, 2006. Investigation of biosorption of Gemazol Turquoise Blue-G reactive dye by dried Rhizopus arrhizus in batch and continuous systems. *Separation and Purification Technology* 48, 24–35.

180. A. Ozer, G. Akkaya, M. Turabik, 2006. The removal of Acid Red 274 from wastewater: Combined biosorption and biocoagulation with Spirogyra rhizopus. *Dyes Pigments* 71, 83–89.

181. A. B. D. Santos, I. A. E. Bisschops, F. J. Cervantes, J. B. van Lier, 2004. Effect of different redox mediators during thermophilic azo dye reduction by anaerobic granular sludge and comparative study between mesophilic (30°C) and thermophilic (55°C) treatments for decolorisation of textile wastewaters. *Chemosphere* 55, 1149–1157.

182. A. B. D. Santos, M. P. Madrid, F. A. M. de Bok, A. J. M. Stams, J. B. van Lier, F. J. Cervantes, 2006. The contribution of fermentative bacteria and methanogenic archaea to azo dye reduction by a thermophilic anaerobic consortium. *Enzyme Microb. Technol.* 39, 38–46.

183. O. K. Mahadwad, P. A. Parikh, R. V. Jasra, C. Patil, 2011. Photocatalytic degradation of Reactive Black-5 dye using TiO_2 impregnated ZSM-5. *Bull. Mater. Sci.* 34(3), 551–556.

184. S. Erdal, and M. Taskin, 2010. Uptake of textile dye Reactive Black-5 by *Penicillium chrysogenum* MT-6 isolated from cement-contaminated soil. *African Journal of Microbiology Research* 4(8), 618–625.

2

Novel Carbon-Based Nanoadsorbents for Removal of Synthetic Textile Dyes from Wastewaters

Shamik Chowdhury[1], Rajasekhar Balasubramanian[1] and Papita Das*,[2]

[1]*Department of Civil and Environmental Engineering, National University of Singapore, Singapore*
[2]*Department of Chemical Engineering, Jadavpur University, Kolkata, West Bengal, India*

Abstract

Adsorption technology is widely recognized as one of the most powerful methods of decolorizing textile effluents. However, the development of efficient inexpensive adsorbent materials remains an ongoing challenge. Nanostructured carbon materials such as carbon nanotubes and graphene are currently attracting immense interest as a promising new class of adsorbent because of their large surface area and exceptional physicochemical properties. This chapter (a) presents a comprehensive update of the latest developments in the use of carbon-based nanoadsorbents for removal of colorants from dye effluents, (b) briefly discusses their characteristics, advantages and limitations, and (c) identifies future research challenges and opportunities in this exciting and emerging research field.

Keywords: Adsorption, carbon nanomaterials, carbon nanotubes, dyes, graphene, textile effluents

Acronyms

AFM: Atomic force microscopy
FTIR: Fourier transform infrared spectroscopy
SEM: Scanning electron microscopy

Corresponding author: papitasaha@gmail.com

Sanjay K. Sharma (ed.) Green Chemistry for Dyes Removal from Wastewater, (35–82)
© 2015 Scrivener Publishing LLC

TGA: Thermogravimetric analysis
XPS: X-ray photoelectron spectroscopy

2.1 Introduction

Textile manufacturing involves several processes (e.g., desizing, scouring, bleaching, rinsing, mercerizing, dyeing and finishing) which generate large volumes of colored effluents [1]. The presence of dyes and pigments in water, even at very low concentrations, is highly undesirable and represents a serious environmental problem due to their negative ecotoxicological effects and bioaccumulation in wildlife [2]. The removal of dyes from textile effluents is thus of prime importance. During the past three decades, several physical, chemical, and biological methods have been developed for decoloration of textile wastewaters with varying levels of success. These include coagulation, flocculation, microfiltration, ultrafiltration, reverse osmosis, adsorption, ion-exchange, irradiation, sonochemical degradation, photocatalytic degradation, oxidation using chlorine, chlorine dioxide, hydrogen peroxide and Fenton's reagent, ozonation, electrocoagulation, electrochemical destruction, aerobic or anaerobic treatment, and microbial degradation [3–5]. Amongst these techniques, adsorption is widely acknowledged as the most promising and efficient method because of its low capital investment, simplicity of design, ease of operation, insensitivity to toxic substances and complete removal of pollutants even from dilute solutions [6–11]. Adsorption treatment also does not result in any harmful substances and produces a high quality treated effluent.

Activated carbon, a crude form of graphite, is undoubtedly the most preferred adsorbent because of its highly porous structure and large surface area [10,12]. However, its widespread use is restricted due to economic considerations [13]. Attempts have thus been made by many researchers to find inexpensive alternative substitutes to activated carbon. Most research undertaken for this purpose has focused on the use of waste/byproducts from industries (e.g., fly ash, bottom ash, steelplant slag, red mud, metal hydroxide sludge) and agricultural operations (e.g., rice husk, orange peel, banana peel, sawdust, soybean hull), natural materials (e.g., bentonite, kaolinite, diatomite, zeolites, dolomite), or microbial and non-microbial biomass [4,10,14–19]. Nevertheless, these low-cost adsorbents have been largely criticized for their low adsorption capacities and potential disposal problems and have thus not been applied at an industrial scale [16]. Therefore, the exploration of new promising adsorbents with higher adsorption capacity and better regeneration ability is still in progress.

In recent years, nanotechnology has introduced a myriad of novel nanomaterials that can have promising outcomes in environmental clean-up and remediation [20]. Particularly, carbon-based nanomaterials such as carbon nanotubes and graphene are being intensively studied as new types of adsorbents for removal of toxic pollutants from aquatic systems. This extraordinary interest stems from their unique morphology, nanosized scale and novel physicochemical properties [21–24]. Both carbon nanotubes and graphene offer chemically inert surfaces for physical adsorption, and their high specific surface areas stand comparison with those of activated carbon. They also have far more well-defined and uniform structure than activated carbon and most other adsorbent materials reported till date. In addition, these nanoadsorbents are not only capable of sequestering contaminants with varying molecular size, but also have considerably high adsorption capacities and can be easily regenerated for repeated use. With the aforementioned, this chapter reviews the potential of carbon-based nanoadsorbents (CBNAs), viz., carbon nanotubes and graphene, in treating colored effluents. The advantages and limitations of CBNAs in adsorptive treatment of dye wastewaters are evaluated. To highlight their treatment performance, selected information such as optimum pH, reaction time and temperature, the initial concentration of dye pollutant, and adsorption capacity is also presented. We further emphasize the role of surface modification of these nanosystems, and discuss its effects on the adsorption properties of CBNAs. Finally, the chapter concludes by highlighting the key challenges involved in the development of these novel nanoadsorbents to help identify future research directions for this emerging field to continue to grow.

2.2   Basic Properties of Carbon Nanoadsorbents

2.2.1   Carbon Nanotubes

Ever since their discovery in 1991 by Ijima, carbon nanotubes (CNTs) have been the focus of considerable research because of their unprecedented physical and chemical properties [25]. CNTs are cylinder-shaped macromolecules made up of a hexagonal lattice of carbon atoms analogous to the atomic planes of graphite (i.e., graphene), normally capped at their ends by one half of a fullerene-like molecule [26]. They can be either single walled (SWCNTs) with diameters as small as 0.4 nm and length of up to centimeters, or multi-walled (MWCNTs) consisting of nested tubes with outer diameters ranging from 5 to 100 nm and lengths of tens of microns (Figure 2.1a–c) [27,28]. The constituent cylinders within MWCNTs are separated by ~0.35 nm, similar to the basal plane separation in graphite, and may possess different chiral structures (Figure 2.1b). A special case of MWCNTs is the double-walled CNT composed of just two concentric cylinders (Figure 2.1c).

Carbon nanotubes have extremely high tensile strength ($\approx$150 GPa, more than 100 times that of stainless steel), high Young's modulus ($\approx$1 TPa), large aspect ratio, low density (1.1–1.3 g cm^{-3}, one-sixth of that of stainless steel), good chemical and environmental stability, high thermal conductivity (~3000W m^{-1} K^{-1}, comparable to diamond) and high electrical conductivity (comparable to copper) [30]. Thus CNTs are very promising in field emitters [31], nanoelectronic devices [32], electrodes for electrochemical double-layer capacitors [33], hydrogen storage [34], and many more areas.

In environmental engineering, CNTs are envisaged as a promising new class of adsorbent for the remediation of toxic pollutants because of their highly porous and hollow structure, large specific surface area, light mass density, surface functional groups and hydrophobic surfaces [22,35]. Due

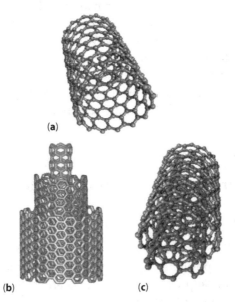

(a)

(b)   (c)

Figure 2.1 Structural representation of (a) a SWCNT, (b) a MWCNT made up of three shells of different chirality and (c) a double-walled CNT. Reprinted from [26]; Copyright 2004, with permission from John Wiley and Sons, and from [29]; Copyright 2009, with permission from the American Chemical Society.

to this unique combination of properties, CNTs are expected to have strong interactions with both organic and inorganic contaminants. There are basically four possible sites in CNT bundles for the adsorption of different pollutants (see Figure 2.2) [36,37]:

i.   *Internal sites*: These sites are found within the hollow interior of individual tubes and are accessible only if the caps are removed and the open ends are unblocked.

ii. *Interstitial channels*: These sites are found in the interior space between individual nanotubes in the bundles and are easily accessible to the adsorbate species.

iii. *External groove sites*: Grooves are present on the periphery of a nanotube bundle and the exterior surface of the outermost nanotubes, where two adjacent parallel tubes meet.

iv. *Outside surface*: The adsorbate can also bind to the curved surface of individual tubes on the outside of the nanotube bundles.

A consensus is that adsorption on close-ended CNTs first takes place in the grooves between adjacent tubes on the perimeter of the bundles, followed by the adsorption on the convex external walls [22]. For purified opened CNT bundles, the first step of adsorption proceeds by the

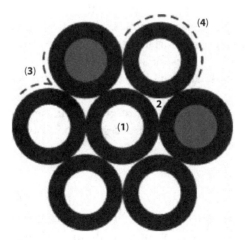

Figure 2.2 Different adsorption sites on a homogeneous bundle of partially open-ended SWCNTs: (1) internal, (2) interstitial channel, (3) external groove site, and (4) external surface. Sites 1 and 2 comprise the internal porous volume of the bundle, whereas sites 3 and 4 are both located on the external surface of the bundle. Reprinted from [38]; Copyright 2006, with permission from the American Chemical Society.

population of the walls inside the opened nanotubes and formation of one-dimensional chains in the grooves at the outer surface of the bundles. The second step can be assigned to the filling of the remaining axial sites inside the nanotubes and the completion of a quasi-hexagonal monolayer on the outer surfaces of the bundles [22]. It is worth noting that the adsorption reaches equilibrium much faster on external sites (grooves and outer surfaces) than on the internal sites (interstitial channels and inside the tube) under the same temperature and pressure conditions.

The overall performance of CNTs in an adsorption treatment process depends on several factors, the most important being the fraction of opened and unblocked nanotubes. Opened CNT bundles provide more adsorption sites than capped CNTs, resulting in fast sorption kinetics and increased saturation capacity. In addition, CNTs are often found mixed with impurities such as carbon-coated catalyst particles, soot, and other forms of carbons, which can significantly decrease their adsorption efficiency. These impurities can be removed by using several treatment methods like acid/base treatment, heat treatment and so forth [23]. Furthermore, the surface functionalities also influence the maximum adsorption capacity of CNTs. CNTs mostly contain the oxygen functional groups such as hydroxyl ($-OH$), carbonyl ($-CO$), and carboxyl ($-COOH$) on their surface, formed during the synthesis procedure and purification process, or can also be intentionally generated by oxidation using various acids, ozone, or plasma [39–44]. These functional groups can change the wettability of CNT surfaces, and make them more hydrophilic and suitable for the adsorption of relatively low molecular weight and polar compounds. They may also increase diffusion resistance and decrease the surface area, which can reduce the accessibility and affinity of CNT surfaces for some organic chemicals [45]. Meanwhile, the functional groups can also block the access to the interior space of the uncapped tubes. By subjecting the nanotubes to a high-temperature treatment under vacuum, these chemical groups can be removed in order to make the internal space of the tube accessible for adsorption [46,47].

2.2.2   Graphene

Graphene is the newest member in the family of carbon allotropes and has emerged as the "celeb" material of the 21st century. Since its successful isolation by the exfoliation of graphite in 2004 by Kostya Novoselov, his fellow Nobel laureate Andre Geim and their group at the University of Manchester, graphene has been attracting enormous attention in both scientific and engineering communities. Graphene is a single atomic layer of sp^2 hybridized carbon atoms, densely packed into an ordered two-dimensional honeycomb network (Figure 2.3) [48]. It is the thinnest, yet strongest material known to man, being both brittle and ductile simultaneously [49,50]. In its purest form, graphene is impermeable to even the smallest gas molecules, including Helium. This one-atom-thick allotrope of carbon can be viewed as the basic structural unit of other carbon allotropes. It can be wrapped into zero-dimensional fullerenes, rolled into one-dimensional CNTs, or can be stacked into three-dimensional graphite (Figure 2.4) [51].

A hexagonal unit cell of graphene comprises two equivalent sub-lattices of carbon atoms, joined together by σ bonds with a carbon–carbon bond length of 0.142 nm [52]. Each carbon atom in the lattice has a π orbital that contributes to a delocalized network of electrons, making graphene sufficiently stable compared to other nanosystems [53]. Theoretical and experimental studies have proved that graphene offers a unique combination of high three-dimensional aspect ratio and large specific surface area, superior mechanical stiffness and flexibility, remarkable optical transmittance, exceptionally high electronic and thermal conductivities, as well as many other supreme properties, as shown in Table 2.1. Because of these fascinating properties, it is not surprising that graphene has the potential to be used in a plethora of applications across many fields. The wide range

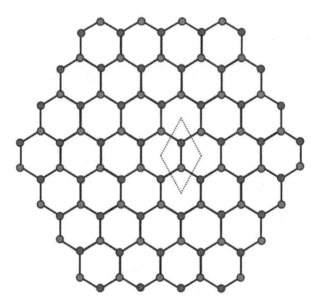

Figure 2.3 Representation of the honeycomb lattice of graphene and its unit cell (indicated by the dashed lines). The unit cell contains two atoms, each one belonging to a different sub-lattice.

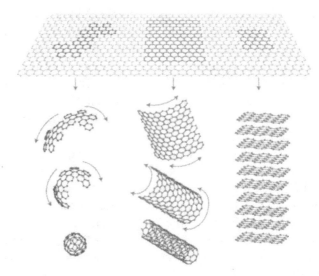

Figure 2.4 Graphene as a basic building block for graphitic materials of all other dimensionalities. It can be wrapped up into 0-dimensional fullerenes, rolled into 1-dimensional nanotubes or stacked into 3-dimensional graphite. Reprinted from [54], Copyright 2007, with permission from Macmillan Publishers Ltd.

Table 2.1 Physical properties of single-layer graphene at room temperature [53, 55–62].

Property	Value
C-C bond length	0.142 nm
Theoretical BET-specific surface area*	2630 $m^2\,g^{-1}$
Young's modulus	1,100 GPa
Tensile strength	125 Gpa
Carrier density	$10^{12}\,cm^{-2}$
Resistivity	$10^{-6}\,\Omega$ cm
Electron mobility	200, 000 $cm^2\,V^{-1}\,s^{-1}$
Thermal conductivity	5,000 $W\,m^{-1}\,K^{-1}$
Optical transparency	97.7%

*BET: Brunauer-Emmett-Teller

of graphene's applications includes: nanoelectronics [63], structural composites [64], conducting polymers [64], battery electrodes [65,66], supercapacitors [67], transport barriers [68,69], printable inks [70], antibacterial papers [67], and biomedical technologies [71,72].

More recently, the multifarious applications of graphene have encouraged not only the development of substrate-bound extended graphene monolayers, but also related materials such as graphene oxide (GO), reduced graphene oxide (RGO), and few-layered graphene oxide (FGO) [73]. The product of chemical exfoliation of graphite, GO is a highly oxidative form of graphene consisting of a variety of oxygen functionalities in its graphitic backbone: carboxyl (—COOH) and carbonyl (—C=O) groups at the sheet edges and epoxy (C—O—C) and hydroxyl (—OH) groups on the basal plane (Figure 2.5) [64,74]. The existence of GO has been known since the middle of the 19th century, but it was only during the last decade that GO has attracted considerable research interest due to its role as a precursor for the cost-effective and mass-production of graphene-based materials [75].

Apart from being a precursor material for preparing graphene, GO itself has many remarkable properties. It can be characterized as an unconventional soft material such as a two-dimensional polymer, anisotropic colloid, soft membrane, liquid crystal, or even an amphiphile [76]. The oxygen-containing functional groups render it a good candidate for application in diverse fields of science and engineering. The functional groups also provide reactive sites for a variety of surface modification reactions to develop functionalized GO and graphene-based materials for a wider range of applications [77]. Additionally, the disruptions of the sp^2 bonding network by these functional groups makes GO electrically insulating [76]. The conductivity can, however, be partially recovered by restoring the π-network through chemical, thermal, or electrochemical reduction of GO to graphene-like sheets, i.e., RGO [75]. Although RGO is more defective and thus less conductive than pristine graphene, it is sufficiently conductive for many applications. The restoration of graphitic network in the basal plane of RGO facilitates its frequent modification by noncovalent physisorption

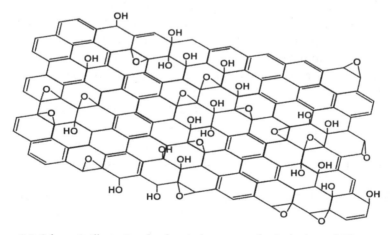

Figure 2.5 Schematic illustrating the chemical structure of a single sheet of GO.

of both polymers and small molecules via π-π stacking or van der Waals interactions [75]. As a result, interest in RGO has also spread across many disciplines.

Undoubtedly the discovery of graphene has brought about enormous opportunities for science and technology. Because of their perfect sp^2 hybrid carbon nanostructure, large specific surface area, and strong interactions with other molecules or atoms, graphene nanomaterials have recently attracted the interest of researchers as a new type of adsorbent for the removal of harmful contaminants from aqueous systems. Large numbers of experimental studies have already been carried out on the adsorption of different dye pollutants by graphene nanomaterials, as will be discussed in the following section.

2.3   Adsorpton of Textile Dyes by Carbon Nanoadsorbents

2.3.1   Adsorption by CNTs and Their Composites

2.3.1.1   CNTs

Numerous investigations have been undertaken to evaluate the adsorption potential of CNTs for treatment of dye-contaminated waters (Table 2.2). Yao *et al.* [78] explored the possibility of using CNTs for removal of Methylene Blue from its aqueous solution. The adsorption equilibrium data were well described by the Langmuir model, implying monolayer coverage of dye molecules onto the adsorbent surface. The maximum dye adsorption capacity increased from 35.4 mg g^{-1} at 273 K to 64.7 mg g^{-1} at 333 K, indicating an endothermic nature of the adsorption process.

Shirmardi *et al.* [79] studied the performance of SWCNTs as adsorbent for the removal of Acid Red 18 from aqueous solutions. The adsorption equilibrium data showed an excellent fit to the Langmuir isotherm model, with a maximum monolayer dye adsorption of 166.66 mg g^{-1}. In another study, Moradi [80] investigated the adsorption of Basic Red 46 onto SWCNTs. Adsorption of Basic Red 46 was strongly dependent on the initial solution pH with maximum uptake occurring at about pH 9. The adsorption process was spontaneous and exothermic in nature but dominated by physisorption. The maximum dye uptake capacity was determined to be 38.35 mg g^{-1} at 298 K.

Kuo *et al.* [81] tested the adsorption efficiency of MWCNTs for removing direct dyes (Direct Yellow 86 and Direct Red 224) from their aqueous

Table 2.2 Reported results of batch adsorption studies on the removal of dyes from water by CNTs.

Adsorbent	Target dye	Conc.	pH	Temp. (K)	Contact time (h)	Adsorption capacity	Isotherm	Kinetic model	Thermodynamics	Ref.
CNTs	Methylene Blue	5–40 mg L⁻¹	7.0	273 298 333	1.5	35.4 mg g⁻¹ 46.2 mg g⁻¹ 64.7 mg g⁻¹	Langmuir	Pseudo-second-order	Endothermic	[78]
SWCNTs	Acid Red 18	25–100 mg L⁻¹	3.0	298	6	166.66 mg g⁻¹	Langmuir	Pseudo-second-order	—	[79]
SWCNTs	Basic Red 46	50–200 mg L⁻¹	9.0	298 308 318 328	3	38.35 mg g⁻¹ 33.12 mg g⁻¹ 30.12 mg g⁻¹ 27.16 mg g⁻¹	Langmuir	Pseudo-second-order	Exothermic	[80]
MWCNTs	Direct Yellow 86	—	—	288 308 328	4	35.8 mg g⁻¹ 54.9 mg g⁻¹ 56.2 mg g⁻¹	Freundlich	Pseudo-second-order	Endothermic	[81]
MWCNTs	Direct Red 224	—	—	288 308 328	4	47.2 mg g⁻¹ 52.1 mg g⁻¹ 61.3 mg g⁻¹	Dubinin-Radushkevich	Pseudo-second-order	Endothermic	[81]
MWCNTs	Procion Red MX-5B	20 mg L⁻¹	6.5	281 291 301 321	24	42.92 mg g⁻¹ 44.64 mg g⁻¹ 39.84 mg g⁻¹ 35.71 mg g⁻¹	Langmuir, Freundlich	Pseudo-second-order	Endothermic	[82]
MWCNTs	Methylene Blue	—	6.0	298	2	59.7 mg g⁻¹	Langmuir, Freundlich	Pseudo-second-order	—	[83]

(Continued)

Table 2.2 (Cont.)

Adsorbent	Target dye	Conc.	pH	Temp. (K)	Contact time (h)	Adsorption capacity	Isotherm	Kinetic model	Thermodynamics	Ref.
MWCNTs	Acid Red 183	—	6.0	298	2	45.2 mg g^{-1}	Langmuir, Freundlich	Pseudo-second-order	—	[83]
MWCNTs	Safranine O	10–50 mg L^{-1}	1.0	Room temp.	0.5	43.48 mg g^{-1}	Temkin	Pseudo-second-order	Endothermic	[84]
MWCNTs	Reactive Red M-2BE	300–600 mg L^{-1}	2.0	298 303 308 313 318 323	1	312.3 mg g^{-1} 319.2 mg g^{-1} 325.1 mg g^{-1} 327.8 mg g^{-1} 330.0 mg g^{-1} 335.7 mg g^{-1}	Liu	Avrami fractional	Endothermic	[85]
MWCNTs	Methyl Orange	10 mg L^{-1}	—	293 303 313	2.5	44.16 mg g^{-1} 47.56 mg g^{-1} 54.09 mg g^{-1}	—	Pseudo-second-order	—	[86]
MWCNTs	Methyl Orange	20–40 mg L^{-1}	—	273 298 318 333	3	50.25 mg g^{-1} 51.784 mg g^{-1} 52.43 mg g^{-1} 52.86 mg g^{-1}	Langmuir	Pseudo-second-order	Endothermic	[87]
MWCNTs	Methylene Blue	—	—	303 313 338	—	54.54 mg g^{-1} 52.51 mg g^{-1} 43.98 mg g^{-1}	Langmuir	—	—	[88]
MWCNTs	Orange II	—	—	303 338	—	66.12 mg g^{-1} 57.34 mg g^{-1}	Langmuir	—	—	[88]

MWCNTs	Acid Blue 161	10–200 mg L⁻¹	3.0	298	1	1000 mg g⁻¹	Freundlich, Temkin	Pseudo-second-order	—	[89]
MWCNTs	Reactive Blue 29	5–30 mg L⁻¹	2.0	298	3	55.7 mg g⁻¹	Freundlich	—	—	[90]
MWCNTs	Arsenazo III	10–40 mg L⁻¹	1.0	293 303 313	0.25	10.2 mg g⁻¹ 10.8 mg g⁻¹ 14.08 mg g⁻¹	Langmuir, Freundlich, Temkin	Pseudo-second-order	Endothermic	[91]
MWCNTs	Alizarin Red S	50–200 mg L⁻¹	1.0	—	0.5	161.29 mg g⁻¹	Langmuir	Pseudo-second-order	Endothermic	[92]
MWCNTs	Morin	50–200 mg L⁻¹	1.0	—	0.5	26.24 mg g⁻¹	Langmuir	Pseudo-second-order	Endothermic	[92]
MWCNTs	Acid Red 183	—	6.0	298 308	2	45.237.4	Langmuir	Pseudo-second-order	Exothermic	[93]
MWCNTs	Reactive Blue 4	—	6.0	298 308	2.3	68.253.0	Langmuir	Pseudo-second-order	Exothermic	[93]
f-SWCNTs	Basic Red 46	50–200 mg L⁻¹	9.0	298 308 318 328	3	49.45 mg g⁻¹ 45.33 mg g⁻¹ 41.23 mg g⁻¹ 39.12 mg g⁻¹	Langmuir	Pseudo-second-order	Exothermic	[80]
f-MWCNTs	Bromothymol Blue	10–70 mg L⁻¹	1.0	293.15	—	55.25 mg g⁻¹	Langmuir	Pseudo-second-order	Endothermic	[94]

(Continued)

Table 2.2 (*Cont.*)

Adsorbent	Target dye	Conc.	pH	Temp. (K)	Contact time (h)	Adsorption capacity	Isotherm	Kinetic model	Thermodynamics	Ref.
f-MWCNTs	Methyl Red	5–60 mg L^{-1}	4.0	Room temp.	0.4	108.69 mg g^{-1}	Langmuir	Pseudo-second-order	Endothermic	[95]
f-MWCNTs	Congo Red	50–400 mg L^{-1}	3.0	—	—	148.08 mg g^{-1}	Langmuir, Freundlich, Temkin	Pseudo-first-order	—	[96]
f-MWCNTs	Reactive Green HE4BD	50–400 mg L^{-1}	5.0	—	—	151.88 mg g^{-1}	Langmuir, Freundlich, Temkin	Pseudo-first-order	—	[96]
f-MWCNTs	Golden Yellow MR	50–400 mg L^{-1}	7.0	—	—	141.61 mg g^{-1}	Langmuir, Freundlich, Temkin	Pseudo-first-order	—	[96]
f-MWCNTs	Malachite Green	25–100 mg L^{-1}	7.0	298	1.3	142.85 mg g^{-1}	Langmuir	Pseudo-first-order	—	[97]
Chitosan/MWCNTs	Congo Red	10–1000 mg L^{-1}	5.0	303	—	450.4 mg g^{-1}	Langmuir, Sips	Pseudo-second-order	—	[98]
Calcium alginate/MWCNTs	Methyl Orange	—	7.0	298 ± 1	2	14.13 mg g^{-1}	Langmuir	—	—	[99]
NiFe$_2$O$_4$/MWCNTs	Toluidine Blue	20–65 mg L^{-1}	—	298 313 323	4.1	25 mg g^{-1} 32.26 mg g^{-1} 38.02 mg g^{-1}	Langmuir	Pseudo-second-order	Endothermic	[100]

Adsorbent	Dye	Concentration	pH	Temperature	Time	q_{max}	Isotherm	Kinetics		Reference
CNTs/Activated carbon fabric	Basic Violet 10	—	—	303	7	103.1 mg g^{-1}	Langmuir, Dubinin-Radushkevich	—	—	[101]
CNTs/A-Fe$_2$O$_3$	Methyl Orange	20–80 mg L^{-1}	—	298	3	28.8 mg g^{-1}	Freundlich	Pseudo-second-order	—	[102]
Fe$_3$O$_4$-MWCNTs	Methylene Blue	10–30 mg L^{-1}	—	298 ± 1	2	48.06 mg g^{-1}	Langmuir	Pseudo-second-order	—	[103]
MWCNTs/γ-Fe$_2$O$_3$	Methylene Blue	20 mg L^{-1}	6.0	298	4	42.3 mg g^{-1}	Freundlich	—	—	[104]
MWCNTs/γ-Fe$_2$O$_3$	Neutral Red	20 mg L^{-1}	6.0	298	4	77.5 mg g^{-1}	Freundlich	—	—	[104]
Magnetic MWCNTs	Crystal Violet	20 mg L^{-1}	7.0	Room temp.	0.25	227.27 mg g^{-1}	Langmuir	—	—	[105]
Magnetic MWCNTs	Thionine	20 mg L^{-1}	7.0	Room temp.	0.25	36.63 mg g^{-1}	Langmuir	—	—	[105]
Magnetic MWCNTs	Janus Green B	20 mg L^{-1}	7.0	Room temp.	0.25	250.00 mg g^{-1}	Langmuir	—	—	[105]
Magnetic MWCNTs	Methylene Blue	20 mg L^{-1}	7.0	Room temp.	0.25	48.08 mg g^{-1}	Langmuir	—	—	[105]
Magnetic MWCNTs	Brilliant Cresyl Blue	1.4–37.4 mg L^{-1}	7.0	298 ± 2	24	6.28 mg g^{-1}	Freundlich	Pseudo-second-order	—	[106]

(Continued)

Table 2.2 (*Cont.*)

Adsorbent	Target dye	Conc.	pH	Temp. (K)	Contact time (h)	Adsorption capacity	Isotherm	Kinetic model	Thermodynamics	Ref.
Magnetic MWCNTs	Methylene Blue	1.4–37.4 mg L^{-1}	7.0	298 ± 2	24	11.86 mg g^{-1}	Freundlich	Pseudo-second-order	—	[106]
Magnetic MWCNTs	Neutral Red	1.4–37.4 mg L^{-1}	7.0	298 ± 2	24	9.77 mg g^{-1}	Freundlich	Pseudo-second-order	—	[106]
Magnetic Polymer MWCNTs	Orange II	5–25 mg L^{-1}	6.2	298 308 318	6	67.57 mg g^{-1} 109.89 mg g^{-1} 83.33 mg g^{-1}	Langmuir	Pseudo-second-order	—	[107]
Magnetic Polymer MWCNTs	Sunset Yellow FCF	5–25 mg L^{-1}	6.2	298 308 318	6	85.47 mg g^{-1} 98.04 mg g^{-1} 76.92 mg g^{-1}	Langmuir	Pseudo-second-order	—	[107]
Magnetic Polymer MWCNTs	Amaranth	5–25 mg L^{-1}	6.2	298 308 318	6	47.39 mg g^{-1} 63.29 mg g^{-1} 57.14 mg g^{-1}	Langmuir	Pseudo-second-order	—	[107]
Magnetic chitosan/ γ-Fe$_2$O$_3$/ MWCNTs	Methyl Orange	5–50 mg L^{-1}	—	297 312 323	5	66.09 mg g^{-1} 60.94 mg g^{-1} 60.50 mg g^{-1}	Langmuir	Pseudo-second-order	Exothermic	[108]
MWCNT-Starch-Iron oxide	Methylene Blue	373.9 mg L^{-1}	—	—	6	93.7 mg g^{-1}	—	Pseudo-second-order	—	[109]

Adsorbent	Dye	Concentration		Temp.		q_{max}	Isotherm	Kinetics	Thermodynamics	Reference
MWCNT-Starch-Iron oxide	Methyl Orange	327.33 mg L⁻¹	—	—	6	135.6 mg g⁻¹	—	Pseudo-second-order	—	[109]
Magnetic MWCNT-Fe₃C	Direct Red 23	20 mg L⁻¹	7.0	303	5.8	172.4 mg g⁻¹	Freundlich	Pseudo-second-order	Endothermic	[110]
Magnetic MWCNT-Fe₃C	Acid Red 88	10–58 mg L⁻¹	7.0	303	4.5	54.4 mg g⁻¹	Freundlich	Pseudo-second-order	Exothermic	[111]
Guar gum-MWCNT-Fe₃O₄	Neutral Red	0.05–0.6 mmol L⁻¹	—	—	6	89.85 mg g⁻¹	Langmuir	Pseudo-second-order	—	[112]
Guar gum-MWCNT-Fe₃O₄	Methylene Blue	0.05–0.6 mmol L⁻¹	—	—	6	61.92 mg g⁻¹	Langmuir	Pseudo-second-order	—	[112]
Polyurethane foam/Diatomite/MWCNTs	Acridine Orange	—	—	Room temp.	3	421.73 µg g⁻¹	Langmuir	—	—	[113]
Polyurethane foam/Diatomite/MWCNTs	Methylene Blue	—	—	Room temp.	3	378.38 µg g⁻¹	Langmuir	—	—	[113]

(Continued)

Table 2.2 (*Cont.*)

Adsorbent	Target dye	Conc.	pH	Temp. (K)	Contact time (h)	Adsorption capacity	Isotherm	Kinetic model	Thermodynamics	Ref.
Polyurethane foam/ Diatomite/ MWCNTs	Ethidium Bromide	—	—	Room temp.	3	296.11 µg g^{-1}	Langmuir	—	—	[113]
Polyurethane foam/ Diatomite/ MWCNTs	Eosin Y	—	—	Room temp.	3	243.2 µg g^{-1}	Langmuir	—	—	[113]
Polyurethane foam/ Diatomite/ MWCNTs	Eosin B	—	—	Room temp.	3	202.71 µg g^{-1}	Langmuir	—	—	[113]

solutions. The dye decolorization potential increased with increasing adsorbent dose and reaction temperature, while it decreased with an increase in initial dye concentration. The equilibrium adsorption data of Direct Yellow 86 were best fitted to the Freundlich isotherm, while that of Direct Red 224 to the Dubinin-Radushkevich isotherm. The MWCNTs exhibited a stronger binding affinity for Direct Red 224 than Direct Yellow 86. Thermodynamic analyses indicated that the adsorption of direct dyes onto MWCNTs was endothermic and spontaneous.

The feasibility of MWCNTs for removal of the reactive dye Procion Red MX-5B was investigated by Wu et al. [82]. Both the Langmuir and the Freundlich models were found suitable to describe the equilibrium dye adsorption data, suggesting that monolayer sorption as well as heterogeneous energetic distribution of active sites on the surface of the adsorbent were possible. The adsorption process followed the pseudo-second-order kinetics with activation energy of 33.35 kJ mol^{-1}, indicating physisorption of Procion Red MX-5B by MWCNTs. The MWCNTs have also been successfully applied to the adsorption of other reactive dyes including Reactive Red M-2BE and Reactive Blue 29 [85,90].

Geyikci [89] studied the adsorption characteristics of Acid Blue 161 onto MWCNTs as a function of reaction time, pH and initial dye concentration. The maximum adsorption yield (91.68%) was obtained at the conditions of pH 3.0, initial dye concentration = 50 mg L^{-1}, adsorbent dose = 0.1 g L^{-1}, temperature = 298 K and contact time = 1 h. The adsorption kinetics was found to obey the pseudo-second-order mechanism. The intraparticle diffusion was not the sole rate-controlling step. Adsorption equilibrium data fitted well to the Freundlich isotherm model as well as the Temkin model. MWCNTs exhibited a significantly high adsorption capacity of 1000 mg g^{-1} for Acid Blue 161.

The efficacy of MWCNTs to adsorb anthraquinone dyes, Alizarin Red S and Morin, from wastewaters has also been investigated [92]. The maximum monolayer adsorption capacity was found to be 161.29 and 26.24 mg g^{-1} for Alizarin Red S and Morin, respectively. The adsorption process followed pseudo-second-order kinetics with involvement of the intraparticle-diffusion mechanism.

Azo dyes have also been successfully removed from aqueous systems using MWCNTs as adsorbent. Yao et al. [87] investigated the adsorption of Methyl Orange onto MWCNTs from aqueous solutions by performing simple batch adsorption experiments. The MWCNTs exhibited a fairly high adsorption capacity of approximately 52.86 mg g^{-1} with an adsorbent loading of 0.3 g L^{-1}. The adsorption followed pseudo-second-order kinetics and the equilibrium data fitted well with the Langmuir isotherm. Additionally, the adsorption process was found to be endothermic and thermodynamically favorable. Adsorption of Methyl Orange by MWCNTs has also been studied by Zhao *et al.* [86]. Ghaedi *et al.* [91] reported the removal of Arsenazo III using MWCNTs as adsorbent. The effects of pH, temperature, initial dye concentration, adsorbent dose, and contact time on the decolorization efficiency were systematically studied. More than 96% dye removal was recorded at pH 2. Kinetic studies showed that adsorption of Arsenazo III onto MWCNTs obeyed the pseudo-second-order model and that intraparticle diffusion was not the sole rate-controlling step.

Adsorption of cationic dye (Methylene Blue) and anionic dye (Acid Red 183) on MWCNTs was examined by Wang *et al.* [83] in single and binary dye systems. The MWCNTs showed higher adsorption capacity towards Methylene Blue in both the systems. The authors explained that Methylene Blue, being a planar molecule, could easily approach MWCNTs via a face-to-face conformation, which is favorable for π-π interactions between the conjugated aromatic chromophore skeleton and the nanotubes. On the contrary, non-planar molecules like Acid Red 183 are kept apart from MWCNTs due to the spatial restriction, resulting in low π-π interactions with the MWCNTs. Similarly, the adsorption of Reactive Blue 4 and Acid Red 183 on MWCNTs was investigated in single and binary dye systems by Wang *et al.* [93]. In single dye systems, adsorption of Reactive Blue 4 and Acid Red 183 followed a pseudo-second-order kinetic mechanism, and the equilibrium data could be well described by the Langmuir isotherm. From the viewpoint of structure and properties of the dyes and MWCNT, the adsorption was attributed to strong electrostatic attraction, as illustrated in Figure 2.6. On the other hand, in binary dye systems, MWCNTs could only adsorb Reactive Blue 4 and did not show any binding selectivity for Acid Red 183. Such adsorption pattern was due to the competition between the dyes for the same electrophilic sites and MWCNT's preference for adsorbing planar molecules (Figure 2.6).

A comparative study of MWCNTs and activated carbon as adsorbents for the removal of the textile dye Safranine O from aqueous solutions was performed by Ghaedi *et al.* [84]. The equilibrium isotherm data best fitted to the Langmuir isotherm model with the maximum dye sorption capacity of MWCNTs being 43.48 mg g^{-1} and that of activated carbon 1.32 mg g^{-1}. In another study, the comparative adsorption of cationic Methylene Blue and anionic Orange II dyes from aqueous solution by using MWCNTs and carbon nanofibers (CNF) as adsorbents was explored in batch experiments by Rodríguez and coworkers [88]. The adsorption of Methylene Blue onto CNF was slightly higher than adsorption onto MWCNTs, while for Orange II, the adsorption capacity of CNF was considerable lower than that of MWCNTs.

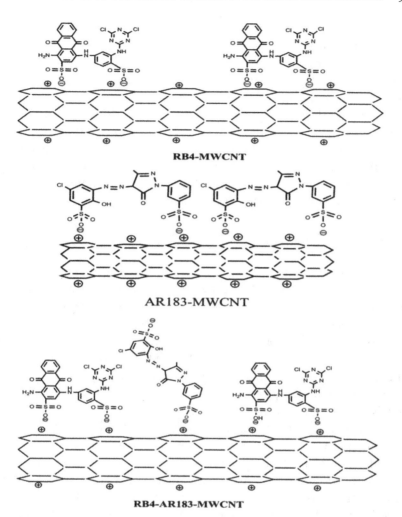

Figure 2.6 Adsorption of Acid Red 183 and Reactive Blue 184 ontoMWCNTs in single and binary dye systems. Reprinted from [93]; Copyright 2012, with permission from the American Chemical Society.

Recently, functionalized CNTs (ƒ-CNTs) are also being intensively studied for treatment of dye bearing effluents. The ƒ-CNTs are hydrophilic in nature, resulting in high adsorption uptake of cationic and anionic dyes from aqueous solution due to the presence of oxygen-containing functional surface groups. Moradi [80] prepared ƒ-SWCNTs having covalent attachments of carboxylic groups (SWCNT–COOH) and tested their potential to adsorb Basic Red 46 from aqueous solutions. It was confirmed that functionalization of SWCNT improved its dye-binding capacity. In another study by Mishra *et al.* [96], *f*-MWCNTs developed by an acid treatment method have been used as an efficient adsorbent material for three different azo dyes: Congo Red, Reactive Green HE4BD and Golden Yellow MR. An extremely fast dye uptake rate was observed with the experimental kinetic data conforming to the pseudo-first-order model. Adsorption isotherms for all the three dyes followed Langmuir, Freundlich and Temkin models. The maximum dye adsorption capacity of *f*-MWCNTs was found to be 148.08, 151.88 and 141.61 mg g^{-1} for Congo Red, Reactive Green HE4BD and Golden Yellow MR, respectively. Similarly, *f*-MWCNTs have also been found effective to remove Bromothymol Blue [94], Methyl Red [95], and Malachite Green [97] from aqueous solutions.

2.3.1.2   CNT-Based Nanocomposites

From a series of discussions we had in the previous section of this chapter, it is evident that CNTs have an extremely high affinity for many classes of dyes. Compared to activated carbon, CNTs offer a number of advantages including fast adsorption rate and small adsorbent dose, which make them attractive alternatives for the selective removal of toxic dyes from industrial wastewaters. However, the separation of CNTs from aqueous medium after treatment is very difficult because of their smaller size and high aggregation property. This problem can be overcome by making composites of CNTs with different materials such as polymers, metal oxide, carbon, etc., which act as a stable matrix to the CNTs, thereby facilitating the easy separation of CNTs from the aqueous environment. These materials also provide additional active sites to CNTs, which in turn makes them better adsorbents compared to their parent nanomaterial. As such, a number of CNT-based composites have been developed and used as a promising adsorbent for the treatment of colored wastewaters in recent years. The adsorption characteristics of these nanocomposites will be discussed in this section.

Chatterjee *et al.* [98] impregnated MWCNTs with chitosan hydrogel beads and used it as adsorbent to remove Congo Red from aqueous solution by batch adsorption process. The kinetic data conformed to the pseudo-second-order kinetic model with intraparticle diffusion playing a significant role during the initial stage of adsorption. The equilibrium data fitted well to the Langmuir isotherm model. The composite exhibited a remarkably high adsorption capacity of 450.4 mg g^{-1}. Zhu *et al.* [99] investigated the adsorption of Methyl Orange onto calcium alginate/MWCNTs composites with respect to MWCNT content, initial dye concentration and pH. The adsorption uptake increased with increasing MWCNT content in the composite, while it decreased with increasing solution pH. Based on the Langmuir isotherm, the maximum equilibrium adsorption capacity of Methyl Orange was 14.13 mgg^{-1}. Recently, Bhagat et al. [100] prepared NiFe$_2$O$_4$/MWCNT composite materials for removal of Toluidine Blue. The amount of dye adsorbed increased with NiFe$_2$O$_4$/MWCNT dosage. The activation energy of the adsorption process calculated using the Arrhenius equation suggested that adsorption of Toluidine Blue onto NiFe$_2$O$_4$/CNTs involved physisorption.

In order to resolve the aggregation and dispersion problem of CNTs, Wang et al. [101] prepared CNTs/activated carbon fabric (CNTs/ACF) composite and tested its efficacy for the removal of Basic Violet 10 from aqueous solutions. The experimental data obtained showed that the dye uptake process was very fast and could be well characterized by three different types of isotherm models: Freundlich, Langmuir, and Dubinin-Radushkevich. The maximum monolayer adsorption of Basic Violet 10 onto CNTs/ACF was 220 mgg^{-1}.

The incorporation of magnetic property in CNTs is another useful approach to separate CNTs from test solutions. The magnetic adsorbent can be well dispersed in the aqueous environment and can also be easily separated magnetically. Madrakian et al. [105] developed magnetic-MWCNTs, by depositing Fe$_2$O$_3$ nanoparticles onto MWCNTs, for removal of cationic dyes Thionine, Crystal Violet, Janus Green B and Methylene Blue. Batch adsorption experiments were conducted to study the effect of different operating parameters including initial pH, adsorbent dosage and contact time on the dye uptake process. Optimum adsorption occurred at pH 7.0 for all the dyes. The maximum monolayer adsorption capacities for Crystal Violet, Thionine, Janus Green B and Methylene Blue dyes were calculated as 227.7, 36.63, 250.0, and 48.08 mg g^{-1}, respectively. Desorption studies were conducted by washing the spent adsorbent with different eluents such as methanol, N,N-dimethyl formamide and acetonitrile. For all the studied dyes, the desorption equilibrium time was 2 min and the desorption efficiency was found in the order: methanol < N,N-dimethyl formamide < acetonitrile. After five successive cycles, no change in the adsorption behavior was observed. It was therefore recommended that magnetic MWCNTs could be used as a good reusable and economical adsorbent for treatment of colored effleunts. Magnetic MWCNTs have also been identified as a suitable adsorbent for removal of Brilliant Cresyl Blue [106], Methylene Blue [103,104,106], Methyl Orange [102], and Neutral Red [104,106] by several other researchers.

Also, a number of ternary nanocomposites have also been developed and tested for the decoloration of textile effluents. Gao et al. [107]

synthesized magnetic polymer MWCNT (MPMWCNT) nanocomposites composed of MWCNTs, poly (1-glycidyl-3-methylimidazoliumchloride) (an ionic liquid-based polyether) and Fe_3O_4 nanoparticles. The adsorption capacity of MPMWCNT for anionic azo dyes, Orange II, Sunset Yellow FCF and Amaranth, were examined. The effect of solution pH was investigated and the adsorption kinetics and isotherms were also studied. It was found that a low pH value favored the adsorption of anionic azo dyes and the adsorption kinetic and equilibrium data fitted well to the pseudo-second-order model and Langmuir isotherm, respectively. It was inferred that the introduction of ionic liquid-based polyether and Fe_3O_4 moieties significantly improved the adsorption and separation performance of MWCNTs.

Yan et al. [112] prepared guar gum grafted MWCNTs/Fe_3O_4(GG/MWCNT/Fe_3O_4) ternary composite for the removal of Neutral Red and Methylene Blue. The pseudo-second-order kinetic model provided a better correlation for the experimental kinetic data in comparison to the pseudo-first-order kinetic model. The adsorption equilibrium could be best represented by the Langmuir isotherm model, with maximum monolayer adsorption capacity of 89.85 and 61.92 mg g^{-1} for Neutral Red and Methylene Blue, respectively. The high adsorption capacity of GG/MWCNT/Fe_3O_4 was ascribed to the hydrophilic property of guar gum, which improved the dispersion of GG/MWCNT/Fe_3O_4 in the solution, and also facilitated the diffusion of dye molecules to the surface of CNTs. GG/MWCNT/Fe_3O_4 also exhibited a high saturation magnetization of 13.3 emu g^{-1}, and could thus be easily separated from the aqueous solution by applying a suitable magnetic field.

Chang et al. [109] developed starch-functionalized MWCTs/iron oxide composites to improve the hydrophilicity and biocompatibility of MWCNTs. The as-synthesized MWCNT–starch–iron oxide was used as an adsorbent for removing Methyl Orange and Methylene Blue from aqueous solutions. The nanocomposite exhibited superparamagnetic properties with a saturation magnetization of 23.15 emug^{-1} and a high removal capacity of 93.7 and 135.6 mg g^{-1} for Methyl Orange and Methylene Blue, respectively.

A novel ternary magnetic composite composed of chitosan wrapped magnetic nanosized γ-Fe_2O_3 and multi-walled carbon nanotubes (m-CS/γ-Fe_2O_3/MWCNTs) was prepared for the removal of Methyl Orange by Zhu et al. [108]. The adsorption capacity of Methyl Orange onto m-CS/γ-Fe_2O_3/MWCNTs was 2.2 times higher than m-CS/γ-Fe_2O_3. The adsorption reaction followed the pseudo-second-order kinetics, while the adsorption isotherm was well described by the Langmuir model with a maximum adsorption capacity of 66.09 mg g^{-1} at 297 K. The results from thermodynamic studies indicated that the adsorption was feasible, spontaneous and exothermic in nature.

Yu and Fugetsu [113] developed a novel nanoadsorbent by inserting MWCNTs into the cavities of dolomite for scavenging of Ethidium Bromide, Acridine Orange, Methylene Blue, Eosin B, and Eosin Y dyes from wastewater. Polyurethane polymer was used as binder to produce the foam line MWCNTs/dolomite adsorbent. The adsorption process reached equilibrium within 30 min for the cationic dyes Acridine Orange, Ethidium Bromide, and Methylene Blue, while it was about 60 min for the anionic dyes, Eosin B and Eosin Y. The Langmuir adsorption isotherm model best fitted to the experimental equilibrium data. To evaluate the possibility of regeneration of the dye-saturated adsorbent, desorption experiments were carried out using used ethanol/water (50/50) solution in an ultrasonic generator. After three adsorption/regeneration cycles, the dye uptake capacity of the ternary composite remained unchanged.

Recently, Konikci *et al.* [110] used magnetic MWCNT-Fe$_3$C nanocomposite (MMWCNTs-Fe$_3$C) as an adsorbent for the removal of Direct Red 23 from aqueous solution. The effects of various parameters such as initial dye concentration (9–54 mg L^{-1}), solution pH (3.7–11.1) and temperature (293–333 K) were investigated through batch adsorption studies. The adsorption kinetics followed the pseudo-second-order model, while a thermodynamic assessment indicated the endothermic and spontaneous nature of adsorption of Direct Red 23 on MMWCNTs-Fe$_3$C. The Freundlich isotherm model showed a good fit to the experimental equilibrium data, implying that adsorption of Direct Red 23 on MMWCNTs-Fe$_3$C was multilayer and applicable to heterogeneous surfaces. More recently, Konikci *et al.* [111] studied the adsorption of Acid Red 88 by MMWCNTs-Fe$_3$C composites. Although quite similar results were obtained, the adsorption of Acid Red 88 onto MMWCNTs was found to be spontaneous and exothermic in nature.

2.3.2   Adsorption by Graphene and Its Related Materials

In recent years, graphene and its related materials, including GO and RGO, as well as their nanocomposites, have all been widely examined as a promising class of adsorbents for removal of synthetic dyes from aqueous environments (Table 2.3). Without going into too much detail, some of the latest important results on the dye adsorption characteristics of graphene materials are discussed in the following sections.

Table 2.3 Reported results of batch adsorption studies on the removal of dyes from water by graphene and its related materials.

Adsorbent	Target dye	Conc.	pH	Temp. (K)	Contact time (h)	Adsorption capacity	Isotherm	Kinetic model	Thermodynamics	Ref.
Graphene	Cationic Red X-GRL	20–140 mg L^{-1}	—	288 313 333	24	217.39 mg g^{-1} 227.27 mg g^{-1} 238.10 mg g^{-1}	Langmuir	Pseudo-second-order	Endothermic	[114]
Graphene	Methylene Blue	20–120 mg L^{-1}	—	293 313 333	—	153.85 mg g^{-1} 185.19 mg g^{-1} 204.08 mg g^{-1}	Langmuir	Pseudo-second-order	Endothermic	[115]
Graphene	Methyl Blue	5 mg L^{-1}	—	303	96	1.52 g g^{-1}	—	—	—	[116]
Graphene sponge	Methylene Blue	2 × 10-4mol L^{-1}	—	298	4	184 mg g^{-1}	—	—	—	[117]
Graphene sponge	Rhodamine B	2 × 10-4mol L^{-1}	—	298	4	72.5 mg g^{-1}	—	—	—	[117]
Graphene sponge	Methyl Orange	2 × 10-4mol L^{-1}	—	298	24	11.5 mg g^{-1}	—	—	—	[117]
GO	Methylene Blue	0.188–1.000 g g^{-1}	6.0	298	1	714 mg mg^{-1}	Freundlich	—	—	[118]
GO	Methylene Blue	0.33–3.3 mg L^{-1}	7.0	293	2	1.939 mg mg^{-1}	Langmuir	—	—	[119]
GO	Methylene Blue	10–50 mg L^{-1}	10.0	—	—	17.3 mg g^{-1}	Langmuir	Pseudo-second-order	—	[120]
GO	Methyl Violet	10–50 mg L^{-1}	6.0	—	—	2.47 mg g^{-1}	Langmuir	Pseudo-second-order	—	[120]

GO	Rhodamine B	1–10 mg L^{-1}	6.0	—	—	1.24 mg g^{-1}	Freundlich	Pseudo-second-order	—	[120]
GO	Acridine Orange	0.1 g L^{-1}	—	Room temp.	3	1428 mg g^{-1}	Langmuir	—	—	[121]
GO	Methylene Blue	40–120 mg L^{-1}	6.0	298	5	240.65 mg g^{-1}	Langmuir	Pseudo-second-order	—	[122]
GO	Methyl Green	—	4.0 5.0 6.0 7.0 9.0	398	1	4.821 mmol g^{-1} 5.496 mmol g^{-1} 6.167 mmol g^{-1} 6.628 mmol g^{-1} 7.613 mmol g^{-1}	Langmuir	Pseudo-second-order	Endothermic	[123]
In situ reduced GO	Acridine Orange	0.1 g L^{-1}	—	Room temp.	3	3333 mg g^{-1}	Langmuir	—	—	[121]
RGO	Orange G	1–60 mg L^{-1}	—	—	—	5.98 mg g^{-1}	Langmuir	Pseudo-second-order	—	[120]
RGO-based hydrogel	Methylene Blue	0.5–10 mg L^{-1}	6.4	298	2	7.85 mg g^{-1}	Freundlich	Pseudo-second-order	—	[124]
RGO-based hydrogel	Rhodamine B	0.5–10 mg L^{-1}	6.4	298	2	29.44 mg g^{-1}	Freundlich	Pseudo-second-order	—	[124]
Graphene/Fe$_3$O$_4$	Fuchsine	20–60 mg L^{-1}	6.6 ± 0.2	298	1	89.4 mg g^{-1}	Langmuir	Pseudo-second-order	—	[125]

(Continued)

Table 2.3 (Cont.)

Adsorbent	Target dye	Conc.	pH	Temp. (K)	Contact time (h)	Adsorption capacity	Isotherm	Kinetic model	Thermodynamics	Ref.
Graphene/Magnetite	Methylene Blue	10–25 mg L^{-1}	–	298	–	43.82 mg g^{-1}	Langmuir	Pseudo-second-order	Endothermic	[126]
Magnetite@Graphene	Congo Red	–	–	298 ± 052	–	33.66 mg g^{-1}	Langmuir	Pseudo-second-order	–	[127]
Magnetite@Graphene	Methylene Blue	–	–	298 ± 0.5	–	45.27 mg g^{-1}	Langmuir	Pseudo-second-order	–	[127]
Graphene/Fe$_3$O$_4$	Pararosaniline	20–60 mg L^{-1}	6.6 ± 0.2	298	1	198.23 mg g^{-1}	Langmuir, Freundlich	Pseudo-second-order	–	[128]
Graphene-SO$_3$H/Fe$_3$O$_4$	Safranine T	20–250 mg L^{-1}	6.0	Room temp.	–	199.3 mg g^{-1}	Langmuir	Pseudo-second-order	–	[129]
Graphene-SO$_3$H/Fe$_3$O$_4$	Neutral Red	20–250 mg L^{-1}	6.0	Room temp.	–	216.8 mg g^{-1}	Langmuir	Pseudo-second-order	–	[129]
Graphene-SO$_3$H/Fe$_3$O$_4$	Victoria Blue	20–250 mg L^{-1}	6.0	Room temp.	–	200.6 mg g^{-1}	Langmuir	Pseudo-second-order	–	[129]
CoFe$_2$O$_4$-Functionalized graphene	Methyl Orange	10 mg L^{-1}	–	–	–	71.54 mg g^{-1}	–	Pseudo-second-order	–	[130]

Adsorbent	Dye	Concentration	pH	Temperature	Time	Capacity	Isotherm	Kinetics	Thermodynamics	Reference
Graphene/CoFe$_2$O$_4$	Methyl Green	50–400 mg L^{-1}	—	298 313 323	—	203.51 mg g^{-1} 258.39 mg g^{-1} 312.80 mg g^{-1}	Langmuir	Pseudo-second-order	Endothermic	[131]
Graphene/Sand	Rhodamine 6G	5 mg L^{-1}	—	303 ± 2	8	55 mg g^{-1}	—	Pseudo-second-order	—	[132]
Graphene-Sand	Rhodamine 6G	—	—	303 ± 2	6	75.4 mg g^{-1}	—	Pseudo-second-order	—	[133]
Graphene-CNT	Methylene Blue	10–30 mg L^{-1}	—	—	3	81.97 mg g^{-1}	Freundlich	Pseudo-second-order	—	[134]
Graphene/c-MWCNT	Rhodamine B	20 mg L^{-1}	—	Room temp.	—	150.2 mg g^{-1}	—	Pseudo-second-order	—	[135]
Graphene/c-MWCNT	Methylene Blue	20 mg L^{-1}	—	Room temp.	—	191.0 mg g^{-1}	—	Pseudo-second-order	—	[135]
Graphene/c-MWCNT	Fuchsine	20 mg L^{-1}	—	Room temp.	—	180.8 mg g^{-1}	—	Pseudo-second-order	—	[135]
Graphene/c-MWCNT	Acid Fuchsine	20 mg L^{-1}	—	Room temp.	—	35.8 mg g^{-1}	—	Pseudo-second-order	—	[135]
PES/GO	Methylene Blue	50–250 µmol L^{-1}	7.0	303 ± 1	60	62.50 mg g^{-1}	Langmuir	Pseudo-second-order	—	[136]

(Continued)

Table 2.3   (Cont.)

Adsorbent	Target dye	Conc.	pH	Temp. (K)	Contact time (h)	Adsorption capacity	Isotherm	Kinetic model	Thermodynamics	Ref.
Polystyrene @ Fe_3O_4@GO	Rhodamine B	0–150 mg L^{-1}	—	Room temp.	24	13.8 mg g^{-1}	—	—	—	[137]
Fe_3O_4-SiO_2-GO	Methylene Blue	—	—	298 318 333	—	97 mg g^{-1} 102.6 mg g^{-1} 111.1 mg g^{-1}	Langmuir	Pseudo-second-order	Endothermic	[138]
Graphene-Chitosan	Methylene Blue	0–80 mg L^{-1}	6.5	294 ± 1	58	390 mg g^{-1}	—	—	—	[139]
Graphene-Chitosan	Eosin Y	0–80 mg L^{-1}	7.0	294 ± 1	36	326 mg g^{-1}	—	—	—	[139]
Magnetic chitosan/GO	Methyl Blue	60–200 mg L^{-1}	5.3	303	1	95.31 mg g^{-1}	Langmuir	Pseudo-second-order	—	[140]
Magnetic chitosan/GO	Methylene Blue	50–100 mg L^{-1}	—	303 ± 0.2	—	180.83 mg g^{-1}	Langmuir	Pseudo-second-order	—	[141]
Magnetic β-cyclodextrin-chitosan/GO	Methylene Blue	—	—	298	—	84.32 mg g^{-1}	Langmuir	Pseudo-second-order	Endothermic	[142]
GO/Calcium alginate	Methylene Blue	30–80 mg L^{-1}	—	298	5	181.81 mg g^{-1}	Langmuir	Pseudo-second-order	Exothermic	[143]

Magnetite/RGO	Rhodamine B	7.0	298	2	13.15 mg g^{-1}	Langmuir, Freundlich	Pseudo-second-order	—	[144]
Magnetite/RGO	Malachite Green	7.0	298	2	22.0 mg g^{-1}	Langmuir, Freundlich	Pseudo-second-order	—	[144]
RGO@ZnO	Rhodamine B	—	293	—	32.6 mg g^{-1}	—	—	—	[145]
RGO-Titanate	Methylene Blue	—	—	—	83.26 mg g^{-1}	—	Pseudo-second-order	—	[146]

2.3.2.1   Graphene

Liu *et al,* [115] reported the use of graphene as an adsorbent for removal of Methylene Blue from its aqueous solution. The adsorption capacity decreased with increasing adsorbent dose, while it increased with increasing temperature. The maximum adsorption capacity increased from 153.85 mg g^{-1} to 204.08 mg g^{-1} with the rise in temperature from 293 K to 333 K. The adsorption kinetics was studied in terms of pseudo-first-order, pseudo-second-order, Elovich and intraparticle diffusion models. It was found that the pseudo-second-order mechanism was predominant and that intraparticle diffusion was not the sole rate-controlling step. Adsorption equilibrium data fitted well to the Langmuir isotherm model rather than the Freundlich model. Estimation of the thermodynamic parameters indicated that adsorption of Methylene Blue onto graphene was a spontaneous, endothermic and physisorption process.

Adsorption of Cationic Red X-GRL onto graphene was studied by Li *et al,* [114]. The maximum dye adsorption capacity obtained from the Langmuir isotherm equation was 153.85 mg g^{-1} at 293 K. Similar to the investigations of Liu *et al.* [115], the adsorption process was spontaneous and endothermic, and obeyed the pseudo-second-order rate equation.

Wu *et al.* [116] investigated the use of graphene for removal of Methyl Blue. The amount of dye adsorbed was found to be strongly dependent on the initial concentration of Methyl Blue, and the adsorption process attained equilibrium after 1 h. Graphene showed a remarkably high dye adsorption capacity of 1.52 g g^{-1}, which was mainly due to π-π stacking interactions as inferred through fluorescence spectroscopy studies. The researchers also investigated the possibility to regenerate and reuse graphene and observed no significant changes in the efficiency of graphene, at least during the first five cycles of the adsorption-desorption process.

In a very interesting study by Zhao *et al.* [117], a new graphene material called "graphene sponge" was developed by hydrothermal treatment of GO sheets in the presence of thiourea. The material had a very porous structure and behaved like a sponge (hence the name). Graphene sponge showed excellent mechanical properties, very good processability and structural stability. The feasibility of using graphene sponge to remove cationic (Methylene Blue, Rhodamine B) and anionic (Methyl Orange) dyes from their aqueous solutions was investigated. Graphene sponge was found to have an affinity for both cationic and anionic dyes. However, the sorption capacity for basic dyes was much higher than for acidic dyes because of the ionic charges on the dyes and surface characteristics of graphene sponge. The maximum adsorption uptake was found to be 184, 72.5 and 11.5 mg g^{-1} for Methylene Blue, Rhodamine B and Methyl Orange, respectively.

2.3.2.2 Graphene Oxide

GO has also proven to be an excellent adsorbent for removal of synthetic dyes from aqueous systems. For instance, Yang et al. [118] demonstrated the removal of Methylene Blue from its aqueous solution by using GO as adsorbent. Almost complete removal of Methylene Blue could be achieved at initial dye concentrations of less than 250 mg L^{-1}. The dye removal efficiency increased with an increase in pH and ionic strength. Lower temperatures and the presence of dissolved organic matter favored the adsorption process. Removal of Methylene Blue by GO has also been explored by several other researchers [119,122].

Ramesha et al. [120] evaluated the sorption capacity of GO for the uptake of three different cationic dyes, viz., Methylene Blue, Methyl Violet, and Rhodamine B, from their aqueous solutions. A significantly high adsorption capacity for all three of the dyes was exhibited by GO. However, the sorption preference increased in the order Methylene Blue > Methyl Violet > Rhodamine B. Such behavioral pattern was attributed to the fact that Methylene Blue and Methyl Violet were positively charged, whereas Rhodamine B had both positive and negative charges associated with its structure and hence the electrostatic interactions between Rhodamine B and GO were considerably weaker. In the same study, the utilization of GO for the removal of an acidic dye, Orange G, was also investigated by Ramesha et al. [120]. However, GO showed a poor binding affinity for Orange G. The two sulfonic groups of Orange G made it negatively charged, resulting in electrostatic repulsion between the dye and the adsorbent, and hence no significant removal was observed. Sharma and Das [123] have also tested GO for the removal of Methyl Green dye. An adsorption capacity in the range of 4.82–7.61 mmol g^{-1} was recorded at different pH (4.0–9.0).

A study on the use of modified GO for removing Acridine Orange from its aqueous solution was conducted by Sun et al. [121]. That research attempted to improve the effectiveness of GO as an adsorbent through in situ reduction with sodium hydrosulfite. Parallel adsorption tests under similar experimental conditions, carried out with pristine GO and in situ reduced GO, showed that the latter had a much higher adsorption capacity (3333 mg g^{-1}) than the former (1428 mg g^{-1}). To identify the mechanism of enhancement in adsorption capacity, the structure and morphology of GO before and after the reduction were studied and compared using modern characterization techniques (SEM, FTIR, AFM, XPS, and TGA). It was concluded that the enhancement was probably due to the reduction of carbonyl groups to hydroxyl groups.

2.3.2.3   Reduced Graphene Oxide

The efficacy of RGO in adsorbing dyes has also been explored in recent years. Ramesha et al. [124] investigated the potential of RGO to remove an acidic dye, Orange G, from its aqueous solutions. An excellent removal efficiency of about 95% was recorded. In another study by Tiwari et al. [129], a three-dimensional RGO-based hydrogel was synthesized by chemical reduction of GO with sodium ascorbate. The prepared hydrogel had a very large surface area with uniform pore size distribution, and was tested as an adsorbent for the removal of Methylene Blue and Rhodamine B from aqueous solutions in a batch system. The results showed that with an adsorbent dose of 0.6 g L^{-1}, extremely high removal efficiencies of up to ~100% for Methylene Blue and ~97% for Rhodamine B could be achieved within 2 h at room temperature. The high uptake capacity was due to adsorption through strong π-π stacking and anion–cation interactions. Furthermore, desorption studies demonstrated that the RGO-based hydrogel could be easily regenerated by using an inexpensive solvent such as ethylene glycol and reused for at least three adsorption-desorption cycles without any significant loss in adsorption capacity.

2.3.2.4   Graphene-Based Nanocomposites

The potential application of graphene materials as adsorbent strongly depends on their homogeneous dispersion in the liquid phase as well as their ability to remove different types of contaminants. However, graphene as a bulk material has the tendency to agglomerate and restack to form graphite during liquid processing [147]. On the other hand, GO has a weak binding affinity for anionic compounds due to strong electrostatic repulsion between them. Additionally, both graphene and GO cannot be easily collected and separated from treated water, leading to serious recontamination.

Chemical functionalization of graphene materials is an effective and practical approach which can actually facilitate the dispersion and stabilize graphene to prevent agglomeration [148]. It also can improve their processability and enhance their interaction with different types of substances. The functional groups attached can be nanosized metal oxides (NMOs) such as nanosized TiO_2, ZnO, MnO_2, SiO_2, Fe_3O_4, and $CoFe_2O_4$, or organic polymers such as chitosan, polystyrene, polyaniline, polyurethane and polycaprolactone [149]. The different types of polymers or nanoparticles can be directly decorated on the graphenic sheets, and no molecular linkers are needed to bridge the polymers/nanoparticles and the graphenic sheets, which may prevent additional trap states along the sheets [77]. The resulting nanocomposite is not merely the sum of the individual components, but instead a new material with new functionalities and properties [150]. The NMO/polymer anchored on graphene two-dimensional structures not only arrests the agglomeration and restacking but also increases the available surface area of the graphene sheet alone, leading to high adsorption activity. The incorporated material also provides high selectivity and strong binding of the desired contaminant depending on its structure, size and crystallinity. In contrast, the graphenic materials provide chemical functionality and compatibility to allow easy processing of the deposited NMO/polymer in the composite. The ultimate goal is to maximize the practical use of the combined advantages of both the components as active materials for improving the adsorption performance and potential.

Wang *et al.* [125] synthesized a graphene-based magnetite nanocomposite (G/Fe_3O_4) by *in situ* chemical co-precipitation of Fe^{2+} and Fe^{3+} in alkaline solution in the presence of graphene and investigated its potential as an adsorbent for the removal of Fuchsine dye from aqueous solution. Adsorption isotherm and kinetics of Fuchsine on G/Fe_3O_4 were studied employing a batch experimental set-up. The effects of adsorbent dose (0.2–0.5 g L^{-1}), pH (3.0–10.0), and ionic strength (0–20% w/v NaCl) on the removal efficiency of Fuchsine were investigated. The experimental data obtained showed that the dye uptake process was very fast; about 96% of the dye was adsorbed within 10 min and 99% of the dye was adsorbed within 30 min. The amount of dye adsorbed increased with an increase in pH from 3.0 to 5.5. Further increase in pH did not significantly change the adsorption yield. Ionic strength did not show any direct influence on the dye removal efficiency. The adsorption equilibrium data was best described by the Langmuir isotherm model, with a maximum monolayer adsorption capacity of 89.4 mg g^{-1}. The adsorption process was found to be a pseudo-second-order reaction. Furthermore, maximum desorption of 94% was achieved at pH 2.0 using ethanol as the eluent. The adsorption capacity of G/Fe_3O_4 for Fuchsine did not show any significant decrease even after five regenerations. Thus, G/Fe_3O_4 proved to be a highly efficient adsorbent for removal of color from dye-bearing wastewater. Graphene/magnetite composites have also been prepared and successfully employed as adsorbent for the removal of Methylene Blue [126,127], Congo Red [127] and Pararosaniline [128].

Wang *et al.* [129] synthesized a magnetic-sulfonic graphene nanocomposite ($G\text{-}SO_3H/Fe_3O_4$) and explored it as adsorbent for the batch removal of three cationic dyes: Safranine T, Neutral Red, Victoria Blue, and three anionic dyes: Methyl Orange, Brilliant Yellow, Alizarin Red, from their aqueous solutions. The $G\text{-}SO_3H/Fe_3O_4$ adsorbent showed excellent adsorption capacity towards cationic dyes compared to anionic dyes. More than 93% of the cationic dyes were removed during the first 10 min of dye-adsorbent contact. The maximum adsorption capacity of G-SO$_3$H/Fe$_3$O$_4$ for cationic dyes, calculated from the Langmuir isotherm, increased in the following order: Safranine T < Victoria Blue < Neutral Red. Furthermore, the nanocomposite could be easily regenerated using ethanol (adjusted to pH 2.0 with 0.1 mol L^{-1}HCl) as eluent and reused for at least six adsorption-desorption cycles without any significant loss in adsorption capacity.

Magnetic CoFe$_2$O$_4$-functionalized graphene sheet (CoFe$_2$O$_4$–FGS) nanocomposites were prepared by Li et al. [130] to remove Methyl Orange. The adsorption process followed pseudo-second-order kinetics and the adsorption capacity was found to be as high as 71.54 mg g^{-1}. Recently, Farghali et al. [131] have also synthesized CoFe$_2$O$_4$–FGS nanocomposites for the removal of Methyl Green from aqueous solution. The adsorption isotherm was well described by the Langmuir model, whereas the adsorption kinetics corresponded to the pseudo-second-order kinetic model. Although intraparticle diffusion was involved in the adsorption process, it was not the rate-controlling step. Also, thermodynamic analyses indicated that adsorption of Methyl Green was a spontaneous, endothermic and a physisorption process.

In another study conducted by Sen Gupta et al. [132], graphene immobilized on sand was used as an adsorbent for the removal of Rhodamine 6G. Graphene was prepared in situ from cane sugar and anchored onto the surface of river sand without the need of any additional binder, resulting in a composite, referred to as graphene-sand composite (GSC). The ability of GSC to remove Rhodamine 6G from its aqueous solution was tested through batch and continuous column experiments. The adsorption process followed pseudo-second-order kinetics and equilibrium was attained in 8 h. The equilibrium adsorption capacity of Rhodamine 6G was 55 mg g^{-1} at 303 ± 2 K. Fixed-bed column experiments were performed to study the practical applicability of the adsorbent and breakthrough curves at different bed depths were obtained. The bed depth service time (BDST) model showed good agreement with the dynamic flow experimental data. Desorption studies revealed that GSC could be regenerated using acetone for multiple uses. In general, the results suggested that GSC should be considered for dye wastewater treatment with appropriate engineering. In another study by the same research group [133], GSC was prepared using asphalt as the carbon source and tested as an adsorbent for removal of Rhodamine 6G. Both batch and continuous column experiments were carried out. Similar to their previous work, the experimental data correlated well with the pseudo-second-order model and an adsorption capacity of

75.4 mg g^{-1} was achieved. Fixed-bed column adsorption studies were also conducted in multiple cycles and the results thus obtained confirmed that GSC could be used for water purification applications.

Ai and Jiang [134] reported the efficient removal of Methylene Blue from its aqueous solution by a self-assembled cylindrical graphene-carbon nanotube hybrid (G-CNT). The hybrid showed good adsorption performance with a maximum adsorption capacity of 81.97 mg g^{-1}. Sui *et al.* [135] fabricated graphene-CNT hybrid aerogels by supercritical CO$_2$ drying of their hydrogel precursors obtained from heating the aqueous mixtures of GO and CNTs with vitamin C without stirring. The CNTs used in the study were either pristine MWCNTs or acid-treated MWCNTs (c-MWCNTs). The resulting hybrid aerogels, i.e., graphene/MWCNT and graphene/c-MWCNT, showed excellent adsorption performance in removal of basic dyes (Rhodamine B, Methylene Blue, Fuchsine) from their aqueous solutions. The adsorption was found to be pseudo-second-order with the binding capacity of graphene/c-MWCNT (150.2, 191.0 and 180.8 mg g^{-1} for Rhodamine B, Methylene Blue and Fuchsine, respectively) being higher than that of graphene/MWCNT (146.0, 134.9 and 123.9 mg g^{-1} for Rhodamine B, Methylene Blue and Fuchsine, respectively).

Zhang *et al.* [136] developed polyethersulfone (PES) enwrapped GO porous particles to remove Methylene Blue from aqueous solutions. Batch isotherm and kinetic studies were carried out to study the effects of contact time, initial dye concentration (50–250 µmol L^{-1}), pH (3.0–11.0), and temperature (288–333 K) on the adsorption phenomena. Unlike other adsorption systems with fast equilibration time, the adsorption of Methylene Blue attained equilibrium after about 60 h. The delayed equilibrium was ascribed to the porous structure of the particles. The interior micropores were abundant for which the diffusion of dye molecules took a long time. Both temperature and pH were found to have a significant influence on the adsorption of Methylene Blue. The experimental adsorption equilibrium data correlated well with the Langmuir isotherm model. The pseudo-second-order kinetic model provided a better correlation for the experimental kinetic data in comparison to the pseudo-first-order kinetic model. The dye uptake process was controlled by intraparticle diffusion. Enwrapping of GO with PES facilitated its easy separation from aqueous environment after adsorption. However, it was found that pure GO had a higher adsorption uptake capacity than the prepared porous particles. After blending with PES, the adsorption capacity of GO decreased. Moreover, the mass of GO in the particles did not show a linear relationship with the adsorption capacity. This was mainly due to the coverage of the adsorption sites by PES and the agglomeration between GO sheets. The investigators ruled out that the adsorption capacity of the porous particles was still significantly higher than that of many other carbon-based adsorbents and should therefore be considered for treatment of textile effluents.

The adsorption of Rhodamine B by core-shell structured polystyrene-Fe_3O_4-GO nanocomposites was studied by Wang et al. [137]. The maximum adsorption capacity was found to be 13.8 mg g^{-1}. The same research group also investigated the removal of Rhodamine B onto RGO/ZnO composite [145]. The composite showed an excellent recycling performance for dye adsorption with up to 99% recovery over four cycles. In another study conducted by Yao et al. [138], Fe_3O_4/SiO_2-GO nanocomposite was synthesized by a covalent bonding technique to remove Methylene Blue. Isotherm data best fitted the Langmuir model with maximum adsorption capacities of 97.0, 102.6, and 111.1 mg g^{-1} at 298, 318, and 333 K, respectively.

Cheng et al. [147] developed a three-dimensional chitosan-graphene nanocomposite with large specific surface area and unique mesoporosity and used it as an adsorbent to remove Reactive Black 5 from its aqueous solution. A removal efficiency of 97.5% was obtained with an initial dye concentration of 1 mg L^{-1}. It is worth mentioning that graphene used in that study was prepared from graphite derived from waste sugarcane bagasse. The potential of GO-chitosan (GO-CS) composite hydrogel to remove acidic (Eosin Y) and basic (Methylene Blue) dyes from water was explored by Chen et al. [139]. The equilibrium adsorption capacities were reported to be 390 and 326 mg g^{-1} for Methylene Blue and Eosin Y, respectively. The investigators also reported that GO-CS hydrogel could be used as a column packing material to fabricate a continuous water purification process.

A novel magnetic chitosan-GO (MCGO) nanocomposite has been developed by covalently binding chitosan on the surface of Fe_3O_4 nanoparticles, followed by covalent functionalization of GO with magnetic chitosan, by Fan et al. [140]. Simple batch adsorption experiments were conducted to estimate the adsorption properties of MCGO for Methyl Blue. The linearized Langmuir isotherm model best represented the experimental equilibrium data. A maximum monolayer adsorption capacity of 95.31 mg g^{-1} was recorded. The dye uptake kinetics followed a pseudo-second-order mechanism. The values of thermodynamic parameters indicated the spontaneous and exothermic nature of the adsorption process. The MCGO could be easily regenerated using 0.5 mol L^{-1} NaOH and the adsorption capacity was about 90% of the initial saturation adsorption capacity after four adsorption-desorption cycles. In a separate study, Fan et al. [141] have found that MCGO also has extraordinary adsorption capacity and fast removal rate for Methylene Blue. The maximum monolayer adsorption capacity was found to be 180.83 mg g^{-1}. The adsorption of Methylene Blue by magnetic β-cyclodextrin–chitosan/GO nanocomposites has also been studied by Fan *et al.* [142]. In this case, as much as 84.32 mg g^{-1} of Methylene Blue could be adsorbed as determined by the Langmuir model.

The possibility of using GO/calcium alginate (GO/CA) composite as an adsorbent for the removal of Methylene Blue from its aqueous solution has recently been explored by Li *et al.* [143]. The optimum pH for dye removal was found to be in the range of 4.5 to 10.2. The adsorption capacity decreased from 163.93 to 140.85 mg g^{-1} with an increase in temperature from 298 to 328 K, indicating the exothermic nature of the adsorption process. A maximum adsorption capacity of 181.81 mg g^{-1} for an adsorbent dose of 0.05 g per 100 mL dye solution was recorded. The adsorption kinetic data could be best described by the pseudo-second-order rate equation.

RGO-based nanocomposites have also been developed and investigated for their dye removal potential. Sun *et al.* [144] prepared magnetite/RGO (MRGO) nanocomposites for the removal of Rhodamine B and Malachite Green dyes from aqueous solutions. The MRGO nanocomposite exhibited excellent dye removal efficiency, with over 91% of Rhodamine B and over 94% of Malachite Green being removed within 2 h of dye-MRGO contact at room temperature. Desorption studies showed that by using an inexpensive eluent such as ethylene glycol, MRGO could be subjected to multiple rounds of recycle and reuse without any significant change in its adsorption efficiency. In order to further evaluate the practical applicability of the prepared adsorbent material, real water samples, including local industrial wastewater and lake water collected from Lake Tai in China, were first contaminated with dyes and then treated using the as-prepared MRGO. It was found that real water samples had little interference with the decolorization efficiency of MRGO. In addition, MRGO also showed excellent removal efficiency for other dye pollutants including Crystal Violet and Methylene Blue from industrial wastewater. Nguyen-Phan *et al.* [146] fabricated RGO–titanate (RGO–Ti) hybrids by incorporating spherical TiO$_2$ nanoparticles within GO layers in the presence of NaOH, followed by solvothermal treatment. The adsorption characteristic of Methylene Blue onto the RGO-Ti hybrids was then investigated. The equilibrium adsorption capacity was reported as 83.26 mg g^{-1} and the kinetics of adsorption indicated the process to be chemisorption.

2.4 Mechanism of Dye Adsorption onto Carbon-Based Nanoadsorbents

Based on experimental evidence, several factors have been identified to affect the overall efficiency of CBNAs. These include the size of adsorbent and adsorbate, charge of the adsorbate and adsorbent surface, temperature and pH of the aqueous medium and so forth. Therefore, the prediction of the actual adsorption mechanism is neither simple nor straightforward. The mechanisms by which adsorption of dyes onto CBNAs occur have been a matter of considerable debate. Different studies have reached different conclusions. It is believed that different kinds of interactions such as hydrophobic interactions, π–π electron donor-acceptor interactions, electrostatic attraction, precipitation and hydrogen bonds act simultaneously during adsorption of dye ions onto CBNAs, the predominant mechanism being different for different types of dyes [22,23]. For example, π–π interactions depend on the size and shape of the aromatic system and the substitution unit of molecules, and increases with the increasing number of aromatic rings in the dye molecule [22].

2.5   Conclusion and Future Perspectives

In this chapter, the latest research work on the use of CBNAs as advanced adsorbent materials for removal of various synthetic dyes from aqueous solutions has been compiled and summarized. It is evident that CBNAs are a promising alternative to activated carbon and other adsorbent materials that are currently being considered for treatment of industrial wastewaters. Their simplicity, flexibility, fast kinetics, and high adsorption capacities make them attractive for the selective removal of dye pollutants. However, research in this area is still in its infancy. There are a number of challenges that have to be overcome before we can realize the full potential of CBNAs for practical applications. Here, we would like to highlight some of those important issues that might help future research.

Firstly, study on the adsorption characteristics of CBNAs is mostly restricted to batch adsorption studies. There are hardly any reports on the adsorption of dyes using fixed-bed dynamic adsorption techniques. Continuous column study is therefore highly recommended since it allows a more efficient utilization of CBNAs and provides a better quality effluent. In addition, future research works should also focus on verifying the performance of CBNAs at the pilot plant scale in order to ascertain their commercial applications.

Secondly, the equilibrium adsorption data, without exception, have been empirically correlated with the conventional isotherm models, viz. Langmuir, Freundlich and a few others. Although a few researchers have proposed adsorption mechanisms, no further attempts have been made to interpret the adsorption equilibrium data using the proposed mechanisms.

Meanwhile, the applications of pseudo-kinetic models, particularly the pseudo-second-order kinetic model, have provided a good description of the batch adsorption kinetic data. However, these models require a pseudo-rate constant that is concentration dependent and also do not account for pH variations. It is thus suggested to apply the well-established surface ionization/complexation model or the double-layer retention model to investigate the effect of pH or ionic strength on the adsorption capacity of CBNAs.

Thirdly, it is necessary to conduct regeneration studies as they help determine the reusability of an adsorbent, which in turn contributes in evaluating the effectiveness and economic feasibility of the adsorbent. Based on the literature, not many adsorbent regeneration studies have been reported and should therefore be performed in detail.

Fourthly, it is extremely essential to investigate the simultaneous removal of many coexisting dye pollutants from multicomponent solutions. Unlike laboratory tests using pure aqueous solutions, real textile effluents contain different types of dye pollutants and other undesirable substances. However, most of the studies reported so far are based on single-solute systems. There is absolutely little or no effort in investigating the competitive adsorption of dyes. Such studies are essential for accurate designing of adsorption systems as the effect of competitive interactions may cause deterioration in the adsorption capacity. Therefore, some future research in this respect should be pursued to provide insights into competitive adsorption and possible interference from other contaminants to target species removed by CBNAs.

Fifthly, although several processes, including ion exchange, electrostatic interactions, π-π electron donor-acceptor interactions, surface adsorption, and complexation, have been suggested to explain the mechanisms involved in the adsorption between water pollutants and CBNAs, mechanistic studies need to be conducted in detail to validate the proposed binding mechanism of aquatic pollutants with CBNAs. Finally, very little information is currently available on the toxicity and biocompatibility of graphene and related materials. Future investigations should, therefore, focus on *in vitro* and *in vivo* interactions between CBNAs and different living systems to effectively evaluate the utilization of CBNAs in dye wastewater remediation.

Undoubtedly, CBNAs hold great potential for being robust materials to address the environmental problems posed by synthetic colorants. However, it is only when we overcome the above-mentioned challenges that the commercial application of CBNAs in dye-wastewater treatment can be successfully realized.

References

1. P. A. Pekakis, N. P. Xekoukoulotakis, and D. Mantzavinos, *Water Research*, Vol. 40, p. 1276, 2006.
2. P. Saha, S. Chowdhury, S. Gupta, I. Kumar, *Chemical Engineering Journal*, Vol. 165, p. 874, 2010.
3. M. Doble, A. Kumar, *Biotreatment of Industrial Effluents*, Elsevier Butterwoth-Heinemann, Burlington, 2005.
4. K. Singh, S. Arora, *Critical Reviews in Environmental Science & Technology*, Vol. 41, p. 807, 2011.
5. M. Rafatullah, O. Sulaiman, R. Hashim, A. Ahmad, *Journal of Hazardous Materials*, Vol. 177, p. 70, 2010.
6. A. Dabrowski, *Advances in Colloid and Interface Science*, Vol. 93, p. 135, 2001.
7. G. Crini, *Progress in Polymer Science*, Vol. 30, p. 38, 2005.
8. G. Crini, P.-M. Badot, *Progress in Polymer Science*, Vol. 33, p. 399, 2008.
9. K. Y. Foo, B. H. Hameed, *Chemical Engineering Journal*, Vol. 156, p. 2, 2010.
10. A. Bhatnagar, M. Sillanpaa, *Chemical Engineering Journal*, Vol. 157, p. 277, 2010.
11. W. S. Wan Ngah, L. C. Teong, M. A. K. M. Hanafiah, *Carbohydrate Polymers*, Vol. 83, p. 1446, 2011.
12. V. K. Gupta, and Suhas, *Journal of Environmental Management*, Vol. 90, p. 2313, 2009.
13. V. K. Gupta, P. J. M. Carrott, M. M. L. Ribeiro Carrott, Suhas, *Critical Reviews in Environmental Science & Technology*, Vol. 39, p. 783, 2009.
14. I. M. Banat, P. Nigam, D. Singh, R. Marchant, *Bioresource Technology*, Vol. 58, p. 217, 1996.
15. T. Robinson, G. McMullan, R. Marchant, P. Nigam, *Bioresource Technology*, Vol. 77, p. 247, 2001.
16. Z. Aksu, *Process Biochemistry*, Vol. 40, p. 997, 2005.
17. G. Crini, *Bioresource Technology*, Vol. 97, p. 1061, 2006.
18. A. Demirbas, *Journal of Hazardous Materials*, Vol. 167, p. 1, 2009.
19. A. Srinivasan, and T. Viraraghavan, *Journal of Environmental Management*, Vol. 91, p. 1915, 2010.
20. M. M. Khin, A. S. Nair, V. J. Babu, R. Murugan, S. Ramakrishna, *Energy & Environmental Science*, Vol. 5, p. 8075, 2012.
21. V. K. K. Upadhyayula, S. Deng, M. C. Mitchell, G. B. Smith, *Science of the Total Environment*, Vol. 408, p. 1, 2009.
22. X. Ren, C. Chen, M. Nagatsu, X. Wang, *Chemical Engineering Journal*, Vol. 170, p. 395, 2011.
23. V. K. Gupta, R. Kumar, A. Nayak, T. A. Saleh, M. A. Barakat, *Advances in Colloid and Interface Science*, Vol. 193–194, p. 24, 2013.
24. H. Wang, X. Yuan, Y. Wu, H. Huang, X. Peng, G. Zeng, H. Zhong, J. Liang, *Advances in Colloid and Interface Science*, Vol. 195–196, p. 19, 2013.
25. S. Ijima, *Nature*, Vol. 354, p. 56, 1991.

26. K. Balasubramanian, and M. Burghard, *Small*, Vol. 1, p. 180, 2005.
27. M. Endo, T. Hayashi, Y. A. Kim, M. Terrones, M. S. Dresselhaus, *Philosophical Transactions: Mathematical, Physical and Engineering Sciences*, Vol. 362, p. 2223, 2004.
28. J. N. Coleman, U. Khan, W. J. Blau, Y. K. Gun'ko, *Carbon*, Vol. 44, p. 1624, 2006.
29. Y.-L. Zhao, and J. F. Stoddart, *Accounts of Chemical Research*, Vol. 42, p. 1161, 2009.
30. K. Dasgupta, J. B. Joshi, S. Banerjee, *Chemical Engineering Journal*, Vol. 171, p. 841, 2011.
31. J. M. Bonard, H. Kind, T. Stockli, L. O. Nilsson, *Solid State Electronics*, Vol. 45, p. 893, 2001.
32. K. Tsukagoshi, N. Yoneya, S. Uryu, *Physica B: Condensed Matter*, Vol. 323, p. 107, 2002.
33. J. H. Chen, W. Z. Li, D. Z. Wang, S. X. Yang, J. G. Wen, Z. F. Ren, *Carbon*, Vol. 40, p. 1193, 2002.
34. A. C. Dillon, K. M. Jones, T. A. Bekkedahl, C. H. Kiang, D. S. Bethune, M. J. Heben, *Nature*, Vol. 386, p. 377, 1997.
35. C. L. Chen, J. Hu, D. D. Shao, J. X. Li, X. K. Wang, *Journal of Hazardous Materials*, Vol. 164, p. 923, 2009.
36. S. M. Gatica, M. J. Bojan, G. Stan, M. W. Cole, *Journal of Chemical Physics*, Vol. 114, p. 3765, 2001.
37. S. Agnihotri, J. P. B. Mota, M. Rostam-Abadi, M. J. Rood, *Langmuir*, Vol. 21, p. 896, 2005.
38. S. Agnihotri, J. P. B. Mota, M. Rostam-Abadi, M. J. Rood, *The Journal of Physical Chemistry B*, Vol. 110, p. 7640, 2006.
39. J. Liu, A. G. Rinzler, H. J. Dai, J. H. Hafner, R. K. Bradley, P. J. Boul, A. Lu, T. Iverson, K. Shelimov, C. B. Huffman, F. Rodriguez-Macias, Y. S. Shon, T. R. Lee, D. T. Colbert, R. E. Smalley, *Science*, Vol. 280, p. 1253, 1998.
40. M. L. Toebes, E. M. P. van Heeswijk, J. H. Bitter, A. J. van Dillen, K. P. de Jong, *Carbon*, Vol. 42, p. 307, 2004.
41. O. Byl, J. Liu, J. T. Yates Jr., *Langmuir*, Vol. 21, p. 4200, 2005.
42. L. V. Liu, W. Q. Tian, Y. A. Wang, *The Journal of Physical Chemistry B*, Vol. 110, p. 13037, 2006.
43. M. L. Sham, J. K. Kim, *Carbon*, Vol. 44, p. 768, 2006.
44. C. L. Chen, B. Liang, A. Ogino, X. K. Wang, M. Nagatsu, *The Journal of Physical Chemistry C*, Vol. 113, p. 7659, 2009.
45. G. Onyestyák, Z. Ötvös, J. Valyon, I. Kiricsi, L. V. C. Rees, *Helvetica Chimica Acta*, Vol. 87, p. 1508, 2004.
46. A. Kuznetsova, J. T. Yates Jr., J. Liu, R. E. Smalley, *The Journal of Chemical Physics*, Vol. 112, p. 9590, 2000.
47. A. Kuznetsova, D. B. Mawhinney, V. Naumenko, J. T. Yates Jr., J. Liu, R. E. Smalley, *Chemical Physics Letters*, Vol. 321, p. 292, 2000.
48. A. L. Ivanovskii, *Russian Chemical Reviews*, Vol. 81, p. 571, 2012.

49. Y. Chen, B. Zhang, G. Liu, X. Zhuang, E.-T. Kang, *Chemical Society Reviews*, Vol. 41, p. 4688, 2012.

50. F. Adar, *Spectroscopy*, Vol. 26, p. 16, 2011.

51. D. A. C. Brownson, D. K. Kampouris, C. E. Banks, *Journal of Power Sources*, Vol. 196, p. 4873, 2011.

52. P. Avouris, C. Dimitrakopoulos, *Materials Today*, Vol. 15, p. 86, 2012.

53. Y. Zhu, S. Murali, W. Cai, X. Li, J. W. Suk, J. R. Potts, R. S. Ruoff, *Advanced Materials*, Vol. 22, p. 3906, 2010.

54. A. K. Geim, K. S. Novoselov, *Nature Materials*, Vol. 6, p. 183, 2007.

55. S. V. Tkachev, E. Y. Buslaeva, S. P. Gubin, *Inorganic Materials*, Vol. 47, p. 1, 2011.

56. S. Basu, P. Bhattacharya. *Sensors and Acutators B: Chemical*, Vol. 173, p. 1, 2012.

57. M. D. Stoller, S. Park, Y. Zhu, J. An, R. S. Ruoff, *Nano Letters*, Vol. 8, p. 3498, 2008.

58. C. Lee, X. Wei, J. W. Kysar, J. Hone, *Science*, Vol. 321, p. 385, 2008.

59. W. Choi, I. Lahiri, R. Seelaboyina, Y. S. Kang, *Critical Reviews in Solid State and Material Sciences*, Vol. 35, p. 52, 2010.

60. K. I. Bolotin, K. J. Sikes, Z. Jiang, M. Klima, G. Funderberg, J. Hone, P. Kim, H. L. Stromer, *Solid State Communications*, Vol. 146, p. 351, 2008.

61. A. A. Balandin, S. Ghosh, W. Bao, I. Cailzo, D. Teweldebrhan, F. Miao, C. N. Lau, *Nano Letters*, Vol. 8, p. 902, 2008.

62. S. Park, R. S. Ruoff, *Nature Nanotechnology*, Vol. 4, p. 217, 2009.

63. R. Ruoff, *Nature Nanotechnology*, Vol. 3, p. 10, 2008.

64. S. Stankovich, D. A. Dikin, G. H. B. Dommett, K. M. Kohlhass, E. J. Zimney, E. A. Stach, R. D. Piner, S. T. Nguyen, R. S. Ruoff, *Nature*, Vol. 442, p. 282, 2006.

65. S. M. Paek, E. Yoo, I. Honma, *Nano Letters*, Vol. 9, p. 72, 2009.

66. F. Y. Su, C. You, Y. B. He, W. Lv, W. Cui, F. Jin, B. Li, Q. H. Yang, F. Kang, *Journal of Materials Chemistry*, Vol. 20, p. 9644, 2010.

67. D. A. Dikin, S. Stankovich, E. J. Zimney, R. D. Piner, G. H. Dommett, G. Evmenenko, S. T. Nguyen, R. S. Ruoff, *Nature*, Vol. 448, p. 457, 2007.

68. J. S. Bunch, S. S. Verbridge, J. S. Alden, A. M. van der Zande, J. M. Parpia, H. G. Craighead, P. L. McEuen, *Nano Letters*, Vol. 8, p. 2458, 2008.

69. O. C. Compton, S. Kim, C. Pierre, J. M. Torkelson, S. T. Nguyen, *Advanced Materials*, Vol. 22, p. 4759, 2010.

70. S. Wang, P. K. Ang, Z. Wang, A. L. Tang, J. T. Thong, K. P. Loh, *Nano Letters*, Vol. 10, p. 92, 2010.

71. L. Zhang, J. Xia, Q. Zhao, L. Liu, Z. Zhang, *Small*, Vol. 6, p. 537, 2010.

72. S. R. Ryoo, Y. K. Kim, M. H. Kim, D. H. Min, *ACS Nano*, Vol. 4, p. 6587, 2010.

73. V. C. Sanchez, A. Jachak, R. H. Hurt, A. B. Kane, *Chemical Research in Toxicology*, Vol. 25, p. 15, 2012.

74. J. Kim, J. L. Cote, F. Kim, W. Yuan, K. R. Shull, J. J. Huang, *Journal of the American Chemical Society*, Vol. 132, p. 8180, 2010.

75. D. R. Dreyer, S. Park, C. W. Bielawski, R. S. Ruoff, *Chemical Society Reviews*, Vol. 39, p. 228, 2010.
76. D. Krishnan, F. Kim, J. Luo, R. Cruz-Silva, L. J. Cote, H. D. Jang, J. Huang, *Nano Today*, Vol. 7, p. 137, 2012.
77. V. Singh, D. Joung, L. Zhai, S. Das, S. I. Khondaker, S. Seal, *Progress in Materials Science*, Vol. 56, p. 1178, 2011.
78. Y. Yao, F. Xu, M. Chen, Z. Xu, Z. Zhu, *Bioresource Technology*, Vol. 101, p. 3040, 2010.
79. M. Shirmardi, A. H. Mahvi, A. Mesdaghinia, S. Nasseri, R. Nabizadeh, *Desalination and Water Treatment*, doi: 10.1080/19443994.2013.793915, 2013.
80. O. Moradi, *Fullerenes, Nanotubes, and Carbon Nanostructures*, Vol. 21, p. 286, 2013.
81. C.-Y. Kuo, C.-H. Wu, J.-Y. Wu, *Journal of Colloid and Interface Science*, Vol. 327, p. 308, 2008.
82. C.-H. Wu, *Journal of Hazardous Materials*, Vol. 144, p. 93, 2007.
83. S. Wang, C. W. Ng, W. Wang, Q. Li, Z. Hao, *Chemical Engineering Journal*, Vol. 197, p. 34, 2012.
84. M. Ghaedi, S. Haghdoust, S. Nasiri Kokhdan, A. Mihandoost, R. Sahraie, A. Daneshfar, *Spectroscopy Letters*, Vol. 45, p. 500, 2012.
85. F. M. Machado, C. P. Bergmann, T. H. M. Fernandes, E. C. Lima, B. Royer, T. Calvete, S. B. Fagan, *Journal of Hazardous Materials*, Vol. 192, p. 1122, 2011.
86. D. Zhao, W. Zhang, C. Chen, X. Wang, *Procedia Environmental Sciences*, Vol. 18, p. 890, 2013.
87. Y. Yao, B. He, F. Xu, X. Chen, *Chemical Engineering Journal*, Vol. 170, p. 82, 2011.
88. A. Rodriguez, G. Ovejero, J. L. Sotelo, M. Mestanza, J. Garcia, *Journal of Environmental Science and Health Part A*, Vol. 45, p. 1642, 2010.
89. F. Geyikci, *Fullerenes, Nanotubes, and Carbon Nanostructures*, Vol. 21, p. 579, 2013.
90. M. H. Dehghani, A. Naghizadeh, A. Rashidi, E. Derakhshani, *Desalination and Water Treatment*, doi: 10.1080/19443994.2013.791772, 2013.
91. M. Ghaedi, A. Shokrollahi, H. Tavallali, F. Shojaiepoor, B. Keshavarz, H. Hossainian, M. Soylak, M. K. Purkait, *Toxicological & Environmental Chemistry*, Vol. 93, p. 438, 2011.
92. M. Ghaedi, A. Hassanzadeh, S. N. Kokhdan, *Journal of Chemical & Engineering Data*, Vol. 56, p. 2511, 2011.
93. S. Wang, C. W. Ng, W. Wang, Q. Li, L. Li, *Journal of Chemical & Engineering Data*, Vol. 57, p. 1563, 2012.
94. M. Ghaedi, H. Khajehsharifi, A. H. Yadkuri, M. Roosta, A. Asghari, *Toxicological & Environmental Chemistry*, Vol. 94, p. 873, 2012.
95. M. Ghaedi, S. N. Kokhdan, *Desalination and Water Treatment*, Vol, 49, p. 317, 2012.

96. A. K. Mishra, T. Arockiadoss, S. Ramaprabhu, *Chemical Engineering Journal*, Vol. 162, p. 1026, 2010.

97. M. Shirmardi, A. H. Mahvi, B. Hashemzadeh, A. Naeimabadi, G. Hassani, M. V. Niri, *Korean Journal of Chemical Engineering*, Vol. 30, p. 1603, 2013.

98. S. Chatterjee, M. W. Lee, S. J. Woo, *Bioresource Technology*, Vol. 101, p. 1800, 2010.

99. H. Y. Zhu, R. Jiang, L. Xiao, G. M. Zeng, *Bioresource Technology*, Vol. 101, p. 5063, 2010.

100. M. Bhagat, A. A. Farghali, W. M. A. El Rouby, M. H. Khedr, *Fullerenes, Nanotubes, and Carbon Nanostructures*, Vol. 22, p. 454, 2014.

101. J.-P. Wang, H.-S. Yang, C.-T. Hsieh, *Separation Science and Technology*, Vol. 46, p. 340, 2011.

102. F. Yu, J. Chen, L. Chen, J. Huai, W. Gong, Z. Yuan, J. Wang, J. Ma, *Journal of Colloid and Interface Science*, Vol. 378, p. 175, 2012.

103. L. Ai, C. Zhang, F. Liao, Y. Wang, M. LI, L. Meng, J. Jiang, *Journal of Hazardous Materials*, Vol. 198, p. 282, 2011.

104. S. Qu, F. Huang, S. Yu, G. Chen, J. Kong, *Journal of Hazardous Materials*, Vol. 160, p. 643, 2008.

105. T. Madrakian, A. Afkhami, M. Ahmadi, H. Bagheri, *Journal of Hazardous Materials*, Vol. 196, p. 109, 2011.

106. J.-L. Gong, B. Wang, G.-M. Zeng, C.-P. Yang, C.-G. Niu, Q.-Y. Niu, W.-J. Zhou, Y. Liang, *Journal of Hazardous Materials*, Vol. 164, p. 1517, 2009.

107. H. Gao, S. Zhao, X. Cheng, X. Wang, L. Zheng, *Chemical Engineering Journal*, Vol. 223, p. 84, 2013.

108. Y. Li, K. Sui, R. Liu, X. Zhao, Y. Zhang, H. Liang, Y. Xia, *Energy Procedia*, Vol. 16, p. 863, 2012.

109. P. R. Chang, P. Zheng, B. Liu, D. P. Anderson, J. Yu, X. Ma, *Journal of Hazardous Materials*, Vol. 186, p. 2144, 2011.

110. W. Konicki, I. Pelech, E. Mijowska, I. Jasinska, *Chemical Engineering Journal*, Vol. 210, p. 87, 2012.

111. W. Konicki, I. Pelech, E. Mijowska, I. Jasinska, *Clean Soil Air Water*, Vol. 42, p. 284, 2014.

112. L. Yan, P. R. Chang, P. Zheng, X. Ma, *Carbohydrate Polymers*, Vol. 87, p. 1919, 2012.

113. H. Yu, and B. Fugetsu, *Journal of Hazardous Materials*, Vol. 177, p. 138, 2010.

114. Y. H. Li, T. Liu, Q. Du, J. Sun, Y. Xia, Z. Wang, W. Zhang, K. Wang, H. Zhu, D. Wu, *Chemical & Biochemical Engineering Quarterly*, Vol. 25, p. 483, 2011.

115. T. Liu, Y. Li, Q. Du, J. Sun, Y. Jiom G. Yang, Z. Wang, Y. Xia, W. Zhang,K. Wang, H. Zhu, D. Wu, *Colloids and Surfaces B: Biointerfaces*, Vol. 90,p. 197, 2012.

116. T. Wu, X. Cai, S. Tan, H. Li, J. Liu, W. Yang, *Chemical Engineering Journal*, Vol. 173, p. 144, 2011.

117. J. Zhao, W. Ren, H.-M. Cheng, *Journal of Materials Chemistry*, Vol. 22, p. 20197, 2012.
118. S.-T. Yang, S. Chen, Y. Chang, A. Cao, Y. Liu, H. Wang, *Journal of Colloid and Interface Science*, Vol. 359, p. 24, 2011.
119. W. Zhang, C. Zhou, W. Zhou, A. Lei, Q. Zhang, Q. Wan, B. Zou, *Bulletin of Environmental Contamination and Toxicology*, Vol. 87, p. 86, 2011.
120. G. K. Ramesha, A. V. Kumara, H. B. Muralidhara, S. J. Sampath, *Journal of Colloid and Interface Science*, Vol. 361, p. 270, 2011.
121. L. Sun, H. Yu, B. Fugetsu, *Journal of Hazardous Materials*, Vol. 203–204, p. 101, 2012.
122. Y. Li, Q. Du, T. Liu, X. Peng, J. Wang, J. Sun, Y. Wang, S. Wu, Z. Wang, Y. Xiz, L. Xiz, *Chemical Engineering Research and Design*, Vol. 91, p. 361, 2013.
123. P. Sharma, and M. R. Das, *Journal of Chemical & Engineering Data*, Vol. 58, p. 151, 2013.
124. J. N. Tiwari, K. Mahesh, N. H. Le, K. C. Kemp, R. Timilsina, R. N. Tiwari, K. S. Kim, *Carbon*, Vol. 56, p. 173, 2013.
125. C. Wang, C. Feng, Y. Gao, X. Ma, A. Wu, Z. Wang, *Chemical Engineering Journal*, Vol. 173, p. 92, 2011.
126. L. Ai, C. Zhang, Z. Chen, *Journal of Hazardous Materials*, Vol. 192, p. 1515, 2011.
127. Y. Yao, S. Miao, S. Liu, L. P. Ma, H. Sun, S. Wang, *Chemical Engineering Journal*, Vol. 184, p. 326, 2012.
128. Q. Wu, C. Feng, C. Wang, Z. Wang, *Colloids and Surfaces B: Biointerfaces*, Vol. 101, p. 210, 2013.
129. S. Wang, J. Wei, S. Lv, Z. Guo, F. Jiang, *Clean Soil Air Water*, Vol. 41, p. 992, 2013.
130. N. Li, M. Zheng, X. Chang, G. Ji, H. Lu, L. Xue, L. Pan, J. Cao, *Journal of Solid State Chemistry*, Vol. 184, p. 953, 2011.
131. A. A. Farghali, M. Bahgat, W. M. A. El Rouby, M. H. Khedr, *Journal of Alloys and Compounds*, Vol. 555, p. 193, 2013.
132. S. Sen Gupta, S. Sreeprasad, S. M. Maliyekkalm S. K. Das, T. Pradeep, *ACS Applied Materials & Interfaces*, Vol. 4, p. 4156, 2012.
133. T. S. Sreeprasad, S. Sen Gupta, S. M. Maliyekkal, T. Pradeep, *Journal of Hazardous Materials*, Vol. 246–247, p. 213, 2013.
134. L. Ai, J. Jiang, *Chemical Engineering Journal*, Vol. 192, p. 156, 2012.
135. Z. Sui, Q. Meng, X. Zhang, R. Ma, B. Cao, *Journal of Materials Chemistry*, Vol. 22, p. 8767, 2012.
136. X. Zhang, C. Cheng, J. Zhao, L. Ma, S. Sun, C. Zhao, *Chemical Engineering Journal*, Vol. 215–216, p. 72, 2013.
137. J. Wang, B. Tang, T. Tsuzuki, Q. Liu, X. Hou, L. Sun, *Chemical Engineering Journal*, Vol. 204–206, p. 258, 2012.
138. Y. Yao, S. Miao, S. Yu, L. P. Ma, H. Sun, S. Wang, *Journal of Colloid and Interface Science*, Vol. 379, p. 20, 2012.

139. Y. Chen, L. Chen, H. Bai, L. Li, *Journal of Materials Chemistry A*, Vol. 1, p. 1992, 2013.
140. L. Fan, C. Luo, X. Li, F. Lu, H. Qiu, M. Sun, *Journal of Hazardous Materials*, Vol. 215–216, p. 272, 2012.
141. L. Fan, C. Luo, M. Sun, X. Li, F. Lu, H. Qiu, *Bioresource Technology*, Vol. 114, p. 703, 2012.
142. L. Fan, C. Luo, M. Sun, H. Qiu, X. Li, *Colloid and Surfaces B: Biointerfaces*, Vol. 103, p. 601, 2013.
143. Y. Li, Q. Du, J. Sun, Y. Wang, S. Wu, Z. Wang, Y. Xia, L. Xia, *Carbohydrate Polymers*, Vol. 95, p. 501, 2013.
144. H. Sun, L. Cao, L. Lu, *Nano Research*, Vol. 4, p. 550, 2011.
145. J. Wang, T. Tsuzuki, B. Tang, X. Hou, L. Sun, X. Wang, *ACS Applied Materials & Interfaces*, Vol. 4, p. 3084, 2012.
146. T.-D. Nguyen-Phan, V. H. Pham, E. J. Kim, E.-S. Oh, S. H. Hur, J. S. Chung, B. Lee, E. W. Shin, *Applied Surface Science*, Vol. 258, p. 4551, 2012.
147. J.-S. Cheng, J. Du, W. Zhu, *Carbohydrate Polymers*, Vol. 88, p. 61, 2012.
148. T. Kuilla, S. Bhadra, D. Yao, N. H. Kim, S. Bose, J. H. Lee, *Progress in Polymer Science*, Vol. 35, p. 1350, 2010.
149. X. Huang, X. Qi, F. Boey, H. Zhang, *Chemical Society Reviews*, Vol. 41, p. 666, 2012.
150. Z.-S. Wu, G. Zhou, L.-C. Yin, W. Ren, F. Li, H.-M. Cheng, *Nano Energy*, Vol. 1, p. 107, 2012.

3

Advanced Oxidation Processes for Removal of Dyes from Aqueous Media

Süheyda Atalay and Gülin Ersöz

Ege University, Faculty of Engineering, Chemical Engineering Department, İzmir, Turkey

Abstract

Dyes are synthetically aromatic compounds used in various industries such as textile, plastics, pharmaceutical, paper, paint and food. The quality and quantity of wastewaters from these industries are a matter of concern for human health and environmental pollution.

Different conventional methods have been developed for dye removal. However, due to the complex structure and recalcitrant nature of dyes, these methods appear to be ineffective in their complete removal.

Advanced Oxidation Processes (AOPs) have been considered as a promising alternative to treat wastewater containing dye residues. These processes are based on the production of highly reactive radicals, especially hydroxyl radicals, which promote destruction of the target pollutant until mineralization.

This chapter addresses AOPs for removal of dyes from aqueous media by presenting background on advanced oxidation processes. The AOPs will be investigated in two categories: Nonphotochemical and Photochemical Processes. Among these processes, the main focus will be on ozonation, Fenton's reagent oxidation, wet air oxidation and photocatalytic oxidation, with only brief information given on electrochemical oxidation, ultraviolet irradiation and hydrogen peroxide (UV/H_2O_2) and photocatalytic ozonation (UV/O_3) processes.

Keywords: Removal of dyes, advanced oxidation processes, non-photochemical advanced oxidation processes, photochemical advanced oxidation processes, ozonation, Fenton's reagent oxidation, wet air oxidation, electrochemical oxidation, photocatalytic oxidation

Corresponding authors: suheyda.atalay@ege.edu.tr; suheyda.atalay@gmail.com

Sanjay K. Sharma (ed.) Green Chemistry for Dyes Removal from Wastewater, (83–117)
© 2015 Scrivener Publishing LLC

3.1 Introduction

Dyes have a synthetic origin and complex aromatic molecular structures. Based on the chemical structure or chromophore, approximately 30 different groups of dyes are present. Anthraquinone, phthalocyanine, triarylmethane and azo dyes are quantitatively the most important groups of dyes [1].

Azo dyes are the most widely used commercial reactive dyes and characterized by nitrogen to nitrogen double bond. The color of dyes is due to azo bond and associated chromophores [2–4]. Within the overall category of dyestuffs, azo dyes represent the largest class of dyes [3,5,6] and therefore constitute a significant portion of the dye pollutants.

Dyes are widely used in various industrial applications such as textile, plastics, pharmaceutical, paper, paint and food. Wastewater from these dye processing or dye using industries is one of the major environmental problems due to the characteristics of wastewater such as strong color, high chemical oxygen demand (COD) derived from additives and surfactants and a low biodegradability. These dyes are the most problematic pollutants of industrial wastewaters. For example, in the literature it is reported that more than 15% of the textile dyes is lost in wastewater stream during dyeing operation [1,7–9].

Chemical, physical and biological methods have been used to treat dye-containing effluents, however these conventional treatment technologies were found to be unsatisfactory [1] due to the complex structure and synthetic nature of the dyes. The dye containing effluents are chemically and photolytically stable, toxic and mostly nonbiodegradable and also contain many organic compounds, which are not easy to degrade; in other words, they are resistant to destruction by conventional treatment methods. The inappropriate disposal of dyes is a big challenge as it can disturb the ecosystem and constitutes potential environmental and health problems such as reducing light penetration and photosynthesis [10,11]. Consequently, it is necessary to find an effective treatment technology that leads to complete degradation and decolorization. Hence removal of dyes from aqueous media has been the cause of much concern to societies, researchers and regulation authorities around the world.

Recently, alternative to conventional treatment methods, the Advanced Oxidation Processes (AOPs), through which highly oxidizing species like hydroxyl radicals are produced, have been of major concern. These radicals can be produced by means of oxidizing agents such as O_3, H_2O_2, ultraviolet irradiation, ultrasound, and catalysts (homogeneous or heterogeneous).

The AOPs are generally classified in terms of whether UV source i
the process: Nonphotochemical Processes and Photochemical P
The most common AOPs are ozonation, photocatalytic degradation,
Fenton's reagent (H_2O_2/Fe^{2+}), photo-Fenton, and wet air oxidation (WAO).

The objective of this chapter is to provide the readers with knowledge
about AOPs.

In view of the importance of the removal of dyes from aqueous media,
it is the purpose of this chapter:

- to present an overview of AOPs by investigating them in two
  categories such as nonphotochemical and photochemical
  processes;
- to introduce the most important aspects of various AOPs
  such as ozonation, Fenton's reagent oxidation, wet air oxidation and photocatalytic degradation processes;
- to discuss the key factors influencing the AOPs' efficiency;
  and
- to present a perspective to the readers on the research performed on the removal of dyes from aqueous media by AOPs
  which is found in literature.

3.2   Advanced Oxidation Processes

Advanced oxidation processes (AOPs) are defined as a set of chemical
treatment processes designed to remove organic compounds in wastewater
by several oxidation reactions.

In order to generate highly reactive intermediates, •OH radicals, the
AOPs may proceed along one of the two routes given below [12]:

- Use of high energy oxidants such as ozone and H_2O_2 and/
  or photons. The generation of hydroxyl radical might possi-
  bly be by the use of UV, UV/O_3, UV/H_2O_2, Fe^{+2}/H_2O_2, $TiO_2/$
  H_2O_2 and a number of other processes [13].
- Use of O_2 in temperature ranges between ambient conditions
  and those found in incinerators, such as in wet air oxidation
  (WAO) processes in the region of 1–20 MPa and 200–300°C.

They are mainly all characterized by the same chemical feature, which is
the generation of powerful non-selective oxidizing agent, hydroxyl radicals

or another species of similar reactivity such as sulfate radical anion ($\bullet SO_4^-$), which can destroy even the recalcitrant pollutants [14].

The oxidation potential of some important oxidizing agents are summarized in Table 3.1 [15].

The hydroxyl radical has the second highest oxidizing potential and is the most preferred choice for use in oxidation processes. The hydroxyl radicals are highly reactive species that initiate a sequence of reactions resulting in destruction and removal of organic pollutants.

In general, AOPs involve two stages of oxidation:

Stage 1: the formation of strong oxidants (mainly hydroxyl radicals)

Stage 2: the reaction of these oxidants with organic contaminants in water

The AOPs have common principles in terms of the participation of hydroxyl radicals that are assumed to be operative during the reaction. Due to the instability of $\bullet OH$ radical, it has to be generated continuously *in situ* through chemical or photochemical reactions [16].

The technologies that can be classified as AOPs fall under two general categories in terms of whether UV source is used in the process. Table 3.2 shows the classification of AOPs [17].

In recent years, the research found in literature has shown that toxic and refractory organics in wastewater could be destroyed by most AOPs, such as ozonation, photocatalytic degradation, Fenton's reagent (H_2O_2/Fe^{2+}), photo-Fenton, and wet air oxidation, and have proven to be very effective for dye-containing effluents [18–23]. Various combinations of these AOPs may also provide efficient removal of dyes from aqueous media.

Of the many AOPs developed, four have garnered the most study and use in scientific research: ozonation, Fenton's reagent oxidation, catalytic wet air oxidation and photocatalytic oxidation. Consequently, this chapter

Table 3.1 Relative oxidation power of some oxidizing species.

Name of oxidant	Oxidation potential (eV)
Fluorine	3.06
Hydroxyl radical	2.80
Sulfate	2.60
Ozone	2.07
Hydrogen peroxide	1.77
Chlorine	1.36

Table 3.2 Classification of Advanced Oxidation Processes.

Non-photochemical	Photochemical
Ozonation	Photocatalytic oxidation, UV/Catalyst
Ozonation with hydrogen peroxide (O_3/H_2O_2)	UV/H_2O_2
Fenton (Fe^{2+} or Fe^{3+}/H_2O_2)	UV/O_3
Wet air oxidation (WAO)	$UV/O_3/H_2O_2$
Electrochemical oxidation	Photo-Fenton ($Fe^{3+}/H_2O_2/UV$)

will mainly focus on these processes and only brief information will be given on electrochemical oxidation, ultraviolet irradiation and hydrogen peroxide (UV/H_2O_2) and photocatalytic ozonation (UV/O_3) processes.

3.2.1   Nonphotochemical Advanced Oxidation Processes

There are several well known methods for generating hydroxyl radicals without using light energy. These are ozonation, Fenton's reagent oxidation, wet air oxidation and electrochemical oxidation. In the following sections, information on these processes will be given.

3.2.1.1   Ozonation

Ozone and ozone-based advanced oxidation processes are powerful, environmentally friendly technologies capable of degrading a variety of organic pollutants [24].

Ozonation is the method which is widely preferred for removal of colored substances since the chromophore groups with conjugated double bonds, which are responsible for color, can be easily broken down by ozone, either directly or indirectly forming smaller molecules [25].

It is a promising treatment process due to its unique features such as decolorization and degradation that occurs in one step; increases dissolved oxygen; reduces oxygen demanding matter, no remaining sludge; little space is required and it is easily installed on a site [26,27]. In contrast, there are some disadvantages of ozonation such as high capital cost, high electric consumption and its being highly corrosive.

The ozonation process is realized in three steps:

1. Generation of ozone
2. Dissolution of ozone in the wastewater
3. Oxidation of organic matter

Depending on the pH, the oxidation may follow different pathways, which are direct (molecular ozone attack) and indirect (free radical mechanism) radical oxidation, resulting in different oxidation products, degrees of mineralization and color removal efficiencies [28].

Direct ozonation involves degradation of organics by ozone molecule under acidic conditions [29]. In this method, an ozone molecule attacks an unsaturated bond and its dipolar structure breaks up the bond [30]. However, in direct oxidation only some parts of the organic compound can be degraded, byproducts are produced and generally additional treatment may be required to remove them.

Indirect ozonation considers the degradation mechanism of organics throughout hydroxyl radicals under basic conditions [29]. The hydroxyl radicals which are generated in radical chain reactions react with organic compound according to the following equations:

$$O_3 + OH^- \rightarrow \bullet O_3 + \bullet OH \qquad (3.1)$$

$$\bullet O_3^- \rightarrow \bullet O^- + O_2 \qquad (3.2)$$

$$\bullet O^- + H + \rightarrow \bullet OH \qquad (3.3)$$

The hydroxyl radical decomposes most organic compounds into CO_2 and H_2O [29,30].

The ozonation of reactive dyes is known to produce mainly two types of byproducts. These are carboxylic acids with hydroxyl functional groups and aldehydes. Lopez *et al.* described the main ozonation byproducts of azo dyes, as formaldehyde, acetaldehyde, glyoxal, acetone, acetic, formic, oxalic and carbonic acids. Since these degradation products of azo dyes are known to have extremely complex characteristics, their identification requires efficient analytical methods that combine high separation efficiency and rich structural information. The techniques such as MS, HPLC, GC, GC/MS, HPLC/MS are particularly recommended for analysis [31,32].

In the case where ozone cannot completely oxidize organic compounds to CO_2 and H_2O due to the formation of partial oxidation products relatively unreactive towards ozone, the combination of ozone with homogeneous or heterogeneous catalysts, with or without metallic phases, is recommended for the enhancement of degradation of organic compounds [33].

Up till now, numerous studies have demonstrated that dyes can be effectively degraded by ozone and some published studies are summarized in Table 3.3.

Table 3.3 Summary of some results in the removal of dyes by ozonation.

Dye	Time (min)	Ozone Amount (mg/min)	Catalyst	pH	Color Removal (%)	COD Removal (%)	TOC Removal (%)	References
Textile Dyeing Factory wastewater	360	–	No catalyst		90	–		[45]
C. I. Reactive Blue 5		–	LaCoO$_3$				100	[56]
Acid Red B	60	30	Fe–Cu oxide	6.8	90	70	–	[48]
ReactiveBlue 19	90	–	No catalyst			55	17	[41]
Procion red MX-5B	30	4.3	Metal ions		97	–	75	[27]
RR198	9	–	MgOnanocrystal catalysts	8	100	–	–	[44]
C. I. Acid Blue 113, C. I. Reactive Yellow 3, C. I. Reactive Blue 5	less than 10	–	activated carbon, cerium oxide, ceria-activated carbon composite		100	–	–	[57]
Reactive Blue 19	10	1.2 L/min	No catalyst	–	BOD/COD = 0.33	–	–	[22]
Bomaplex Red CR-L dye	10–30	–	No catalyst		100	35–56	–	[26]
Acid, Direct and Reactive	180	–	No catalyst		100	66	–	[58]
C. I. Direct Black 22	160	–	No catalyst		70–83	–	33	[4]
Red X-GLR	120	–	No catalyst		100	5.7–35	–	[59]
Remazol Black 5	360	–	No catalyst	–	–	40	25	[60]

3.2.1.1.1   Factors Influencing the Ozonation Efficiency

Researchers have demonstrated that the rate of decomposition of ozone is correlated with initial pH and inlet ozone concentration, reaction time, temperature, catalyst and H_2O_2 concentration.

pH

The initial pH reaction is an important parameter influencing the ozonation process [27] since the oxidizing agent generated depends on the pH of the solution; ozone decomposition in water is strongly pH dependent and occurs faster with an increase of pH. As mentioned before, degradation of dyes is either due to direct oxidation of ozone or radical oxidation by •OH radical, and this mainly depends on the pH of the solution. Consequently, the oxidizing agent is known to be ozone in acidic solution, whereas in neutral and basic solution it is the hydroxyl radicals [14,29,34,35]. Since the oxidation potential of •OH radicals is higher than that of the ozone molecule, the decomposition reaction is faster in basic conditions [27,36–39].

Inlet Ozone Concentration

The ozone concentration that is used in the reaction is an important parameter affecting the extent of oxidation. The effect of the amount of ozone concentration on decolorization and degradation of different dyes has been studied extensively by other researchers. They have concluded that an increase in the inlet ozone concentration considerably increases the removal efficiency [40,41].

However, it is very important to calculate the optimum ozone concentration before the oxidation process in order to avoid overdosing. Ozone is a gas that is very unstable in the liquid phase and hence it is difficult to obtain saturation in the regions where mass transfer is limiting. This results in a steady state concentration that often is much lower than the saturation concentration. It is therefore important to carefully determine optimum ozone concentration. If ozonation is carried out at ambient temperature and atmospheric pressure, the concentration of ozone in aqueous phase can be calculated according to Henry's Law [42,43].

Reaction Time

The studies reported in literature have shown that that the reaction time in the ozonation process has a positive effect on the treatment efficiency. The researchers observed that removal of both color and COD increased with an increase in reaction time [44,45].

Temperature

Ozonation reactions are recommended to be carried out at ambient temperature. Even though higher temperatures generally increase the reaction rate, at the same time they reduce the solubility of O_3 in water and hence the amount of O_3 available for reaction will decrease [42,46].

Using Catalyst in Ozonation

Recently, catalyst-based ozonation processes have been reported as a powerful technology capable of degrading a variety of organic pollutants.

Using transition metals such as Fe, Ni, Mn, Cu and Zn in ozonation enhances the oxidizing power of ozone. The usage of these as catalysts leads to a faster removal of organic pollutants and reduces the ozone amount necessary [27,39,44,47–51].

As can be easily seen from Table 3.3, the ozonation processes required a shorter amount of time when there was a catalyst in the medium.

Using H_2O_2 in Ozonation (O_3/H_2O_2)

In the O_3/H_2O_2 system, hydroxyl radicals are generated by a radical chain mechanism by interaction between the ozone and the hydrogen peroxide.

The global reaction is as follows [52]:

$$H_2O_2 + 2O_3 \rightarrow 2OH^- + 3O_2 \qquad (3.4)$$

Although H_2O_2 reacts very slowly with the ozone molecule in water, its conjugate base (HO_2-) can rapidly react with molecular ozone, thereby initiating the formation of hydroxyl radicals [53]. H_2O_2 initiates O_3 decomposition by electron transfer [54,55]; alternatively, the reaction can be described as the activation of H_2O_2 by ozone.

The efficiency of this process can also be improved by introducing UV radiation [52].

Upon investigation of some of the research on ozonation of dyes that has been reported in literature (Table 3.3), the following conclusions can be reached:

- due to different experimental results it can be generally concluded that ozonation is effective for color removal,
- however it is only moderately efficient for chemical oxygen demand (COD) or total organic compound (TOC) removal,
- the dye solutions are decolorized or degraded in a shorter reaction time when a catalyst is used, and
- studies are generally performed in alkaline solutions, since ozone decomposition is accelerated in high pH.

3.2.1.2   Fenton's Reagent Oxidation

Fenton's reagent oxidation is a catalytic oxidation process using a mixture of strong chemical oxidizer, hydrogen peroxide, and ferrous ions as a catalyst and an acid as an optimum pH adjuster. This particular reagent mixture is called Fenton's Reagent.

The main advantages of Fenton's reagent oxidation compared to other AOPs are its simplicity, the lack of toxicity of the reagents and the cost-effective source of hydroxyl radicals it offers, since the chemicals are always available at moderate cost and there is no need for special equipment [61].

The general mechanism using Fenton's reagents, via which the hydroxyl radicals are produced, is a number of cyclic reactions, which utilize the Fe^{2+} or Fe^{3+} ions as a catalyst to decompose the H_2O_2 [62].

There are numerous parallel and consecutive reactions, but are generally described by the following two dominant reactions [63]:

$$Fe^{2+} + H_2O_2 \rightarrow Fe^{3+} + \bullet OH + OH^- \qquad (3.5)$$

$$Fe^{3+} + H_2O_2 \leftrightarrow Fe^{2+} + \bullet OOH + H^+ \qquad (3.6)$$

The principal active component of Fenton's reagent is the hydroxyl free radical which is produced by catalytic chemical reaction (Reaction 3.5) between hydrogen peroxide and ferrous iron under an optimum pH condition [64,65]. In this reaction, the ferrous ion initiates and catalyses the decomposition of H_2O_2 and hydroxyl radicals are generated [63]. In addition to the oxidation, the iron (III) ions generated during the oxidation stage promote the removal of other pollutants by coagulation and sedimentation. Reaction 3.6 is known to be several orders of magnitude slower than Reaction 3.5 [67]. The slow regeneration of Fe^{3+} to Fe^{2+} is the rate-determining step of the overall reaction. Thus, in AOP, rate of dye degradation is fast in the beginning due to high initial concentration of Fe^{2+}. However, subsequently the rate is drastically reduced due to the drop in the concentration of Fe^{2+} and poor rate of its regeneration [68].

The Fenton oxidation process can be operated both homogeneously or heterogeneously under various combinations.

The classical homogeneous Fenton oxidation may be the most extensively studied method. However, there are two major drawbacks of the process. The process is limited by the acidic pH required (pH = 2–4) and it needs high amounts of iron ions in the solution, and the iron ions need to be separated from the system at the end of the reaction, which means an additional removal process [69–75].

To overcome these disadvantages, increasing attention has been paid to research on the heterogeneous Fenton systems. The use of heterogeneous Fenton and Fenton-like catalysts has recently received much attention, such as with zeolites,clays and activated carbon (from organic waste) loaded with the iron is recommended [69,76–80].

In the heterogeneous phase, the physical steps in addition to chemical changes take place on the surface of the catalyst at the active sites where mass tranfer limited adsorption of reactant molecules occurs [75,81]. Also heterogeneous solid catalysts can support Fenton-like reactions over a wide range of pH values, because the iron (III) species in solid catalysts is immobilized within the structure and the catalyst can maintain its ability to generate hydroxyl radicals from H_2O_2. And the catalyst can easily be recovered after the reaction. A wide range of solid materials, such as transition metal exchanged zeolites, and iron oxide minerals have been reported in the literature as heterogeneous catalysts for the oxidative removal of dyes through Fenton oxidation [69,82].

The efficiency of the treatment can be considerably increased when ultraviolet light is simultaneously irradiated in the Fenton process, the so-called photo-Fenton's process [83,84]. Under irradiation, ferric ion complexes produce extra •OH radicals and the regeneration of Fe (II), which will further react with more H_2O_2 molecules in Fenton reaction.

Some studies performed on Fenton's Reagent oxidation of dye pollutants are illustrated in Table 3.4.

3.2.1.2.1   Factors Influencing the Fenton's Reagent Oxidation Efficiency
The oxidation efficiency of the Fenton process depends on several variables, namely pH, iron (Fe^{2+}) concentration, H_2O_2 concentration, and temperature for a given wastewater.

pH
Fenton oxidation is known as a highly pH-dependent process since pH plays an important role in the mechanism of •OH production in the Fenton's reaction [62]. The efficiency of the Fenton reaction system is mainly a function of pH. It directly affects the generation of hydroxyl radicals and thus the oxidation efficiency. The optimum pH recommended is between pH = 3 and pH = 6.

For the homogeneous Fenton reaction, the solution pH drastically influences the efficiency, whereas this effect in heterogeneous Fenton reactions is generally only moderate. In heterogeneous Fenton systems, the pH can potentially affect surface adsorption of the organic contaminant, the complex formation of H_2O_2 with active sites, and changes in iron (II, III) recycling due to the pH-dependent oxidation [85].

A change in pH of the solution involves a variation of the concentration of Fe^{2+}, and therefore the rate of production of •OH radicals responsible for oxidation dyes is restricted [86].

For pH values less than 3, the reaction between hydrogen peroxide and Fe^{2+} is affected, leading to a reduction in hydroxyl radical production, and

Table 3.4 Summary of some results in the removal of dyes by Fenton's Oxidation.

Dye	Time (min)	T (°C)	[H$_2$O$_2$] (mM)	Catalyst	pH	Color Removal (%)	COD Removal (%)	TOC Removal (%)	References
Reactive Black 5	–	30	4.0	iron (III) impregnated on rice husk ash	3	89.18	–	–	[69]
Alcian Blue	–	50	30.0		2.5	93.2	–	54.1	[70]
Reactive Black 5	–	30	4.0	iron (III) impregnated on AC	3	89.18	–	–	[82]
Reactive Blue 19	5	–	8.3	–	3	80	–	–	[99]
C. I. Acid Red 14	6	80	8.7	Fe (II)-Y Zeolite	5.96	99	–	–	[77]
Reactive Black 5	150	–	12.0	Montmorillonite K10	2.5	99	–	–	[100]
Acid Red 1	120	30	8.0	iron ions loading on rice husk	2.0	96	–	–	[81]
Direct Blue 15	50	30	2.8	–	4.0	100	–	–	[101]
Reactive Yellow 145	30	50	–	–	3	91.4	52.27	–	[102]
Reactive Black 5	15	–	–	–	3	97.5	21.6	–	[1]

in addition to this, $\bullet OH$ radical formation is inhibited due to the small amount of soluble iron (Fe^{3+}) at pH below 3 [1,87].

A decrease in efficiency for pH values in alkaline conditions attributed to the precipitation of iron hydroxide (Fe (OH)$_3$) occurs, decreasing the concentration of dissolved Fe^{3+} and consequently of Fe^{2+} species [88,89]. In addition, in such conditions hydrogen peroxide is less stable and hence decomposes into water and oxygen [88,89]; therefore fewer hydroxyl radicals are formed, reducing the process efficiency.

Iron (Fe^{2+}) Concentration
It is well known that dye degradation efficiency by Fenton process is influenced by the concentration of Fe^{2+} ions, which catalyze hydrogen peroxide decomposition resulting in $\bullet OH$ radical production and hence the degradation of organic molecule. According to the literature, it is recommended to determine an optimum concentration for iron [62].

In the absence of iron, there is no evidence of hydroxyl radical formation and hence there is no removal or the removal efficiency is considerably lower.

But as the concentration of iron is increased, dye removal accelerates until a point is reached where further addition of iron becomes inefficient [62].

When Fe^{2+} concentration is increased, the catalytic effect also increases accordingly, and when its concentration is higher, a great amount of Fe^{3+} is produced. Fe^{3+} undergoes a reaction with hydroxyl ions to form Fe (OH)$^{2+}$, which has a strong absorption band causing higher absorbance and results in a decrease in $\bullet OH$ generation and hence removal efficiency decreases [91–93].

H$_2$O$_2$ Concentration
In the Fenton process, hydrogen peroxide plays an important role as a source of hydroxyl radical generation. Generally, it can be concluded that increasing H_2O_2 concentration improves the degradation and decolorization of dyes.

Previous studies on Fenton oxidation have highlighted the existence of an optimum in the H_2O_2 concentration with respect to achievable rates of removal.

In many studies performed by Fenton oxidation, the researchers concluded that when a low amount of H_2O_2 is added, the intermediate compounds of degradation are still present in the reaction mixture even after enough reaction time.

When a higher amount of H_2O_2 is used, it will generate more hydroxyl radicals, which are available for oxidation.

However, when H_2O_2 is overdosed, this does not improve the degradation efficiency. The excess H_2O_2 reacts with the hydroxyl radicals and inhibits the oxidation reaction, which might cause decomposition of H_2O_2 to oxygen and water, and the recombination of •OH radicals. And in addition to this, higher concentrations of hydrogen peroxide might act as free-radical scavenger itself [62].

The reason for its being an optimum value can be concluded as follows: At moderate concentrations, the •OH radicals attack the dye molecules, whereas at excess H_2O_2 concentration the scavenging of •OH radicals may occur and hence the degradation may decrease [69,81,94].

In addition to this, besides considering the process efficiency, the selection of an optimum H_2O_2 concentration is also recommended due to environmental aspects and the cost of H_2O_2[94,96].

Temperature

Increasing the temperature has a positive effect on the removal of dyes. Fenton's reaction can be accelerated by raising the temperature, which improves the generation rate of •OH and therefore enhances the removal of dyes [97].

However, as temperatures increase above a certain value (generally 40–50°C), the efficiency of H_2O_2 utilization decreases. This is due to the possible decomposition of H_2O_2 into oxygen and water. As a practical matter, temperatures between 20 and 40°C are recommended for most applications of Fenton's reagent [93].

3.2.1.3   Wet Air Oxidation

The wet air oxidation (WAO) process, which was first patented by Zimmerman over 50 years ago, removes organic compounds in the liquid phase by oxidizing them using an oxidant such as oxygen or air at high temperatures (120–300°C) and pressures (0.5–20 MPa) [103,104].

With the help of WAO processes, the organic contaminants dissolved in water are in turn partially degraded by means of an oxidizing agent into biodegradable intermediates or mineralized into innocuous inorganic compounds such as carbon dioxide, water and inorganic salts, which remain in the aqueous phase [105].

In this method, by setting an appropriate temperature and pressure and injecting determined values of oxidant (air, oxygen, hydrogen peroxide, ozone, etc.), the oxidation operation of organic matter is performed [106].

The severe operating conditions and high costs sometimes limit its application in wastewater treatment. Catalytic wet air oxidation (CWAO) introduces a catalyst to the system of traditional WAO, thus it enables mild reaction conditions, increases the oxidation capacity of oxidants, improves the formation of hydroxyl radicals and shortens reaction time, thereby reducing the investment in operating costs [105,107–109]. The reaction pathway is different for the WAO and CWAO reactions, being that the intermediates formed during the catalytic tests were less refractory to oxidation than those obtained with WAO [110]. In particular, azo dyes are readily converted into hazardous aromatic amines under anoxic conditions [111,112].

The catalysts included may be homogenous or heterogeneous types. The key problems that need to be solved are recycling of homogenous catalysts and better stability of heterogeneous catalysts [106]. Although the homogenous catalysts such as copper and iron salts, and the heterogeneous catalysts such as noble metals like (Pt, Pd, Ru, Rh) and transition metal oxide catalysts (oxides of Cu, Ni, Mn, Fe, Co) are effective in CWAO, they have some disadvantages like the need for an additional separation step to remove or recover the metal ions from the treated effluent and deactivation of the catalyst [113,114].

Recently, various carbon materials like activated carbon have been used in catalytic wet oxidation processes effectively due to the stability, high porosity and surface areas of the carbon materials [82].

The wet air oxidation mechanism is divided into two steps:

1. A physical step, which involves the transfer of oxygen from the gas phase to the liquid phase, usually considers that oxygen diffuses rapidly within the gas phase, and the transfer of carbon dioxide to the gas from the liquid.
2. A chemical step, which involves the reaction between the organic matter and dissolved oxygen in the liquid phase, producing carbon dioxide.

Usually the mechanism includes initiation, propagation and termination reactions. Numerous reactions have been proposed, and the main reactions in the presence of organic compounds (RH) are listed [104].

$$RH + O_2 \rightarrow HO_2\bullet + R\bullet \qquad (3.7)$$

$$H_2O + O_2 \rightarrow HO_2\bullet + \bullet OH \qquad (3.8)$$

$$R\bullet + O_2 \rightarrow ROO\bullet \qquad (3.9)$$

$$ROO\bullet + RH \rightarrow ROOH + R\bullet \qquad (3.10)$$

$$\bullet OH + RH \rightarrow R\bullet + H_2O \qquad (3.11)$$

$$HO_2\bullet + \bullet OH \rightarrow H_2O + O_2 \qquad (3.12)$$

This method has been successfully applied to the removal of many organic contaminants, including azo dyes [113,115–120].

Studies on the WAO of some model dyes are listed in Table 3.5.

Table 3.5 Summary of some results in removal of dyes by Catalytic Wet Air Oxidation.

Dye	Time (min)	P (MPa)	T (°C)	Catalyst	pH	Color Removal (%)	COD Removal (%)	TOC Removal (%)	Reference
Reactive Black 5	90	–	70	Commercial activated carbons (AquaSorb 5000 P)	3	92	–		[82]
Methyl Orange Direct Brown Direct Green	120	–		$CuO/c\text{-}Al_2O_3$	–	99	–	70	[3]
BasicYellow 11		–	150	Ni/MgAlO	–	–	–	–	[121]
Acid Orange 7	160	0.5	150	$NaNO_2/FeCl_3$	2.6	–	–	56	[127]
Reactive Dye Solution	120	1.0	135 165	$CoAlPO_4$-5 CeO_2	–	95 100	90 95	–	[127]
Orange II	100	0.6–3.0	260	$H_4SiW_{12}O_{40}$, $Na_2HPW_{12}O_{40}$	–	–	–	90	[128]
Orange II	150	1.0	230	catalyst comprising ZnO, CuO, and Al_2O_3	–	–	–	70	[129]

3.2.1.3.1 Factors Influencing the Catalytic Wet Air Oxidation Process Efficiency

The key features that influence the free-radicals formation, and thus the pollutant degradation rate, in CWAO are believed to be pressure, temperature, initial dye concentration and catalyst loading. Because these parameters determine the overall removal efficiency, it is important to understand the mutual relationships between the parameters in terms of free-radical formation, degradation and decolorization. The degradation effects are generally evaluated by the extent of removal of color, chemical oxygen demand (COD) and total organic carbon (TOC).

Pressure

The application of pressure requires knowledge of some thermodynamic properties, like solubilities of gases in liquids at pressures higher than those for which such data is ordinarily available.

The dissolved oxygen concentration in the liquid phase is proportional to the partial pressure of oxygen in the gas phase (Henry's law). The high oxygen pressure in the gas phase ensures sufficient dissolved oxygen concentration providing a strong driving force for oxidation, which is required for the dye oxidation. In other words, the concentration of free radicals increases with oxygen pressure and then enhances the organic compounds degradation [105,110]. Elevated pressures are also required to keep water in the liquid state [121].

Temperature

Temperature is one of the most important variables in WAO and CWAO reactions. Considering the Arrhenius law in the kinetic regime, the oxidation reaction in terms of dye degradation is expected to accelerate when increasing the temperature [121,122].

The studies in literature show that dye removal is very sensitive to temperature in wet air oxidation. The mass transfer of oxygen from the gas phase to the liquid phase depends on the operating temperature. Higher temperatures lead to a higher production of free radicals which are responsible for the initiation of the oxidation reaction. This explains why the performances obtained for reactions carried out at higher temperatures are better [123,124].

Initial Dye Concentration

Decolorization and degradation efficiency depends on the initial dye concentration. With an increase in initial concentration of the dye, when the loading of catalyst is kept constant, more dye molecules are adsorbed onto the surface of the catalyst, limiting the generation of hydroxyl radicals, and so the degradation rate decreases. Thus, an increase in the number of substrates being accommodated in the lattice of the catalyst inhibits the action of the catalyst with oxygen, thereby decreasing the degradation efficiency [123]. Hence, it can be concluded that the oxidation process requires more catalyst loading and longer reaction times for higher dye concentrations [124].

Catalyst Loading
Catalyst is known to increase the applicability of the wet oxidation process using dedicated catalysts, which potentially promote oxidation in a shorter reaction period under milder operating conditions [125].

When the loading of the catalyst is increased, the number of dye molecules adsorbed also increase. However, above a certain level, the dye molecules available may not be sufficient for adsorption, hence the additional catalyst will not be involved in the catalysis activity [126].

Some studies on the WAO of dyes are summarized in Table 3.5. These studies show that the removal of dyes has attracted a lot of interest by showing good potential for dealing with dye wastewater. The following points can be concluded:

- The WAO process is suitable for a wide range of dyes.
- A high degree of decolorization can be achieved.
- CWAO processes have proven to be extremely efficient in TOC removal from wastewaters.
- The use of catalyst not only improves treatment efficiency, but also reduces the severity of reaction conditions.

3.2.1.4   Electrochemical Oxidation

Electrochemical oxidation is another method that belongs to the advanced oxidation methods. The electrochemical oxidation method is considered an environmentally friendly technology that is able to electrogenerate *in situ* hydroxyl radicals [130–132]. One of the main advantages of the method is that on the surface of electrodes only electrons are produced and consumed.

Electrochemical methods oxidize and reduce pollutants in wastewater by means of electrode reactions. The electrodes needed are available in various shapes (bar, plate, porous and fiber) and are made of various materials; processes are influenced significantly by the anode material. The requirements for an ideal anode material include acceptable efficiency, cost-effectiveness and stability in severe conditions [133].

Various anodes have been tested for dye-containing wastewater treatment and boron-doped diamond has proven to be the most efficient for dyes [133–136].

Generally, oxidation of organic matter by electrochemical treatment can be classified as direct oxidation on the surface of the anode and indirect oxidation distant from the anode surface.

In a direct anodic oxidation process, the pollutants are first absorbed on the anode surface and then destroyed by the anodic electron transfer reaction. Direct anodic oxidation yields rather poor removal [137].

In an indirect oxidation process, during electrolysis, the oxidants are regenerated by the electrochemical reactions. The pollutants are then destroyed in the bulk solution by the oxidation reaction of regenerated oxidants. All the oxidants are generated *in situ* and are used immediately [137].

Some researchers have defined the steps which the energy supplied to an electrode undergoes in four sections: Firstly, transportation of electroactive particle from the bulk solution to the electrode surface is realized. Then electroactive particle is adsorbed on the surface of the electrode. Electron transfer between the bulk and the electrode is realized. And finally, the reacted particle is either transported to the bulk solution (desorption) or deposited on the electrode surface [138].

This method has been recently used for decolorizing and degrading dyes from aqueous solutions [133–137]. The results of these investigations indicated that the operating parameters such as cell voltage and pH play an important role on the electrochemical oxidation of organic pollutants, and controlling these parameters leads to an efficient treatment.

The cell voltage significantly influences both the electrochemical and adsorption catalytic process. It is recommended to study at an optimum cell voltage even though a higher cell voltage causes faster removal efficiency. Because increasing cell voltage increases the overpotential required for the generation of oxidants, the energy consumption becomes higher as larger cell voltage is applied. At high cell voltage, the service life of electrodes is shortened [139].

It has been reported that the influent pH is an important operating factor influencing the performance of the electrochemical process [140]. Although the treatment performance depends on the nature of the pollutants, with all pollutants the best removal efficiency has been found near a pH of 7. The power consumption is, however, higher at neutral pH due to the variation of conductivity. When conductivity is high, the pH effect is not significant [140].

3.2.2 Photochemical Advanced Oxidation Processes

To produce photochemical changes in a molecule, irradiation of light in the UV-visible range must occur within the system. The visible spectrum covers wavelengths between 400 and 800 nm [55].

The most common sources of UV light are high-pressure mercury vapor lamps and pulsed-UV (P-UV) xenon arc lamps with good emission [55].

Even though photochemical processes are known as some of the most efficient, feasible and kinetically favorable types, light-mediated AOPs are not always adequate for the removal of mixtures of substances of high absorbance, or containing high amounts of solids in suspension, because the quantum efficiency decreases through loss of light, dispersion and/or by competitive light absorption [55].

In the following sections, among the photochemical processes, information on photocatalytic oxidation, ultraviolet irradiation and hydrogen peroxide (UV/H_2O_2) and photocatalytic ozonation (UV/O_3) will be given.

3.2.2.1 Photocatalytic Oxidation

Direct photolysis involves the interaction of light with molecules to bring about their dissociation into fragments, with the following pathway:

$$R\text{-} + h\nu \rightarrow \text{Intermediates} \qquad (3.13)$$

$$\text{Intermediates} + h\nu \rightarrow CO_2 + H_2O + R\text{-} \qquad (3.14)$$

This process appears to be less effective than other processes where radiation is combined with hydrogen peroxide or ozone, or where homogeneous, heterogeneous catalysis or photocatalysis are employed.

Photocatalytic oxidation is a very efficient and promising advanced oxidation process in which various types of organic compounds can be broken down to CO_2, water and mineral salts, and complete mineralization can be achieved by the help of activated oxygen species such as the hydroxyl radical and super oxide radicals [141].

The photocatalytic process is based on the irradiation, usually UV ($\lambda = 320$–400 nm), in the presence of a catalyst, usually a semiconductor that creates electron donor and acceptor sites allowing the formation of a highly oxidizing agent, hydroxyl radical. This powerful reactive species attacks most of the organic molecules with a relatively high rate of reaction and nonselectivity [142–144].

The photocatalytic oxidation is mainly realized in four steps: The transferring of the reactants onto the surface of catalyst, adsorption of the reactant on the surface, photocatalytic reaction in the adsorbed phase, and finally, desorption and removal of the final products [145].

The basic mechanism for generating •OH involves the adsorption of a photon, the generation of an electron-hole pair ($e_{cb}^{-} + h_{vb}^{+}$) and subsequently the production of •OH and superoxide radical anions.

When photocatalyst absorbs ultraviolet radiation from a light source, it will produce pairs of electrons and holes. The electron of the valence band of photocatalyst becomes excited when illuminated by light. The excess energy of this excited electron promotes the electron to the conduction band of photocatalyst, thereby creating the negative-electron (e–) and positive-hole (h+) pair. The energy difference between the valence band and the conduction band is known as the "band gap." The positive-hole of photocatalyst breaks apart the water molecule to form hydrogen gas and hydroxyl radical. The negative-electron reacts with oxygen molecule to form superoxide anion. This cycle continues when light is available [146].

The mechanism of the photocatalytic degradation of organic compounds and dyes can be summarized as follows [147,148]:

Absorption of photons by photocatalyst and production of photoholes and electron pairs:
In this step, semiconductor photocatalysis is initiated by electron-hole pairs after band gap excitation. When a photocatalyst is illuminated by light with energy equal to or greater than band-gap energy, the valence band electrons can be excited to the conduction band, leaving a positive hole in the valence band:

$$\text{(photocatalyts)} + h\nu \rightarrow h_{vb}^{+} + e_{cb}^{-} \tag{3.15}$$

Formation of superoxide radical anion (oxygen ionosorption):

$$(O_2)_{ads} + e_{cb}^{-} \rightarrow O_2^{\circ}- \tag{3.16}$$

Neutralization of OH⁻ by photoholes:

$$(H_2O \leftrightarrow H^{+} + OH-)_{ads} + h_{vb}^{+} \rightarrow H^{+} + \bullet OH \tag{3.17}$$

It has been suggested that the hydroxyl radical (•OH) and superoxide radical anions ($O_2^{\circ}-$) are the primary oxidizing species in the photocatalytic oxidation processes. Both species are strongly oxidizing and capable of degrading organic pollutants. These oxidative reactions would results in the degradation of the pollutants as shown in the following Equations 3.18–3.19;

oxidation of the organic pollutants via successive attack by •OH radicals

$$R + \bullet OH \rightarrow R^{\bullet\prime} + H_2O \tag{3.18}$$

or by direct reaction with holes

$$R + h^+ \rightarrow R^{\bullet+} \rightarrow \text{degradation products} \qquad (3.19)$$

The benefits of this method are mineralization of organic compounds, no additional wastewater problem and operating in mild pressure and temperature [149].

In literature, there are many studies on degradation and decolorization of azo dyes using advanced oxidation methods comprising the photocatalysis in the presence of various types of photocatalysts. Some of these studies are summarized in Table 3.6.

3.2.2.1.1   Factors Influencing Photocatalytic Degradation

In photocatalytic degradation of dyes in wastewaters, the main operating parameters which affect the process efficiency are UV intensity, photocatalyst type and loading, initial dye concentration and pH. These parameters will be considered one after the other as they influence the photocatalytic processes of the degradation of dyes in wastewaters.

UV Intensity

Generally it is expected that the treatment efficiency steadily increases with increased UV intensity since UV intensity determines the amount of photon absorbed by the catalyst [143].

Under the higher intensity of light irradiation, the removal is expected to be higher because the electron–hole formation is predominant and, hence, electron–hole recombination is negligible. In contrast to this, at lower light intensity, electron–hole pair separation competes with recombination resulting in a decrease in the formation of free radicals, causing less of an effect on the percentage degradation of the dyes [150,151].

Photocatalyst Type and Loading

The use of semiconductors such as TiO_2, ZnO, Fe_2O_3, and CdS as photocatalysts is inevitable for the degradation of organic pollution. Due to its optical and electrical properties, low cost, remarkable photocatalytic activity, chemical stability, nontoxicity and photocorrosion resistance, nano-titanium dioxide is preferred as a common photocatalyst [152,153]. The catalytic activity of TiO_2 catalyst has been raised by using transition metals. Recently, photocatalytic degradation of organic pollutants has been improved by immobilization of the transition metals onto solid matrices such as activated carbon, zeolite, clay, sand, glass, stainless steel and various biomaterials [154,155]. These supporting materials provide a large surface area for adsorption, prevent leaching and allow the catalyst recovery [141,153].

Many studies in literature suggest that the amount of catalyst plays a major role in the degradation of dyes in photocatalytic systems. It is recommended to determine the optimum loading for an efficient removal of dye.

Generally the rate of photocatalytic degradation increases with an increase in the amount of catalyst. By increasing photocatalyst concentration the exposed surface area increases, which absorbs more numbers of photons, and as a result the rate of photocatalytic degradation of the dyes increases [156].

However, after a certain amount of photocatalyst concentration the treatment efficiency may keep constant or decrease. This can be explained in terms of availability of active sites on photocatalyst surface and the light penetration of photoactivating light. Because by using the photocatalyst concentration more than necessary, the light penetration through the solution becomes difficult, and hence, the photoactivated volume of the suspension is reduced. In such a condition, part of the catalyst surface probably becomes unavailable for photon absorption and dye adsorption [149,157].

Initial Dye Concentration
The studies in literature reveal that increasing dye concentration will reduce photocatalytic degradation [143,158,159]. This can be attributed to the quantity of intermediates increased with the increasing dye concentration. At higher dye concentration, a significant amount of UV light might be absorbed by dye molecules rather than the catalyst, reducing the photodegradation efficiency [160,161].

Sacco *et al.* concluded in their study that at constant reaction times, the increase of initial dye concentration will lead to a decrease in photocatalytic activity because the increase in the color intensity of the solution will reduce the light penetration into the aqueous medium, meaning that the path length of photons inside the solution will decrease [157].

Another point is that the active site of the catalyst surface area is fixed. If the concentration of dye increases, since a limited number of dye molecules can attach at the active site of the catalyst, the rate of degradation decreases. Competitions between dye molecules to attach to active site also affect the rate of degradation [162].

pH
The interpretation of pH effects on the efficiency of dye photocatalytic oxidation is a difficult and interesting task because of its several roles.

Previous studies indicate that pH may affect photocatalytic oxidation efficiency in a number of ways [143]. For example, the charge of the dye molecules with ionizable functional groups and the surface of the catalyst that can be used are both pH dependent. Acid-base properties of the metal oxide surfaces can have considerable implications upon their photocatalytic activity [163]. For example, in the case of using TiO_2 as photocatalyst the point of zero charge of the TiO_2 is at pH 6.8 [164]. Thus, the TiO_2 catalyst surface will be positively charged in acidic media (pH < 6.8), whereas it is negatively charged under alkaline conditions (pH > 6.8) [149]. Consequently, pH changes can thus influence the adsorption of dye molecules onto the catalyst surfaces [165]. In addition to this, the catalyst particles (especially TiO_2) might tend to agglomerate under acidic conditions and the surface area available for dye adsorption and photon absorption would be less [165]. Another obvious point is that the formation of hydroxyl radicals is more responsible for photocatalytic degradation and that hydroxyl radicals can be formed by the reaction between hydroxide ions and positive holes. The positive holes are considered as the major oxidation species at low pH, whereas hydroxyl radicals are considered as the predominant species at neutral or high pH levels [162]. An alkaline condition would thus favor •OH formation and enhance degradation.

From the existing literature (Table 3.6), it can be seen that generally:

- the photocatalytic oxidation process is suitable for a wide range of dyes,
- a high degree of decolorization can be achieved,
- pH may affect photocatalytic oxidation efficiency, and
- the effect of temperature is not considered that much.

3.2.2.2   Ultraviolet Irradiation and Hydrogen Peroxide (UV/H_2O_2)

Currently gaining more attention is the UV/H_2O_2 system, which is an advanced oxidation process in which hydrogen peroxide (H_2O_2) is added in the presence of ultraviolet (UV) light to generate hydroxyl radicals (•OH).

As reported, hydrogen peroxide is a strong oxidant for reducing low levels of pollutants present in wastewaters [68]. However, the individual use of H_2O_2 is not always efficient in oxidizing more complex pollutants. The use of H_2O_2 becomes more effective when it acts in conjunction with other reagents or energy sources capable of dissociating it to generate hydroxyl radicals.

With UV irradiation in wavelengths shorter than 300 nm, H_2O_2 can decompose and generate hydroxyl radicals. In the combined ultraviolet irradiation and hydrogen peroxide system, because the rate of dye removal is mainly dependent on the concentration of H_2O_2, it is necessary to determine the optimum H_2O_2 concentration to avoid an excess that could retard

Table 3.6 Summary of some results in the removal of dyes by Photocatalytic Oxidation.

Dye	Time (min)	T (°C)	UV Intensity mW/cm²	Catalyst Type	pH	Color Removal (%)	COD Removal (%)	TOC Removal (%)	References
Reactive Blue 2	–	–	32 Watt	Nano-titania		95.22	94.05	92.52	[166]
Acid Yellow	–	–	25 mW/cm²	TiO$_2$ and activated carbon	–	–	–	–	[167]
Azo dye containing wastewater	–	–		Iron or Cesium doped nano-titaina films	–	–	–	–	[168]
Acid Red 27	–	room	–	Nano ZnO	7	–	100	–	[169]
Remazol Black 5 Remazol Brilliant Orange	–	–	–	TiO$_2$	2	–	–	–	[170]
Alizarin Cyanine Green	90	–	–	TiO$_2$-Activated carbon -TiO$_2$		–	–	–	[172]
Methyl Orange	–	–	–	Anatase and rutile phase TiO$_2$ supported on spherical activated carbon	5.7	98		–	[153]

(Continued)

Table 3.6 (*Cont.*)

Dye	Time (min)	T (°C)	UV Intensity mW/cm^2	Catalyst Type	pH	Color Removal (%)	COD Removal (%)	TOC Removal (%)	References
Methylene blue (MB) Real textile wastewater (TW)	120	–	–	Silver doped titania	6.8	–	99 (MB) 98 (TW)	–	[172]
Basic Red 18 (BR18) Basic Red 46 (BR46	–	–	–	Titania nanoparticle/ activated carbon	–	–	81.2 (BR18) 76.2 (BR46)	–	[173]
Procion Red MX-5B	20	23	17	TiO$_2$	5	100			[143]

the degradation [55]. For high removal efficiencies, the H_2O_2 concentration must be above the stoichiometric demand and sufficient reaction of time under UV radiation must be applied [174].

The main advantage of using the combined ultraviolet irradiation and hydrogen peroxide process is that H_2O_2 is totally soluble in water, there is no mass transfer limitation, and there is no need for a separation process after treatment [17,46,55].

Studies on the removal of dyes using UV/H_2O_2 have shown that it is a promising method to treat dilute aqueous solutions of azo dyes [175,176].

3.2.2.3   Photocatalytic Ozonation (UV/O_3)

Ozonation combined with ultraviolet (UV) radiation is deemed as a more effective process to remove organics; compared to sole ozonation, a larger quantity of •OH radicals can be formed. UV radiation is commonly employed to enhance ozone decomposition, yielding more free radicals and resulting in a higher ozonation rate [177].

The photodecomposition of ozone leads to two hydroxyl radicals per ozone molecule [178]:

$$H_2O + O_3 + h\nu \rightarrow 2 \bullet OH + O_2 \qquad (3.20)$$

$$2 \bullet OH \rightarrow H_2O_2 \qquad (3.21)$$

The key parameters for the success of ozone/UV system are ozone dosage, UV irradiation level and pH.

The main difference between combined photocatalytic ozone decomposition and ozone decomposition alone in aqueous solution is the initiation of the reaction. The starting radical is formed photochemically by an electron transfer from photocatalyst to oxygen and not by the reaction of OH^- ion with ozone [40].

In the literature there are various promising studies using ozonation combined with ultraviolet (UV) radiation in the treatment of dyes [179–181].

3.3   Concluding Remarks

The use of conventional wastewater treatment processes for dye removal has become increasingly challenging because of the complex structure and recalcitrant nature of dyes and the identification of more and more intermediates. Consequently, advanced oxidation processes have emerged as powerful technologies and are widely used in dye-containing wastewater.

The AOPs can be classified in terms of whether UV source is used in the process as non-photochemical and photochemical processes. If an overview of several AOPs is performed and their effectiveness in the degradation of all kinds of dyes is compared, there are mainly four treatment technologies that hold the greatest promise: ozonation, Fenton's reagent oxidation, wet air oxidation and photocatalytic oxidation.

The principal mechanism of the AOPs function is the generation of highly reactive free radicals. A combination of two or more AOPs generally enhances free radical generation, which eventually leads to higher oxidation rates.

Much research has been performed within the field of advanced oxidation of dyes and there is no doubt that these methods work and are efficient for elimination of dyes in wastewaters. However each method has its own advantages and disadvantages. The different methods may result in different intermediates since toxicity patterns for the applied treatments are substantially different. They may also differ in the optimum operating parameters with respect to the wastewater quality and quantity. Consequently, it is the researchers' responsibility to choose the most appropriate treatment method for their cases.

References

1. M. S. Lucas, and J. A. Peres, *Dyes and Pigments*, Vol. 71(3), p. 236, 2006.
2. M. A. Behnajady, N. Modirshahla, R. Hamzavi, *J. Hazard. Mater.* Vol. 133, p. 226, 2006.
3. L. Hua, H. Ma, and L. Zhang, *Chemosphere*, Vol. 90(2), p. 143, 2013.
4. H.-Y. Shu, and M.-C. Chang, *J. Hazard. Mater.*, Vol. 121(1–3), p. 127, 2005.
5. H. Kusic, N. Koprivanac, and L. Srsan, *J. Photochem. Photobiol. A: Chem.*, Vol. 181, p. 195, 2006.
6. P. A. Ramalho, M. H. Cardoso, A. Cavaco-Paulo, and M. T. Ramalho, *Appl. Environ. Microbiol.*, Vol. 70, p. 2279, 2004.
7. R. Gong, X. Zhang, H. Liu, Y. Sun, and B. Liu, *Bioresour. Technol.*, Vol. 98, p. 1319, 2007.
8. M. Greluk, and Z. Hubicki, *Chemical Engineering Journal*, Vol. 162(3), p. 919, 2010.
9. H. Park, and W. Choi, *Journal of Photochemistry and Photobiology A: Chemistry*, Vol. 159, p. 241, 2003.
10. J. G. Montano, F. Torrades, L. A. Perez Estrada, I. Oller, S. Malato, M. I. Maldonado, and J. Peral, *Environ. Sci. Technol.*, Vol. 42, p. 6663, 2008.
11. W. T. Tsai, C. Y. Chang, M. C. Lin, S. F. Chien, H. F. Sun, and M. F. Hsieh, *Chemosphere*, Vol. 45, p. 51, 2001.

12. R. Munter, *Proc. Estonian Acad. Sci. Chem.*, Vol. 50, p. 59, 2001.
13. A. Mandal, K. Ojha, A. K. De, and S. Bhattacharjee, *Chemical Engineering Journal*, Vol. 102, p. 203, 2004.
14. C. R. Da Silva, M. G. Maniero, S. Rath, and J. R. Guimarães, *J. Adv. Oxid. Technol.*, Vol. 14, p. 106, 2011.
15. Ullmann's *Encyclopedia of Industrial Chemistry*, Germany, VCH Verladsgesellschaft, 5 ed., p. 415, 1991.
16. J. Yoon, Y. Lee, and S. Kim, *Water Sci. Technol.*, Vol. 44(5), p. 15, 2001.
17. A. L. N. Mota, L. F. Albuquerque, L. T. C. Beltrame, O. Chiavone-Filho, A. Machulek Jr., and C. A. O. Nascimento, *Brazilian Journal of Petroleum and Gas*, Vol. 2(3), p. 122, 2008.
18. I. Arslan, I. A. Balcioglu, T. Tuhkanen, *Environ. Technol.*, Vol. 20, p. 921, 1999.
19. R. B. M. Bergamini, E. B. Azevedo, L. R. R. Araújo, *Chem. Eng. J.*, Vol. 149, p. 215, 2009.
20. T. Y. Chen, C. M. Kao, A. Hong, C. E. Lin, S. H. Liang, *Desalination*, Vol. 249, p. 1238, 2009.
21. L. G Devi, S. G. Kumar, K. M Reddy, C. Munikrishnappa, *J. Hazard. Mater.*, Vol. 164, p. 459, 2009.
22. J. M. Fanchiang, D. H. Tseng, *Chemosphere*, Vol. 77, p. 214, 2009.
23. A. Muhammad, A. Shafeeq, M. A. Butt, Z. H. Rizvi, M. A. Chughtai, S. Rehman, *Braz. J. Chem. Eng.*, Vol. 25, p. 453, 2008.
24. K. E. O'Shea, and D. Dionysiou, *J. Phys. Chem. Lett.*, Vol. 3, p. 2112, 2012.
25. P. Sukanchan, *Res. J. Chem. Environ.*, Vol. 16(2), p. 83, 2012.
26. E. Oğuz, B. Keskinler, and Z. Çelik, *Dyes and Pigments*, Vol. 64(2), p. 101, 2005.
27. K. Pachhade, S. Sandhya, and K. Swaminathan, *J. Hazard. Mater.*, Vol. 167 (1–3), p. 313, 2009.
28. A. C. Silva, J. S. Pic, G. L. Sant'Anna Jr., and M. Dezotti, *Journal of Hazardous Materials*, Vol. 169(1–3), p. 965, 2009.
29. S. Palit, *Journal of Environmental Research and Development*, Vol. 6(3), p. 575, 2012.
30. J. B. Parsa and S. H. Negahdar, *Separation and Purification Technology*, Vol. 98, p. 315, 2012.
31. G. Lopez, G. Ricco, and G. Mascolo, *Water Sci. Technol.*, Vol. 38, p. 239, 1998.
32. F. Zhang, A. Yediler, and X. Liang, *Chemosphere*, Vol. 67(4), p. 712, 2007.
33. B. Kasprzyk-Hordern, M. M Ziółek, and J. Nawrocki, *Appl. Catal. B: Environ.*, Vol. 46, p. 639, 2003.
34. A. R. Freshour, S. Mawhinney, and D. Bhattacharyya, *Water Res.*, Vol. 30, p. 1949, 1996.
35. A. Reife, H. S. Freeman, *Environmental Chemistry of Dyes and Pigments*, New York, John Wiley & Sons, Inc., 1996.
36. S. Baig, and P. A. Liechti, *Water Sci. Technol.*, Vol. 43, p. 197, 2001.
37. I. A. Balcioglu, and I. Arslan, *Water Sci. Technol.*, Vol. 43, p. 221, 2001.

38. F. Zhang, A. Yediler, X. Liang, and A. Kettrup, *Dyes Pigments*, Vol. 60, p. 1, 2004.
39. S. Zhang, D. Wang, S. Zhang, X. Zhang, and P. Fan, *Procedia Environmental Sciences*, Vol. 18, p. 493, 2013.
40. T. E. Agustina, H. M. Ang, and V. K. Vareek, *Journal of Photochemistry and Photobiology C: Photochemistry Reviews*, Vol. 6(4), p. 264, 2005.
41. A. R. Tehrani-Bagha, N. M. Mahmoodi, and F. M. Menger, *Desalination*, Vol. 260(1–3), p. 34, 2010.
42. I. Karat, Advanced oxidation processes for removal of COD from pulp and paper mill effluents: A Technical, Economical and Environmental Evaluation. Master of Science Thesis, Stockholm. Presented at Industrial Ecology, Royal Institute of Technology, 2013.
43. J. A. Roth, and D. E. Sullivan, *Ind. Eng. Chem. Fundam.*, Vol. 20(2), p. 137, 1981.
44. G. Moussavi, and M. Mahmoudi, *Chemical Engineering Journal*, Vol. 152(1), p.1, 2009.
45. S. Wijannarong, S. Aroonsrimorakot, P. Thavipoke, C. Kumsopa, and S. Sangjan, *APCBEE Procedia*, Vol. 5, p. 279, 2013.
46. P. R. Gogate, and A. B. Pandit, *Advances in Environmental Research*, Vol. 8, p. 501, 2004.
47. B. Legube, and N. K. V. Leitner, *Catal. Today*, Vol. 53, p. 61, 1999.
48. X. Liu, Z. Zhou, G. Jing, and J. Fang, *Separation and Purification Technology*, Vol. 115, p. 129, 2013.
49. K. Singh, and S. Arora, *Crit. Rev. Env. Sci. Tec.*, Vol. 41, p. 807, 2011.
50. C. Tizaoui, and N. Grima, *Chem. Eng. J.*, Vol. 173, p. 463, 2011.
51. A. O. Yıldırım, S. Gul, O. Eren, and E. Kusvuran, *CLEAN-Soil Air Water*, Vol. 39, p. 795, 2011.
52. S. Esplugas, J. Gimenez, S. Contreras, E. Pascual, M. Rodriguez, *Water Research*, Vol. 36, p. 1034, 2002.
53. W. H. Glaze, *Environ. Sci. Technol.*, Vol. 21, p. 224, 1987.
54. C. P. Huang, C. Dong, and Z. Tang, *Waste Manag.*, Vol. 13, p. 361, 1993.
55. M. Litter, *Handbook Env. Chem.*, Vol. 2, p. 325, 2005.
56. C. A. Orge, J. J. M. Órfão, M. F. R. Pereira, A. M. D. de Farias, and M. A. Fraga, *Chemical Engineering Journal*, Vol. 200–202, p. 499, 2012.
57. P. C. C. Faria, J. J. M. Órfão, and M. F. R. Pereira, *Applied Catalysis B: Environmental*, Vol. 88(3–4), p. 341, 2009.
58. F. Gökçen, and T. A. Özbelge, *Chem. Eng. J.*, Vol. 123(3), p. 109, 2005.
59. W. R. Zhao, H. X. Shi, and D. H. Wang, *Chemosphere*, Vol. 57(9), p. 1189, 2004.
60. C. Wang, A. Yediler, D. Lienert, and Z. Wang, *Chemosphere*, Vol. 52(7), p. 1225, 2003.
61. S. M. Arnold, W. J. Hickey, and R. F. Harris, *Environmental Science and Technology*, Vol. 29, p. 2083, 1995.
62. S. A. Abo-Farha, *Journal of American Science*, Vol. 6(10), p. 128, 2010.

63. J. J. Pignatello, *Environ. Sci. Technol.*, Vol. 26, p. 944, 1992.
64. W. G. Kuo, *Water Research*, Vol. 26(7), p. 881, 1992.
65. C. Walling, *Accounts of Chemical Research*, Vol. 31(4), p. 155, 1998.
66. R. Venkatadri, and R. W. Peters, *Hazardous Waste and Hazardous Materials*, Vol. 10(2), p.107, 1993.
67. F. Velichkova, C. Julcour-Lebigue, B. Koumanova, and H. Delmas, *Journal of Environmental Chemical Engineering*, Vol. 1(4), p. 1214, 2013.
68. E. Neyens, and J. Baeyens, *Journal of Hazardous Materials B*, Vol. 98, p. 33, 2003.
69. G. Ersöz, *Applied Catalysis B: Environmental*, Vol. 147, p. 353, 2014.
70. Z. Bıçaksız, G. Aytimur, S. Atalay, *Water Environment Research*, Vol. 80(6), p. 540, 2008.
71. M. M. Cheng, W. H. Ma, J. Li, Y. P. Huang, and J. C. Zhao, *Environ. Sci. Technol.*, Vol. 38, p. 1569, 2004.
72. F. Ji, C. Li, J. Zhang, and L. Deng, *Desalination*, Vol. 269, p. 284, 2001.
73. S. F. Kang, C. H. Liao, and M. C. Chen, *Chemosphere*, Vol. 46, p. 923, 2002.
74. R. M. Liu, S. H. Chen, M. Y. Hung, C. S. Hsu, and J. Y. Lai, *Chemosphere*, Vol. 59, p. 117, 2005.
75. H. Hassan, and B. H. Hameed, *Int. J. Environ. Sci. Dev.*, Vol. 2, p. 218, 2011.
76. A. Guzman-Vargas, B. Delahay, C. E. Lima, P. Bosch, J. C. Jumas, *Catal. Today*, Vol. 107–108, p. 94, 2005.
77. R. Idel-Aouad, M. Valiente, A. Yaacoubi, B. Tanouti, and M. López-Mesas, *J. Hazard. Mater.*, Vol. 186(1), p. 745, 2011.
78. O. A. Makhotkina, E. V. Kuznetsova, and S. V. Preis, *Appl. Catal. B: Environ.*, Vol. 68, p. 85, 2006.
79. M. Neamtu, A. Yediler, I. Siminiceanu, and A. Kettrup, *Journal of Photochemistry and Photobiology A: Chemistry*, Vol. 161, 87, 2003.
80. Y. Zhan, X. Zhou, B. Fu, Y. Chen, *Journal of Hazardous Materials*, Vol. 187 (1–3), p. 348, 2011.
81. N. K. Daud, and B. H. Hameed, *J. Hazard. Mater.*, Vol. 176, p. 938, 2010a.
82. G. Ersöz, A. Napoleoni, and S. Atalay, *Journal of Environmental Engineering*, Vol. 139, p. 1462, 2013.
83. M. Neamtu, A. Yediler, I. Siminiceanu, M. Macoveanu, and A. Kettrup, *Dyes and Pigments*, Vol. 60, p. 61, 2004.
84. J. Pignatello, E. Oliveros, and A. MacKay, *Crit. Rev. Environ. Sci. Technol.*, Vol. 36, p. 1, 2006.
85. A. Rusevova, R. Köferstein, M. Rosell, H. H. Richnow, F.-D. Kopinke, and A. Georgi, *Chemical Engineering Journal*, Vol. 239, p. 322, 2014.
86. C. Bouasla, M. El-Hadi Samar, and F. Ismail, *Desalination*, Vol. 254, p. 35, 2010.
87. J. A. S. Peres, L. H. M. Carvalho, R. A. R. Boaventura, and C. A. V. Costa, *Journal of Environmental Science and Health: Part A*, Vol. 11–12, p. 2897, 2004.
88. R. Oliveira, M. F. Almeida, L. Santos, L. M. Madeira, *Ind. Eng. Chem. Res.*, Vol. 45, p. 1266, 2006.

89. C. S. D. Rodrigues, L. M. Madeira, R. A. R. Boaventura, *Journal of Hazardous Materials*, Vol. 164(2–3), p. 987, 2009.
90. J. Herney-Ramirez, M. Lampinen, M. A. Vicente, C. A. Costa, L. M. Madeira, *Ind. Eng. Chem. Res.*, Vol. 47 p. 284, 2008.
91. N. Modirshahla, M. A. Behnajady, and F. Ghanbary, *Dyes Pigments*, Vol. 73, p. 305, 2007.
92. R. Li, C. Yang, H. Chen, G. Zeng, G. Yu, and J. Guo, *J. Hazard. Mater.*, Vol. 167, p. 1028, 2009.
93. *Fenton's Reagent General Chemistry*, http://www.h2o2.com/industrial/fentons-reagent.aspx?pid = 143&name = General-Chemistry-of-Fenton-s-Reagent, 2014.
94. S. K. Singh, and W. Z. Tang, *Waste Manage.*, Vol. 33, p. 81, 2013.
95. J. Fernandez, J. Bandara, A. Lopez, P. Buffar, and J. Kiwi, *Langmuir*, Vol. 15, p. 185, 1999.
96. C. L. Hsueh, Y. H. Huang, C. C. Wang, and C. Y. Chen, *Chemosphere*, Vol. 58, p. 1409, 2005.
97. J.-H. Sun, S.-P. Sun, G.-L. Wang, and L.-P. Qiao, *Dyes Pigments*, Vol. 74, p. 647, 2007.
98. F. Duarte, V. Morais, F. J. Maldonado-Hódar, and L. M. Madeira, *Chemical Engineering Journal*, Vol. 232, p. 34, 2013.
99. J. R. Guimarães, M. G. Maniero, and R. N. de Araújo, *Journal of Environmental Management*, Vol. 110, p. 33, 2012.
100. N. K. Daud, and B. H. Hameed, *J. Hazard. Mater.*, Vol. 176(1–3), p. 1118, 2010b.
101. J.-H. Sun, S.-H. Shi, Y.-F. Lee, and S.-P. Sun, *Chemical Engineering Journal*, Vol. 155(3), p. 680, 2009.
102. A. Mahmood, S. Ali, H. Saleem, and T. Hussain, *Asian Journal of Chemistry*, Vol. 23(9), p. 3875, 2011.
103. S. T. Kolaczkowski, P. Plucinski, F. J. Beltran, F. J. Rivas, D. B. McLurgh, *Chem. Eng. J.*, Vol. 73, p. 143,1999.
104. M. Zhou, and J. He, *Electrochimica Acta*, Vol. 53, p. 1902, 2007.
105. J. Levec, and A. Pintar, *Catal. Today*, Vol. 124, p. 172, 2007.
106. J. Fu, and G. Z. Kyzas, *Chinese Journal of Catalysis*, Vol. 35(1), p. 1, 2014.
107. Y. Liu, and D. Sun, *Appl. Catal. B: Environ.*, Vol. 72, p. 205, 2007.
108. X. Hu, L. Lei, G. Chen, and P. L. Yue, *Water Res.*, Vol. 35, p. 2078, 2001.
109. D. J. Chang, I.-P. Chen, M.-T. Chen, and S.-S. Lin, *Chemosphere*, Vol. 52, p. 943–949, 2003.
110. G. Ovejero, A. Rodríguez, A. Vallet, and J. García, *Chemical Engineering Journal*, Vol. 215–216, p. 168, 2013.
111. P. Ekici, G. Leupold, H. Parlar, *Chemosphere*, Vol. 44, p. 721, 2001.
112. R. Ganesh, G. D. Boardman, D. Michelsen, *Water Res.*, Vol. 28, p. 1367, 1994.
113. S. Keav, A. Martin, J. Barbier Jr., and D. Duprez, *Catal. Today.*, Vol. 151, p. 143, 2010.

114. A. Quintanilla, J. A. Casas, J. A. Zazo, A. F. Mohedano, J. J. Rodríguez, *Applied Catalysis B: Environmental*, Vol. 62(1–2), p. 115, 2006.
115. F. J. Benitez, J. García, J. L. Acero, F. J. Real, and G. Roldan, *Process Saf. Environ.*, Vol. 89, p. 334, 2011.
116. G. Ersöz, and S. Atalay, *Industrial and Engineering Chemistry Research*, Vol. 49, p. 1625, 2010.
117. G. Ersöz, and S. Atalay, *Industrial and Engineering Chemistry Research*, Vol. 50, p. 310, 2011.
118. O. Türgay, G. Ersöz, S. Atalay, J. Forss, and U. Welander, *Separation and Purification Technology*, Vol. 79(1), p. 26, 2011.
119. K. H. Kim, and S. Ihm, *J. Hazard. Mater.*, Vol. 186, p. 16, 2011.
120. Y. Zhang, D. Li, Y. Chen, X. Wang, and S. Wang, *Appl. Catal. B: Environ.*, Vol. 86, p. 182, 2009.
121. Y. Peng, D. Fu, R. Liu, F. Zhang, X. Liang, *Chemosphere*, Vol. 71(5), p. 990, 2008.
122. A. M. T. Silva, J. Herney-Ramirez, U. Söylemez, and L. M. Madeira, *Applied Catalysis B: Environmental*, Vol. 121, p. 10–19, 2012.
123. G. Ovejero, A. Rodriguez, A. Vallet, and J. Garcia, *Coloration Technology*, Vol. 127(1), p. 10, 2010.
124. Y. Zhan, H. Li, Y. Chen, *J. Hazard. Mater.*, Vol. 180, p. 481, 2010.
125. C. Hung, and W. Lin, *Sustain. Environ. Res.*, Vol. 20(4), p. 251, 2010.
126. R. Aravindhan, N. N. Fathima, J. R. Rao, and B. U. Nair, *J. Hazard. Mater.*, Vol. 138, p. 152, 2006.
127. D. Chang, I. Chen, M. Chen, S. Lina, *Chemosphere*, Vol. 52, p. 943, 2003.
128. I. A. Alaton, and J. L. Ferry, *Dyes Pigments*, Vol. 54(1), p. 25, 2002.
129. J. Donlagic, and J. Levec., *Appl. Catal., B*, Vol. 17(1–2), p. L1, 1998.
130. E. Brillas, I. Sirés, and M. A. Oturan, *Chem. Rev.*, Vol. 109, p. 6570, 2009.
131. A. Mhemdi, M. A. Oturan, N. Oturan, R. Abdelhédi, and S. Ammar, *Journal of Electroanalytical Chemistry*, Vol. 709, p. 111, 2013.
132. M. A. Oturan, *J. Appl. Electrochem.*, Vol. 30, p. 477, 2000.
133. M. Zhou, H. Särkkä, and M. Sillanpää, *Separation and Purification Technology*, Vol. 78(3), p. 290, 2011.
134. P. Canizares, A. Gadri, J. Lobato, B. Nasr, R. Paz, M. A. Rodrigo, and C. Saez, *Ind. Eng. Chem. Res.*, Vol. 45, p. 3468, 2006.
135. X. M. Chen, and G. H. Chen, *Sep. Purif. Technol.*, Vol. 48, p. 45, 2006.
136. M. F. Elahmadi, N. Bensalah, A. Gadri, *J. Hazard. Mater.*, Vol. 168(2–3), p. 1163, 2009.
137. N. Nordin, S. F. M. Amir, Riyanto, M. R. Othman, *Int. J. Electrochem. Sci.*, Vol. 8, p. 11403, 2013.
138. C. Feng, N. Sugiura, S. Shimada, and T. Maekawa, *Journal of Hazardous Materials*, Vol. 103(1–2), p. 65, 2003.
139. L. Yue, K. Wang, J. Guo, J. Yang, X. Luo, J. Lian, L. Wang, *Journal of Industrial and Engineering Chemistry*, Vol. 20(2), p. 725, 2014.

140. G. H. Chen, *Separation and Purification Technology*, Vol. 38(1), p. 11, 2004.
141. O. K. Mahadwad, P. A. Parikh, R. V. Jasra, and C. Patil, *Environmental Technology*, Vol. 33(3), p. 307, 2012.
142. M. R. Hoffmann, S. T. Martin, W. Choi, and D. W. Bahnemann, *Chem. Rev.*, Vol. 95(1), p. 69. 1994
143. C. M. So, M. Y. Cheng, J. C. Yu, and P. K. Wong, *Chemosphere*, Vol. 46(6), p. 905, 2002.
144. K. Soutsas, V. Karayannis, I. Poulios, A. Riga, K. Ntampegliotis, X. Spiliotis, and G. Papapolymerou, *Desalination*, Vol. 250(1), p. 345, 2010.
145. M. A. Rauf, M. A. Meetani, and S. Hisaindee, *Desalination*, Vol. 276, p.13, 2011.
146. http://www.mchnanosolutions.com/mechanism.html, 2014
147. J. Šíma, P. Hasal, *Chemical Engineering Transactions*, Vol. 32, p. 79, 2013.
148. M. Vautier, C. Guillard, J. M. Herrmann, *Journal of Catalysis*, Vol. 201, p. 46, 2001.
149. I. K. Konstantinou, and T. A. Albanis, *Appl. Catal., B: Environ.*, Vol. 49, p. 1, 2004.
150. D. Bahnemann, Photocatalytic degradation of polluted waters, in: P. Boule, ed., *The Handbook of Environmental Chemistry. 2. Part L: Environmental Photochemistry*, Springer, Berlin, 1999.
151. B. Neppolian, H. C. Choi, S. Sakthivel, B. Arabindoo, V. Murugesan, *Journal of Hazardous Materials*, Vol. B89 p. 303, 2002.
152. D. F. Ollis, E. Pelizzetti, and N. Serpone, *Environ. Sci. Technol.*, Vol. 25, p. 1522, 1991.
153. J. Yoon, M. Baek, J. Hong, C. Lee, and J. Suh, *Korean J. Chem. Eng.*, Vol. 29(12), p. 1722, 2012.
154. G. Puma, A. Bono, D. Krishnaiah, and J. Collin, *Journal of Hazardous Materials*, Vol. 157, p. 209, 2008.
155. E. M. R. Rocha, V. J. P. Vilar, A. Fonseca, I. Saraiva, and R. A. R. Boaventura, *Sol. Energy*, Vol. 85, p. 46, 2011.
156. H. S. Sharma, et al., *Int. J. ChemTech Res.*, Vol. 3(2), p. 1008–1014, 2011.
157. O. Sacco, M. Stoller, V. Vaiano, P. Ciambelli, A. Chianese, and D. Sannino, *International Journal of Photoenergy*, Vol. 2012, 8 pages, 2012.
158. R. J. Davis, J. L. Gainer, G. O'Neal, and I. Wu, *Water Environment Research*, Vol. 66, p. 50, 1994.
159. J. Lea, and A. A. Adesina, *Journal of Photochemistry and Photobiology A: Chemistry*, Vol. 118, p. 111, 1998.
160. L. Karimi, S. Zohoori, M. E. Yazdanshenas, *Soc.: J. Saudi Chem.*, 2011. (In press).
161. R. H. Mills, D. Davis, Worsley, *Chemical Society Reviews*, Vol. 22, p. 417, 1993.
162. M. Swati, Munesh, and R. C. Meena, *Arch. Appl. Sci. Res.*, Vol. 4(1), p. 472, 2012.

163. D. W. Bahnemann, J. Cunningham, M. A. Fox, E. Pelizzetti, P. Pichat, and N. Serpone, in: *Aquatic Surface Photochemistry*, R. G. Zepp, G. R. Heltz, D. G. Crosby, eds., Lewis Publishers, p. 261, 1994.

164. I. Poulios, and I. Tsachpinis, *J. Chem. Technol. Biotechnol.*, Vol. 71, p. 349, 1999.

165. M. A. Fox, and M. Dulay, *Chemical Reviews*, Vol. 93, p. 341, 1993.

166. S. Rashidi, M. Nikazar, A. V. Yazdi, and R. Fazaeli, *Journal of Environmental Science and Health, Part A*, Vol. 49, p. 452, 2014.

167. M. Zeng, Y. Li, M. Ma, W. Chen, and L. Li, *Nonferrous Met. Soc. China*, Vol. 23, p. 1019, 2013.

168. H. J. Song, L. Zhu, and X. Wang, *Applied Mechanics and Materials*, Vol. 295, p. 331, 2013.

169. M. Stanthi, and V. Kuzhalosai, *Indian Journal of Chemistry*, Vol. 51A, p. 428, 2012.

170. C. Y. Chen, M. C. Cheng, and A. H. Chen, *Journal of Environmental Management*, Vol. 102, p. 25, 2012.

171. P. Muthirulan, C. N. Devi, and M. M. Sundaram, *Arabian Journal of Chemistry*, 2013 (In press).

172. C. Sahoo, A. K. Gupta, and I. M. S. Pillai, *Journal of Environmental Science and Health, Part A*, Vol. 47, p. 1428, 2012.

173. N. M. Mahmoodi, *Desalination*, Vol. 279, p. 332, 2011.

174. O. M. Alfano, R. J. Brandi, A. E. Cassano, *Chem. Eng. J.*, Vol. 82, p. 209, 2001.

175. N. Daneshvar, D. Salari, and A. R. Khataee, *J. Photochem. Photobiol. A: Chem.*, Vol. 157, p. 111, 2003.

176. A. M. El-Dein, J. A. Libra, and U. Wiesmann, *Chemosphere*, Vol. 52(6), p. 1069, 2003.

177. H. W. Prengle, *Environ. Sci. Technol.*, Vol. 17, p. 743, 1983.

178. R. Peyton, and W. H. Glaze, *Environmental Science and Technology*, Vol. 22, p. 761, 1988.

179. N. M. Mahmoodi, M. Arami, and J. Zhang, *Journal of Alloys and Compounds*, Vol. 509, p. 4754, 2011.

180. N. M. Mahmoodi, M. Bashiri, S. J. Moeen, G. R. Peyton, W. H. Glaze, *Environ. Sci. Technol.* Vol. 22, p. 761, 1988.

181. J. Sun, X. Yan, K. Lv, S. Sun, K. Deng, D. Du, *Journal of Molecular Catalysis A: Chemical*, Vol. 367, p. 31, 2013.

4

Photocatalytic Processes for the Removal of Dye

Pankaj Chowdhury*[,1], Ali Elkamel[2] and Ajay K. Ray[3]

[1]*University of Calgary, Dept. of Chemical and Petroleum Engineering, Calgary, Alberta, Canada*
[2]*University of Waterloo, Dept. of Chemical Engineering, Waterloo, Ontario, Canada*
[3]*University of Western Ontario, Dept. of Chemical and Biochemical Engineering, London, Ontario, Canada*

Abstract
The heterogeneous photocatalysis process has huge potential as a tertiary treatment method for the degradation of numerous organic compounds and dyes in wastewater. The process will be more economical and sustainable once solar-light-driven photocatalysis techniques are established.

The main objective of this chapter is to demonstrate the basic theory and mechanism of photo-oxidation processes followed by application of solar light in dye degradation. The design features and efficiencies of various solar photoreactors with their current application status will be discussed. Also reviewed are the effects of several factors such as photocatalyst loading, initial dye concentration, solution pH, light intensity, and electron scavenger for photo-oxidation of dye molecule.

Keywords: Dye, degradation, photocatalysis, photoreactor, solar

4.1 Introduction

Despite the fact that almost two-thirds of the earth's surface is covered by water, mankind is still facing challenges in acquiring safe drinking water. The main reason behind this is the contamination of water resources (lakes, rivers and oceans) due to the process of industrialization. Out of the

---

*Corresponding author: pankajchowdhury.ca@gmail.com

Sanjay K. Sharma (ed.) Green Chemistry for Dyes Removal from Wastewater, (119–137)
© 2015 Scrivener Publishing LLC

various sources of water pollution, some main sources are oil spills, sewage, industrial waste, nuclear waste, and agricultural waste [1]. According to a World Health Organization (WHO) report, in 2008 about 884 million people still relied on inferior water sources, out of which 84% live in rural areas. Many of them are suffering from severe waterborne diseases due to their daily usage of microbiologically unsafe water [2]. Dye is one of the most important classes of pollutants which results in colored wastewater that is sometimes hard to degrade because of its complex structure [3]. Dye-containing wastewater generates from textile, tannery, dying, pulp and paper, and paint industries. The majority of dyes arise from dyeing and finishing processes in textile industries. The colored wastewater is not only unpleasant on aesthetic grounds, but also is characterized by high biochemical and chemical oxygen demands (BOD$_5$: 80–6000 mg. L^{-1}; COD: 150–12000 mg. L^{-1}) [4].

Estimated dye production from different sources is reported in the range of 7×10^5 to 1×10^6 tons/annum. A significant amount of dye (10–15% of the produced quantity) enters into the environment. A dye molecule has two main components: i) chromophores, and ii) auxochromes. Chromophore is responsible for color, whereas auxochromes facilitate attachment towards fibers [3]. There are fourteen different types of textile dyes among which six classes (acid dye, direct dye, azonic dye, dispersed dye, sulfur dye, and fiber reactive dye) are classified by the US Environmental Protection Agency (EPA) as major toxic elements [5].

Dye pollutants result in several health hazards such as skin and eye-related diseases. Most dyes are toxic in nature and pose a threat to aquatic living organisms. To comply with strict environmental regulations, several conventional methods such as adsorption, membrane processes, biological processes, electrocoagulation, etc., have been utilized for the removal of dyes from water and wastewater. Table 4.1 shows a comparative review of conventional dye removal techniques with their advantages/disadvantages. Conventional oxidation processes are inferior in oxidizing dyestuff with complex structures, and thus advanced oxidation processes (AOPs) (UV/O$_3$, UV/H$_2$O$_2$, photo-Fenton, etc.) are introduced for dye degradation [4]. Among the AOPs, TiO$_2$/UV-based photocatalytic oxidation processes have received significant attention in recent years as an alternative method for water detoxification [6]. To make the TiO$_2$-based photo-oxidation process economical, solar light can be used as a potential replacement for commercial UV lamps. This again requires modification of TiO$_2$ and TiO$_2$-based photocatalysts, for utilization of solar visible spectra [1].

This chapter summarizes photocatalytic oxidation process for dye degradation under both UV and visible light, application of solar light and solar

Table 4.1 Comparison of Different Dye Removal Technologies.

Technology	Target Dye	Significant factors	Results	Advantages/ Disadvantages	References
Membrane processes: reverse osmosis (RO), nanofiltration (NF)	acid red, reactive black, reactive blue	RO: membrane pore size (0.5 nm), operating pressure (7–100 bars); NF: membrane pore size (0.5–2 nm), operating pressure (5–40 bars);	i) NF method consumes half of the electrical power compared to RO; ii) NaCl in dye solution increases dye removal efficiency; iii) pH, TDS, applied pressure and dye concentration show positive effect on dye removal; iv) feed temperature shows negative effect on dye removal;	Advantages: higher removal potential with lower effective cost; Disadvantages: production of concentrated sludge;	(7)
Electro-coagulation	reactive red 198	voltage, electrode connection mode, interelectrode distance, electrical energy consumption, dye concentration and electrolyte concentration;	i) dye removal follows first order kinetics; ii) 98.6% dye removal and 84% COD removal;	Advantages: requires smaller space, easy operation, ecofriendly process, cost effective method; Disadvantages: sludge generation;	(8)

(Continued)

Table 4.1 (Cont.)

Technology	Target Dye	Significant factors	Results	Advantages/ Disadvantages	References
Advanced oxidation processes (AOPs)	Several textile dyes	Ozone (O_3), hydrogen peroxide (H_2O_2), wavelength of ultraviolet (UV) light, O_3/UV, H_2O_2/UV, O_3/H_2O_2/UV, solution pH, dye concentration, and temperature;	i) in O_3/UV process hydroxyl radical ($HO^{\cdot}$) is produced after activation of O_3 by UV ($\lambda = 254$ nm); ii) H_2O_2/UV also produces $HO^{\cdot}$ radical for dye degradation; iii) among all AOPs O_3/H_2O_2/ UV shows highest efficiency for dye degradation;	Advantages: complete mineralization achieved in presence of $HO^{\cdot}$; no sludge generation, appropriate process for recalcitrant dyes; Disadvantages: several by-product formations, less economic feasibility;	(4)

Biological processes: Green algae (Cosmarium species)	Malachite green	Dye concentration, algal concentration, solution pH, and temperature;	i) dye decolorization rate is described with Michaelis-Menten model; ii) optimum pH is 9; temperature (5–45°C) has positive effect on decolorization rate; iii) optimal kinetic parameters: v_{max} (maximum specific decolorization rate) is 7.63 mg dye g cell^{-1}h^{-1} and K_m (dissociation constant) is 164.57 ppm;	Advantages: alga are available all over world in different kinds of habitats, good capability for verity of dye decolorization, economically feasible, publicly acceptable process; Disadvantages: slow process, needs to be performed under optimal favorable conditions;	(9, 10)
Activated carbon (AC) adsorption	Methylene blue, basic red 22, basic yellow 21	AC surface area, surface chemistry, pore structure; adsorbent dose, and solution pH;	i) three adsorption isotherm models (Langmuir, Freundlich and Redlich-Peterson) have been used, Redlich-Peterson model gave best fit; ii) alkaline pH and low pore size of AC favored dye adsorption;	Advantages: very effective adsorbent, high adsorption capacity; Disadvantages: non-destructive process, AC regeneration is very expensive;	(9, 11)

(Continued)

Table 4.1 (Cont.)

Technology	Target Dye	Significant factors	Results	Advantages/Disadvantages	References
Adsorption with non-conventional adsorbents: Fly ash from thermal power plant	Basic violet, basic fuchsin	Adsorbent particle size, adsorbent dose, solution pH, and temperature;	i) Langmuir and Freundlich adsorption isotherm have been used, nonlinear Freundlich model gave better fit; ii) negative free energy value and positive enthalpy value suggest spontaneous and endothermic process; iii) dye adsorption follow 1st order kinetics;	Advantages: low cost adsorbent, high adsorption capacity; Disadvantages: non-destructive process, expensive regeneration process;	(9, 12)

photoreactor in dye degradation, and finally discusses the dependence of different parameters (pH, photocatalyst loading, initial dye concentration, electron scavenger, light intensity) on dye degradation.

4.2   Photocatalysis – An Emerging Technology

The heterogeneous photocatalysis process has shown huge potential for water and wastewater treatment over the last few decades. The process basically combines semiconductor material with either high energy UV photon (300-400 nm) or low energy visible photon (400–700 nm). In this process, the semiconductor photocatalyst generates electron-hole pair (e^-/h^+) after electronic excitation and finally produces hydroxyl radials (HO·) in the system. Hydroxyl radical is the second highest oxidizing species (Table 4.2) after fluorine, which reacts non-selectively with most of the organic pollutants in wastewater and produces carbon dioxide, water, and mineral acid [1,6].

Photocatalysis for water decontamination was first recognized during the 1980s when photocataltyic mineralization was successfully conducted for different halogenated hydrocarbons [13,14]. Substantial research for photocatalytic degradation of various organic compounds, heavy metals, etc., with both UV and solar radiation was carried out at a later stage [15].

Table 4.2 Oxidation Potentials of Some Oxidants in Descending Order [6].

Species	Oxidation Potential (V)
Fluorine	3.03
Hydroxyl radical	2.80
Atomic oxygen	2.42
Ozone	2.07
Hydrogen peroxide	1.78
Perhydroxyl radical	1.70
Permanganate	1.68
Hypobromous acid	1.59
Chlorine dioxide	1.57
Hypochlorous acid	1.49
Hypoiodous acid	1.45
Chlorine	1.36
Bromine	1.09
Iodine	0.54

4.3   Photo-Oxidation Mechanism

The mechanisms of photocatalytic processes have been well discussed by several authors, but the most interesting explanation of photocatalysis was given by Chen *et al.* [6], who elucidated the existence of light source (electronic phase) as the fourth phase along with solid (photocatalyst), liquid (aqueous or organic solvent) and gaseous phases (O_2).

Photo-oxidation mechanism for dye molecule can be different under different light sources depending on the energy of the incident photons. In the presence of UV light, the conventional photo-oxidation mechanism will be applied with different photocatalysts such as TiO_2, ZnO, $SrTiO_3$, ZnS, WO_3, etc. The reaction mechanism for UV photo-oxidation is shown below [16]:

$$TiO_2 \xrightarrow{h\vartheta} TiO_2\left(h^+ + e^-\right)$$

$$O_2 + e^- \rightarrow O_2^-$$

$$HO^- + h^+ \rightarrow HO^{\cdot}$$

$$HO^{\cdot} + RH \rightarrow R^{\cdot} + H_2O \rightarrow CO_2 + H_2O + mineral\ acid$$

The reaction mechanism would be similar to this in the presence of visible light if metal or nonmetal doped semiconductor materials (TiO_2, ZnO, etc.) are used as photocatalysts. However, dye-sensitization mechanism would play a major role under visible light in the presence of sensitizing materials (dye, polymer, commercial pigment, etc.) [17,18,19].

Chatterjee *et al.* [20] discussed the degradation of a non-sensitizing dye (Acid Blue 1 (AB1)) with the help of TiO_2 and a sensitizing dye (eosin Y(EY)) via dye-sensitization mechanism. The degradation process generated $^{\cdot}O_2^-/^{\cdot}HO_2$ as the reactive species which were primarily responsible for the reactions. At first, EY absorbed visible photon and formed an excited state (EY*). Then EY* injected electron into the conduction band of TiO_2 and transformed to oxidized species (EY+). Finally, EY+ was reduced to ground state (EY) by AB1, which acted as electron donor producing oxidized dye AB1+ (Figure 4.1).

4.4   Solar Photocatalysis/Photoreactors

The earth's surface receives 1.5×10^{18} kWh of sunlight per year. Two major issues regarding the use of solar light for photocatalysis are: i) incident solar light is relatively diluted before reaching the earth surface and needs proper collectors for concentration of the solar light, and ii) the inability of using the entire solar spectra, especially the visible one, as most photocatalysts are active under UV radiation (300-400 nm).

Solar radiation can be classified into direct and diffuse radiation. Direct radiation reaches the earth's surface without being absorbed or scattered by anything, while diffuse radiation is dispersed before reaching the earth's surface. Global radiation is basically the sum of direct and diffuse radiation (Figure 4.2) [1].

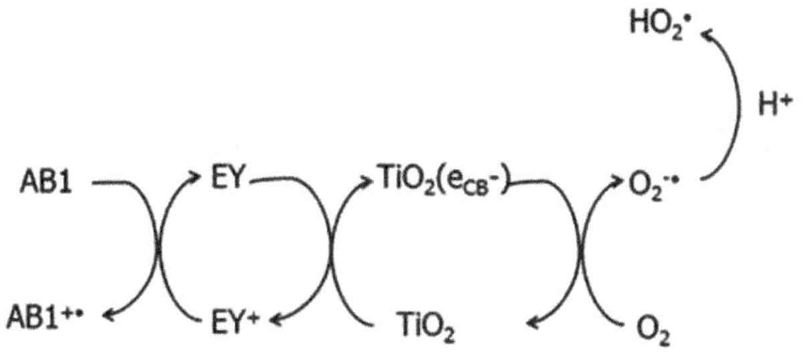

$$AB1/ AB1^+ + O_2^{-\bullet}/HO^{\bullet}_2 \to Products$$

$$EY/ EY^+ + O_2^{-\bullet}/HO^{\bullet}_2 \to Products$$

Figure 4.1 Dye-sensitization Mechanism for Dye Degradation [20].

Figure 4.2 Direct and Global Solar Radiations [1].

In solar photoreactors, solar radiation is collected and distributed inside the photoreactor in such a way that maximizes photocatalytic efficiency. The collectors are of two types: i) non-concentrating collectors, and ii) concentrating collectors. Non-concentrating collectors are capable of using both direct and diffuse radiation and provide high optical and quantum efficiency. However, their limitations are low mass transfer, reactant contamination, reactant vaporization, etc. On the contrary, concentrating collectors have certain advantages such as providing excellent mass transfer, zero vaporization of the reactant, and small reactor tube area. But they experience difficulties in using diffuse solar radiation and thus have low quantum and optical efficiencies [21].

4.5   Solar Photoreactor for Degradation of Different Dyes

This section discusses three different types of solar photoreactors which have been utilized for photodegradation of several dyes. Compound parabolic collector (CPC) allows both direct and diffuse sun light for reflection onto the absorber surface, increasing the number of photons available and making the setup active on an overcast day in the absence of direct sunlight [22].

Augugliaro et al. [23] have installed CPCs at Plataforma Solar de Almeria (PSA, Spain) for the photodegradation of two azo-dyes (methyl-orange and orange II) with aqueous suspension of polycrystalline TiO_2. Two photoreactors with three CPC modules were used in series which provided 3.08 m^2 of irradiation surface. Solar radiation was useful in degrading both dyes, although the mineralization process was slow with the formation of NO_3^-, SO_4^{2-} and carbon dioxide. In another study, Augugliaro et al. [24] applied solar photo-oxidation reaction in combination with membrane technology for the degradation of lincomycin and separation of photocatalyst particles. Montano et al. [25] used photo-Fenton oxidation with a CPC solar photoreactor for the degradation of two reactive dyes (Procion Red H-E7B and Cibacron Red FN-R) in both bench scale and pilot scale. Here, the photo-oxidation method was also used as a pretreatment method along with an anaerobic biological treatment in an immobilized biomass reactor. The combination of photo-oxidation and biological processes worked more efficiently than individual processes in bench scale.

The CPC photoreactor follows a similar principle as that of the solar thermal process, which is deficient in a few areas such as photocatalyst handling, effective photocatalyst illumination, and mass transfer between photocatalyst and reactant fluid. Inclined plate collector (IPC) is another solar collector where all of the above parameters are properly addressed. In IPC, the photocatalysts are supported on the inclined plate and the reactant fluid flows through the catalyst allowing the transfer of photon to photocatalyst via reactant fluid. IPCs do not include any photon concentrator or reflective surface and the design is pretty simple. Some researchers have used the term thin-film fixed-bed reactor (TFFBR) instead of IPC [22,26]. Roselin *et al.* [26] discussed the use of ZnO mediated photocatalytic degradation of reactive dye 22 in a TFFBR under solar radiation. They reported almost complete mineralization of the dye along with the formation of NO_3^-, SO_4^{2-} and NH_4^+ ions. Another simple solar photoreactor setup known as double skin sheet (DSS) reactor has shown better performance compared to TFFBR [22]. Arslan *et al.* [27] performed a comparative study with two different photoreactors (TFFBR and DSS) and three commercial TiO_2 (Degussa P25, Millennium PC 500 and Sachtleben Mikroanatas) for the degradation of simulated dyehouse effluent. Degussa P25 was reported as the best photocatalyst and DSS reactor offered better efficiency even under a lower UV-A dose. Diagrams of the three solar reactors are presented in Figures 4.3–4.5 [28,29].

4.6   Dependence of Dye Degradation on Different Parameters

The photocatalytic degradation of several dyes (acid orange 7, reactive blue 4, red 120, acid brown 14, methylene blue, remazol brill blue R,

Figure 4.3 Compound Parabolic Collector Reactor (Copyright 2005 Elsevier) [28].

Figure 4.4 Thin Film Fixed Bed Reactor (Copyright 2004 Elsevier) [29].

Figure 4.5 Double Skin Sheet Reactor (Copyright 2004 Elsevier) [29].

orange G, naphthol blue black) have been carried out in natural sunlight using TiO_2, ZnO and a few other photocatalysts such as CdS, ZrO_2, WO3, etc. [5]. The TiO_2-based photocatalytic oxidation process became popular mainly because of the following reasons: i) the processes can be carried out under ambient temperature and pressure conditions, ii) the oxidant is strong and less selective, resulting in complete mineralization, iii) the process does not consume any expensive oxidizing chemical, and iv) the photocatalysts are less expensive, non-hazardous, stable and reusable [1,30,31].

UV-light-assisted dye degradation is a very popular technique, but recent studies have also found the use of natural sunlight for dye degradation as a promising and sustainable route. Direct solar radiation was used as a light source (between 9 am and 4 pm) with different ranges of solar light intensities at different locations throughout the year [32–37]. Both non-concentrating and concentrating solar reactors were utilized for the degradation of different dyes under several experimental conditions [32–37].

There are five main parameters that affect photo-oxidation of dye molecules such as: i) photocatalyst loading, ii) initial dye concentration, iii) solution pH, iv) light intensity, and v) electron scavenger. These will be discussed in the next sections.

4.6.1   Effect of Photocatalyst Loading

Mehrotra *et al.* [30, 38] categorized photocatalytic reaction into two distinct regions depending on the catalyst concentration. At a low catalyst concentration (<0.05 g L^{-1}), the overall rate is entirely controlled by reaction kinetics and is known as kinetic regime. However, at a high catalyst concentration (0.05–2 g L^{-1}), the reaction rate depends on several factors such as: i) external mass transfer, ii) catalyst agglomeration, iii) light shielding via catalyst, among others. In the kinetic regime, the initial rate is directly proportional to the catalyst concentration. That means, by using twice the photocatalyst mass the conversion also doubles, and the reaction rate remains unchanged. As mentioned by Konstantinou and Albanis [39], the dye degradation rate increases in the catalyst concentration range of 400–500 mg/L, whereas reduced degradation rate is observed above 2000 mg/L. In the photocatalytic process, there is an optimum concentration level of the photocatalyst beyond which the reaction becomes independent of the photocatalyst mass. At optimum catalyst loading, the complete photocatalyst surfaces would be illuminated with the incident light. Once the catalyst concentration exceeds the optimum level, a fraction of the photocatalyst surface would become unavailable for light absorption and dye adsorption, thus shielding the effect of the particle on the light by excess particles [39,6].

4.6.2   Effect of Initial Dye Concentration

Initial concentration of the pollutant would affect the degradation rate, and thus the effect of initial concentration is always studied. Photocatalytic degradation kinetics of different dyes over TiO$_2$ followed the Langmuir-Hinshelwood (L-H) model as described by Zhou and Ray [40].

$$r_{rxn} = \left(-\frac{dC_b}{dt}\right)\left(\frac{V_L}{V_R}\right) = \frac{k_r K C_s}{1 + K C_s} \qquad (4.1)$$

where V_L is the total volume of liquid treated (m³), V_R is the volume of the reactor (m³), (dC_b/dt) is the observed rate (mol.m⁻³.s⁻¹), k_r is the reaction rate constant (mol.m⁻³.s⁻¹), K is the adsorption-desorption equilibrium constant (m³.mol⁻¹), C_s is the concentration of the reactant on the catalyst surface (mol.m⁻³) in equilibrium with the actual surface concentration [40].

A Langmuir-type kinetic equation is more appropriate for photocatalyst slurry system; in case of immobilized system it is difficult to predict the kinetics without knowing the pollutant concentration on photocatalyst surface. Zhou and Ray [40] considered both external and internal mass-transfer resistance for photocatalytic degradation of Eosin B on immobilized TiO₂ surface to find out the true kinetic parameters. They found out the intrinsic kinetic parameters from the following equation:

$$\frac{1}{k_o} = \frac{1}{k_r'} + \frac{1}{k_{m,int}k} + \frac{1}{k_{m,ext}k} \qquad (4.2)$$

where k_r' is the first-order reaction rate constant (s⁻¹), $k_{m,ext}$ is external mass-transfer coefficient, and k is the specific surface area (m⁻¹).

4.6.3   Effect of Solution pH

The pH of the solution significantly affects a photocatalytic process as it occurs on the surface of the photocatalyst. Various surface properties of photocatalyst such as: i) surface charge, and ii) band edge position, are influenced by pH. Degussa P25 TiO₂ shows point of zero charge (pHzpc) at pH 6.8, and thus in alkaline medium, the TiO₂ surface becomes negatively charged. Therefore in alkaline pH, cationic dye (e.g., methylene blue) is well adsorbed [40,41,42].

$$Ti^{IV} - OH + HO^- \rightarrow Ti^{IV} - O^- + H_2O$$

$$Ti^{IV} - OH + H^+ \rightarrow Ti^{IV} - OH_2^+$$

Zhou and Ray [40] performed a kinetic study for photocatalytic degradation of an anionic dye (Eosin B) with TiO₂. The adsorption of Eosin B was very low in alkaline pH because of the formation of anionic species on negatively charged TiO_2 surface [40]. Ramirez et al. [32] reported the solar-assisted degradation of acid orange 7 (anionic dye) in aqueous solution by cerium-doped TiO_2. Photocatalytic degradation was highest at pH 7 and lowest at alkaline pH. Sakthivel et al. [35] compared solar photodegradation of an azo dye (acid brown 14) with ZnO and TiO_2 photocatalyst. The ZnO has a zero point charge (zpc) of 9.0 ± 0.3 and the dye degradation was higher at alkaline pH.

4.6.4 Effect of Light Intensity

Normally the rate of photo-oxidation of different model compounds increases with increasing light intensity. At a high light intensity, electron-hole formation rate exceeds electron-hole recombination rate and the photo-oxidation rate is found to be proportional to the square root of light intensity. However, at low light intensity, electron-hole formation and recombination rates compete with each other and the photo-oxidation rate is directly proportional to the light intensity [34,35].

Park et al. [34] studied the degradation of Red 120 (reactive dye) under natural sunlight between 10 am and 4 pm. Their results showed that during a clear day, the UV light intensity changed from 0.9 mW cm^{-2} to 0.7 mW cm^{-2}, whereas under thick cloudy skies the UV light intensity changed from 0.11 mW cm^{-2} to 0.3 mW cm^{-2}. The dye degradation rate under clear skies was 13 times higher than that of thick cloudy skies.

Zhou and Ray [40] performed experiments for Eosin B degradation with immobilized photocatalysts at varying light intensities and established a relationship among reaction rate constant (k_r), photocatalyst loading w (kg/m^2), and light intensity I (W/m^2). Their experimental data followed an empirical equation previously proposed by Ray and Beenackers [43] as shown below:

$$k_r = \frac{k_s a w^n I^\beta}{1 + a w^n I^\beta} \qquad (4.3)$$

where k_r is rate constant, w is photocartalyst mass, I is light intensity, and n and β are constants.

4.6.5 Effect of Electron Scavenger

Photocatalytic oxidation reactions require the use of electron scavenger to subdue its recombination with hole. Usually dissolved oxygen/air is used as the electron scavenger. According to Wang *et al.* [44], the photocatalytic reaction could be terminated in the absence of dissolved oxygen and the rate of photocatalytic decomposition of organic compound would be affected by the steady state concentration of dissolved oxygen. Zhou and Ray [40] performed an Eosin B photo-oxidation experiment in oxygen atmosphere (saturated). They have used the Langmuir-Hinshelwood equation to describe the dependence of dye-degradation rate constant on the concentration of dissolved oxygen as follows:

$$k_p \propto \frac{K_{O_2} p_{O_2}}{1 + K_{O_2} p_{O_2}} \tag{4.4}$$

where k_p is the kinetic constant for organic compound degradation, K_{O_2} is the adsorption constant of dissolved oxygen on TiO_2 catalyst, and p_{O_2} is the partial pressure of dissolved oxygen.

4.7   Conclusions

Huge amounts of dye-containing wastewater released from textile industries contaminate water resources every year. Conventional treatment methods are capable of color removal from wastewater but are incapable of complete mineralization.

UV-light-driven photocatalysis is an emerging technique for the mineralization of dye-containing wastewater. Normal band gap excitation and photo-oxidation process occur under UV light. In the presence of visible light, dye degradation is also possible via dye-sensitization pathway in the presence of a sensitizing dye.

As UV light is costly and hazardous, today solar-light-assisted photocatalysis is becoming popular around the world. There are several photoreactors such as CPC, TFFBR, DSS, etc., that have shown potential for dye degradation (mineralization) under natural sunlight. Like other photo-oxidation processes, dye photo-oxidation also depends on different parameters such as photocatalyst dose, solution pH, light intensity, and electron scavenger.

Acknowledgement

The First author would like to thank the Natural Sciences and Engineering Research Council of Canada (NSERC), for financial support.

References

1. P. Chowdhury, Solar and visible light driven photocatalysis for sacrificial hydrogen generation and water detoxification with chemically modified TiO_2. University of Western Ontario – Electronic Thesis and Dissertation Repository. Paper 702, 2012.

2. WHO, Part I: Health related millennium development goals, World Health Statistics, Publications of the World Health Organization 20 Avenue Appia, 1211 Geneva 27, Switzerland, 2011.

3. V. K. Gupta, and Suhas, *Journal of Environmental Management*, Vol. 90, p. 2313, 2009.

4. A. A. Kdasi, A. Idris, K. Saed, and C. T. Guan, *GLOBAL NEST: the International Journal*, Vol. 6, p. 222, 2004.

5. T. E. Agustina, AOPs application on dyes removal, in: S. K. Sharma, and R. Sanghi, eds., *Wastewater Reuse and Management*, Springer, pp. 353–372, 2013.

6. D. Chen, M. Sivakumar, and A. K. Ray, *Developments in Chemical Engineering and Mineral Processing*, Vol. 8, p. 505, 2000.

7. M. F. Abid, M. A. Zablouk, and M. A. Alameer, *Iranian Journal of Environmental Health Science & Engineering*, Vol. 9, p. 1, 2012.

8. A. Dalvand, M. Gholami, A. Joneidi, and N. M. Mahmoodi, *Clean – Soil, Air, Water*, Vol. 39, p. 665, 2011.

9. G. Crini, *Bioresource Technology*, Vol. 97, p. 1061, 2006.

10. N. Daneshvar, M. Ayazloo, A. R. Khataee, and M. Pourhassan, *Bioresource Technology*, Vol. 98, p. 1176, 2007.

11. E. N. E. Qada, S. J. Allen, and G. M. Walker, *Chemical Engineering Journal*, Vol. 135, p. 174, 2008.

12. D. Mohan, K. P. Singh, G. Singh, and K. Kumar, *Industrial and Engineering Chemistry Research*, Vol. 41, p. 3688, 2002.

13. A. L. Pruden, and D. F. Ollis, *Journal of Catalysis*, Vol. 82, p. 404, 1983.

14. D. F. Ollis, C. Y. Hsiao, L. Budiman, and C. L. Lee, *Journal of Catalysis*, Vol. 82, p. 404, 1984.

15. A. K. Ray, Photocatalytic reactor configuration for water purification: Experimentation and modelling, in: H. I. de Lasa, and B. S. Rosales, eds., *Advances in Chemical Engineering, Photocatalytic Technologies*, Vol. 36, Elsevier, pp. 145–183, 2009.

16. D. Chen, F. Li, and A. K. Ray, *AIChE Journal*, Vol. 46, p. 1034, 2000.

17. P. Chowdhury, H. Gomaa, and A. K. Ray, Dye-sensitized photocatalyst: A breakthrough in green energy and environmental detoxification, in: N. Shamim, and V. K. Sharma, eds., *Sustainable Nanotechnology and the Environment: Advances and Achievements*, Vol. 1124, pp. 231–266, 2013.

18. D. Zhang, T. Yoshida, and H. Minoura, *Advanced Materials*, Vol. 15, p. 814, 2003.

19. X. Shang, B. Li, T. Zhang, C. Li, X. Wang, *Procedia Environmental Science*, Vol. 18, p. 478, 2013.

20. D. Chatterjee, S. Dasgupta, R. S. Dhodapkar, and N. N. Rao, *Journal of Molecular Catalysis A: Chemical*, Vol. 260, p. 264, 2006.

21. S. Malato, Wastewater treatment by advanced oxidation processes (solar photocatalysis in degradation of industrial contaminants), Course on: Innovative processes and practices for wastewater treatment and re-use, Ankara University, Turkey, 2007.

22. R. J. Braham, and A. T. Harris, *Industrial and Engineering Chemistry Research*, Vol. 48, p. 8890, 2009.

23. V. Augugliaro, C. Baiocchi, A. B. Prevot, E. G. Lopez, V. Loddo, S. Malato, G. Marci, L. Palmisano, M. Pazzi, and E. Pramauro, *Chemosphere*, Vol. 49, p. 1223, 2002.

24. V. Augugliaro, E. G. Lopez, V. Loddo, S. M. Rodriguez, I. Maldonado, G. Marci, R. Molinari, and L. Palmisano, *Solar Energy*, Vol. 79 p. 402, 2005.

25. J. G. Montano, L. P. Estrada, I. Oller, M. I. Maldonado, F. Torrades, and J. Peral, *Journal of Photochemistry and Photobiology A: Chemistry*, Vol. 195, p. 205, 2008.

26. L. S. Roselin, G. R. Rajarajeswari, R. Selvin, V. Sadasivam, B. Sivasankar, and K. Rengaraj, *Solar Energy*, Vol. 73, p. 281, 2002.

27. I. Arslan, I. A. Balcioglu, and D. W. Bahnemann, *Water Science and Technology*, Vol. 44, p. 171, 2001.

28. P. Fernandez, J. Blanco, C. Sichel, and S. Malato, *Catalysis Today*, Vol. 101, p. 345, 2005.

29. D. Bahnemann, *Solar Energy*, Vol. 77, p. 445, 2004.

30. K. Mehrotra, G. S. Yablonsky, and A. K. Ray, *Industrial & Engineering Chemistry Research*, Vol. 42, p. 2273, 2003.

31. M. N. Chong, B. Jin, C. W. K. Chow, and C. Saint, *Water Research*, Vol. 44, p. 2997, 2010.

32. R. J. Ramirez, J. V. Sanchez, and S. S. Martinez, *Mexican Journal of Scientific Research*, Vol. 1, p. 42, 2012.

33. B. Neppolian, H. C. Choi, S. Sakthivel, B. Arabindoo, and V. Murugesan, *Chemosphere*, Vol. 46, p. 1173, 2002.

34. J. H. Park, I. H. Cho, and Y. G. Kim, *Journal of Environmental Science and Health, Part A: Toxic/Hazardous Substances and Environmental Engineering*, Vol. A39, p. 159, 2004.

35. S. Sakthivel, B. Neppolian, M. V. Shankar, B. Arabindoo, M. Palanichamy, and V. Murugesan, *Solar Energy Materials & Solar Cells*, Vol. 77, p. 65, 2003.

36. K. Nagaveni, G. Sivalingam, M. S. Hegde, and G. Madras, *Applied Catalysis B: Environmental*, Vol. 48, p. 83, 2004.

37. B. Krishnakumar, and M. Swaminathan, *Desalination and Water Treatment*, Vol. 51, p. 6572, 2013.

38. K. Mehrotra, G. S. Yablonsky, and A. K. Ray, *Chemosphere*, Vol. 60, p. 1427, 2005.

39. I. K. Konstantinou, and T. A. Albanis, *Applied Catalysis B: Environmental*, Vol. 49 p. 1, 2004.

40. S. Zhou, and A. K. Ray, *Industrial and Engineering Chemistry Research*, Vol. 42, p. 6020, 2003.
41. P. Chowdhury, J. Moreira, H. Gomaa, and A. K. Ray, *Industrial and Engineering Chemistry Research*, Vol. 51, p. 4523, 2012.
42. A. Houas, H. Lachheb, M. Ksibi, E. Elaloui, C. Guillard, and J. M. Herrmann, *Applied Catalysis B: Environmental*, Vol. 31, p. 145, 2001.
43. A. K. Ray, and A. A. C. M. Beenackers, *AIChE Journal*, Vol. 43, p. 2571, 1997.
44. C. M. Wang, A. Heller, and H. Gerischer, *Journal of the American Chemical Society*, Vol. 114, p. 5230, 1992.

5

Removal of Dyes from Effluents Using Biowaste-Derived Adsorbents

Pejman Hadi[1], Sanjay K. Sharma[2] and Gordon McKay[*,1]

[1]Chemical and Biomolecular Engineering Department, Hong Kong University of
Science and Technology, Clear Water Bay Road, Hong Kong SAR
[2]Green Chemistry & Sustainability Research Group, Department of Chemistry,
JECRC University, Jaipur, India

Abstract

The removal of dyes from aqueous media has always been the focus of attention since the invention of the dyeing industry. Dyes not only disturb the visual appearance of the aquatic environment, but can also negatively affect human health due to their toxicological, carcinogenic and mutagenic nature.

The hazards associated with these materials have inspired scientists to look for different technologies to remove these toxic substances from effluents. It is now well-established that adsorption/ion exchange is a much preferred technique over the traditional wastewater treatment methods owing to its ease of process, effectiveness and no sludge formation.

Notwithstanding the very promising dye removal efficiency of activated carbon materials, their high cost has incited debate on the economic aspects of these materials. In the last few decades, researchers have employed many different types of waste materials to develop various porous carbonaceous and siliceous materials to act as adsorbents for dye removal.

This chapter will primarily provide an overview of the relevant studies reported in literature undertaken to modify waste materials and use them for dye adsorption. Furthermore, the advantages and disadvantages of the waste-produced adsorbents, their comparative uptake capacities as well as the relevant challenges will be addressed.

Keywords: Dye removal, adsorption, ion exchange, waste materials, activated carbon, wastewater treatment

---

*Corresponding author: kemckayg@ust.hk

Sanjay K. Sharma (ed.) Green Chemistry for Dyes Removal from Wastewater, (139–201)
© 2015 Scrivener Publishing LLC

139

5.1   Introduction

Over the last few decades, the awareness of society towards the protection of the environment has vastly increased. In this regard, the chemical industry has become one of the foremost targets of environmentalists, since its potential adverse impacts can have irreparable damage to the ecosystem. Closely related to the environment is the presence of chemicals, which may seriously jeopardize human health if not treated properly [1,2]. Dyes and pigments are one of the most criticized such chemicals that pollute the environment, although they are not as toxic as other pollutants such as heavy metals. This can be related to the visible nature of these contaminants, which not only causes toxicity in aqueous media, but also is at the forefront of aesthetic pollution even at very low concentrations, thus receiving significant public concern. These substances can also hinder the penetration of light into the water streams, endangering aquatic life. Since the degradation of dye molecules by aquatic microorganisms is almost negligible, they potentially accumulate in the medium [3,4].

This is why governments have ratified stringent legislations regarding the discharge of dye-containing wastewater into the environment and more specifically into the water bodies. This has obliged industries to look for technologies to remove such contaminants from aqueous streams. Some of the physicochemical and biological technologies employed for this purpose are ozonation, coagulation-flocculation, photodecomposition, flotation, membranes, electrochemical destruction, adsorption and ion exchange. However, these technologies have been impeded due to several drawbacks such as low efficiency, generation of secondary byproducts, short half-life of oxidizing agents, high cost, low removal rate, high sludge formation, technical constraints and high energy consumption [5-12].

Adsorption, among others, has been proved to be one of the most promising and efficient dye removal techniques in industry [13,14]. Nevertheless, its high cost has been known to be a major disadvantage [15-17], which adds financial burden to the treatment of wastewater and might appear in the form of higher final product price. Therefore, in practice although consumers are the ultimate beneficiaries of pollution remediation, the onus of the financial responsibility is imposed on them [18]. This, in turn will probably influence the overall demand of consumers. All these interactions have obliged industries to endeavor to reduce the cost of the adsorption process while maintaining the high efficiency. In this regard, the application of low cost carbonaceous or siliceous waste materials is found to be highly attractive with its significant cost reduction despite the challenges encountered in the chemical or physical modification of these substances to increase their efficiency [10,16,19–23].

The adsorption capacity of an adsorbent can be greatly influenced by two well-known factors; high surface area, which assures a high amount of adsorption sites for the adsorbates and abundant surface functional groups, which can form complexes with the adsorbate molecules, enhancing the selectivity of the adsorbent [24]. Depending upon the modification process and the activating agent applied, the aforementioned properties of the adsorbent material can be systematically monitored.

There are two common types of activation methodologies basically aimed at increasing the surface area of the material and/or conjugating functional moieties to the surface of the adsorbent. *Physical activation* involves a two-step process in which the carbonaceous precursor is pyrolyzed in an inert atmosphere at elevated temperatures in order to remove the non-carbon elements and then the resultant char is exposed to a controlled oxidizing atmosphere, usually steam or carbon dioxide. In this process, the carbonaceous material reacts with the oxidizing agent and the gaseous reaction volatiles are stripped off, building up a porous structure. It is worth noting that some literature has reported the combination of these two steps and the development of a single-stage activation process with comparable surface areas. On the other hand, *chemical activation* of a siliceous or carbonaceous substance includes a simultaneous pyrolysis and activation using a dehydrating agent, such as zinc chloride and phosphoric acid, at relatively high temperatures [25–28]. Chemical activation process has several superior advantages over the physical activation technique which includes lower reaction temperatures, less reaction time, higher yield and higher surface area [29,30].

The mechanism of adsorption is usually governed chiefly by the binding energy of the adsorbate to the adsorbent surface. *Physisorption* involves the weak van der Waals interaction between the adsorbate and the surface of the adsorbent. Since no chemical bond forms, the chemical nature and the electronic orbital patterns of the adsorbent and adsorbate molecules are not perturbed. In contrast, a chemical reaction between the adsorbate molecules and the adsorption sites on the surface of the adsorbent drives the *chemisorption*, so that the identity of the sorbed molecule might be altered by the chemical bond [31–33].

This chapter will primarily provide an overview of the relevant literature undertaken to modify the waste carbonaceous and siliceous materials for use in dye adsorption purposes. It has been endeavored to compare different activation conditions and their effects on the dye removal efficiency of the resulting materials. Furthermore, the advantages and disadvantages of such waste-based adsorbents and the relevant challenges will be addressed in addition to their comparative dye uptake capacities.

5.2 Agro-Based Waste Materials as Dye Adsorbents

Agricultural waste materials are abundantly available, renewable, inexpensive and nontoxic [34]. All these specifications account for the high popularity of these waste substances, primarily as pollution sources, to be widely used as a carbonaceous source for the production of inexpensive, high-value activated carbon [23,27]. The economic value of these lingocellulosic materials coupled with their high potential in the economically-viable sequestration of pollutants renders them particularly attractive to be used in eco-friendly wastewater treatment.

Although the basic components of agricultural waste materials are cellulose, hemicelluloses and lignin [23], the appropriate choice of these materials and proper activation techniques is a critical factor affecting their ultimate performance. Herein, an exhaustive literature review on the subject of the application of various waste agro-based materials for the production of activated carbons with different surface areas, nanopore structures and functional groups has been summarized. Also, it has been endeavored to examine the influence of the original precursor type and the modification parameters on the dye adsorption properties of the resulting activated carbon.

5.2.1 Rice Husk

Rice husk is the outermost protective coating of a rice kernel which comprises significant amounts of silica and carbon [35]. During the rice harvesting season, the rice is milled and the husk is separated from the grain in a so-called husking stage. Annually, over 500 million tons of paddy is produced globally over 20 wt% of which is assumed to be husk [36]. This amount is steadily increasing due to the cultivation growth as well as an increase in the production yield per unit area.

Due to the large quantities of rice husk produced in developing countries, rice husk used to be either dumped into the landfill or open-burnt in order to decrease the space required to landfill this voluminous waste. However, with the swift progress in technology and increasing public environmental awareness, this problematic waste has turned out to be a popular focus of research for energy recovery [37], fuel production [38], electrodes in batteries and capacitors [39], catalyst [40], hydrogen storage [41] and adsorbent for pollutant sequestration [42]. The latter has been one of the most attractive applications of this waste and will be reviewed in detail.

Han *et al.* [43] used rice husk without any treatment for the adsorption of Congo red dye from effluents in a fixed-bed column. They found that the pH, initial dye concentration, flow rate of the effluent and bed depth have significant effects on the adsorption capacity of rice husk. They observed that an increase in the pH level of the solution would lead to a decrease in the adsorption efficiency of the adsorbent. They attributed this effect to the change in the surface charge of the adsorbent and a competition between the dye molecules and OH^- ions at higher pH values. Also it has been pointed out that the existence of salt in the effluent would enhance the removal capacity of the adsorbents due to the reduced repulsive forces between the surface functional groups of the rice husk and dye molecules. However, it should be noted that the maximum capacity of the untreated rice husk was measured to be only 3 mg.g^{-1}, which is a very low value. The adsorption equilibrium and kinetics of two types of dye molecules by unmodified rice husk have been investigated by Safa *et al.* [44]. They considered the effect of various parameters such as initial dye concentration, pH, adsorbent dose and adsorbent particle size on its removal efficiency. Similar to the study by Han *et al.* [43], they also observed a decrease in the removal percent by an increase in the pH value. Moreover, the adsorption capacity of the adsorbent decreases by increasing the rice husk dose, which was ascribed to the aggregation of the adsorbent particles at high loadings and the availability of fewer binding sites.

In order to increase the adsorption capacity of the rice husk, some researchers have investigated its chemical and/or physical modification. Chemical modification using sodium hydroxide has been carried out for this purpose [45,46]. However, the maximum adsorption capacity of the NaOH-modified rice husk for Malachite Green was found to be around 16 mg.g^{-1}, whereas the same adsorbent material has a capacity of around 45 mg.g^{-1} for Crystal Violet. These results show that this modification procedure is not quite useful to increase the adsorption capacity of rice husk. In contrast to the literature studies conducted by Han *et al.* [43], an increase in the pH value of the dye solution resulted in a drastic increase in the removal amount of the dye molecules (see Figure 5.1). This behavior was rationalized by the fact that the surface of the adsorbent was positively-charged at very low pH values and thus electrostatic repulsion occurred between the positively-charged dye molecules and the surface of the adsorbent. As the pH of the solution increased, deprotonation of the adsorbent surface sites took place and subsequent attraction between the negatively-charged surface sites and positively-charged dye molecules led

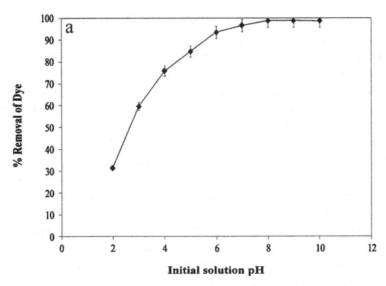

Figure 5.1 Effect of pH on adsorption of CV using treated rice husk [46].

to an increase in the dye removal amount. Also the effect of adsorbent dose was considered as a prominent parameter that strongly affects the adsorption process. It was shown that the removal percent increased, but the maximum dye uptake decreased by increasing the adsorbent loading. They attributed this phenomenon to the increasing number of adsorption sites as the adsorbent loading increases to a certain extent. However, further increase in the adsorbent dose results in the aggregation of the adsorbent particles, which in turn would lead to a decrease in the surface area of the adsorbent and a reduction in the removal percent. Furthermore, Figure 5.2 illustrates that the removal percentage of the dye molecules was decreased by raising the temperature. It was related to the weakening of the bonds between the surface functional groups of the adsorbent and the dye molecules. The adsorption mechanism of crystal violet (CV) dye molecules was perceived to consist of several steps; dissociation of the dye molecules into CV^+ and Cl^-, dissociation of surface hydroxyl groups of the adsorbent, migration of dye molecules from the bulk of the solution to the adsorbent surface, diffusion of the dye molecules from the boundary layer to the surface of the adsorbent, hydrogen bonding between the nitrogen atoms of the dye molecule and hydroxyl moieties of the adsorbent surface sites via dye-hydrogen ion exchange process. The possible adsorption mechanism of CV onto rice husk has been illustrated in Figure 5.3.

Many authors have used various acid treatment methods in order to prepare activated carbon from rice husk for wastewater treatment purposes.

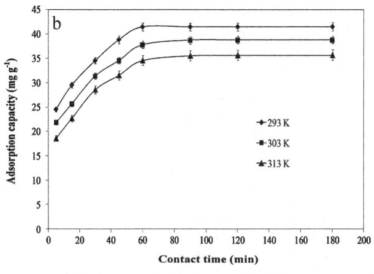

Figure 5.2 Time profiles for adsorption of CV by rice husk at different temperatures [46].

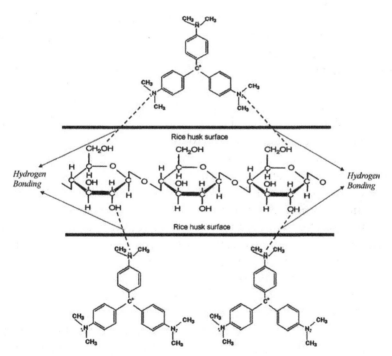

Figure 5.3 Possible adsorption mechanism of CV onto rice husk [46].

Mohamed [47] modified the rice husk using phosphoric acid as activating agent. They impregnated the rice husk in phosphoric acid at 353 K overnight and then activated the impregnated sample in a reactor at 773 K for 2.5 h. Although the resultant activated carbon had an almost high surface area and total pore volume (S_{BET} = 352 $m^2.g^{-1}$ and 0.42 $ml.g^{-1}$, respectively) with a mean pore radius size of 2.4 nm, its adsorption capacities for acid blue and acid yellow are only 38 $mg.g^{-1}$ and 10 $mg.g^{-1}$, respectively. Since no information is available about the surface functionality of the activated carbon, this low adsorption capacity cannot be explained. In a study by Rahman et al. [48], rice husk was treated with both an acid and a base and the removal percentage of malachite green dye was compared for these samples. Acid-activation was performed by impregnating the rice husk with 10% and 20% phosphoric acid and subsequent carbonization at three different temperatures, namely 400°C, 500°C and 650°C. Sodium hydroxide was used for base-activation, where 10% NaOH was mixed with rice husk at 30°C and 100°C for an hour. Then the sodium hydroxide was washed out and the resulting material was carbonized at 500°C. Since mass loss of all the samples remained constant after 30 min, it was assumed to be the optimum carbonization time. It was shown that base-activation led to the removal of silica from the rice husk while acid-activation left the silica content unchanged. The adsorption capacity of the acid-treated material was the highest when it was carbonized at 500°C. This behavior was attributed to the pore blockage at lower temperatures due to the decomposition of the organic constituents and higher silica content at high temperatures. The adsorption experiments showed that acid-activated rice husk had a higher removal efficiency compared to the base-activated rice husk.

A comprehensive study on the production of activated carbon from rice husk and its application for malachite green and Rhodamine B removal was conducted by Guo et al. [49,50]. The rice husk was carbonized at 450°C under nitrogen atmosphere and the resultant material was soaked in a caustic solution and was further heated for an hour at 400°C. Then the temperature was raised to 650°C, 700°C and 750°C for activation. The activation process altered the surface area of the rice husk considerably. Depending on the activation time and temperature and the activating chemical reagent used, the surface areas of the activated samples ranged from 1400 to 2700 $m^2.g^{-1}$ with significantly high pore volumes (see Table 5.1). The adsorption experiments revealed very promising adsorption capacities (above 550 $mg.g^{-1}$ for malachite green and 450 $mg.g^{-1}$ for Rhodamine B) for the prepared activated carbons at a high adsorption rate (equilibrium time, 90 min). It has been shown that as the temperature increases, the adsorption amount also increases, which was attributed to the increase in the intraparticle diffusion rate of the dye molecules into the activated carbon pores. It was also found that the addition of salt barely had any effect on the adsorption capacity of the activated carbon when the solution concentration is relatively low. As the solution concentration increased, the effect of salt addition became more pronounced (see Figure 5.4). At such high concentrations, the addition of salt was believed to result in a partial neutralization of the surface positive charge on the activated carbon by Cl^- ions and consequent compression of the electric double layer, enhancing the adsorption capacity of the activated carbon. Also, they hypothesized

Table 5.1 Summary of porous structure of the rice husk-based activated carbons prepared at various conditions [49,50].

Activation condition	BET SA $(m^2 g^{-1})$	Pore vol. $(cc.g^{-1})$	Micropore SA $(m^2 g^{-1})$	Av. Pore Size $(Å)$
750°C, 30 min	1886	0.98	721	20.00
750°C, 60 min	1987	1.32	785	23.47
750°C, 90 min	2721	1.88	1044	25.79
650°C, 120 min	1392	0.70	955	20.20
700°C, 60 min	1759	0.79	1735	17.89
750°C, 60 min	1930	0.97	1090	20.02

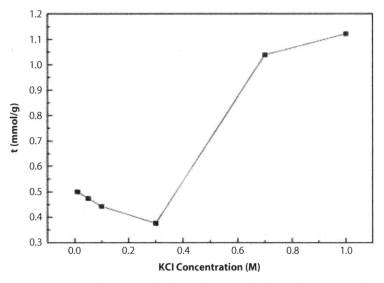

Figure 5.4 Adsorption of RB on rice husk-based activated carbon as a function of KCl [50]. (Conditions: contact time 2 h; adsorbent dosage 0.8 g/L; [RB] 1.2 mmol/L; temperature 25°C)

that the Cl^- present in the solution would pair with the dye molecules and reduce the repulsion forces between the adjacent adsorbed dye molecules. The adsorption behavior of the activated carbon for Rhodamine B dye removal at different pH values was interesting. According to Figure 5.5, as the pH was increased to pH value of 7, the adsorption capacity decreased and further increase in pH level enhanced the removal capability of the activated carbon. They attributed the dye uptake reduction at moderate pH values to the aggregation of monomeric dye molecules and formation of larger molecules, dimers, which can hinder the diffusion of dye molecules into the pores. Further increase in the pH level will increase the OH^- ions, which generates a competition between the functional moieties of the dye molecule and decrease the aggregation.

The physical activation of rice husk using steam has been carried out by Malik [51]. The rice husk was first carbonized at 400°C for an hour under air atmosphere and then it was activated at 600°C for 1 h by means of steam. The surface area obtained was 272 m²/g. The adsorption capacity of the steam-activated carbon for the removal of Acid Yellow 36 was determined to be around 30 $mg.g^{-1}$. Hameed and El-Khaiary [52] carbonized the rice straw at 700°C for 2 h under nitrogen atmosphere. The obtained material had an uptake capacity of around 150 $mg.g^{-1}$ for malachite green. Table 5.2 presents a comparison of the adsorption capacities of the rice husk-based materials together with the modification procedures.

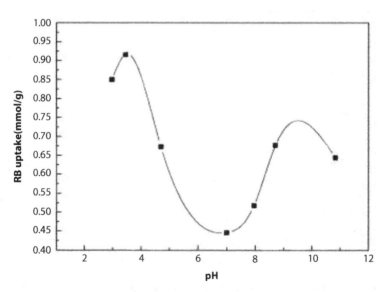

Figure 5.5 Adsorption of RB on rice husk-based adsorbent as a function of pH [50]. (Conditions: contact time 2 h; adsorbent dosage 0.8 g/L; [RB] 1.2 mmol/L; temperature 25°C)

Table 5.2 Comparison of the adsorption capacities of the rice husk-based materials.

Modification	Surface Area ($m^2 g^{-1}$)	Dye Type	pH	Adsorption Capacity ($mg·g^{-1}$)	Ref.
N/A	...	Methylene Blue	7	40	[120]
N/A	...	Direct Red-31	3	130	[44]
		Direct Orange-26		66.7	
N/A	...	Direct Red-31	2	20.5	[121]
NaOH	116	Malachite Green	7	10	[45]
		Crystal Violet	8	45	[46]
NaOH	0.47	Methylene Blue	5.2	8.3	[122]
H_3PO_4	352	Acid Blue	...	38	[47]
H_3PO_4	...	Malachite Green	...	92.6	[48]
Carbonization – NaOH – Activation	1015	Methylene Blue	12	396.4	[35]
NaOH – H_2SO_4 – Carbonization – Activation	2028	Methylene Blue	10	552	[123]
Carbonization	...	Malachite Green	5	146.5	[52]
Carbonization – NaOH – Activation	1886	Rhodamine B	3.5	480	[50]
		Malachite Green	6.6	547	[49]

5.2.2   Bagasse

Bagasse is a lingocellulosic residue from the processing of sugarcane after the sucrose extraction from crushed cane. It consists of approximately 40–45% cellulose and 30–35% hemicelluloses and 20–30% lignin [53]. Out of the numerous applications of bagasse, adsorption of various pollutants has been found to be one of the most attractive topics which have gained the attention of researchers due to the low ash content of this material.

Ho and McKay [54] carried out an exhaustive study on the kinetics of dye removal with bagasse pith. The adsorption capacity of the bagasse pith was found to be 82.8 $mg.g^{-1}$ for Maxilon Red and for Erionyl Red. Effects of several parameters, such as adsorbent dose, initial concentration of the dye solution, particle size of the adsorbent and temperature have been extensively studied. Also, different kinetics models have been employed to fit the experimental data, including pseudo-first-order, pseudo-second-order and intraparticle diffusion model. They concluded that the sorption of dyes onto bagasse pith is best described by the pseudo-second-order model. It is notable that McKay *et al.* have also conducted research on the application of pore diffusion model, which is out of the scope of this chapter [55]. The adsorption of several dyes by ball-milled sugarcane bagasse has been studied by Zhang *et al.* [56,57]. By investigating the effect of initial dye concentration, pH and adsorbent dosage, the adsorption capacity of the bagasse was determined to be 38 $mg.g^{-1}$ for Congo red, 43.5 $mg.g^{-1}$ for Rhodamine B and 27.8 $mg.g^{-1}$ for Basic Blue 9. It has been observed that an increase in the surface area of the adsorbent resulted in an enhanced dye removal due to the provision of more active sites (see Figure 5.6). The mechanism of adsorption has been well-described by conducting Fourier transform infrared (FTIR) analysis. It was shown that the interactions between the carboxyl and hydroxyl groups of bagasse with sulfonic acid groups of Congo red and amine groups of Rhodamine B and Basic Blue 9 were responsible for the uptake of the dye molecules. Tsai *et al.* [58] impregnated the bagasse with zinc chloride followed by pyrolysis at 500°C under nitrogen atmosphere. The modification process has resulted in the appearance of oxygen complexes, probably carboxyl and phenolic hydroxyl groups, on the surface of the activated material, which would lead to its hydrophilicity and surface acidity. The surface area analysis showed that a large amount of micropores had been developed and the surface area increased substantially when the impregnation ratio was raised to more than 50 wt% (see Table 5.3). Although sufficiently high microporosity was developed, the adsorption capacity remained very low (~5 $mg.g^{-1}$). Valix *et al.* [59] successfully prepared very high surface area activated carbons with high dye uptake capacity. They carbonized the bagasse at a

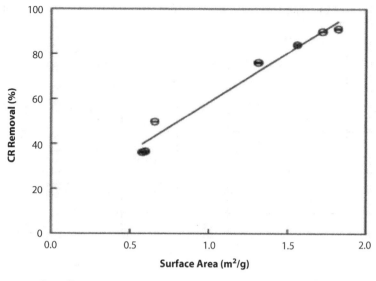

Figure 5.6 Effect of bagasse surface area on Congo red adsorption [56].

Table 5.3 Effect of impregnation ratio on the textural characteristics of bagasse-based activated carbons [58].

Impregnation Ratio	Langmuir SA $(m^2 g^{-1})$	Pore vol. $(cc.g^{-1})$	Av. Pore Diameter (Å)
25	4.86	0.003	27.5
50	287	0.107	14.9
75	526	0.194	14.8
100	790	0.288	14.6

high temperature of 900°C under nitrogen atmosphere. Then the atmosphere was changed to carbon dioxide for various periods of time (1–15 h) for activation. Depending on the activation time, the surface area ranged from 400–1400 $m^2 g^{-1}$. More details regarding the physical and chemical properties of the prepared activated carbons have been given in Table 5.4. Also, as illustrated in Figure 5.7, as the surface area increased, the acid blue uptake capacity of the activated carbon was enhanced, where the activated carbon with a surface area of 1433 $mg.g^{-1}$ had an adsorption capacity of 385 $mg.g^{-1}$. They applied two mathematical models, Langmuir and Freundlich, to fit their experimental data and found that the Freundlich equation was the best-fit model. By reviewing the Freundlich parameters, it was concluded that the steric hindrance associated with the size of the dye molecules accounts for the observed effect of pore size and surface

Table 5.4 Physical and chemical properties of activated carbons from activated bagasse [59].

Carbon ID	Gasification Time	Burn-off	BET SA $(m^2\ g^{-1})$	Pore vol. $(cc.g^{-1})$	Av. Pore Width (Å)	Ash (wt%)	Carbon pH
CC–0	0	0.00	403	0.013	3.66	10.4	2.6
CC–1	1	0.17	614	0.31	0.49	25	6.4
CC–3	3	0.20	737	0.39	0.54	28.7	6.6
CC–5	5	0.27	860	0.46	0.60	34.8	6.8
CC–7	7	0.38	1146	0.67	0.83	44.3	7.1
CC–10	10	0.51	1165	0.70	1.05	46.7	7.4
CC–15	15	0.57	1433	0.91	1.16	61.1	7.4

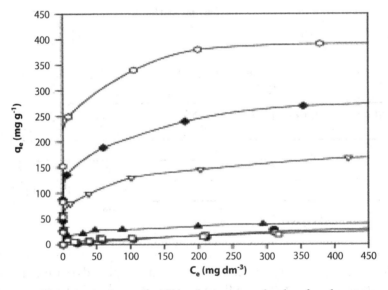

Figure 5.7 Adsorption isotherm of acid blue dye in activated carbon from bagasse (•: CC-1; □: CC-3; ▲: CC-5; ▽: CC-7; ♦: CC-10; ○: CC-15) [59].

area on the adsorption capacity of the activated carbon. It was also inferred that an increase in the surface carbon pH will immensely enhance the adsorption capacity of acid blue dye. It was related to the dissociation of the dye molecule with a negative charge and its enhanced adsorption capability on the more basic carbon surfaces. Kadam *et al.* [60] have carried out several modification processes using $CaCl_2$, NaOH, NH_4OH and steam and concluded that the sample treated with $CaCl_2$ showed the highest textile effluent dye removal efficiency (~88 $mg.g^{-1}$). Succinylated and EDTA dianhydride-modified sugarcane bagasse were prepared by Gusmão et al. [61,62] and were successfully applied for the adsorption of methylene blue and gentian violet from wastewater. The functionalization was conducted by adding succinic anhydride or EDTA dianhydride to the bagasse and heating the slurry under reflux condition for a long period of time. The introduction of carboxylate functional groups was believed to account for the successful modification of the precursor. The surface area of the EDTA-modified sample was found to be very low (1.4 $m^2\ g^{-1}$), but had a larger average pore diameter (7.8 nm) compared to activated carbons. Since the pH of zero charge point, pH_{PZC}, for the prepared materials was 7.5, the adsorbent becomes negatively-charged when the pH of the dye solution is higher than pH_{PZC}. The maximum adsorption capacity of the succinylated and EDTA-modified sugarcane bagasse was determined to be 478 and 202 $mg.g^{-1}$ for methylene blue and 1273 and 328 $mg.g^{-1}$ for gentian violet, respectively, whilst adsorption capacity of unmodified bagasse was 48.5 and 85.6 $mg.g^{-1}$ for methylene blue and gentian violet, respectively. In another attempt, the tetraethylenepentamine-modified sugarcane bagasse (TEPA-SB) was successfully used to remove eosin Y from the effluent [63]. Due to the existence of amine functional groups on the surface of the modified material, a high adsorption capacity of 349.3 $mg.g^{-1}$ for this dye was obtained. The adsorption capacities of sugarcane bagasse modified with various methods have been compared in Table 5.5.

5.2.3   Peat

Peat, consisting of a complex array of organic matter (lignin, cellulose and humic substances), is formed through incomplete decomposition of dead plant matters under limited oxygen and excess humidity conditions. The existence of a variety of functional groups on this organic material, including carboxylic acid, phenolic hydroxide, ester, aldehyde and ketone, in addition to its polarity and porosity renders it a potential adsorbent to bind with different types of pollutants in wastewater [64].

Ho and McKay [65] have explored the adsorption of two dyes onto peat. It was found that the adsorption capacity of the peat for Basic Blue 69 (195 $mg.g^{-1}$) is much higher than the Acid Blue 25 (12.7 $mg.g^{-1}$). This behavior was attributed to the surface characteristics of peat and the ionic charges of the dye molecules. They observed that increasing the agitation speed significantly enhances the adsorption capacity of the adsorbent due to the decrease in the boundary layer resistance to mass transfer from the bulk of the solution to the adsorbent surface. As shown in Figure 5.8, the

Table 5.5   Comparison of the adsorption capacities of the bagasse-based materials.

Modification	Surface Area ($m^2 g^{-1}$)	Dye Type	pH	Adsorption Capacity ($mg \cdot g^{-1}$)	Ref.
N/A	...	Acid Blue 25	...	21.7	[55]
		Acid Red 114		22.9	
		Basic Red 22		76.6	
		Basic Blue 69		157.4	
N/A	...	Maxilon Red	...	82.8	[54]
		Erionyl Red		17.6	
N/A	1.8	Rhodamine B	7	43.5	[57]
		Basic Blue 9	7	27.8	
		Congo Red	5	38.2	[56]
Succinic anhydride	...	Methylene Blue	8	478.5	[62]
		Gentian violet	8	1273.2	
EDTA dianhydride	1.46	Methylene Blue	8	202.4	[61]
		Gentian violet		327.9	
H_2SO_4 – Carbonization – Activation	1433	Acid Blue	...	92.6	[59]
$ZnCl_2$	790	Acid Orange 10	...	4.9	[58]
$CaCl_2$	...	Textile effluent	...	88	[60]
NaOH – TEPA	...	Eosin Y	6	349.3	[63]

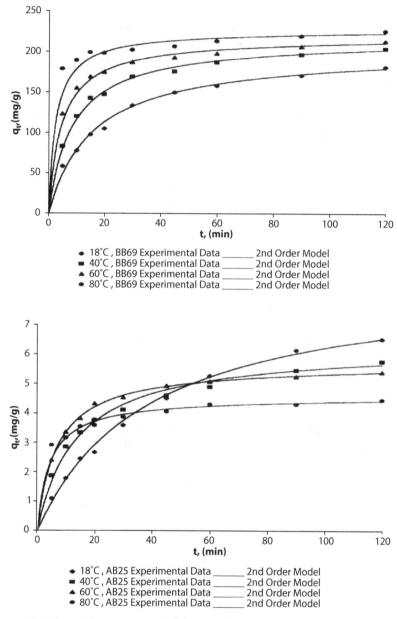

Figure 5.8 Effect of temperature on the sorption of (a) Basic Blue 69 and (b) Acid Blue 25 onto peat [65].

sorptions of Acid Blue 25 and Basic Blue 69 are favored in lower and higher temperatures, respectively. They attributed this finding to the fact that at higher temperatures, an increase in the free volume occurs due to the increased mobility of the solute in the former case, whereas in the case of acidic dye, an elevation in temperature increases the escaping tendency of the dye from the interface. A very low surface area peat was employed by Ip *et al.* for the removal of Reactive Black 5 from an aqueous solution [66]. By measuring the acidic and basic surface groups as well as the pH_{PZC}, it was revealed that the peat surface is predominantly acidic due to the presence of carboxylic groups of humic acid and fulvic groups. This surface acidity would exert a repulsive force for the acidic dye molecules, resulting in a low uptake capacity of the peat. It was noted that although the adsorption capacity of the peat for this dye is quite low (7 $mg.g^{-1}$), this capacity was considerably increased when a salt was introduced into the solution (see Table 5.6). This phenomenon was assigned to the reduced repulsive forces between the dye molecules and the surface of the peat. Fernandes *et al.* [67] obtained reasonably high uptake capacities by peat for methylene blue (303 $mg.g^{-1}$), as seen in Figure 5.9. Such high adsorption capability coupled with fast equilibrium time was attributed to the cation exchange between the H^+ ion of the carboxylic group associated with the humic substances and the cationic groups of the dye. Also, an increase in the solution temperature resulted in a slight improvement of the adsorption capacity. Two opposing factors are created when increasing the temperature; this may lead to an increase in the tendency of the adsorbed molecules to escape from the pores and it can also enhance the intraparticle diffusion rate of the dye molecules. Depending on the net effect, the removal capability of the adsorbent will either be enhanced or decreased. Allen *et al.* [68] used five different isotherm models together with several error analysis functions to fit the adsorption of three dyes onto the peat. It was shown that the

Table 5.6 The effect of salts on Reactive Black dye adsorption by peat [66].

Initial Dye Concentration (ppm)	Type of Salt	Adsorption Capacity (mg.g⁻¹)
400	N/A	3.16
	NaCl	24.28
	Na₂PO₄	51.45
1000	N/A	5.44
	NaCl	29.82
	Na₂PO₄	55.27

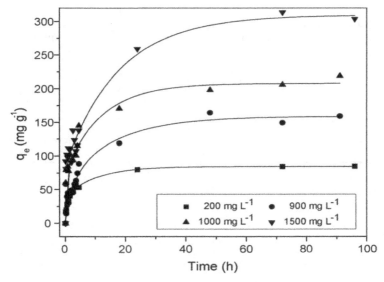

Figure 5.9 Progressive removal of MB from aqueous solutions by peat, for different initial concentrations, at 35°C [67].

Redlich-Peterson isotherm model yields the best fit for all three dyes. Also, the linear transform model provided the highest regression coefficient for the case of Redlich-Peterson isotherm, whereas the Freundlich isotherm was better represented by HYBRID error function. The comparison of the uptake capacities of peat has been given in Table 5.7.

5.2.4   Bamboo

Bamboo is a type of abundant perennial plant mostly found in Asia, Africa and Latin America. Bamboo has been traditionally utilized for handicrafts, construction materials, pulping and papermaking. The rapid growth of bamboo and its excessive amount have created some problems in its usage and disposal. Therefore, more practical usage of this material has been the focus of attention for researchers. Due to its unique characteristics in chemical composition, this lingocellulosic biomass has been extensively used for the production of bioethanol and as a green biofuel. In addition, the usage of bamboo-based activated carbon as an adsorbent for the removal of dyes, heavy metals, organic pollutants and humidity has recently attracted the attention of many researchers [69–71].

Various activation methods have been used to produce activated carbon from waste bamboo. Mui *et al.* [72] carried out the pyrolysis of bamboo under nitrogen atmosphere at various temperatures and observed that as

Table 5.7 Comparison of adsorption capacities of peat.

Modification	Surface Area ($m^2 g^{-1}$)	Dye Type	pH	Adsorption Capacity ($mg.g^{-1}$)	Ref.
N/A	...	Acid Blue 25	...	12.7	[65]
		Basic Blue 69		195	
N/A	0.7	Reactive Black 5	7	7.0	[66]
N/A	...	Methylene Blue	3	303	[67]
H_2SO_4 – Polyvinylalcohol and formaldehyde	9–11	Basic Magenta	...	67.6	[124]
		Basic Green 4		69.0	
N/A	32.5	Basic Blue 3	...	512.9	[68]
		Basic Red 22		270.9	
		Basic Yellow 21		655.8	

Table 5.8 Effect of temperature on textural characteristics of bamboo chars [72].

Temperature (K)	Yield (%)	BET SA ($m^2 g^{-1}$)	Pore vol. ($cc.g^{-1}$)	Micropore vol. ($cc.g^{-1}$)	Mesopore vol. ($cc.g^{-1}$)
673	30.8	110	0.044	0.018	0.026
773	25.3	126	0.103	0.025	0.078
873	25.1	156	0.079	0.065	0.014
973	24.0	194	0.102	0.083	0.019
1073	23.0	263	0.145	0.097	0.048
1173	22.6	327	0.185	0.140	0.045

the pyrolysis temperature increased from 673 to 1173 K, the surface area also increased from 110 to 327 $mg.g^{-1}$, according to Table 5.8. This was attributed to the pore development with the removal of volatiles at higher temperatures, which also accounts for the reduction in the production yield of activated carbon. The very rapid increase of surface area at temperatures of higher than 1073 K was related to the decomposition of the lignin, whereas cellulose and hemicellulose decompositions occurred at a much lower temperature (673 K). Very prolonged furnace holding time decreased the surface area due to the realignment of carbon atoms and consequent contraction of char. Also, by changing the heating rate different surface areas revealed the temperature gradient inside the particles. It is worth noting that the highest surface area was obtained at a heating rate of 5 $K.min^{-1}$. The adsorption capacity of the pyrolyzed bamboo activated carbon was determined to be 16.9, 35.3 and 319.2 $mg.g^{-1}$ for Acid Blue 25, Acid Yellow 117 and methylene blue, respectively. The more favorable adsorption of methylene blue was attributed to the molecular sieve effect of the activated carbon with narrow micropore sizes.

A number of studies have been conducted to activate the bamboo by orthophosphoric acid. Chan *et al.* [73] impregnated bamboo with phosphoric acid and activated it in a two-step process, where the impregnated slurry was first subjected to 150°C and the temperature was raised to either 400 or 600°C under flowing nitrogen. The results showed that the surface area increased considerably with an increase in both the acid-to-bamboo ratio and the temperature. The highest surface area and total pore volume obtained were determined to be 2517 $m^2.g^{-1}$ and 1.4 $cm^3.g^{-1}$ (see Figure 5.10). The highest adsorption capacities of the bamboo-based activated carbon were measured as 785.3 $mg.g^{-1}$ for Acid Blue 25 and 111.2 $mg.g^{-1}$ for Acid Yellow 117. The higher adsorption capacity for Acid Blue 25 compared to Acid Yellow 117 was assigned to the smaller molecular size of the former. The same reason accounts for the preferential adsorption of the former dye to the latter in a multicomponent system. Using the hybrid fractional error function, it was shown that the Sips model was the best fit model for the adsorption of dyes by bamboo-derived activated carbons [74,75]. A similar activation method was employed to modify the bamboo to remove Reactive Black 5 [76]. Two bamboo activated carbons with impregnation ratio of 2 and 6, namely BACX2 and BACX6, were prepared. It was found that the activated carbon with lower impregnation ratio gives higher surface area and micropore volume, whereas the one with higher impregnation ratio has higher mesopore volume. The adsorption capacities of BACX2 and BACX6 adsorbents were found to be 281.2 and 441.7 $mg.g^{-1}$, respectively, for Reactive Black 5. The higher adsorption capacity of the activated carbon with lower surface area (BACX6) was ascribed to the pore widening at higher impregnation ratio, enabling the dye molecules to penetrate the pores more effectively. Ahmed and Hameed have determined the surface functional groups of phosphoric acid-activated bamboo as phenols, carboxylic acid and carbonyl [77]. A high surface area bamboo-based activated carbon prepared by similar activation method showed a very tiny adsorption capacity for the Disperse Red 167 [78]. Choy and McKay [79] prepared bamboo charcoal at 500°C under a nitrogen atmosphere. Also, activated bamboo was produced by mixing the raw material with sulfuric acid at 160°C and carbonizing the resulting material at 750°C. These adsorbents were used to study the multilayer dye adsorption behavior. The single-component adsorption study indicated that bamboo, bamboo

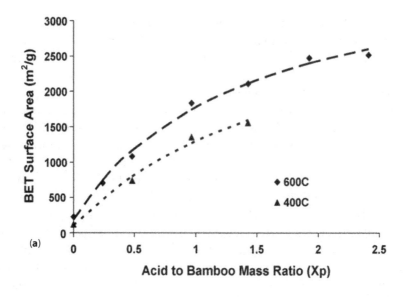

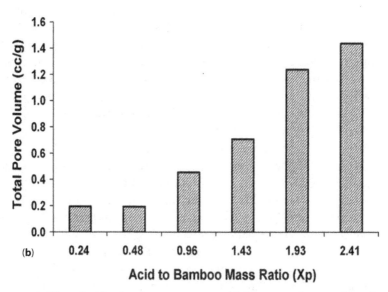

Figure 5.10 Effect of acid to bamboo mass ratio on (a) the BET surface area ($m^2\,g^{-1}$) at different activation temperatures and (b) total pore volume at 600°C [73].

charcoal and activated bamboo had adsorption capacities of 57.2, 75.4 and 113.0 $mg.g^{-1}$ for methylene blue, respectively. Despite the lack of any acid yellow adsorption for bamboo, this dye molecule can be adsorbed on the methylene blue-saturated surface, which is indicative of the interaction between the acidic and basic dye molecules and the synergistic multilayer

Table 5.9 Comparison of adsorption capacities of bamboo-based materials.

Modification	Surface Area $(m^2 g^{-1})$	Dye Type	pH	Adsorption Capacity $(mg.g^{-1})$	Ref.
H_3PO_4 – Carbonization – Activation	2517	Acid Blue 25	...	785.3	[73]
		Acid Yellow 117		111.2	
		Reactive Black		441.7	[76]
Carbonization	327	Acid Blue 25	...	16.9	[72]
		Acid Yellow 117		35.3	
		Methylene Blue		319.2	
H_2SO_4 – Carbonization	518.6	Methylene Blue	...	113.0	[79]
H_3PO_4 – Carbonization	747	Disperse Red 167	1	3.0	[78]

effect. The adsorption capacities of various bamboo-based adsorbents have been tabulated in Table 5.9.

5.2.5   Date Pits

Date pits are the seeds of the fruit of the date palm tree. Pit, which constitutes around 10 wt% of the whole date, is considered a byproduct of the date fruit processing industry. Considering the large production amount of date pits, which was equivalent to 750 thousand tons in 2011, a more systematic approach is required to utilize this large volume of waste. Currently, the major fraction of this waste is used as feed for various livestock or as a soil organic additive [80]. However, date pits can be considered as a valuable source of functional carbon for various applications.

The adsorption of methylene blue by unmodified date pits has been studied by two research groups. Belala *et al.* [81] characterized the surface functional groups of date pits and found that this waste material consists of various functional groups, such as hydroxyl, carboxylic and carbonyl, and may be a potential adsorbent for the removal of dyes from wastewater. They found an adsorption capacity of 43.5 $mg.g^{-1}$ for date pit as adsorbent (particle size of 0.5–1 mm) and methylene blue as adsorbate. In order to measure the adsorption capacity, the pH of the dye solution and the adsorbent dosage were fixed at 6.3 and 10 $g.L^{-1}$, respectively. Al-Ghouti *et al.* [82] succeeded in enhancing the adsorption capacity of unmodified waste date pit by altering the experimental conditions. They found that as the adsorbent particle size increased, the adsorption capacity of the date pits for methylene blue decreased. Also, it has already been mentioned that as the adsorbent dosage decreases, the uptake capacity is enhanced [65]. Accordingly, an enhanced methylene blue uptake of 281 $mg.g^{-1}$ for date pit with a surface area of 80.5 $m^2.g^{-1}$ was obtained when the adsorbent dosage, pH and particle size were 1 $g.L^{-1}$, 4 and 125–300 μm. Under the same experimental conditions, when the adsorbent particle size increased to 500–710 μm, the uptake capacity was dramatically reduced to 131 $mg.g^{-1}$ (see Figure 5.11). They proposed the hydrogen bonding mechanism between the hydroxyl groups of the adsorbent and nitrogen and sulfur atoms of adsorbate, as illustrated in Figure 5.12, which was verified by the shift of relevant FTIR peaks. Furthermore, lack of desorption of methylene blue molecules from the surface of the adsorbent confirmed the strong hydrogen bonds between the dye molecules and the adsorbent. Banat *et al.* [83] studied the adsorption of methylene blue by raw and modified date pit. When the particle size was 125–212 μm, the adsorption capacity of raw date pit was determined to be 80.3 $mg.g^{-1}$. Different adsorption capacities from the abovementioned literature could be because of the different adsorbent dosage (5 $g.L^{-1}$) used in this work. The decrease in the adsorption capacity by increasing the particle size was attributed to the lower surface area of the adsorbent at larger particle size. Raising the solution temperature resulted in a decrease in the adsorption capacity due to the tendency of the dye molecules to escape from the adsorbent pores and surface with an increase in the solution. They also converted the date pits to

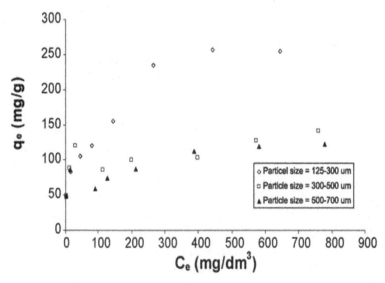

Figure 5.11 Effect of particle size on the methylene blue adsorption capacity of date pit-based activated carbon [82].

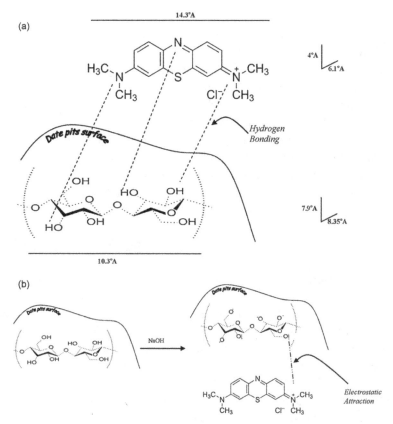

Figure 5.12 Schematic representation of (a) hydrogen bonding between nitrogen atoms of MB and hydroxyl groups on the adsorbent surface, cellulose unit, and (b) electrostatic attraction between MB and the adsorbent surface, cellulose unit [82].

activated carbon by first carbonizing them at 700°C for 2 h under nitrogen atmosphere and then activating them at either 500 or 900°C for half an hour under carbon dioxide flush. However, the adsorption results under the same experimental conditions as the raw date pit revealed that the adsorption capacity drastically decreased to 12.9 and 17.3 $mg.g^{-1}$ for the date pit activated at 500 and 900°C, respectively. They postulated that either the destruction of functional groups or the smaller size of the pores generated during the activation was responsible for the lower adsorption capacity of the activated carbons. Ahmed and Dhedan [84] endeavored to activate the date pit with $ZnCl_2$ in a two-stage process. The precursor was mixed with the activating agent at an impregnation ratio of 2 and then it was carbonized at 500°C for an hour. The resultant activated carbon with a surface area of 1045.6 $m^2.g^{-1}$ showed a promising adsorption capacity of

398.2 $mg.g^{-1}$ for methylene blue at pH value of 7 and adsorbent dose of 0.5 $g.L^{-1}$. Also, the high iodine number of the prepared activated carbon (1008.86 $mg.g^{-1}$) was due to the ability of zinc chloride to produce a high amount of micropores. Theydan and Ahmed [85] conducted a similar activation process, where ferric chloride was used as the activating agent at an impregnation ratio of 1.5 and carbonization temperature of 700°C. The yield and other characteristics of the prepared activated carbon have been summarized in Table 5.10. This table shows a very high yield compared to literature which use phosphoric acid as activating agent. It has been related to the tendency of the ferric chloride to produce more microporous structure, whereas phosphoric acid tends to produce more macropores. The surface area of the resultant activated carbon and its adsorption capacity for methylene blue at a pH value of 7 and adsorbent dose of 0.5 $g.L^{-1}$ were found to be 780.1 $m^2.g^{-1}$ and 259.2 $mg.g^{-1}$. This dye uptake capacity is equivalent to 90% of that of the prepared activating carbon using zinc chloride as dehydrating agent. In order to determine the rate-limiting step, the adsorption process was divided into 4 steps, namely transport of dye molecules from the bulk to the boundary layer surrounding the adsorbent; surface diffusion, which is defined as the transport of the dye molecules to the external surface of the adsorbent; transport of the dye molecules from the surface to the intraparticle active sites, or so-called pore diffusion, and; adsorption of the dye molecules by the adsorbent surface sites. The first and last steps are so fast that the probability of their being limiting steps is ruled out. Also, because the linear regression coefficient for the intraparticle diffusion model was low and the plot of q_t versus $t^{1/2}$ did not pass through the origin, intraparticle diffusion was also dismissed. The feasibility of the application of microwave-assisted activation was studied by Foo

Table 5.10 Effect of temperature on textural characteristics of bamboo chars [85].

Characteristic	Value
Yield (%)	47.08
Bulk density ($g.ml^{-1}$)	0.271
Ash content (%)	6.62
Moisture content (%)	12.54
Iodine number ($mg.g^{-1}$)	761.40
Surface area ($m^2.g^{-1}$)	780.06
Micropore volume ($cm^3.g^{-1}$)	0.468
Mesopore volume ($cm^3.g^{-1}$)	0.105

and Hameed [86], where the crushed date pits were initially carbonized at 700°C and then the resultant char was soaked in potassium hydroxide and activated in a microwave oven at a power of 600 W for an irradiation time of 8 min. The obtained activated carbon demonstrated a well-developed porous structure with BET surface area of 856 $m^2.g^{-1}$ and total pore volume of 0.47 $cm^3.g^{-1}$. As shown in Figure 5.13, the SEM images of the char and the activated carbon clearly illustrate the creation of pores after the activation process. The adsorption capacity of the activated carbon for methylene blue was determined as 296.5 $mg.g^{-1}$ under the operating conditions of pH 6 and adsorbent dose 1 $g.L^{-1}$. The strong dependence of the adsorption capacity to the pH level was ascribed to the protonation of the dye molecule in the acidic medium and presence of excess H^+ ions which can

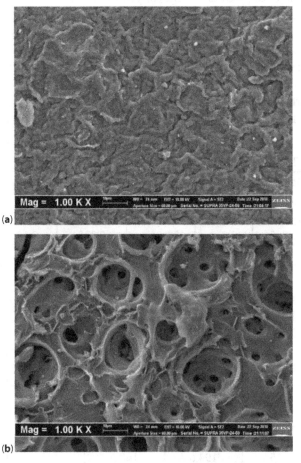

Figure 5.13 SEM micrographs of (a) date stone char and (b) date stone activated carbon [86].

compete with the adsorbate molecules to reach the adsorption sites on the surface of the adsorbent. The steam activation of date pits carried out by Ashour involved the carbonization of the precursor at 600°C followed by an activation stage at 950°C under partial steam environment to gain two activated carbons with 20% and 50% burn-off, DS2 and DS5, respectively [87]. The nitrogen adsorption-desorption graph indicated the existence of a hysteresis loop at DS5 indicative of the formation of mesopores in contrast to DS2 which was predominantly microporous. As tabulated in Table 5.11, the surface areas were calculated to be 705 and 1040 $m^2.g^{-1}$ for DS2 and DS5, respectively. Since the surface pH values of the obtained carbons were considerably higher than 7, their surfaces were mainly basic. Both of these activated carbons revealed high efficiency for the adsorption of methylene blue (MB) and remazol yellow (RY). It is interesting to note that contrary to the previously-discussed literature, the adsorption capacity of the activated carbon for the dye molecules decreases as the pH level increases (see Table 5.12). This is related to the basic nature of the steam-activated carbon surfaces, where at lower pH values, the activated carbon takes positive charge by adsorbing H^+ ion and can uptake the anionic dye owing to the strong electrostatic attraction between the adsorbent and adsorbate. Reddy et al. [88] employed two activation methods to increase

Table 5.11 Textural properties of date pit-based activated carbons [87].

Carbon	BET SA $(m^2\,g^{-1})$	Pore vol. $(cc.g^{-1})$	Mean Pore Radius (nm)	Micropore vol. $(cc.g^{-1})$	Mesopore vol. $(cc.g^{-1})$	Surface pH
DS2	705	0.361	1.03	0.228	0.133	8.3
DS5	1040	0.884	1.70	0.365	0.519	9.5

Table 5.12 Adsorption capacities of date pit-based activated carbons for MB and RY [87].

Carbon	Solution pH	q_e $(mg.g^{-1})$, MB / 27°C	q_e $(mg.g^{-1})$, MB / 37°C	q_e $(mg.g^{-1})$, RY / 27°C	q_e $(mg.g^{-1})$, RY / 37°C
DS2	3	84.6	97.6	42.1	55.0
	5	68.1	88.8	37.4	50.4
	7	64.0	76.9	33.6	44.0
DS5	3	220.0	200.0	164.0	152.2
	5	194.6	168.5	147.6	136.3
	7	174.1	148.6	129.9	123.1

the surface area of the date pits. In the physical activation method, the precursor was carbonized at 800°C for 1 hour at a nitrogen flow followed by the activation at 971°C for 56 min under CO_2 atmosphere. These conditions were chosen based on the reported optimum conditions for maximizing the yield and BET surface area of date palm pits by Reddy *et al.* [89]. The chemical activation method involves the impregnation of the raw material in phosphoric acid and its subsequent activation at 400°C under self-generated gas atmosphere. The activated carbons obtained via these two methods exhibited considerably different behaviors where H_3PO_4-activated and CO_2-activated samples had surface areas of 725 and 666 $m^2 g^{-1}$, respectively. The microporosities of these adsorbents were another major distinction. It was observed that CO_2-activated carbon was mainly microporous, whereas in the case of the sample activated by H_3PO_4, only 31% of the total volume was related to the micropores. The existence of a distinct hysteresis verified the presence of mesoporosity in the latter activated carbon. The preparation conditions and the textural characteristics of the optimized and the S_{BET}-maximized date pit-based activated carbons have been compiled in Table 5.13. A surface functional group study of the raw date pit was compared with the two activated carbons. It was shown that some of the functional groups disappeared upon activation. It was also observed that other than the hydroxyl group peak, the rest of the bands vanished for CO_2 – activated carbon compared to the H_3PO_4 – activated carbon. The adsorption capacity of the two activated carbons were determined to be 110 and 345 mg methylene blue per gram of CO_2– and H_3PO_4– activated carbon, respectively. The higher adsorption capacity of the latter to the former adsorbent was related to the higher average pore size of the H_3PO_4 – activated carbon and easier diffusion of the large dye molecules

Table 5.13 Physical properties of the activated carbons derived from date palm pits with different activation procedures [88].

Activation	T, °C	t, min	BET SA $(m^2 g^{-1})$	Pore vol. $(cc.g^{-1})$	v_{micro}/v_t	D_p, nm	Yield (%)
Physical (optimized)	971	56	666	0.41	0.92	1.56	14.8
Chemical (optimized)	400	58	725	1.26	0.31	2.90	44.0
Physical (maximized)	1063	68	980	0.61	0.93	1.57	9.5
Chemical (maximized)	450	75	952	1.38	0.36	2.91	41.0

into the pores. Table 5.14 summarizes the different adsorption capacities of date pit-based adsorbents for dye molecules.

5.2.6   Palm Tree Waste

Palm tree is one of the most commonly grown trees in tropical regions. In 2012, the total world area of oil palm plantations exceeded 14 million hectares, nearly 80% of which is located in Malaysia and Indonesia. A variety of waste is generated either directly from the tree, such as flower, trunk and branches, or from the residues of the oil palm industry. The large volume of waste is due to the fact that the productive life of this tree is 25 years and replantation should be done after this period [90]. Cellulose, hemicellulose and lignin are the major components of palm-based waste and can be used as a precursor for many different types of applications, such as bio-oil production [91], energy generation [92], filler [93], biolubricant synthesis [94], bioethanol production [95] and so on. In addition, a couple of researchers have recently investigated the feasibility of pollutant removal by this waste.

Table 5.14 Comparison of the adsorption capacities of date pit-based materials.

Modification	Surface Area $(m^2 g^{-1})$	Dye Type	pH	Adsorption Capacity $(mg.g^{-1})$	Ref.
N/A	...	Methylene Blue	6.3	43.5	[81]
N/A	80.5	Methylene Blue	4	281	[82]
N/A	...	Methylene Blue	...	80.3	[83]
Carbonization – CO_2 Activation				17.3	
$ZnCl_2$ – Carbonization	1045.6	Methylene Blue	7	398.2	[84]
$FeCl_3$ – Carbonization	780.1	Methylene Blue	7	259.2	[85]
Carbonization – KOH – Microwave	856	Methylene Blue	6	296.5	[86]
Carbonization – Steam activation	1040	Methylene Blue	3	220.0	[87]
		Remazol Yellow		164.0	
Carbonization – CO_2 Activation	666	Methylene Blue	...	109.9	[88]
H_3PO_4 – Carbonization	725			344.8	

The surface functional groups of palm tree waste, studied by Belala *et al.* [81], illustrated the existence of various moieties, such as hydroxyl, carbonyl and carboxyl groups, representative of lingocellulosic materials. The maximum uptake capacity of unmodified palm tree waste for methylene blue at a pH value of 6.3 and adsorbent loading of 10 $g.L^{-1}$ was 39.5 $mg.g^{-1}$. Palm kernel fiber is another waste product of this category. The functional group characterization by Ofomaja revealed that in addition to the widely-seen moieties of hydroxyl, carboxyl and carbonyl, amine functional groups were also present on the surface of the fiber [96]. He also found that the point of zero charge pH value (pH_{PZC}) for this waste material is 2.76 (see Figure 5.14). The point of zero charge pH is a pivotal parameter in which the net electrical charge density on the surface of the adsorbent is zero. Hence, in the case of this fiber, the adsorbent surface will, above the pH value of 2.76, exhibit acidic behavior, donate protons to the aqueous solution and become negatively-charged. Therefore, the fiber surface will favor the adsorption of methylene blue at pH values higher than pH_{PZC}. The effect of pH level on the adsorption capability of palm kernel fiber has been illustrated in Figure 5.15. The adsorption capacity of the palm kernel fiber for methylene blue was determined to be 655.9 $mg.g^{-1}$ at an initial pH value of 7. As shown in Table 5.15, as the initial dye concentration increased, the final pH of the solution decreased. Ofomaja related this to the ion exchange sorption mechanism in which more H^+ ions were exchanged with the dye molecules with an increase in the initial

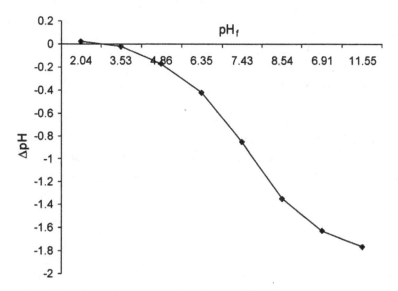

Figure 5.14 Point of zero charge curve for palm kernel fiber [96].

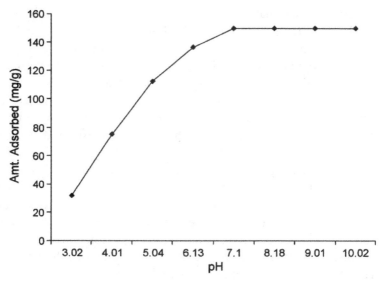

Figure 5.15 Effect of pH on methylene blue adsorption from solution onto palm kernel fiber [96].

Table 5.15 Relationship between methylene blue amount adsorbed (q_e) and changes in hydrogen ion concentration [96].

Initial Dye Concentration ($g.L^{-1}$)	Initial Solution pH	Final Solution pH	ΔpH ($mol.L^{-1}$)	q_e ($mg.g^{-1}$), MB
500	7.01	5.02	$9.45 \times 10{-3}$	671.78
400	7.01	5.14	$7.15 \times 10{-3}$	546.27
300	7.01	5.31	$4.80 \times 10{-3}$	418.57
200	7.01	5.56	$2.66 \times 10{-3}$	283.77

dye concentration. The adsorption of an anionic dye by palm kernel fiber has also been reported by Ofomaja and Ho [97]. The maximum monolayer adsorption capacity of the fiber for this anionic dye (38.6 $mg.g^{-1}$) was much lower than that observed for methylene blue as the cationic dye. As the mass of the adsorbent was increased, the number of the unsaturated adsorption sites was believed to increase, resulting in the reduction in the adsorbent capacity, as seen in Figure 5.16. They also postulated that this reduction in capacity could be due to the aggregation of the particles and a decrease in the surface area of the adsorbent. Furthermore, they showed that temperature increase had a reverse effect on the uptake capacity of the fiber, which was ascribed to the weakening of the interaction between

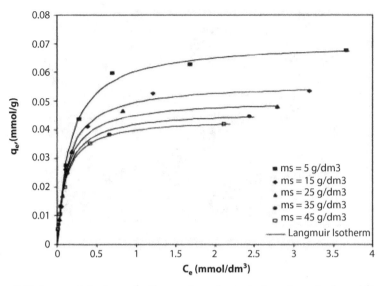

Figure 5.16 Langmuir isotherm for the sorption of anionic dye using palm kernel fiber at various doses [97].

the adsorbent active sites and adsorbate and thus enhanced desorption rates at higher temperatures. A multi-stage process has been designed in a similar study in order to minimize the adsorbent mass and contact time, the two parameters which are critical in the design and cost analysis of a larger scale [98]. In order to enhance the adsorption properties of the oil palm fiber, Tan *et al.* [99] carbonized the precursor at 700°C under a pure nitrogen flow. The resultant char was impregnated with KOH and activated at 850°C under CO_2 atmosphere. This process resulted in the development of a honeycomb-shaped porous structure, probably due to the intercalation of metallic potassium into the carbonaceous network and expansion of the carbon material (see Figure 5.17). The surface area of the resultant activated carbon was determined to be 1354 $m^2.g^{-1}$ with an average pore diameter of 2.3 nm. The study of the adsorption of methylene blue by the prepared activated carbon indicated that the maximum uptake capacity was 276 $mg.g^{-1}$ at a pH of 6.5 and an adsorbent loading of 1 $g.L^{-1}$. As shown in Figure 5.18, the effect of solution temperature was not pronounced for low initial dye concentrations, whereas at higher concentrations, a considerable increase in the adsorption capacity was observed by raising the temperature. When the temperature of the solution was increased from 30 to 50°C, the uptake capacity of the adsorbent was changed from 276 to 384 $g.g^{-1}$, which is quite substantial. This enhancement was attributed to either the increased intraparticle diffusion rate,

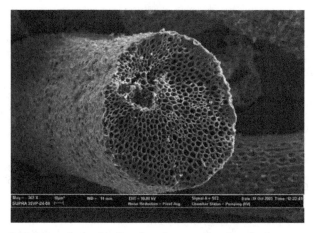

Figure 5.17 SEM image of oil palm fiber-based activated carbon [99].

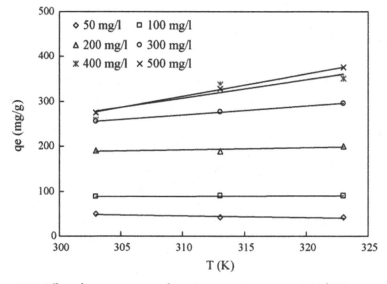

Figure 5.18 Effect of temperature on adsorption capacity at various initial MB concentrations [99].

chemical interaction of adsorbent active sites and the dye molecules or creation of new adsorption sites. Since the Langmuir isotherm was the best-fit model for this system, the homogeneity of the active sites on the surface of the activated carbon with equal activation energy and the formation of adsorbate monolayer coverage on the surface of the adsorbent was postulated. These promising results encouraged Hameed *et al.* [100] to conduct a comprehensive research on the optimization of the activated carbon yield and methylene blue uptake by varying the preparation parameters of activated carbon. The results indicated that the impregnation ratio had the greatest effect on the methylene blue uptake capacity followed by the activation temperature. As the activation temperature was increased, more carbon was gasified by CO_2 and the devolatilization was enhanced, leading to the development of more pores or widening of the existing pores, thereby enhancing the dye uptake capacity. Also, as the activation time and temperature and impregnation ratio were increased, the burn-off and devolatilization were accelerated, resulting in a decrease in the carbon yield. Obviously, the goal is to have the highest yield and adsorption efficiency for the activated carbon. Since the activated carbon preparation parameters had opposite effects on these two factors, an optimization was carried out in order to find a trade-off to get reasonably high yield and sufficient adsorption capacity. The optimum condition was found at the activation time and temperature of 1 h and 862°C, respectively, and the impregnation ratio of 3.1. Applying these conditions to the activated carbon preparation procedure, methylene blue uptake capacity of 204 $mg.g^{-1}$ and carbon yield of 16.5% were predicted. The previous studies inspired Foo and Hameed to use the same carbonization and impregnation as Tan et al. [99], but with activation of the char by microwave irradiation with a power of 360W for 5 min [101]. The obtained BET surface area of 707.8 $m^2.g^{-1}$ and pore volume of 0.38 $m^3.g^{-1}$ were lower than the ones obtained for the activated carbon samples with CO_2-activation method (1354 $m^2.g^{-1}$ and 0.78 $m^3.g^{-1}$, respectively). However, the methylene blue adsorption capacity of the microwave-irradiated carbon was found to be 312.5 $g.g^{-1}$, which is comparatively higher than that of the CO_2-activated carbon (276 $mg.g^{-1}$). Figure 5.19 shows the effects of the activated carbon preparation conditions on the surface area and adsorption capacity [102]. It was shown that the increase of the impregnation ratio from 0.25 to 0.75 enhanced both the methylene blue uptake capacity and the carbon yield, whereas further increase in the impregnation ratio had an opposite effect. It was proposed that below impregnation ratio of 0.75, the porosities were not completely developed and thus an increase in the impregnation ratio increased the uptake capacity. However the increase of the impregnation ratio above 0.75 entailed the catalytic oxidation of the carbon, widening of the mesopores and a decrease in both the carbon yield and dye uptake capacity. Increase in the microwave power resulted in continual reduction in the carbon yield due to the severe reaction at higher thermal radiation. Also, as the microwave power increased, more porous structure was developed and consequently the adsorption capacity was drastically enhanced. The maximum adsorption capacity

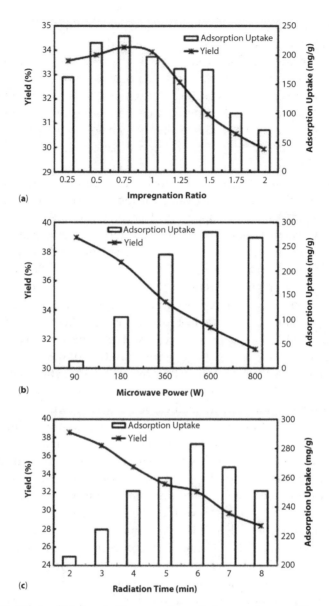

Figure 5.19 Effects of (a) chemical impregnation ratio (preparation conditions: microwave power = 360W; radiation time = 5 min), (b) microwave power (preparation conditions: chemical impregnation ratio = 0.75; radiation time = 5 min) and (c) radiation time (preparation conditions: chemical impregnation ratio = 0.75; microwave power = 600W) on the carbon yield and adsorption uptake [102].

was observed at an irradiation time of 6 min, beyond which the pore walls were destroyed and the efficiency was lowered. The optimum activation conditions led to the production of a mixed micro- and mesostructured activated carbon with a total BET surface area of 1223 $m^2.g^{-1}$, an average pore size of 2.4 nm and surface acidic groups (see Table 5.16). The adsorption capacity of the microwave-assisted activated carbon was found to be 382 $g.g^{-1}$, which is much higher than the capacity of the conventionally-prepared CO_2 activated carbon. The comparison of the dye removal efficiencies of palm tree waste-based adsorbents have been compiled in Table 5.17.

5.2.7   Coconut

Coconut trees are mainly grown in tropical regions because of the high oil content of the endosperm utilized in the food industry. The total global production of the coconut has reached 50 million tons by 2010 and is increasing steadily. A major fraction of this tree, such as husks, shells, leaves and stems, is considered as waste. Traditionally these wastes are either left behind as mulch or used as fertilizer because of their high potash content [103,104]. However, in the last few decades, coconut waste has found a vast variety of applications including animal feedstock [105], concrete filler [106], biodiesel fuel production [105] and thermal insulator [107]. In addition, similar to other lingocellulosic agricultural wastes, coconut waste is also used enormously to sequester pollutants such as dye and heavy metal from wastewater [104].

The removal of dye molecules by coconut bunch waste without any modification has been elaborated by Hameed *et al.* [108]. The characterization

Table 5.16 Porosity structures of oil palm fiber activated carbon [102].

Property	Value
BET surface area ($m^2 g^{-1}$)	1223
Micropore surface area ($m^2 g^{-1}$)	796
External surface area ($m^2 g^{-1}$)	427
Langmuir surface area ($m^2 g^{-1}$)	1827
Total pore volume ($cm^3.g^{-1}$)	0.72
Micropore volume ($cm^3.g^{-1}$)	0.42
Mesopore volume ($cm^3.g^{-1}$)	0.30
Average pore size (Å)	23.57

Table 5.17 Comparison of adsorption capacities of palm waste-based materials.

Modification	Surface Area ($m^2 g^{-1}$)	Dye Type	pH	Adsorption Capacity ($mg.g^{-1}$)	Ref.
N/A	...	Methylene Blue	6.3	39.5	[81]
N/A	...	Methylene Blue	7.1	655.9	[96]
N/A	...	Azo Dye	5	38.6	[97]
N/A	...	Methyl Violet	10	96.3	[98]
N/A	...	Basic Yellow 21	...	327.6	[125]
		Besic Red 22		180.3	
		Basic Blue 3		91.3	
		Basic Red 18		242	[126]
N/A	...	Methylene Blue	7	95.4	[127]
		Crystal Violet		78.9	
Carbonization – KOH – CO_2 Activation	1354	Methylene Blue	6.5	275.7	[99]
Carbonization – KOH – CO_2 Activation	596.2	Methylene Blue	6.5	241	[128]
Carbonization – KOH – CO_2 Activation – HCl	...	Methylene Blue	6.5	303	[129]
Carbonization – KOH – Microwave irradiation	707.8	Methylene Blue	...	312.5	[101]
Carbonization – KOH – Microwave irradiation	1223	Methylene Blue	12	382.3	[102]
Carbonization – KOH – Microwave irradiation	807.5	Methylene Blue	12	344.8	[130]

of this waste showed that it contains a number of functional groups, such as $N - H, C - O - H, C = O, C–N$ and $P –H$. The comparison of the detected peaks before and after the dye adsorption process revealed the presence of amine, carbonyl and carboxyl groups in the adsorption process. The methylene blue uptake capacity of 70.9 $mg.g^{-1}$ was observed at a pH value of 7 and an adsorbent dosage of 1 .L^{-1}, which was further enhanced when the pH level was increased up to 10. Cazetta et al. [109] successfully attempted to produce high surface area activated carbon from coconut shell by a mixture of chemical and physical activation methods. The waste precursor was charred at 500°C under nitrogen atmosphere. Then the prepared char was impregnated at NaOH at various ratios and subsequently activated at

700°C under inert atmosphere. The activation mechanism using NaOH is reported to occur according to the following reaction:

$$6NaOH + 2C \leftrightarrow 2Na + 2Na_2CO_3 + 3H_2 \qquad (5.1)$$

The yield of the prepared activated carbons ranged from 19–30% depending on the impregnation ratio used. As the impregnation ratio increased, more carbon bonds were cleaved, more intense dehydrating reactions occurred and consequently the yield decreased. Also the increase in the impregnation ratio had a dramatic effect on the surface area, pore volume and average pore size of the prepared activated carbons. Figure 5.20 shows the N_2 adsorption-desorption isotherm curves for the activated carbons prepared at impregnation ratios of 1, 2 and 3, labeled as AC-1, AC-2 and AC-3, respectively. A progressive increase in the nitrogen volume uptake is observed by increasing the impregnation ratio. Also, the adsorption and desorption curves show a complete reversibility for all the activated carbons, which is indicative of the existence of micropores and high affinity of the adsorbate and adsorbents. Table 5.18 summarizes the porous characteristics of the activated carbons derived from coconut shell. The highest surface area and pore volume of the activated carbon (2825 $m^2.g^{-1}$ and 1.5 $cm^3.g^{-1}$, respectively) were obtained when the impregnation ratio was the highest. As well as these two parameters, the fraction of mesopores also increased with increasing the impregnation ratio. The increase in the amount of micro- and mesopores with an increase in the amount of dehydrating agent indicates

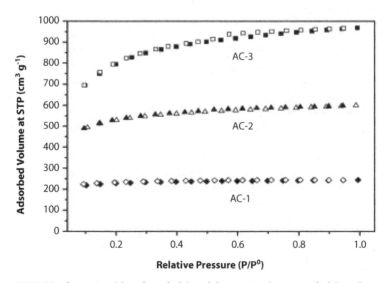

Figure 5.20 N_2 adsorption (closed symbols) and desorption (open symbols) isotherms at 77 K for the AC-1, AC-2 and AC-3 [109].

that not only new pores are created on the activated carbon, but also a portion of the micropores are merged to develop mesopores. The Boehm titration method was employed to quantify the chemical functional groups. It was found that the increase in the impregnation ratio does not have any impact on the basic functional groups (carbonyl, pyrone and chromene), but significantly increases the amount of acidic functional groups (carboxylic and phenolic groups), and hence the surface of the activated carbon becomes more acidic (see Table 5.19). The adsorption efficiency of the high surface area activated carbon was tested by using methylene blue as adsorbate. At pH value of 6.5, the adsorption capacity of the coconut shell-based activated carbon (AC-3) was determined to be 916.3 $mg.g^{-1}$. The activation of coconut husk by a similar dehydrating agent, potassium hydroxide, also provided promising results [110]. Tan *et al.* carried out the same procedure used in the activation of oil palm fiber, where the precursor was carbonized at 700°C in an inert atmosphere, the resultant char was soaked in KOH and then activated at 850°C under CO_2 atmosphere. Very high BET surface area and pore volume of 1940 $m^2.g^{-1}$ and 1.14 $cm^3.g^{-1}$, respectively, were reported for the honeycomb-shaped activated material with an average pore diameter of 2.4 nm. Figure 5.21 shows the SEM images of the precursor and the derived activated carbon. The creation of many large pores can evidently be seen in these figures. The methylene blue adsorption capacity of the coconut

Table 5.18 Textural characteristics of the activated carbons derived from coconut shells [109].

Carbon	BET SA ($m^2 g^{-1}$)	Pore vol. ($cc.g^{-1}$)	Average Pore Diameter (nm)	Micropore vol. ($cc.g^{-1}$)	Mesopore vol. ($cc.g^{-1}$)	v_{micro}/v_t (%)	Yield (%)
AC–1	783	0.378	1.63	0.356	0.022	94.2	28.9
AC–2	1842	0.927	1.80	0.775	0.152	83.6	23.4
AC–3	2825	1.498	2.27	1.143	0.355	76.3	18.8

Table 5.19 Results of the Boehm and pH drift methods for the AC-1, AC-2 and AC-3 [109].

Carbon	Carboxylic ($mmol.g^{-1}$)	Lactonic ($mmol.g^{-1}$)	Phenolic ($mmol.g^{-1}$)	Acid ($mmol.g^{-1}$)	Basic ($mmol.g^{-1}$)	Total ($mmol.g^{-1}$)	pH drift
AC-1	0.37	0	0.38	0.75	0.73	1.47	6.00
AC-2	0.62	0	0.88	1.5	0.75	2.25	5.09
AC-3	0.75	0	1.00	1.75	0.75	2.50	5.01

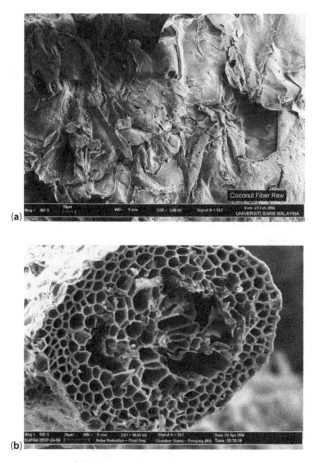

Figure 5.21 SEM images of (a) coconut husk and (b) the derived activated carbon [110].

husk-based activated carbon was relatively large (434.8 $mg.g^{-1}$) when compared with some other literature, which can be attributed to the nature of the precursor used, the activation process applied or a mixture of these two. An increase in the temperature led to a reduction in the adsorption capacity of the material due to the weakening of the adsorbent-adsorbate bonds and the increase in the solubility of the dye. The activation of coconut coir dust was also attempted by zinc chloride [111]. The impregnation of the precursor in $ZnCl_2$ and subsequent activation at 800°C under CO_2 atmosphere resulted in the production of a mesoporous activated carbon with surface area of 1884 $m^2.g^{-1}$ with a variety of functional groups on the surface. The N_2 adsorption-desorption isotherm curve, illustrated in Figure 5.22, depicts a hysteresis loop typical of Type IV isotherm, verifying the development of a mesoporous structure. The mechanism of the dye uptake by the adsorbent

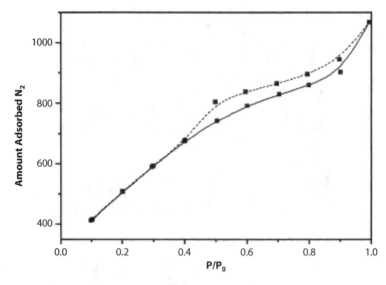

Figure 5.22 Nitrogen adsorption/desorption isotherms of coconut coir dust-derived activated carbon [111].

was suggested to consist of two distinct adsorption profiles, the fast external diffusion with no barrier in which the dye molecules migrate to reach the adsorbent surface and the internal diffusion in which the penetration of the dye molecules into the pores of the adsorbent occurs. These two regimes can be clearly observed in Figure 5.23. The adsorption capacities of the prepared activated carbon were determined as 15.2 and 176.1 $mg.g^{-1}$ for cationic methylene blue and anionic remazol yellow, respectively. It has been discussed that the internal diffusion is an endothermic process, whereas the adsorption processes generally show exothermic behavior owing to the entropy decrease. Nevertheless, the endothermic effect, or in other words the increase in the entropy, can be explained by the displacement of one dye molecule with more than one adsorbed water molecule. The electrostatic interaction of dye molecules with acidic and basic sites of the activated carbon and $\pi - \pi$ dispersive interaction between the conjugated bonds of the dye molecules and active sites of the adsorbent were the two plausible adsorption mechanisms proposed. The comparison of the adsorption capacities of the coconut-based adsorbents have been listed in Table 5.20.

5.2.8   Tea and Coffee

Tea and coffee are two of the most important food commodities in the world. According to the International Coffee Organization, the production of coffee only amounted to about 680 million tons in 2008 [112]. Due to

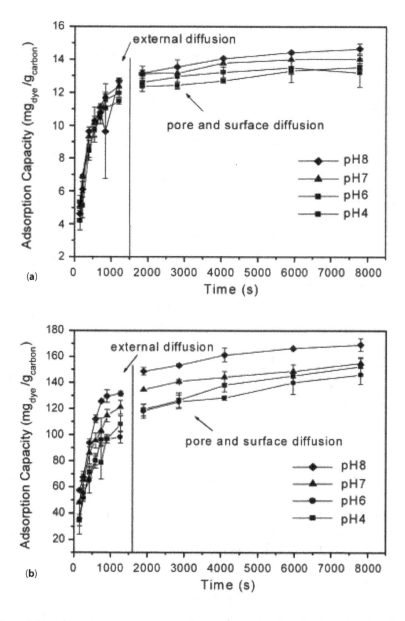

Figure 5.23 Adsorption capacity curves of coconut coir dust-based activated carbon from aqueous dye solutions: (a) MB and (b) RY [111].

the large global consumption of these products, they can be considered as two main waste streams that should be handled carefully. Although several studies have been carried out on the utilization of waste tea and coffee in various applications, this research is still in its infancy. Tea and coffee wastes represent two underutilized carbon resources with the potential

Table 5.20 Comparison of adsorption capacities of coconut waste-based materials.

Modification	Surface Area $(m^2\,g^{-1})$	Dye Type	pH	Adsorption Capacity $(mg.g^{-1}))$	Ref.
N/A	...	Methylene Blue	7	70.9	[108]
N/A	...	Methylene Blue	6	29.5	[131]
H_2SO_4 – Carbonization	556.3	Crystal Violet	6	85.8	[132]
H_3PO_4 – Carbonization	328.2			60.4	
		Reactive Red		189.9	[133]
$ZnCl_2$ – CO_2 Activation	1884	Methylene Blue	...	15.2	[111]
		Remazol Yellow		176.1	
Carbonization – NaOH – Activation	2825	Methylene Blue	6.5	916.3	[109]
Carbonization – KOH – CO_2 Activation	1940	Methylene Blue	6.5	434.8	[110]

to be used as niche adsorbents for the removal of heavy metals, dyes and organics from wastewater.

Tea and coffee wastes have been reported to have good dye adsorption capacity by several researchers. These wastes were used as adsorbents with or without physical and/or chemical modification. The adsorption of methylene blue by spent tea leaves has been studied by Hameed [113]. The tea leaves were thoroughly washed to remove any color or other impurities and were used without any further modification. The maximum monolayer adsorption capacity was computed with Langmuir isotherm model, which showed a relatively high adsorption capacity of 300 $mg.g^{-1}$ for methylene blue. Also, Nasuha et $al.$ [114] obtained a methylene blue adsorption capacity of 147 $mg.g^{-1}$ for rejected tea, although the surface area of the adsorbent was very low (4.2 $m^2.g^{-1}$). This relatively high dye uptake was attributed to the presence of various functional groups on the surface of the rejected tea. The shift of the hydroxyl, carbonyl and –C–C– groups to lower frequencies and the disappearance of aromatic nitro compounds, as exhibited in Table 5.21, demonstrated their involvement in the

Table 5.21 FTIR spectral characteristics of rejected tea before and after adsorption [114].

IR Peak	Frequency (cm^{-1})			Assignment
	Before adsorption	After adsorption	Difference	
1	3430	3414	-16	Bonded O-H groups
2	–	2920		Aliphatic C-H groups
3	–	1733		C=O stretching
4	1630	1600	-30	C=O stretching
5	1509	–		Aromatic nitro compound
6	1383	1385	+2	Symmetric bending of CH_3
7	–	1333		Aromatic nitro compound
8	–	1247		S=O stretching
9	1033	1035	+2	C-O stretching
10	–	884		C-C stretching
11	608	557	-51	-C-C- group

adsorption process. Similar surface area and surface functional groups on Pu-erh tea were reported elsewhere [115]. Also, untreated coffee residues have been proven to have a high uptake capacity. The characterization of this waste illustrated that, similar to tea waste, the surface area of the coffee waste was also very low [116]. The Boehm titration method illustrated the presence of 0.94 mmol/g carboxylic and 0.91 mmol/g basic functional groups, where small amounts of phenolic and lactonic moieties were also observed. The experimental titration curves for point of zero charge (PZC) determination are presented in Figure 5.24. The PZC value was observed between 3.2–3.4, above which the adsorbent surface exhibits a predominantly negative surface charge. The PZC value played an important role in explaining the mechanism of the dye adsorption. The study of the pH effect showed that the adsorption of the Remazol Blue (RB) was decreased with an increase in the pH level, whereas an opposite trend was observed for Basic Blue (BB). The dissociation of the RB in the solution was presented as follows:

$$D - SO_3 Na \rightarrow D - SO_3^- + Na^+ \qquad (5.2)$$

Also, the adsorbent surface was protonated at low pH values. Then electrostatic interaction could occur between the positive surface of the adsorbent and the negatively-charged adsorbate. As the pH value

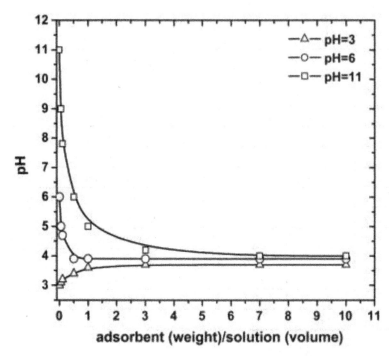

Figure 5.24 Determination of the point of zero charge (PZC) of the untreated coffee residues [116].

was increased to higher than the PZC, the surface of the adsorbent was negatively-charged and consequently, the electrostatic interaction between the adsorbent surface and the dye molecule was decreased, resulting in lower adsorption capacities. The fact that at high pH values, dye removal can still be observed, although in lower percentages, confirms the existence of other interactions, such as van der Waals forces and hydrogen bonding. In extremely high pH values, the dissociation of chlorine atoms in the monochlorotriazinyl anchor of the dye molecule results in the creation of a positive charge on the dye molecule which can interact, to some extent, with the negatively-charged adsorbent. As shown in Figure 5.25, the trend of the BB adsorption capacity change with pH level was opposite to that of the RB dye, i.e., the BB adsorption capacity was enhanced by increasing the pH value. This is due to the positive charge of the dye molecule after dissociation in the solution which can interact with negatively-charged surface of the adsorbent at pH values higher than the PZC. The maximum adsorption capacities of the adsorbent for RB and BB at 25°C in pH values of 2 and 10, respectively, were reported to be 179 and 295 $mg.g^{-1}$, respectively. Higher adsorption capacities were observed at higher

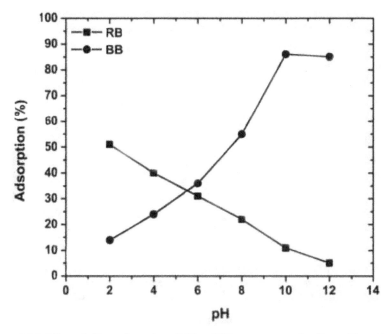

Figure 5.25 Effect of pH on adsorption of RB and BB onto untreated coffee residues [116].

adsorption temperatures for both of the dyes due to the increased penetration of the dye molecules in the porous structure of the adsorbent. Also, the agitation rate was found to have a considerable impact on the removal capacity of the adsorbent. According to Figure 5.26, as the agitation speed increased, the boundary layer resistance decreased, the mobility of the dye molecules increased and the removal percent was increased. Desorption experiments for RB and BB were carried out at extreme alkaline and acidic media, respectively. The results showed the RB and BB desorption of 90 and 94%, respectively, in the aforementioned media. In order to investigate the possibility of the reuse of the adsorbent, the adsorption-desorption experiments were conducted in many cycles and the reduction in adsorption percentage was revealed to be only 7% for BB and 10% for RB (see Figure 5.27). This reduction is sufficiently low enough to make it a suitable adsorbent for further reuse. The decrease in the adsorption capacity after several cycles was attributed to a progressive saturation of the surface active sites of the adsorbent, pore blockage by the impurities or the degradation of the material at extreme pH conditions.

Reffas *et al.* [117] used a chemical activation method to prepare activated carbon from coffee grounds. The washed coffee grounds were impregnated with different amounts of phosphoric acid as the activating

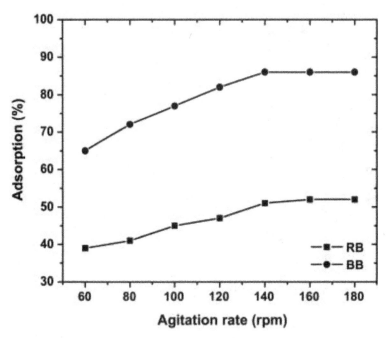

Figure 5.26 Effect of agitation rate on adsorption of RB and BB onto untreated coffee residues [116].

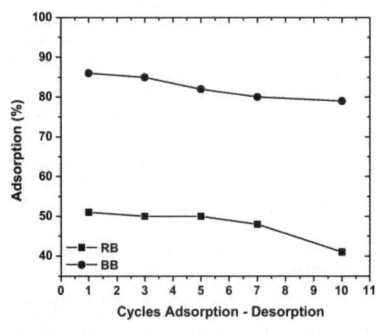

Figure 5.27 Cycles of adsorption–desorption for the reuse of untreated coffee residues as adsorbents for RB and BB removal from single-dye solutions [116].

agent and heated up to 450°C in an air atmosphere. The increase in the amount of activating agent had a drastic effect on the porous structure of the activated carbon. With an increase in the amount of phosphoric acid, the surface area of the activated carbon was increased steadily from 514 to 925 $m^2.g^{-1}$, as given in Table 5.22. This also resulted in the mergence of the micropores favoring the formation of more mesopores on the activated carbon. However, the yield of the carbon was affected insignificantly when the impregnation ratio was changed. The Boehm titration of the activated carbon with the lowest impregnation ratio revealed that its surface mainly consisted of phenolic and carboxylic groups, whereas the carbon with the highest impregnation ratio showed a decrease in the amount of the carboxylic groups and increase in the carbonyl groups (see Table 5.23). These findings were in total agreement with the PZC study results which confirmed the PZC of 3.7 and 7.4 for the CGAC30 and CGAC180, respectively. The thermogravimetric analysis illustrated four steps in the mass loss, mainly due to the dehydration of the physisorbed water and the carbonization of the samples. The FTIR analysis not only confirmed the existence of the functional groups detected by the titration method, but also indicated the development of phosphorous groups on the surface of the

Table 5.22 Textural properties of activated carbons derived from coffee waste obtained by N_2 adsorption/desorption studies [117].

Property	Calculation Method	CGAC30*	CGAC60*	CGAC120*	CGAC 180*
BET surface area $(m^2 g^{-1})$	BET	514	618	745	925
Micropore surface area $(m^2 g^{-1})$	t-Plot	311	243	158	60
External surface area $(m^2 g^{-1})$	t-Plot	202	376	587	864
Mesopore surface area $(m^2 g^{-1})$	BJH Desorption	103	205	417	821
Total pore volume $(cm^3.g^{-1})$	Single-Point Adsorption	0.280	0.338	0.446	0.718
Micropore volume $(cm^3.g^{-1})$	t-Plot	0.169	0.143	0.100	0.046
Mesopore volume $(cm^3.g^{-1})$	BJH Desorption	0.066	0.124	0.279	0.666
Yield (%)		37.7	37.3	35.2	32.0

*CGAC(n) represents the activated carbon with an impregnation ratio of "n" wt%.

Table 5.23 Surface chemical characteristics of the activated carbons derived from coffee waste [117].

Functional group	Calculation Method	CGAC30*	CGAC60*	CGAC120*	CGAC180*
Carboxylic ($meqg^{-1}$)	BET	0.466	0.390	0.266	0.266
Lactonic (m^2g^{-1})	t-Plot	0.200	0.200	0.234	0.014
Phenolic (m^2g^{-1})	t-Plot	0.709	0.700	1.033	0.786
Carbonyl (m^2g^{-1})	BJH Desorption	0.125	0.133	0.124	0.434
Total oxygenated ($cm^3.g^{-1}$)	Single-Point Adsorption	1.500	1.312	1.657	1.500
Total basic (m^2g^{-1})	t-Plot	0.000	0.000	0.000	0.000
pH	BJH Desorption	3.7	3.6	3.6	6.9
pH_{PZC}		3.7	3.6	3.9	7.4

*CGAC(n) represents the activated carbon with an impregnation ratio of "n" wt%.

activated carbons which verified the successful activation of the carbon (see Figure 5.28). As the impregnation ratio increased, the adsorption of the methylene blue was enhanced. The activated carbons with the lowest and highest impregnation ratios showed methylene blue uptake capacities of 96 and 182 $mg.g^{-1}$, respectively. The dye-covered surface areas have been compared with the BET surface area and also mesoporous surface area of the activated carbons in Table 5.24. The MB coverage ratio is virtually independent of impregnation ratio, whereas mesopore volume increases significantly with an increase in the impregnation ratio. This indicates that the MB molecules do not completely fill the mesopores of CGAC120 and CGAC180. Akar et al. [118] prepared an activated carbon from spent tea leaves by impregnating the raw material in dilute NaOH and carbonizing the material at 450°C under nitrogen atmosphere. The BET and micropore surface areas of the carbon after the activation process were increased to 134 and 21.5 $m^2.g^{-1}$, respectively, while the untreated material showed a BET surface area of 30 $m^2.g^{-1}$ with no microporosity. The BJH pore diameter was 22.5 Å for the unactivated material, whereas it decreased considerably to 5.1 Å after the activation process. The surface functional group

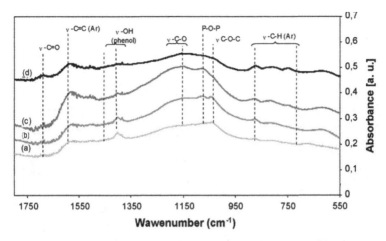

Figure 5.28 FTIR spectra of activated carbons: CGAC30 (a), CGAC60 (b), CGAC120 (c), and CGAC180 (d) [117].

Table 5.24 Langmuir adsorption capacity of methylene blue onto the activated carbons and the surface areas of materials covered by methylene blue molecules [117].

Activated Carbon	Adsorption Capacity $(mg.g^{-1})$	Surface area covered by methylene blue, S_{MB} (%)	Methylene blue coverage ratio, S_{MB}/S_{BET} (%)	Mesopore coverage ratio, S_{ext}/S_{BET} (%)
CGAC30*	96.1	264	51	39
CGAC60*	129.9	357	58	61
CGAC120*	163.9	452	61	79
CGAC180*	181.8	501	54	93

*CGAC(n) represents the activated carbon with an impregnation ratio of "n" wt%.

characterization showed that most of the carbons were bonded together as aromatic rings for the activated carbon, while these aromatic rings were not observed in the raw material. Furthermore, the appearance of C–N stretching vibrations for the dye-loaded adsorbent confirmed the presence of the dye molecules on the surface of the adsorbent. The FTIR spectra of the original material, activated carbon and the dye-loaded adsorbent have been illustrated in Figure 5.29. The maximum adsorption capacity of the activated carbon for the malachite green was found to be 244 $g.g^{-1}$, which was comparable with many commercial and laboratory-prepared activated carbons. The effect of the addition of salt in the adsorption capacity of the prepared activated carbon is shown in Figure 5.30. Since the adsorbent and

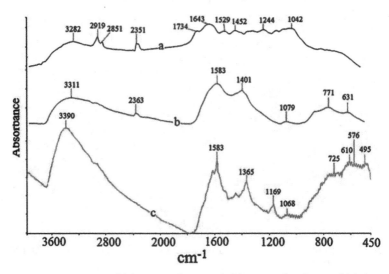

Figure 5.29 FTIR spectra of (a) untreated material, (b) activated carbon and (c) dye-loaded adsorbent [118].

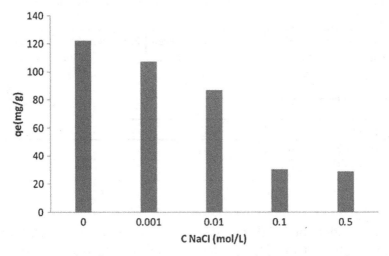

Figure 5.30 The effect of ionic strength on sorption of malachite green onto the activated carbon [118].

the adsorbate are oppositely charged at the experimental pH level (pH 3), the addition of salt has resulted in the reduction of electrostatic interaction between the dye molecules and the activated carbon surface, leading to a drastic decrease in the adsorption capacity of the adsorbent. A similar activation method using potassium acetate was employed by Auta and Hameed [119]. The tea waste was impregnated in potassium acetate and carbonized at 800°C under purified nitrogen atmosphere. A surface area of 820 $m^2.g^{-1}$ with a pore volume of 0.219 $cm^3.g^{-1}$ was obtained for the activated carbon. The average pore diameter of the activated carbon was calculated to be 2.46 nm. The increase in the surface area was attributed to the intercalation of the potassium metal ions into the carbon structure. A lot of functional groups were observed on the surface of the activated carbon, namely O–H, N–H, C–O, C = O and Si–O–Si. The presence of extra functional groups on the surface of the dye-loaded activated carbon, such as NH_2, NO_2 and C–N, signified the contribution of the activated carbon for color removal. The titration results confirmed the presence of 1.57 and 2.56 mmol/g acidic and basic functional groups, respectively. A maximum uptake capacity of 203.3 $mg.g^{-1}$ was determined for the Acid Blue 25. It was found that the optimum adsorption capacity was observed at pH values of 7–11 due to the interaction of the carboxyl groups of the activated carbon with the amine groups of the dye molecule. Moreover, the high adsorption at both acidic and basic media was ascribed to the amphoteric nature of the silica and amino groups. The dye uptake capacities of tea and coffee waste-based materials have been compiled in Table 5.25.

Table 5.25 Comparison of adsorption capacities of tea and coffee waste-based materials.

Modification	Surface Area ($m^2 g^{-1}$)	Dye Type	pH	Adsorption Capacity ($m^2 g^{-1}$)	Ref.
N/A	...	Methylene Blue	...	300	[113]
N/A	...	Methylene Blue	6	147	[114]
N/A	...	Methylene Blue	7	173	[134]
N/A	1.71	Methyl Violet	...	294	[115]
N/A	...	Methylene Blue	5	18.7	[135]
N/A	...	Methylene Blue	8	90.1	[136]
N/A	3	Remazol Blue	2	175	[116]
		Basic Blue	10	240	
N/A	...	Methylene Blue	...	85.2	[137]
NaOH	6.5	Methylene Blue	7	242.1	[138]
NaOH	173.2	Malachite Green	...	55.3	[139]
NaOH	134	Malachite Green	...	256.4	[118]
CH_3CO_2K – Activation	854.3	Methylene Blue	...	554.3	[140]
		Acid Blue		453.1	

(Continued)

Table 5.25 (*Cont.*)

Modification	Surface Area ($m^2\,g^{-1}$)	Dye Type	pH	Adsorption Capacity ($m^2\,g^{-1}$)	Ref.
CH_3CO_2K – Activation	1224	Acid Blue	...	203.3	[119]
Carbonization – NaOH – CO_2 Activation	...	Remazol Brilliant Orange	...	66.8	[141]
Carbonization	...	Methylene Blue	5	14.9	[142]
H_3PO_4 – Carbonization	925	Methylene Blue	6	181.8	[117]
		Nylosan Red	4	367	
$(ZnCl_2 + H_3PO_4)$ – Carbonization	640	Acid Blue	...	3.6	[143]
		Basic Yellow		10	
$FeCl_2 + FeCl_3$	...	Crystal Violet	10	113.6	[144]
		Janus Green	10	129.9	
		Methylene Blue	10	119.0	
		Thionine	10	128.2	
		Neutral Red	6	126.6	
		Congo Red	6	82.6	
		Reactive Blue	3	87.7	

References

1. P. Hadi, P. Gao, J. P. Barford, G. McKay, Novel application of the nonmetallic fraction of the recycled printed circuit boards as a toxic heavy metal adsorbent, *J. Hazard. Mater.* 252–253, 166–70, 2013.
2. P. Hadi, J. Barford, G. McKay, Toxic heavy metal capture using a novel electronic waste-based material-mechanism, modeling and comparison, *Environ. Sci. Technol.* 47, 8248–55, 2013.
3. I. Savin, R. Butnaru, Wastewater characterization in textile finishing mills, *Environ. Eng. Manag. J.* 7, 859–864, 2008.
4. M. Solís, A. Solís, H. I. Pérez, N. Manjarrez, M. Flores, Microbial decolouration of azo dyes: A review, *Process Biochem.* 47, 1723–1748, 2012.
5. O. Panasiuk, Phosphorus removal and recovery from wastewater using magnetite, Master of Science Thesis, KTH Royal Institute of Technology, 2010.
6. S. Elumalai, G. Muthuraman, Comparative study of liquid – liquid extraction and bulk liquid membrane for Rhodamine B, *Int. J. Eng. Innov. Technol.* 3, 387–392, 2013.
7. J. Rubio, M. L. Souza, R. W. Smith, Overview of flotation as a wastewater treatment technique, *Miner. Eng.* 15, 139–155, 2002.

8. P. Mavros, A. C. Daniilidou, N. K. Lazaridis, L. Stergiou, Colour removal from aqueous solutions. Part I. Flotation, *Environ. Technol.* 15, 601–616, 1994.

9. L. Pereira, M. Alves, Dyes - Environmental impact and remediation, in: *Environmental Protection Strategies for Sustainable Development, Strategies for Sustainability series*, A. Malik, and E. Grohmann, eds., pp. 111–162, Springer, New York, 2012.

10. P. Hadi, J. Barford, G. Mckay, Electronic waste as a new precursor for adsorbent production, *SIJ Trans. Ind. Financ. Bus. Manag.* 1, 128–135, 2013.

11. P. Hadi, J. Barford, G. McKay, Synergistic effect in the simultaneous removal of binary cobalt–nickel heavy metals from effluents by a novel e-waste-derived material, *Chem. Eng. J.* 228, 140–146, 2013.

12. P. Hadi, J. Barford, G. McKay, Selective toxic metal uptake using an e-waste-based novel sorbent – Single, binary and ternary systems, *J. Environ. Chem. Eng.* 2, 332–339, 2014.

13. M. N. Rashed, Adsorption technique for the removal of organic pollutants from water and wastewater, in: *Organic Pollutants - Monitoring, Risk and Treatment*, pp. 167–194, InTech, 2013.

14. P. Basnet, Y. Zhao, Superior dye adsorption capacity of amorphous WO3 sub-micrometer rods fabricated by glancing angle deposition, *J. Mater. Chem. A.* 2, 911–914, 2014.

15. S. Das, S. Barman, Studies on removal of Safranine-T and Methyl Orange dyes from aqueous solution using NaX zeolite, *Int. J. Sci. Environ. Technol.* 2, 735–747, 2013.

16. M. Xu, P. Hadi, G. Chen, G. McKay, Removal of cadmium ions from wastewater using innovative electronic waste-derived material, *J. Hazard. Mater.* 273, 118–123, 2014.

17. P. Hadi, J. Barford, G. McKay, *Production of novel adsorbents from waste electric and electronic equipments*, presented at AICHE Annual Meeting, Pittsburgh, USA, 2012.

18. S. Kaul, T. Nandy, L. Szpyrkowicz, A. Gautam, D. R. Khanna, Joint wastewater management for tannery complex, in: *Wastewater Manag.* pp. 105–109, 2005.

19. G. Crini, Non-conventional low-cost adsorbents for dye removal: A review., *Bioresour. Technol.* 97, 1061–85, 2006.

20. A. Bhatnagar, M. Sillanpää, Utilization of agro-industrial and municipal waste materials as potential adsorbents for water treatment – A review, *Chem. Eng. J.* 157, 277–296, 2010.

21. M. Rafatullah, O. Sulaiman, R. Hashim, A. Ahmad, Adsorption of methylene blue on low-cost adsorbents: A review., *J. Hazard. Mater.* 177, 70–80, 2010.

22. V. K. Gupta, Suhas, Application of low-cost adsorbents for dye removal – A review, *J. Environ. Manage.* 90, 2313–42, 2009.

23. A. Demirbas, Agricultural based activated carbons for the removal of dyes from aqueous solutions: A review., *J. Hazard. Mater.* 167, 1–9, 2009.

24. D. Li, Y. Wu, L. Feng, L. Zhang, Surface properties of SAC and its adsorption mechanisms for phenol and nitrobenzene, *Bioresour. Technol.* 113, 121–6, 2012.
25. A. C. Deiana, M. F. Sardella, H. Silva, A. Amaya, N. Tancredi, Use of grape stalk, a waste of the viticulture industry, to obtain activated carbon, *J. Hazard. Mater.* 172, 13–19, 2009.
26. D. Mohan, C. U. Pittman, Activated carbons and low cost adsorbents for remediation of tri- and hexavalent chromium from water, *J. Hazard. Mater.* 137, 762–811, 2006.
27. O. Ioannidou, A. Zabaniotou, Agricultural residues as precursors for activated carbon production – A review, *Renew. Sustain. Energy Rev.* 11, 1966–2005, 2007.
28. A. R. Mohamed, M. Mohammadi, G. N. Darzi, Preparation of carbon molecular sieve from lignocellulosic biomass: A review, *Renew. Sustain. Energy Rev.* 14, 1591–1599, 2010.
29. J. Wang, S. Kaskel, KOH activation of carbon-based materials for energy storage, *J. Mater. Chem.* 22, 23710–25, 2012.
30. T. Tay, S. Ucar, S. Karagöz, Preparation and characterization of activated carbon from waste biomass., *J. Hazard. Mater.* 165, 481–5, 2009.
31. F. Rouquerol, J. Rouquerol, K. Sing, *Adsorption by Powders and Porous Solids: Principles, Methodology and Applications*, Academic Press, London, UK, 1999.
32. C. Minot, A. Markovits, Introduction to theoretical approaches to chemisorption, *J. Mol. Struct. THEOCHEM.* 424, 119–134, 1998.
33. J. B. Condon, An overview of physisorption, in: *Surface Area and Porosity Determinations by Physisorption: Measurements and Theory*, pp. 1–27, Elsevier, 2006.
34. D. Sud, G. Mahajan, M. P. Kaur, Agricultural waste material as potential adsorbent for sequestering heavy metal ions from aqueous solutions – A review., *Bioresour. Technol.* 99, 6017–27, 2008.
35. L. Lin, S.-R. Zhai, Z.-Y. Xiao, Y. Song, Q.-D. An, X.-W. Song, Dye adsorption of mesoporous activated carbons produced from NaOH-pretreated rice husks, *Bioresour. Technol.* 136, 437–43, 2013.
36. Ş. Sargın, M. Saltan, N. Morova, S. Serin, S. Terzi, Evaluation of rice husk ash as filler in hot mix asphalt concrete, *Constr. Build. Mater.* 48, 390–397, 2013.
37. E. Natarajan, A. Nordina, A. N. Rao, Overview of combustion and gasification of rice husk in fluidized bed reactors, *Biomass and Bioenergy* 14, 533–46, 1998.
38. F. Vitali, S. Parmigiani, M. Vaccari, C. Collivignarelli, Agricultural waste as household fuel: Techno-economic assessment of a new rice-husk cookstove for developing countries, *Waste Manag.* 33, 2762–70, 2013.
39. Y. Chen, Y. Zhu, Z. Wang, Y. Li, L. Wang, L. Ding, et al., Application studies of activated carbon derived from rice husks produced by chemical-thermal process – A review, *Adv. Colloid Interface Sci.* 163, 39–52, 2011.
40. F. Adam, J. N. Appaturi, A. Iqbal, The utilization of rice husk silica as a catalyst: Review and recent progress, *Catal. Today.* 190, 2–14, 2012.

41. H. Chen, H. Wang, Z. Xue, L. Yang, Y. Xiao, M. Zheng, et al., High hydrogen storage capacity of rice hull based porous carbon, *Int. J. Hydrogen Energy* 37, 18888–18894, 2012.

42. K. Y. Foo, B. H. Hameed, Utilization of rice husk ash as novel adsorbent: A judicious recycling of the colloidal agricultural waste, *Adv. Colloid Interface Sci.* 152, 39–47, 2009.

43. R. Han, D. Ding, Y. Xu, W. Zou, Y. Wang, Y. Li, et al., Use of rice husk for the adsorption of Congo red from aqueous solution in column mode, *Bioresour. Technol.* 99, 2938–46, 2008.

44. Y. Safa, H. N. Bhatti, Kinetic and thermodynamic modeling for the removal of Direct Red-31 and Direct Orange-26 dyes from aqueous solutions by rice husk, *Desalination* 272, 313–322, 2011.

45. S. Chowdhury, R. Mishra, P. Saha, P. Kushwaha, Adsorption thermodynamics, kinetics and isosteric heat of adsorption of malachite green onto chemically modified rice husk, *Desalination* 265, 159–168, 2011.

46. S. Chakraborty, S. Chowdhury, P. Das Saha, Adsorption of Crystal Violet from aqueous solution onto NaOH-modified rice husk, *Carbohydr. Polym.* 86, 1533–1541, 2011.

47. M. M. Mohamed, Acid dye removal: Comparison of surfactant-modified mesoporous FSM-16 with activated carbon derived from rice husk, *J. Colloid Interface Sci.* 272, 28–34, 2004.

48. I. A. Rahman, B. Saad, S. Shaidan, E. S. Sya Rizal, Adsorption characteristics of malachite green on activated carbon derived from rice husks produced by chemical-thermal process, *Bioresour. Technol.* 96, 1578–83, 2005.

49. Y. Guo, S. Yang, W. Fu, J. Qi, R. Li, Z. Wang, et al., Adsorption of malachite green on micro- and mesoporous rice husk-based active carbon, *Dye. Pigment.* 56, 219–229, 2003.

50. Y. Guo, J. Zhao, H. Zhang, S. Yang, J. Qi, Z. Wang, et al., Use of rice husk-based porous carbon for adsorption of Rhodamine B from aqueous solutions, *Dye. Pigment.* 66, 123–128, 2005.

51. P. K. Malik, Use of activated carbons prepared from sawdust and rice-husk for adsorption of acid dyes: A case study of Acid Yellow 36, *Dye. Pigment.* 56, 239–249, 2003.

52. B. H. Hameed, M. I. El-Khaiary, Kinetics and equilibrium studies of malachite green adsorption on rice straw-derived char, *J. Hazard. Mater.* 153, 701–8, 2008.

53. P. K. Rai, S. P. Singh, R. K. Asthana, S. Singh, Biohydrogen production from sugarcane bagasse by integrating dark- and photo-fermentation, *Bioresour. Technol.* 152, 140–146, 2014.

54. Y. S. Ho, G. McKay, A kinetic study of dye sorption by biosorbent waste product pith, *Resour. Conserv. Recycl.* 25, 171–193, 1999.

55. G. McKay, M. El Geundi, M. M. Nassar, Pore diffusion during the adsorption of dyes onto bagasse pith, *Process Saf. Environ. Prot.* 74, 277–288, 1996.

56. Z. Zhang, L. Moghaddam, I. M. O'Hara, W. O. S. Doherty, Congo red adsorption by ball-milled sugarcane bagasse, *Chem. Eng. J.* 178, 122–128, 2011.

57. Z. Zhang, I. M. O'Hara, G. A. Kent, W. O. S. Doherty, Comparative study on adsorption of two cationic dyes by milled sugarcane bagasse, *Ind. Crops Prod.* 42, 41–49, 2013.

58. W. T. Tsai, C. Y. Chang, M. C. Lin, S. F. Chien, H. F. Sun, M. F. Hsieh, Adsorption of acid dye onto activated carbons prepared from agricultural waste bagasse by ZnCl2 activation, *Chemosphere.* 45, 51–58, 2001.

59. M. Valix, W. H. Cheung, G. McKay, Preparation of activated carbon using low temperature carbonisation and physical activation of high ash raw bagasse for acid dye adsorption, *Chemosphere* 56, 493–501, 2004.

60. A. A. Kadam, H. S. Lade, S. M. Patil, S. P. Govindwar, Low cost CaCl2 pretreatment of sugarcane bagasse for enhancement of textile dyes adsorption and subsequent biodegradation of adsorbed dyes under solid state fermentation, *Bioresour. Technol.* 132, 276–84, 2013.

61. K. A. G. Gusmão, L. V. A. Gurgel, T. M. S. Melo, L. F. Gil, Adsorption studies of methylene blue and gentian violet on sugarcane bagasse modified with EDTA dianhydride (EDTAD) in aqueous solutions: Kinetic and equilibrium aspects, *J. Environ. Manage.* 118, 135–43, 2013.

62. K. A. G. Gusmão, L. V. A. Gurgel, T. M. S. Melo, L. F. Gil, Application of succinylated sugarcane bagasse as adsorbent to remove methylene blue and gentian violet from aqueous solutions – Kinetic and equilibrium studies, *Dye. Pigment.* 92, 967–974, 2012.

63. G.-B. Jiang, Z.-T. Lin, X.-Y. Huang, Y.-Q. Zheng, C.-C. Ren, C.-K. Huang, et al., Potential biosorbent based on sugarcane bagasse modified with tetraethylenepentamine for removal of eosin Y, *Int. J. Biol. Macromol.* 50, 707–12, 2012.

64. V. K. C. Lee, J. F. Porter, G. Mckay, Modified design model for the adsorption of dye onto peat, *Food Bioprod. Process.* 79, 21–26, 2001.

65. Y. Ho, G. Mckay, Sorption of dye from aqueous solution by peat, *Chem. Eng. J.* 70, 115–124, 1998.

66. A. W. M. Ip, J. P. Barford, G. McKay, Reactive Black dye adsorption/desorption onto different adsorbents: Effect of salt, surface chemistry, pore size and surface area, *J. Colloid Interface Sci.* 337, 32–8, 2009.

67. A. N. Fernandes, C. A. P. Almeida, C. T. B. Menezes, N. A. Debacher, M. M. D. Sierra, Removal of methylene blue from aqueous solution by peat, *J. Hazard. Mater.* 144, 412–419, 2007.

68. S. J. Allen, G. McKay, J. F. Porter, Adsorption isotherm models for basic dye adsorption by peat in single and binary component systems, *J. Colloid Interface Sci.* 280, 322–333, 2004.

69. R. Wang, Y. Amano, M. Machida, Surface properties and water vapor adsorption–desorption characteristics of bamboo-based activated carbon, *J. Anal. Appl. Pyrolysis.* 104, 667–674, 2013.

70. S.-F. Lo, S.-Y. Wang, M.-J. Tsai, L.-D. Lin, Adsorption capacity and removal efficiency of heavy metal ions by Moso and Ma bamboo activated carbons, *Chem. Eng. Res. Des.* 90, 1397–1406, 2012.

71. P. Liao, Z. Malik Ismael, W. Zhang, S. Yuan, M. Tong, K. Wang, et al., Adsorption of dyes from aqueous solutions by microwave modified bamboo charcoal, *Chem. Eng. J.* 195–196, 339–346, 2012.
72. E. L. K. Mui, W. H. Cheung, M. Valix, G. McKay, Dye adsorption onto char from bamboo, *J. Hazard. Mater.* 177, 1001–5, 2010.
73. L. S. Chan, W. H. Cheung, S. J. Allen, G. McKay, Separation of acid-dyes mixture by bamboo derived active carbon, *Sep. Purif. Technol.* 67, 166–172, 2009.
74. L. S. Chan, W. H. Cheung, S. J. Allen, G. McKay, Error analysis of adsorption isotherm models for acid dyes onto bamboo derived activated carbon, *Chinese J. Chem. Eng.* 20, 535–542, 2012.
75. L. S. Chan, W. H. Cheung, G. McKay, Adsorption of acid dyes by bamboo derived activated carbon, *Desalination* 218, 304–312, 2008.
76. A. W. M. Ip, J. P. Barford, G. McKay, A comparative study on the kinetics and mechanisms of removal of Reactive Black 5 by adsorption onto activated carbons and bone char, *Chem. Eng. J.* 157, 434–442, 2010.
77. A. A. Ahmad, B. H. Hameed, Reduction of COD and color of dyeing effluent from a cotton textile mill by adsorption onto bamboo-based activated carbon, *J. Hazard. Mater.* 172, 1538–43, 2009.
78. L. Wang, Application of activated carbon derived from "waste" bamboo culms for the adsorption of azo disperse dye: Kinetic, equilibrium and thermodynamic studies, *J. Environ. Manage.* 102, 79–87, 2012.
79. K. K. H. Choy, G. Mckay, Synergistic multilayer adsorption for low concentration dyestuffs by biomass, *Chinese J. Chem. Eng.* 20, 560–566, 2012.
80. S. Suresh, N. Guizani, M. Al-Ruzeiki, A. Al-Hadhrami, H. Al-Dohani, I. Al-Kindi, et al., Thermal characteristics, chemical composition and polyphenol contents of date-pits powder, *J. Food Eng.* 119, 668–679, 2013.
81. Z. Belala, M. Jeguirim, M. Belhachemi, F. Addoun, G. Trouvé, Biosorption of basic dye from aqueous solutions by Date Stones and Palm-Trees Waste: Kinetic, equilibrium and thermodynamic studies, *Desalination* 271, 80–87, 2011.
82. M. A. Al-Ghouti, J. Li, Y. Salamh, N. Al-Laqtah, G. Walker, M. N. M. Ahmad, Adsorption mechanisms of removing heavy metals and dyes from aqueous solution using date pits solid adsorbent, *J. Hazard. Mater.* 176, 510–20, 2010.
83. F. Banat, S. Al-Asheh, L. Al-Makhadmeh, Evaluation of the use of raw and activated date pits as potential adsorbents for dye containing waters, *Process Biochem.* 39, 193–202, 2003.
84. M. J. Ahmed, S. K. Dhedan, Equilibrium isotherms and kinetics modeling of methylene blue adsorption on agricultural wastes-based activated carbons, *Fluid Phase Equilib.* 317, 9–14, 2012.
85. S. K. Theydan, M. J. Ahmed, Adsorption of methylene blue onto biomass-based activated carbon by FeCl3 activation: Equilibrium, kinetics, and thermodynamic studies, *J. Anal. Appl. Pyrolysis.* 97, 116–122, 2012.
86. K. Y. Foo, B. H. Hameed, Preparation of activated carbon from date stones by microwave induced chemical activation: Application for methylene blue adsorption, *Chem. Eng. J.* 170, 338–341, 2011.

87. S. S. Ashour, Kinetic and equilibrium adsorption of methylene blue and rem-azol dyes onto steam-activated carbons developed from date pits, *J. Saudi Chem. Soc.* 14, 47–53, 2010.

88. K. S. K. Reddy, A. Al Shoaibi, C. Srinivasakannan, A comparison of micro-structure and adsorption characteristics of activated carbons by CO2 and H3PO4 activation from date palm pits, *New Carbon Mater.* 27, 344–351, 2012.

89. K. S. K. Reddy, A. Al Shoaibi, C. Srinivasakannan, Activated carbon from date palm seed: Process optimization using response surface methodology, *Waste and Biomass Valorization* 3, 149–156, 2012.

90. K. Szymona, P. Borysiuk, P. S. H'ng, K. L. Chin, M. Mamiński, Valorization of waste oil palm (Elaeis guineensis Jacq.) biomass through furfurylation, *Mater. Des.* 53, 425–429, 2014.

91. F. Abnisa, A. Arami-Niya, W. M. A. Wan Daud, J. N. Sahu, I. M. Noor, Utilization of oil palm tree residues to produce bio-oil and bio-char via pyrolysis, *Energy Convers. Manag.* 76, 1073–1082, 2013.

92. Y. L. Chiew, T. Iwata, S. Shimada, System analysis for effective use of palm oil waste as energy resources, *Biomass and Bioenergy* 35, 2925–2935, 2011.

93. S. M. Mirmehdi, F. Zeinaly, F. Dabbagh, Date palm wood flour as filler of linear low-density polyethylene, *Compos. Part B.* 56, 137–141, 2014.

94. S. Syahrullail, B. M. Zubil, C. S. N. Azwadi, M. J. M. Ridzuan, Experimental evaluation of palm oil as lubricant in cold forward extrusion process, *Int. J. Mech. Sci.* 53, 549–555, 2011.

95. H. T. Tan, K. T. Lee, A. R. Mohamed, Second-generation bio-ethanol (SGB) from Malaysian palm empty fruit bunch: Energy and exergy analyses, *Bioresour. Technol.* 101, 5719–27, 2010.

96. A. E. Ofomaja, Kinetics and mechanism of methylene blue sorption onto palm kernel fibre, *Process Biochem.* 42, 16–24, 2007.

97. A. Ofomaja, Y. Ho, Equilibrium sorption of anionic dye from aqueous solu-tion by palm kernel fibre as sorbent, *Dye. Pigment.* 74, 60–66, 2007.

98. A. E. Ofomaja, E. E. Ukpebor, S. A. Uzoekwe, Biosorption of Methyl vio-let onto palm kernel fiber: Diffusion studies and multistage process design to minimize biosorbent mass and contact time, *Biomass and Bioenergy* 35, 4112–4123, 2011.

99. I. A. W. Tan, B. H. Hameed, A. L. Ahmad, Equilibrium and kinetic studies on basic dye adsorption by oil palm fibre activated carbon, *Chem. Eng. J.* 127, 111–119, 2007.

100. B. H. Hameed, I. A. W. Tan, A. L. Ahmad, Optimization of basic dye removal by oil palm fibre-based activated carbon using response surface methodol-ogy, *J. Hazard. Mater.* 158, 324–32, 2008.

101. K. Y. Foo, B. H. Hameed, Microwave-assisted preparation of oil palm fiber acti-vated carbon for methylene blue adsorption, *Chem. Eng. J.* 166, 792–795, 2011.

102. K. Y. Foo, B. H. Hameed, Adsorption characteristics of industrial solid waste derived activated carbon prepared by microwave heating for methylene blue, *Fuel Process. Technol.* 99, 103–109, 2012.

103. T. S. Anirudhan, S. S. Sreekumari, Adsorptive removal of heavy metal ions from industrial effluents using activated carbon derived from waste coconut buttons, *J. Environ. Sci.* 23, 1989–1998, 2011.
104. A. Bhatnagar, V. J. P. Vilar, C. M. S. Botelho, R. A. R. Boaventura, Coconut-based biosorbents for water treatment – A review of the recent literature, *Adv. Colloid Interface Sci.* 160, 1–15, 2010.
105. S. Sulaiman, A. R. Abdul Aziz, M. K. Aroua, Reactive extraction of solid coconut waste to produce biodiesel, *J. Taiwan Inst. Chem. Eng.* 44, 233–238, 2013.
106. P. Shafigh, H. Bin Mahmud, M. Z. Jumaat, M. Zargar, Agricultural wastes as aggregate in concrete mixtures – A review, *Constr. Build. Mater.* 53, 110–117, 2014.
107. M. V. Madurwar, R. V. Ralegaonkar, S. A. Mandavgane, Application of agro-waste for sustainable construction materials: A review, *Constr. Build. Mater.* 38, 872–878, 2013.
108. B. H. Hameed, D. K. Mahmoud, A. L. Ahmad, Equilibrium modeling and kinetic studies on the adsorption of basic dye by a low-cost adsorbent: Coconut (Cocos nucifera) bunch waste, *J. Hazard. Mater.* 158, 65–72, 2008.
109. A. L. Cazetta, A. M. M. Vargas, E. M. Nogami, M. H. Kunita, M. R. Guilherme, A. C. Martins, et al., NaOH-activated carbon of high surface area produced from coconut shell: Kinetics and equilibrium studies from the methylene blue adsorption, *Chem. Eng. J.* 174, 117–125, 2011.
110. I. A. W. Tan, A. L. Ahmad, B. H. Hameed, Adsorption of basic dye on high-surface-area activated carbon prepared from coconut husk: Equilibrium, kinetic and thermodynamic studies, *J. Hazard. Mater.* 154, 337–46, 2008.
111. J. D. S. Macedo, N. B. da Costa Júnior, L. E. Almeida, E. F. D. S. Vieira, A. R. Cestari, I. D. F. Gimenez, et al., Kinetic and calorimetric study of the adsorption of dyes on mesoporous activated carbon prepared from coconut coir dust, *J. Colloid Interface Sci.* 298, 515–22, 2006.
112. S. L. Ching, M. S. Yusoff, H. A. Aziz, M. Umar, Influence of impregnation ratio on coffee ground activated carbon as landfill leachate adsorbent for removal of total iron and orthophosphate, *Desalination* 279, 225–234, 2011.
113. B. H. Hameed, Spent tea leaves: A new non-conventional and low-cost adsorbent for removal of basic dye from aqueous solutions, *J. Hazard. Mater.* 161, 753–9, 2009.
114. N. Nasuha, B. H. Hameed, A. T. M. Din, Rejected tea as a potential low-cost adsorbent for the removal of methylene blue, *J. Hazard. Mater.* 175, 126–32, 2010.
115. P. Li, Y.-J. Su, Y. Wang, B. Liu, L.-M. Sun, Bioadsorption of methyl violet from aqueous solution onto Pu-erh tea powder, *J. Hazard. Mater.* 179, 43–48, 2010.
116. G. Z. Kyzas, N. K. Lazaridis, A. C. Mitropoulos, Removal of dyes from aqueous solutions with untreated coffee residues as potential low-cost adsorbents: Equilibrium, reuse and thermodynamic approach, *Chem. Eng. J.* 189–190, 148–159, 2012.

117. A. Reffas, V. Bernardet, B. David, L. Reinert, M. B. Lehocine, M. Dubois, et al., Carbons prepared from coffee grounds by H3PO4 activation: Characterization and adsorption of methylene blue and Nylosan Red N-2RBL, *J. Hazard. Mater.* 175, 779–88, 2010.

118. E. Akar, A. Altinişik, Y. Seki, Using of activated carbon produced from spent tea leaves for the removal of malachite green from aqueous solution, *Ecol. Eng.* 52, 19–27, 2013.

119. M. Auta, B. H. Hameed, Preparation of waste tea activated carbon using potassium acetate as an activating agent for adsorption of Acid Blue 25 dye, *Chem. Eng. J.* 171, 502–509, 2011.

120. P. Sharma, R. Kaur, C. Baskar, W.-J. Chung, Removal of methylene blue from aqueous waste using rice husk and rice husk ash, *Desalination* 259, 249–257, 2010.

121. Y. Safa, H. N. Bhatti, Biosorption of Direct Red-31 and Direct Orange-26 dyes by rice husk: Application of factorial design analysis, *Chem. Eng. Res. Des.* 89, 2566–2574, 2011.

122. M. S. Ur Rehman, I. Kim, J.-I. Han, Adsorption of methylene blue dye from aqueous solution by sugar extracted spent rice biomass, *Carbohydr. Polym.* 90, 1314–22, 2012.

123. Y. Chen, S.-R. Zhai, N. Liu, Y. Song, Q.-D. An, X.-W. Song, Dye removal of activated carbons prepared from NaOH-pretreated rice husks by low-temperature solution-processed carbonization and H3PO4 activation, *Bioresour. Technol.* 144, 401–9, 2013.

124. Q. Sun, L. Yang, The adsorption of basic dyes from aqueous solution on modified peat-resin particle, *Water Res.* 37, 1535–44, 2003.

125. M. M. Nassar, Y. H. Magdy, Removal of different basic dyes from aqueous solutions by adsorption on palm-fruit bunch particles, *Chem. Eng. J.* 66, 223–226, 1997.

126. M. M. Nassar, M. F. Hamoda, G. H. Radwan, Adsorption equilibria of basic dyestuff onto palm-fruit bunch particles, *Water Sci. Technol.* 32, 27–32, 1995.

127. G. O. El-Sayed, Removal of methylene blue and crystal violet from aqueous solutions by palm kernel fiber, *Desalination* 272, 225–232, 2011.

128. I. A. W. Tan, A. L. Ahmad, B. H. Hameed, Adsorption of basic dye using activated carbon prepared from oil palm shell: Batch and fixed bed studies, *Desalination* 225, 13–28, 2008.

129. I. A. W. Tan, A. L. Ahmad, B. H. Hameed, Enhancement of basic dye adsorption uptake from aqueous solutions using chemically modified oil palm shell activated carbon, *Colloids Surfaces A.* 318, 88–96, 2008.

130. K. Y. Foo, B. H. Hameed, Preparation of oil palm (Elaeis) empty fruit bunch activated carbon by microwave-assisted KOH activation for the adsorption of methylene blue, *Desalination* 275, 302–305, 2011.

131. U. J. Etim, S. A. Umoren, U. M. Eduok, Coconut coir dust as a low cost adsorbent for the removal of cationic dye from aqueous solution, *J. Saudi Chem. Soc.* 2012.

132. S. Senthilkumaar, P. Kalaamani, C. V. Subburaam, Liquid phase adsorption of Crystal violet onto activated carbons derived from male flowers of coconut tree, *J. Hazard. Mater.* 136, 800–8, 2006.

133. S. Senthilkumaar, P. Kalaamani, K. Porkodi, P. R. Varadarajan, C. V. Subburaam, Adsorption of dissolved Reactive red dye from aqueous phase onto activated carbon prepared from agricultural waste, *Bioresour. Technol.* 97, 1618–25, 2006.

134. M. Giahi, R. Rakhshaee, M. A. Bagherinia, Removal of methylene blue by tea wastages from the synthesis waste waters, *Chinese Chem. Lett.* 22, 225–228, 2011.

135. A. S. Franca, L. S. Oliveira, M. E. Ferreira, Kinetics and equilibrium studies of methylene blue adsorption by spent coffee grounds, *Desalination* 249, 267–272, 2009.

136. L. S. Oliveira, A. S. Franca, T. M. Alves, S. D. F. Rocha, Evaluation of untreated coffee husks as potential biosorbents for treatment of dye contaminated waters, *J. Hazard. Mater.* 155, 507–12, 2008.

137. M. T. Uddin, M. A. Islam, S. Mahmud, M. Rukanuzzaman, Adsorptive removal of methylene blue by tea waste, *J. Hazard. Mater.* 164, 53–60, 2009.

138. N. Nasuha, B. H. Hameed, Adsorption of methylene blue from aqueous solution onto NaOH-modified rejected tea, *Chem. Eng. J.* 166, 783–786, 2011.

139. M.-H. Baek, C. O. Ijagbemi, S.-J. O, D.-S. Kim, Removal of Malachite Green from aqueous solution using degreased coffee bean, *J. Hazard. Mater.* 176, 820–8, 2010.

140. M. Auta, B. H. Hameed, Optimized waste tea activated carbon for adsorption of Methylene Blue and Acid Blue 29 dyes using response surface methodology, *Chem. Eng. J.* 175, 233–243, 2011.

141. M. A. Ahmad, N. K. Rahman, Equilibrium, kinetics and thermodynamic of Remazol Brilliant Orange 3R dye adsorption on coffee husk-based activated carbon, *Chem. Eng. J.* 170, 154–161, 2011.

142. A. A. Nunes, A. S. Franca, L. S. Oliveira, Activated carbons from waste biomass: An alternative use for biodiesel production solid residues, *Bioresour. Technol.* 100, 1786–92, 2009.

143. A. Namane, A. Mekarzia, K. Benrachedi, N. Belhaneche-Bensemra, A. Hellal, Determination of the adsorption capacity of activated carbon made from coffee grounds by chemical activation with ZnCl2 and H3PO4, *J. Hazard. Mater.* 119, 189–94, 2005.

144. T. Madrakian, A. Afkhami, M. Ahmadi, Adsorption and kinetic studies of seven different organic dyes onto magnetite nanoparticles loaded tea waste and removal of them from wastewater samples, *Spectrochim. Acta A.* 99, 102–9, 2012.

6

Use of Fungal Laccases and Peroxidases for Enzymatic Treatment of Wastewater Containing Synthetic Dyes

Keisuke Ikehata

Pacific Advanced Civil Engineering, Inc., Fountain Valley, California, USA

Abstract

Synthetic organic dyes such as azo dyes in wastewater discharged by dye man-ufacturing and textile industries constitute a major environmental concern due to their poor biodegradability and toxicity; as well as affecting aesthetics and inhibiting photosynthesis and other biological activities in receiving water. Most synthetic dyes contain characteristic multiple aromatic rings that are fused and/or linked by various C–C, C–O, and C–N linkages and substituted by multiple func-tional groups such as amino, nitro, hydroxyl, sulfonate, carbonyl, and carboxylate groups. Because of the similarity of dye molecules and lignin, a natural polymer of aromatic alcohols, many of the synthetic dyes can be degraded and decolor-ized by lignin-degrading white rot fungi such as *Phanerochaete chrysosporium* and *Trametes versicolor* and their oxidoreductase systems including peroxidases and laccases. This chapter discusses the sources and application of these lignino-lytic enzymes for the treatment of synthetic dyes, reaction mechanisms, and other developments and challenges associated with this enzymatic process.

Keywords: Anthraquinone dye, azo dye, dye-decolorizing peroxidase, enzymes, white rot fungus, indigoid dye, laccase, lignin peroxidase, manganese-dependent peroxidase, triarylmethane dye, versatile peroxidase, wastewater treatment

6.1 Introduction

Synthetic dyes are widely used in fabrics, including textiles, paper making, and inks and other coloring agents in many industries [1]. Most of the

Corresponding author: kikehata@pacewater.com

Sanjay K. Sharma (ed.) Green Chemistry for Dyes Removal from Wastewater, (203–260)

synthetic dyes are organic compounds with multiple aromatic rings, either fused or connected by covalent bonds, and modified by various hydrophilic functional groups, such as amine, carbonyl, carboxylic acid, and hydroxyl groups, to produce desired colors, increase the affinity to the materials being dyed, and improve the solubility in solvents, such as water, during dyeing processes. Being synthetic organics, those dyes are often resistant to biodegradation in engineered treatment processes such as activated sludge, as well as in the natural aquatic environment [1–3]. As a result, the pollution of the aquatic environment by residual synthetic dyes discharged to water bodies via industrial effluents has been a major environmental and public health issue in many countries such as India, Pakistan, Indonesia, and China, where the textile industry has been growing steadily [4–7]. Such dye pollution in the rivers and lakes tends to intensify especially in small townships and villages where many smaller factories are present, and pollution prevention measures, such as wastewater treatment, have not been implemented because effective regulations are absent [8,9].

Colored water due to untreated and partially treated industrial effluents containing synthetic dyes is not only aesthetically unappealing and unpleasant, but can also cause many undesired adverse impacts on the receiving environment and public health. While many synthetic dyes are relatively nontoxic, some of their degradation byproducts, such as anilines, are often more toxic and sometimes carcinogenic than the parent compounds [5,10,11]. In addition to the toxicity, residual organic dyes can also exert various hazards on the aquatic ecosystem by depleting dissolved oxygen, blocking sunlight penetration, and inhibiting photosynthesis and growth of aquatic organisms. Therefore, effective treatment methods for textile wastewater containing synthetic dyes have been sought over the past several decades [2].

A number of physical, chemical, and biological treatment processes have been investigated for the removal and/or destruction of synthetic dyes. For example, chemical oxidation, such as ozonation and Fenton processes, electrocoagulation/electro-oxidation, adsorptive treatment using granular activated carbon and other plant-derived materials, aerobic and anaerobic processes, such as activated sludge, and membrane filtration, such as reverse osmosis and nanofiltration, have been found to be effective [1,12,13]. The enzymatic process using ligninolytic enzymes, such as laccases and peroxidases, is a relatively new emerging technology for the degradation of xenobiotics, including synthetic dyes in textile wastewater [14–16]. This unique process employs a hybrid of chemical and biological oxidation using a combination of crude or purified enzymes from plant materials or fungal cultures as a biocatalyst and dissolved molecular oxygen or hydrogen peroxide as a chemical oxidant. This enzymatic process has a number of advantages over conventional physical, chemical and biological processes [15–17].

This chapter provides a comprehensive literature review on the enzymatic treatment of various synthetic dyes and discusses the recent progress and challenges associated with this technology. In addition, the fungal treatment of synthetic dyes and contaminated effluents, as well as the enzymology of the key ligninolytic enzymes, are covered in this chapter to explore the important roles of fungal enzymes in synthetic dye decolorization. More than 70 different synthetic dyes with various chemical structures, including anthraquinone, azo (including diazo, triazo, and poly-azo), heterocyclic, indigoid, triarylmethane, and phtharocyanine dyes, are covered. The chemical structures of most of the dyes discussed in this chapter are drawn and presented in a consistent and legitimate format to aid the readers' comprehension of different enzyme-substrate (i.e., dye) reactions. In addition, the basic characteristics of textile dyes, including nomenclature, chemical structures, and environmental impacts are briefly discussed.

6.2   Textile Dyes – Classifications, Chemical Structures and Environmental Impacts

6.2.1   Classification of Dyes

More than 10,000 different dyes with a variety of colors and chemical structures are commercially available around the world. A majority of the dyes currently used are synthetic organic compounds, while some are of natural origins and inorganic pigments based on minerals. These synthetic and natural dyes may be classified in many ways such as their modes of application, chemical structures, and colors [1]. Like any other commercial chemicals such as solvents, pesticides, and pharmaceuticals, a dyestuff may have many different names, ranging from common chemical names and the International Union of Pure and Applied Chemistry (IUPAC) name to generic and trade names. A comprehensive database of commercial dyes and pigments called Colour Index International is maintained by the Society of Dyers and Colourists and the American Association of Textile Chemists and Colorists. Colour Index International contains 27,000 individual products under 13,000 Colour Index (C. I.) Generic Names and Colour Index Constitution Numbers, according to the Colour Index International's website (http://www.colour-index.com/). The C. I. generic name of a dye consists of its recognized usage class (e.g., direct, reactive, vat, acid, and basic), its hue, and a serial number [18]. For example, an anthraquinone dye Remazol Brilliant Blue R is also called Reactive Blue 19 and C. I. 61200 (Figure 6.1). For more information, consult Colour Index International's website. In this chapter, the C. I. generic names, if known, are used primarily in the text and tables.

6.2.2   Chemical Structures

As an organic compound, a dye has several common moieties for its functions. There are two critical moieties, namely:

- Chromophore – A functional group or a group of functional groups that is responsible for visible light absorption and color, typically multiple aromatic rings combined with alkene, carbonyl ($>C=O$), azo ($-N=N-$), and nitro ($-NO_2$) groups that form a long conjugated pi-bond system.
- Auxochrome – A functional group that modifies the ability of chromophore to absorb light and change the wavelength and intensity of the color, such as hydroxyl ($-OH$), amino ($-NH_3$), and sulfonate ($-SO_3^-$) groups. Auxochromes also make the dye more hydrophilic and water soluble and provide binding capacity to the dyes onto the textile materials via hydrogen bonds.

Several common chromophores used in synthetic textile dyes are listed in Table 6.1. In this chapter, synthetic dyes are mostly classified by their chromophores. Figures 6.1–6.8 present the chemical structures of the dyes covered in this chapter. Common names (such as Orange G and Congo red) used in the reviewed references are shown below the chemical structures and the C. I. generic names with brackets. Please note that the list shown in Table 6.1 is not exhaustive and there are several other chemical classes of dyes, such as diazo, triazo, and formazan dyes. Interested readers should consult with C. I. International, as well as other chapters of this book and literature such as Zaharia and Suteu [1].

Some auxochromes such as chlorotriazole and vinyl sulfonyl groups are vital for fixation of reactive dyes that are commonly used in dyeing of cotton, flax, wool, and nylon. See Figure 6.6 for a chlorotriazole group in Reactive Blue 15 and Figures 6.1 and 6.3 for vinyl sulfonyl group precursors in Reactive Blue 19 and Reactive Black 5, respectively. In the case of cellulosic textiles, such as cotton and flax, these functional groups can be

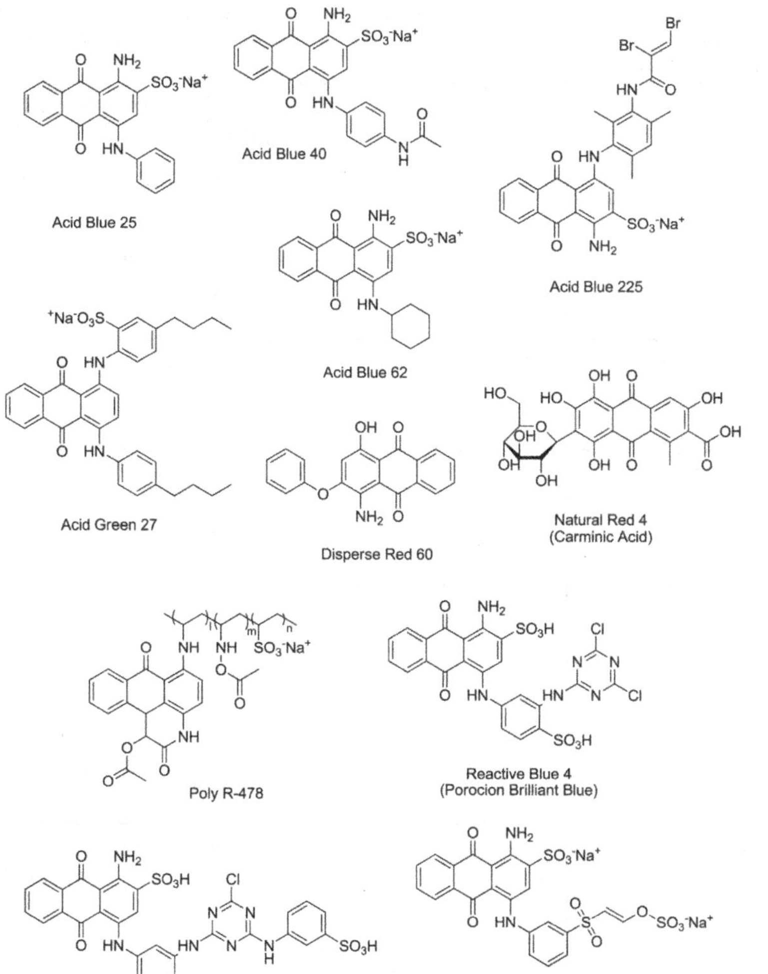

Figure 6.1 Anthraquinone dyes evaluated for fungal and enzymatic treatment.

activated under alkaline conditions and attached to the hydroxyl group of cellulose and form a covalent bond [1]. However, activated vinyl sulfone (R-SO$_2$-CH=CH$_2$) group undergoes hydrolysis with water, which results in discharging the hydrolyzed byproducts into the wastewater [19]. Depending on the fixation efficiency, as much as 50% of reactive dyes can be lost in the effluent [20]. Similar levels of loss of other types of dyes, such as direct and acid dyes, also occur and are inevitable.

Table 6.1 Examples of common chromophores in synthetic textile dyes [1].

Chemical Class	Base Structure	Examples
Anthraquionone	(anthraquinone structure)	Reactive Blue 19 (Remazol Brilliant Blue R), Acid Blue 225, Poly R-478
Azo	(azo structure)	Acid Orange 7 (Orange II), Acid Orange 10 (Orange G), Acid Red 2
Heterocyclic (Phenothiazine)	(phenothiazine structure)	Basic Blue 9 (methylene blue), azure B
Indigoid	(indigoid structure)	Acid Blue 74 (Indigo Carmine)
Phthalocyanine	(phthalocyanine structure)	Pigment Blue 15:2 (Phthalocyanine Blue BN), Reactive Blue 15
Triarylmethane	(triarylmethane structure)	Basic Violet 3 (Crystal Violet), Basic Red 9, bromophenol blue

6.2.3 Environmental Impacts

Owing to their complex chemical structures, synthetic dyes are often recalcitrant and very difficult to be decomposed or removed by conventional wastewater treatment processes, as well as in the aquatic environment [1]. In fact, these synthetic dyes are created as improved alternatives to natural dyes with longer lasting color and wider applicability to a variety of consumer products. Therefore, these compounds are chemically very stable by

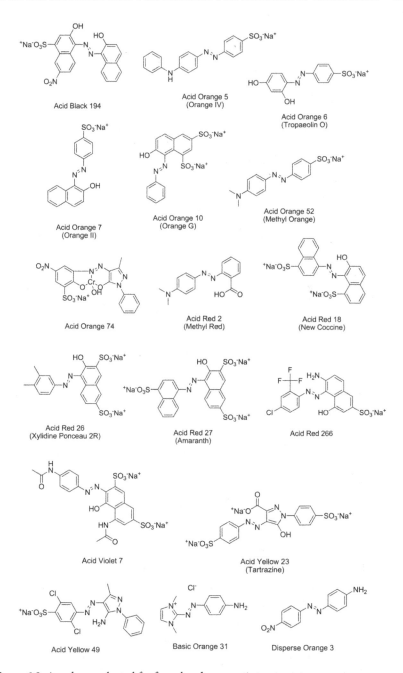

Figure 6.2 Azo dyes evaluated for fungal and enzymatic treatment.

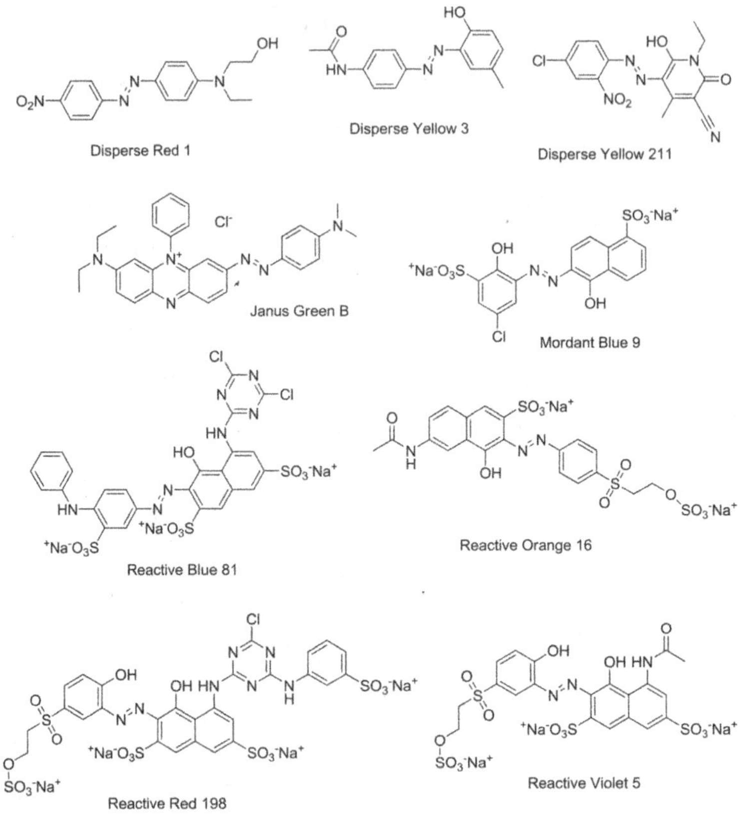

Figure 6.2 Azo dyes evaluated for fungal and enzymatic treatment. (Cont'd)

nature. The major impacts of residual dyes in the receiving water bodies are as follows:

- Visual and aesthetic effect – Unnatural color in river and lake water provokes public attention and protest.
- Depletion of dissolved oxygen due to high organics loading [7] – This disturbs the aquatic ecosystem such as fish, aquatic animals, and macro- and microinvertebrates.
- Prevention of sunlight penetration – Residual dye absorbs visible light and prevents the growth and photosynthesis of algae and plants, disturbing the aquatic ecosystem.
- Toxicity – Many dyes and their breakdown products are acutely and chronically toxic to fish and aquatic organisms [5,6]. Aromatic amines are carcinogenic.

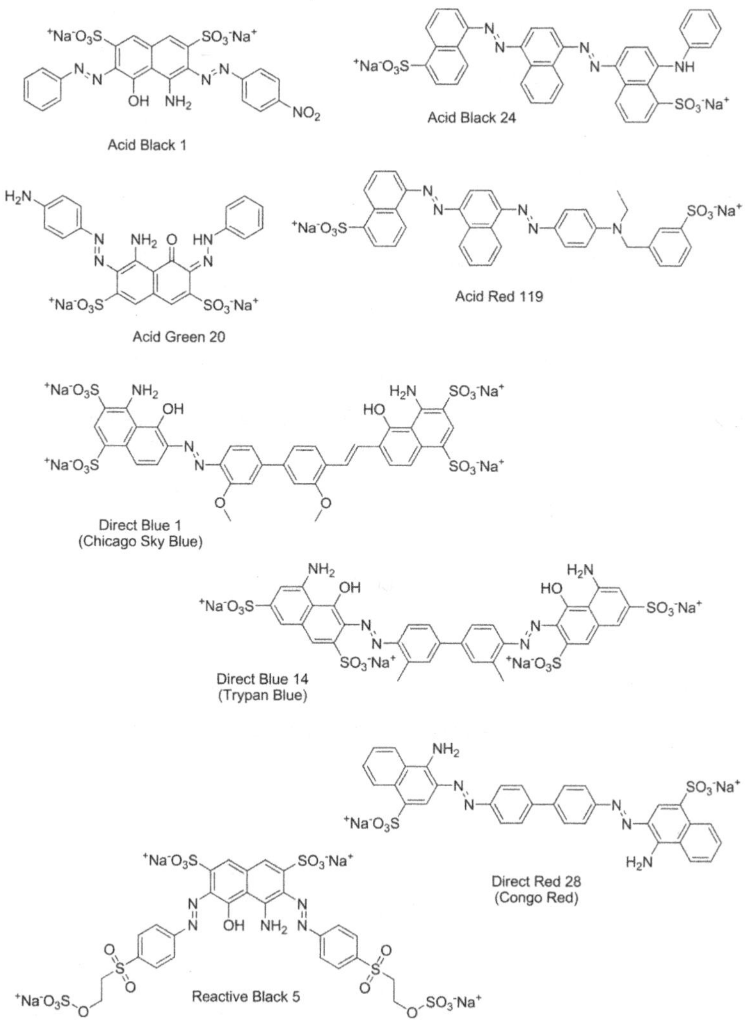

Figure 6.3 Diazo dyes evaluated for fungal and enzymatic treatment.

- Auxiliary chemicals in textile dyeing and processing, such as mineral acid, alkali, salts, oils, fats, surfactants, metals and oxidants – These affect pH and salinity and contribute to BOD and toxicity [7].

Cooper [21] identified several acidic azo dyes that have higher persistence in activated sludge process, including Acid Yellow 17, Acid Yellow 23, Acid Yellow 49, Acid Yellow 151, Acid Orange 10, Direct Yellow 4, Acid Red 14, Acid Red 18, Acid Red 337, and Acid Black 1. It has been suggested

Direct Blue 71

Direct Red 80

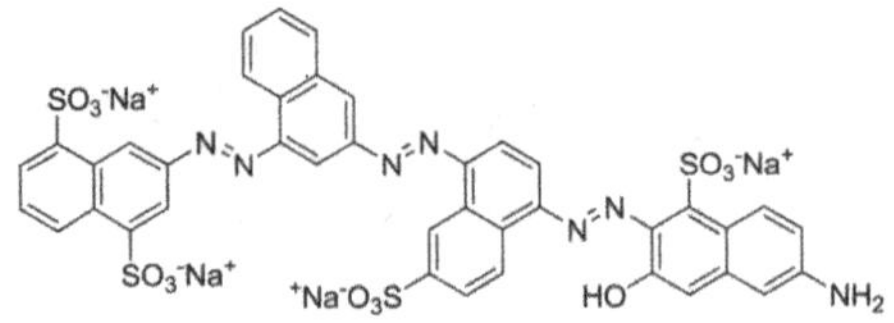

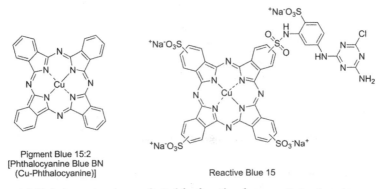

Figure 6.4 Triazo and poly-azo dyes evaluated for fungal and enzymatic treatment.

Acid Blue 74
(Indigo Carmine)

Vat Blue 1
[Indigo (Insoluble)]

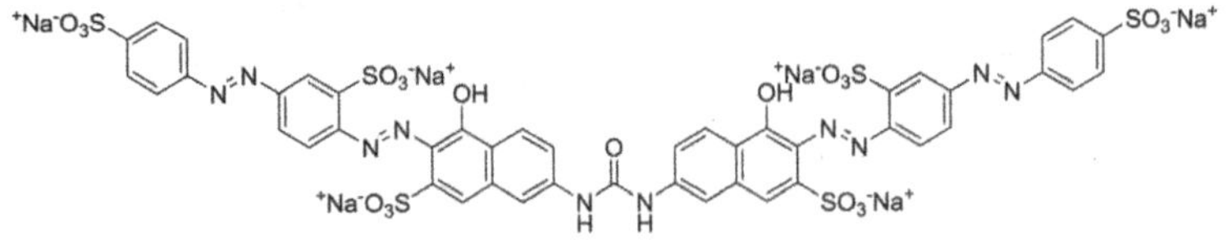

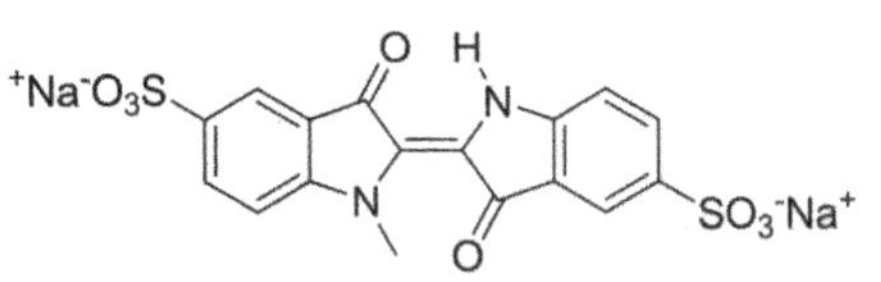

Figure 6.5 Indigoid dyes evaluated for fungal and enzymatic treatment.

Pigment Blue 15:2
[Phthalocyanine Blue BN
(Cu-Phthalocyanine)]

Reactive Blue 15

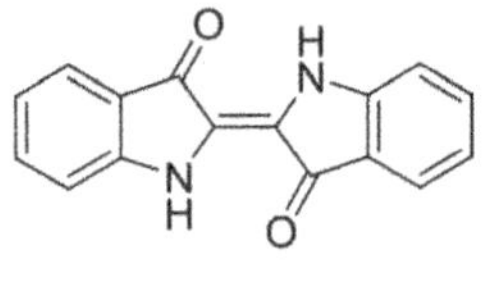

Figure 6.6 Phthalocyanine dyes evaluated for fungal and enzymatic treatment.

that these dyes contain multiple sulfonate groups, which enhance the water solubility and limit the assimilability by bacteria [1]. In addition, partial degradation of azo, diazo, and triazo dyes may result in the release of aromatic amines that are carcinogenic [10,11]. Examples of such degradation byproducts include benzidine (4,4'-diaminobiphenyl) and 2-nitroaniline (Figure 6.9).

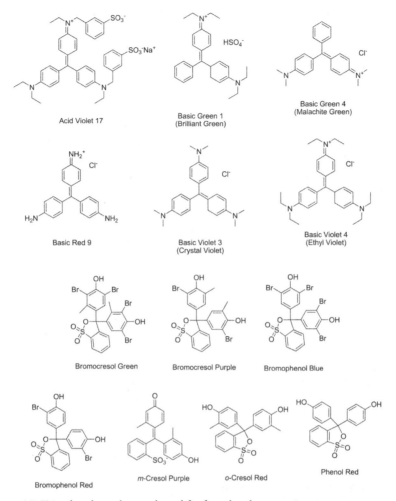

Figure 6.7 Triarylmethane dyes evaluated for fungal and enzymatic treatment.

6.3 Biodegradation of Synthetic Dyes by White Rot Fungi

The ability of white rot fungi, such as *P. chrysosporium* and *T. versicolor*, to decolorize synthetic dye is very well known [22–25]. White rot fungi are one of two groups of wood decaying Basidiomycota along with brown rot fungi. Unlike brown rot fungi that degrade only easily degradable carbohydrate in decaying wood and other plant materials, white rot fungi degrade both lignin and carbohydrate and leave white "bleached" cellulose fibers in

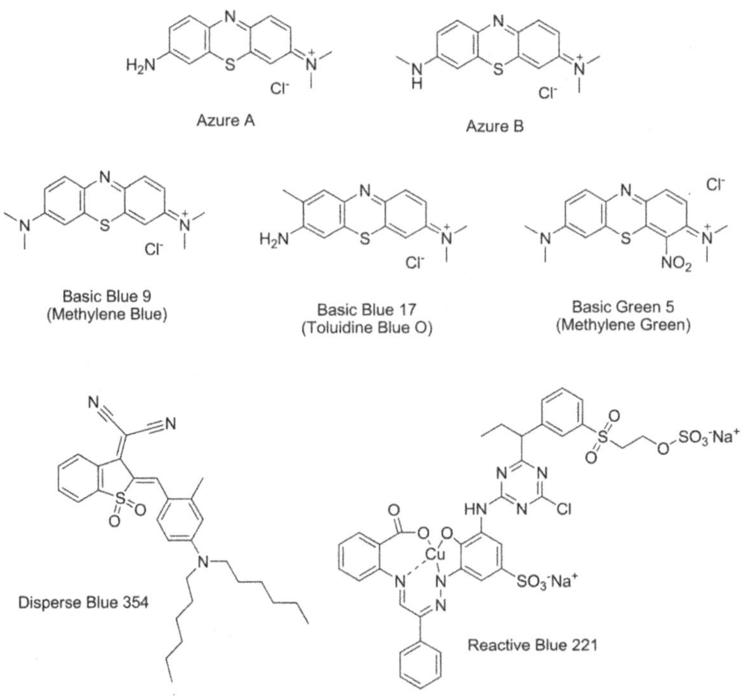

Figure 6.8 Heterocyclic (phenothiazine-structure) dyes and other types of dyes evaluated for fungal and enzymatic treatment.

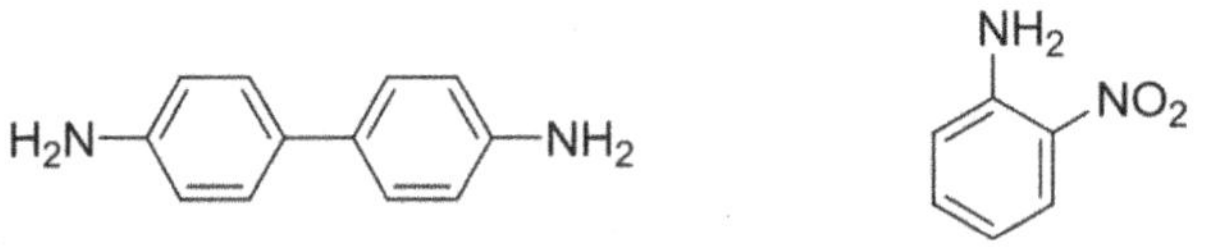

Figure 6.9 Examples of carcinogenic degradation byproducts of azo dyes (including diazo and triazo dyes; left: benzidine, right: 2-nitroaniline).

rotted wood (Figure 6.10). This ability is attributed to secretion of lignin-oxidizing and degrading enzymes, including laccases, lignin peroxidase (ligninases, LiP), and manganese-dependent peroxidase (MnP) by these white rot fungi [26–29]. Lignin is the second most abundant organic polymer composed of aromatic alcohols called monolignols [30,31]. Lignin exerts yellow to dark brown color owing to its very complex, heterogeneous aromatic structure with many crosslinks, as shown in Figure 6.11. It is also known that lignin is highly resistant to bacterial biodegradation and becomes part of natural organic matter found in soil and water.

Because of the structural similarity of synthetic dyes and lignin, a majority of the white rot fungi are capable of decolorizing natural and synthetic

Figure 6.10 White rot fungi (left: Schizoporaceae, right: *Phellinus* sp.). These photos were taken by the author in Edmonton, Alberta, Canada.

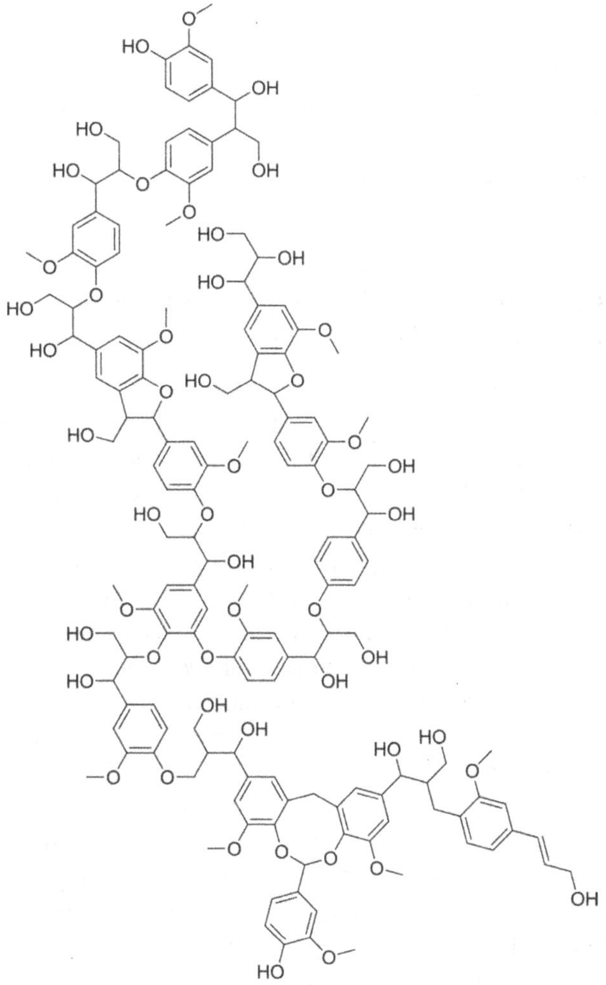

Figure 6.11 One of the structural models of softwood lignin (after Alder [31]).

dyes. In fact, some polymeric dyes such as Poly R-478 (Figure 6.1) and Poly B-411 have been used extensively to study and screen white rot fungi for their ability to degrade lignin, as well as recalcitrant xenobiotics, such as polycyclic hydrocarbons (PAHs) [32–34]. This outstanding ability of white rot fungi to degrade a wide range of recalcitrant organic compounds makes their use in biodegradation and bioremediation very attractive to researchers who look for more environmentally friendly, green treatment technology alternatives [35]. Numerous reports have been published on the use of these fungi for the degradation of pesticides [36,37], chlorinated organics like dioxins and polychlorinated biphenyls (PCBs) [38,39], PAHs [40], pharmaceuticals and endocrine disruptors [41], as well as synthetic dyes [12,22,23]. In this section, research on fungal decolorization of synthetic dyes is briefly reviewed, as this can serve as an excellent introduction to enzymatic treatment. More detailed reviews on this topic can be found elsewhere. For example, earlier works (i.e., prior to 2003) were reviewed by Wesenberg *et al.* [42], Fu and Viraraghavan [43], and Shah and Nerud [14], while more recent studies were reviewed by Asgher *et al.* [44], Kaushik and Malik [45], and Rodríguez-Couto [46].

6.3.1   Earlier Fungal Decolorization Studies with *Phanerochaete chrysosporium*

Since the pioneering work of Glenn and Gold [24] in the early 1980s, a large number of research papers have been published on the decolorization of synthetic dyes using white rot fungi in the last few decades. In their paper published in 1983, Glenn and Gold [24] demonstrated that a model lignin-degrading white rot fungus *P. chrysosporium* could degrade and decolorize three polymeric dyes, namely Poly B-411, Poly R-481, and Poly Y-606. Similar to the cases of lignin biodegradation, they noticed an inhibitory effect of nitrogen in the culture media and suggested the involvement of ligninolytic enzymes in dye decolorization by *P. chrysosporium*. Bumpus and Brock [22] also reported the degradation of another synthetic dye, Basic Violet 3 (crystal violet), and six other triarylmethane dyes including Basic Red 9, cresol red, bromophenol blue, Basic Violet 4, Basic Green 4, and Basic Green 1 by *P. chrysosporium*. Unlike the case of polymeric dyes, crystal violet was degraded in nitrogen-sufficient, non-ligninolytic conditions. Bumpus and Brock [22] also demonstrated that purified LiP from *P. chrysosporium* was able to catalyze *N*-demethylation of Basic Violet 3. Spadaro *et al.* [23] reported the degradation of various azo dyes, such as Disperse Yellow 3 and Disperse Orange 3, by this fungus. Mineralization of substituted aromatic rings was also observed. Cripps *et al.* [47] also showed decolorization of other azo dyes, including Acid Orange 7, Acid Orange 6, and Direct Red 28, as well as an heterocyclic dye Azure B using *P. chrysosporium.*

6.3.2   Other White Rot Fungi

In addition to *P. chrysosporium,* a large number of white rot fungi have been tested for their ability to decolorize synthetic dyes in the 2000s. Such fungi as *T. (Coriolus) versicolor* [25,48–53], *T. hirsuta* [54–56], *Bjerkandera adusta* [25,57–60], *Bjerkandera* sp. [61], *Phanerochaete sordida* [62], *Pleurotus ostreatus* [34,63–66], *P. florida* [55,67], *P. sajor-caju* [68,69], *P. flabellatus* [70], *Pycnoporus sanguineus* [69,71], *Funalia trogii* [52], *Irpex lacteus* [34,72], *I. flavus* [73], *Lentinula edodes* [74,75], *Phlebia* spp. [73], *Ganoderma lucidum* [66], and *Ganoderma* sp. [76] have been evaluated for this purpose. Many authors have reported that some of these fungi, such as *B. adusta* and *P. ostreatus,* showed better performance than *P. chryso-sporium* in synthetic dye degradation and decolorization because of their independence from nitrogen regulation as well as their so-called "hybrid" enzyme (also called versatile peroxidase and "dye-decolorizing peroxidase" (DyP) [77]) that has a wider substrate specificity [61,78]. See Section 6.5 for more description of these enzymes. Several anthraquinone dyes, such as Poly R-478 and Reactive Blue 19, were often used as a standard model dyes to evaluate the decolorization capacity of white rot fungi in many studies [34,51,60,61,73,79], while decolorization of other classes of dyes, such as azo [25,52,59,80–82], phthalocyanine [25], triarylmethane [81], and indigoid dyes [69,81], was also successfully demonstrated.

In addition to the strains that had been maintained in culture collections such as the American Type Culture Collection (ATCC), many new wild strains were isolated more recently in many countries such as India [67,76,80], Pakistan [66], Zimbabwe [83], Tanzania [70], Argentina [84], Brazil [69], and Singapore [51], and were tested for their dye-decolorizing capability. As with other microorganisms, the white rot fungi are distributed globally and universally. Therefore, there is still a possibility of discovering more efficient species and strains that can be used in this application.

6.3.3   Bioreactors and Real Wastewater Treatment

A majority of these fungal decolorization studies were performed *in vitro* or in small batch reactors (i.e., culture flasks) to prove the concept. However, a considerable effort has been made to develop bioreactors for continuous treatment of dye-containing effluents and also treat real wastewater from the textile industry. Zhang *et al.* [85] tested three bioreactors, namely continuous packed-bed bioreactor, fed batch fluidized-bed bioreactor, and continuous fluidized-bed bioreactor, for the decolorization of an azo dye Acid Orange 7 at a rate of 1 g L^{-1} d^{-1} using an unidentified wood-rotting basidiomycota. The working liquid volume of these bioreactors was 1 L. They found that fungal mycelia immobilized by sodium alginate could be reused in the fed batch fluidized-bed bioreactor over a two-month period. Continuous decolorization of Acid Orange 7 (>95%) was achieved in other bioreactors for a period of two months. Mielgo *et al.* [86] constructed a 167-mL capacity continuous packed-bed bioreactor with *P. chrysosporium* that could remove Acid Orange 7 at a dye loading rate of 0.2 g L^{-1} d^{-1}. Polyurethane foam was used as inert support for fungal biomass immobilization. More than 95% of added dye was decolorized at a hydraulic retention time (HRT) of 24 h and a temperature of 37°C. The reactor was fed with 250 mg L^{-1} h^{-1} of glucose and 0.6 mg L^{-1} h^{-1} of ammonia as carbon and nitrogen sources, respectively. The decolorization was improved significantly with purified oxygen gas as compared with air as an oxygen source, which was accompanied by an increased production of MnP.

Kim *et al.* [50] used a microfiltration (MF)-based membrane bioreactor (MBR) with *T. versicolor* to decolorize three reactive dyes including Reactive Blue 19 (anthraquinone), Reactive Blue 49 (anthraquinone), and Reactive Black 5 (diazo). The MF-based MBR was coupled with nanofiltration/reverse osmosis membrane filtration to evaluate the impact of the fungal pretreatment on organics and dye removal. Blánquez *et al.* [48] conducted a study using a fluidized bioreactor with *T. versicolor* to decolorize a mixture of metal complexed dyes called Grey Lanaset G that contains chromium and cobalt. The bioreactor was fluidized by 0.8 L/min of air pulses and was operated at 25°C and pH 4.5 with an HRT of 120 h. They found a good color removal (>90%) using this bioreactor; however, unlike other dye decolorization, the laccase activity did not seem to be related to the decolorization of this metal complexed dye. Rodríguez-Couto *et al.* [56] compared four different carrier materials for the immobilization of *Trametes hirsuta*. They found that stainless steel sponge was the best carrier to improve laccase production (up to 2,200 U/L) and accelerate decolorization of Lanaset Blau as compared with other materials, including alginate beads, polyurethane foam, and nylon sponge, in two batch bioreactors with capacities of 180 mL and 1 L.

A number of research papers have been published that explore the possibility of using white rot fungi to treat real wastewater containing synthetic dyes and related colored byproducts, as reviewed by Rodríguez-Couto [46]. In many cases, the wastewater was supplemented and/or diluted with nutrient solutions, such as culture media containing carbon and nitrogen sources and other trace nutrients. Knapp and Newby [53] investigated the treatment of an effluent from the manufacture of nitrated stilbene sulfonic acids that contain diazo-linked aminostilbene-2,2'-disulfonic acid units with nitro groups using five strains of white rot fungi including *P. chrysosporium*, *T. versicolor*, *P. ostreatus*, and *Piptoporus betulinus*. They found that up to 80% of color removal was achieved by all strains and that a strain of *T. versicolor* showed the best performance. Selvam *et al.* [80] reported somewhat modest decolorization of a dyeing industry effluent using both mycelia of *Thelephora* sp. and purified laccase from this fungus. Up to 61% and 15% of initial color was removed by the mycelia and the purified enzyme in a batch mode, respectively. Nelsson *et al.* [70] treated a real wastewater from a textile industry in Tanzania using a packed bed reactor with natural luffa sponge inoculated with *Pleurotus fabellatus*. The textile wastewater was autoclaved and mixed with a nutrient solution. The pH of the wastewater was adjusted to 3 to 4. The working volume of the bioreactor was 1.5 L and the wastewater was introduced at a flow rate of 59 mL/h that corresponded to an HRT of 25.4 h. Sathiya *et al.* [55] treated a dye house effluent in India with *T. hirsuta* and *Pleurotus florida* in a batch mode. They found that pH adjustment (from 11 to 6) and glucose addition (up to 2%) could improve the decolorization of the effluent by these two fungi. Revankar and Lele [76] and Krishnaveni and Kowsalya [67] reported the decolorization of textile industry effluents in India using *Ganoderma* sp. and *P. florida*, respectively. Shin [72] reported the decolorization of a textile industry effluent in South Korea without addition of any nutrients in shaking or stationary cultures of *I. lacteus*.

6.4   Fungal Decolorization Mechanisms and Involvement of Ligninolytic Enzymes

Virtually all the fungi discussed in Section 6.3 are well-known ligninolytic species that produce high levels of ligninolytic enzymes including laccase, LiP, MnP, and/or manganese-independent peroxidase (versatile peroxidase and DyP) [16], and the major role of those ligninolytic enzymes in dye decolorization has been suggested since the early days [22,24,32,47]. Other significant decolorization mechanisms include sorption to mycelia and biodegradation other than the ligninolytic mechanism [22,43,45], although their contributions are relatively minor. Therefore, many researchers who investigated fungal dye decolorization also measured the activities of those ligninolytic enzymes to explore the dye decolorization mechanisms. Table 6.2 summarizes such studies. It is interesting to note that many of these studies (except for the earlier studies such as Bumpus and Brock [22]) suggested the importance of laccase, MnP, and manganese-independent peroxidase in dye decolorization, while LiP appeared to be less important. The involvement of so-called Remazol Brilliant Blue R (RBBR) oxidase in dye decolorization by certain white rot fungi has been suggested [63]. Because of the requirement of hydrogen peroxide in the catalysis of peroxidase enzymes, hydrogen peroxide-producing oxidases such as aryl-alcohol oxidase and glyoxal oxidase are also important in the fungal decolorization [14,46].

6.5   Classification and Enzymology of Ligninolytic Enzymes

This section briefly describes the classification and enzymology of important ligninolytic enzymes that have been studied for dye decolorization. Those enzymes include LiP, MnP, versatile peroxidase, DyP, and laccase. More detailed literature reviews on the structure, functions, and production of these enzymes can be found elsewhere [16,27,38,97].

6.5.1   Peroxidases

Peroxidases are a group of oxidoreductases that catalyze the reduction of peroxide such as hydrogen peroxide to water and the oxidation of a variety of organic and inorganic compounds [98]. The term peroxidase represents a group of specific enzymes such as NADH peroxidase (EC 1.11.1.1) and glutathione peroxidase (EC 1.11.1.9), and iodide peroxidase (EC 1.11.1.8), as well as a group of nonspecific enzymes. Currently, there are 13 major groups of peroxidase enzymes that can be subdivided into more than 60 classes, according to PeroxiBase (http://peroxibase.toulouse.inra.fr/index.php).

The general catalytic cycle of peroxidases can be described as follows [99,100]:

$$\text{Resting State} + H_2O_2 \rightarrow \text{Compound I} + H_2O \qquad (6.1)$$

$$\text{Compound I} + AH_2 \rightarrow \text{Compound II} + AH\cdot \qquad (6.2)$$

$$\text{Compound II} + AH_2 \rightarrow \text{resting state} + AH\cdot + H_2O \qquad (6.3)$$

Table 6.2 Ligninolytic enzymes involved in fungal dye decolorization studies.

Fungus	Dyes	Enzymes	Notes	Reference
Bjerkandera adusta	Anthraquinone (Reactive Blue 19, Poly R-478), azo (Acid Orange 10, Acid Red 27), phthalocyanine (Pigment Blue 15:2)	Laccase, MnP	Better decolorization than P. chrysosporium and P. ostreatus	[59]
	Anthraquinone (Poly R-478, Natural Red 4)	Horseradish peroxidase (HRP)-like peroxidase, MnP	A mutant that cannot HRP-like peroxidase and MnP showed increased LiP and laccase activities. The mutant showed a lower decolorization efficiency	[58]
	Cibacron Yellow C-2R, Cibacron Red C-2G, Cibacron Blue C-R, Remazol Black B, Remazol Red RB	Laccase, LiP, MnP	High LiP activity, nitrogen did not impact the decolorization, also tested P. ostreatus and T. versicolor	[57]
Clitocybula dusenii	Textile industry effluent	Laccase, MnP	MnP production increased by the textile effluent	[87]
Debaryomyces polymorphus (yeast)	Diazo (Reactive Black 5), Procion Scharlach H-E3G, Procion Marine H-EXL	MnP	Candida tropicalis also showed MnP production and dye decolorization	[88]
Dichomitus squalens	Anthraquinone (Reactive Blue 19), azo (Acid Orange 10)	Laccase, MnP	Second most efficient strain (see I. resinosum and P. calyptratus)	[79]

(Continued)

Table 6.2 (Cont.)

Fungus	Dyes	Enzymes	Notes	Reference
Funalia trogii	Anthraquinone (Reactive Blue 19), Drimarene Blue CL-BR	Laccase	Highest laccase activity	[89]
Geotrichum sp.	Diazo (Reactive Black 5), Reactive Red 158, Reactive Yellow 27	Laccase, MnP, versatile peroxidase	Other fungi (e.g., *B. adusta, Irpex lacteus, T. versicolor*) could not decolorize Reactive Red 158 and Reactive Yellow 27	[90]
Irpex lacteus	Textile industry effluent (Various reactive, acid, metal complex, and disperse dyes)	Laccase, MnP, versatile peroxidase	No LiP or RBBR oxidase activity	[72]
	Anthraquinone (Reactive Blue 19), diazo (Reactive Black 5)	Laccase, MnP	Reactive Black 5 was more resistant, no LiP, laccase and MnP may not be very important	[91]
Ischnoderma resinosum	Anthraquinone (Reactive Blue 19), azo (Acid Orange 10)	Laccase	Most efficient among three strains tested (see *D. squalens* and *P. calyptratus*)	[79]
Lentinula edodes	Anthraquinone (Poly R-478)	Laccase, MnP	Heavy metals inhibited decolorization and MnP production, laccase activity increased	[75]
	Anthraquinone (Poly R-478, Reactive Blue 19), diazo (Direct Red 28), various azo and triarylmethane dyes	MnP	LiP and laccase activities were low	[74]

Organism	Dye	Enzyme	Notes	Ref
Mycena inclinata (Litter-decomposing fungus)	Anthraquinone (Reactive Blue 19), azo (Reactive Orange 16), diazo (Reactive Black 5)	Laccase, MnP	*Collybia dryophila* also effective	[92]
Phanerochaete chrysosporium	Triarylmethanes (e.g., Basic Violet 3)	LiP	*N*-demethylation by purified LiP	[22]
	Azo (Acid Orange 7)	MnP	Purified oxygen improved MnP production and decolorization	[86]
	Azo (Mordant Blue 9) and polyazo (Direct Red 80)	LiP, MnP	Addition of dye stimulated the production of LiP and MnP	[93]
Pleurotus ostreatus	Anthraquinone (Reactive Blue 19), Drimarene Blue CL-BR	Laccase	Both dyes were toxic to *P. ostreatus*	[89]
	Anthraquinone (Reactive Blue 19)	Laccase, MnP, versatile peroxidase	Hydrogen peroxide-dependent RBBR oxidizing enzyme	[63]
P. calyptratus	Anthraquinone (Reactive Blue 19), azo (Acid Orange 10)	Laccase	Least efficient (see *D. squalens* and *I. resinosum*)	[79]
P. sajor-caju	Azo (Acid Red 27, Acid Orange 10, Acid Red 18)	Laccase, MnP	Acid Yellow 23 could not be decolorized, no LiP or veratryl alcohol oxidase activity	[68]
Phlebia tremellosa	Cibacron Yellow C-2R, Cibacron Red C-2G, Cibacron Blue C-R, Remazol Black B, Remazol Red RB	Laccase, LiP, MnP	High LiP and laccase activity, nitrogen did not impact the decolorization, also tested *P. ostreatus* and *T. versicolor*	[57]

(Continued)

Table 6.2 (Cont.)

Fungus	Dyes	Enzymes	Notes	Reference
Trametes (Coriolus) versicolor	Various anthraquinone, azo, tri-arylmethane dyes	Laccase, MnP; versatile peroxidase	Also studied Fome lignosus	[94]
	Anthraquinone (Reactive Blue 19), Drimarene Blue CL-BR	Laccase	60 mg/L of Drimarene Blue was toxic	[89]
	Diazo (Reactive Black 5)	Laccase, MnP	Enzyme activity rapidly decreased under non-sterile conditions	[95]
	Anthraquinone (Poly R-478, Anthraquinone Blue), diazo (Direct Red 28), indigoid (Acid Blue 74), triarylmethane (Basic Green 4)	Laccase	26 strains were evaluated	[84]
T. cingulata	Anthraquinone (Reactive Blue 19), triarylmethane (Cresol Red, Basic Violet 3, bromophenol blue)	Laccase, MnP	Strains of T. versicolor, T. pocas, Datronia concentrica, and Pycnoporus sanguineus were also evaluated.	[83]
T. hirsuta	Lanaset Blau	Laccase	Stainless steel sponge increased laccase production	[56]
Unknown	Twelve dyes (anthraquinone, azo, and diazo)	MnP	Purified MnP can decolorize tested dyes (17% to 98%)	[96]

where AH_2 and AH· represent a substrate and a free radical product, respectively. The oxidized substrate (AH·) undergoes a variety of spontaneous reactions. Each of the nonspecific peroxidase enzymes has different substrate specificity and unique characteristics.

6.5.1.1   Lignin Peroxidase (LiP) and Manganese-Dependent Peroxidase (MnP)

Two archetypical ligninolytic fungal peroxidases, LiP [1,2-bis(3,4-dimethoxyphenyl)propane-1,3-diol:hydrogen-peroxide oxidoreductase; EC 1.11.1.14] and MnP [Mn(II):hydrogen-peroxide oxidoreductase; EC 1.11.1.13], are class II peroxidases in the non-animal peroxidase group that have been discovered and isolated from extracellular culture medium of *P. chrysosporium* [101–103] and many other white rot fungi. Six isozymes of LiP and four isozymes of MnP were purified and sequenced from the extracellular fluid of *P. chrysosporium*. These *P. chrysosporium* LiP and MnP isozymes share 43% amino acid sequence identity and have very similar active site structures around the heme [104,105]. The published crystal structures of LiP and MnP from *P. chrysosporium* are presented in Figure 6.12 [104,106].

The characteristic reactions catalyzed by LiP include nonstereospecific C_α-C_β cleavage and β-O-4 cleavage in lignin model dimers, aromatic ring opening, oxidation of benzyl alcohols such as veratryl alcohol to corresponding aldehydes or ketones and hydroxylation of benzylic methylene groups [103,107,108]. It is known that veratryl alcohol, a secondary metabolite of white rot fungi, acts as a mediator between LiP and polymeric substances that are less accessible to the enzyme's active site [109,110].

As the name suggests, MnP requires Mn(II) ion, as well as hydrogen peroxide, for its catalytic activity. The enzyme catalyzes the oxidation of Mn(II) to Mn(III) which in turn oxidizes various lignin-related organic compounds, including vanillylacetone, 2,6-dimethyloxyphenol, curcumin, syringic acid, guaiacol, syringaldazine, divanillyacetone, and coniferyl alcohol [111–113]. Glenn *et al.* [111] noted that the complex of Mn(III) with α-hydroxy acid, such as lactate, worked as a diffusive intermediate in the oxidative reactions. It is also known that MnP also has an oxidase activity to generate hydrogen peroxide [113].

6.5.1.2   Versatile Peroxidase and Dye-Decolorizing Peroxidase (DyP)

Since the discovery of LiP and MnP in *P. chrysosporium* in the early 1980s, the activity of these ligninolytic enzymes has been found in the

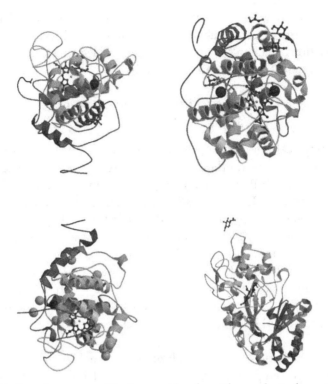

Figure 6.12 Crystal structures of lignin peroxidase from *Phanerochaete chrysosporium* [104] (RCSB PDB Entry: 1LGA, top left), manganese peroxidase from *P. chrysosporium* [106] (RCSB PDB Entry: 3M5Q, top right), versatile peroxidase from *Pleurotus eryngii* [121] (RCSB PDB Entry: 2BOQ, bottom left), and dye-decolorizing peroxidase from *Bjerkandera adusta* [77] (RCSB PDB Entry: 3AFV, bottom right).

extracellular fluid of numerous white rot fungi, including *Pleurotus* spp., *Trametes* (*Coriolus*) spp., and *Bjerkandera* spp. [114]. Later it was found that some isozymes from *P. ostreatus* and *P. eryngii* could exhibit both MnP and LiP activities [115,116]. These enzymes were called by a number of names, such as Mn-independent peroxidase and hybrid peroxidase; however, it is currently described as versatile peroxidase (reactive-black-5:hydrogen-peroxide oxidoreductase; EC 1.11.1.16) [115]. This new enzyme can oxidize a diazo dye Reactive Black 5 (Figure 6.3) that cannot be oxidized by Mn^{3+}, as well as hydroquinones and other dye molecules [116]. Versatile peroxidase is closely related to LiP and MnP and is a member of class II peroxidases in the non-animal peroxidase group.

More recently, another new type of fungal peroxidase called DyP (reactive-blue-5:hydrogen-peroxide oxidoreductase; EC 1.11.1.19) that is able to decolorize a large number of synthetic dyes has been isolated from a plant pathogenic Basidiomycota *Thanatephorus cucumeris* Dec 1 [117]. Interestingly, this *T. cucumeris* Dec 1 was later found to be a white rot fungus *B. adusta*. Although this enzyme was first considered as a class II (fungal) peroxidase because of its origin, the primary structure of the cloned DyP was significantly different from those of other class II peroxidases such as LiP and MnP and those of non-animal peroxidases. Currently, DyP is classified as a member of a separate peroxidase group under heme peroxidase called the DyP-type peroxidase family [98]. DyP also differs from versatile peroxidase because the former enzyme exhibited higher decolorization activities toward anthraquinone dyes such as Reactive Blue 5 than toward azo dyes. In addition to the typical peroxidase reactions, the DyP from *B. adusta* can catalyze the hydrolysis of anthraquinone rings [118,119]. The reported exceptional ability of *B. adusta* and other *Bjerkandera* species on dye decolorization may be due to this DyP enzyme. It is unclear if the previously described high-performance "hybrid" MnP from *B. adusta* [36,116,120] is identical to DyP.

The published crystal structures of versatile peroxidase from *P. eryngii* [121] and DyP from *B. adusta* [77] are shown in Figure 6.12.

6.5.2 Fungal Laccases

Laccases (*p*-diphenol:dioxygen oxidoreductase; EC 1.10.3.2) are copper-containing oxidase enzymes and one of two groups of oxidase enzymes called polyphenol oxidases [122]. Laccases are widely distributed in plant, fungi, and other microorganisms. In particular, laccases are ubiquitous in fungi and have been detected and purified from many species such as white rot fungus *T. versicolor*, edible mushroom *Agaricus bisporus*, and Ascomycota *Neurospora crassa* [16,26,123]. Diphenols such as hydroquinone and catechol are the major substrates of laccases and are oxidized to free radicals by a one-electron reaction catalyzed by laccases. The initial product is usually unstable and may either undergo a second oxidation to quinones or undergo non-enzymatic reactions such as disproportionation and/or polymerization, giving an amorphous insoluble melanin-like product [26]. Laccases are also known to catalyze the oxidation of a range of substituted monophenols, polyphenols, and other aromatic compounds, resulting in demethylation, polymerization and depolymerization of the substrates. A crystal structure of laccase from *T. versicolor* [124] is shown in Figure 6.13.

Although the contribution of laccases to lignin degradation by white rot fungi such as *T. versicolor* has been speculated, its true role in ligninolysis was less clear than those of LiP and MnP, partly because the low

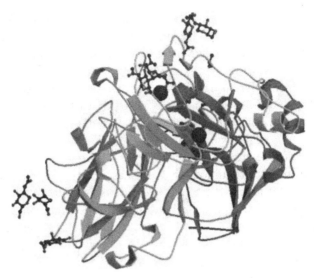

Figure 6.13 Crystal structure of fungal laccase from *Trametes versicolor* [124] (RCSB PDB Entry: 1GYC).

redox potential of laccase enzymes did not seem to be suitable for the oxidation of non-phenolic lignin structures [125]. In addition, not all white rot fungi produce laccases. Various mechanisms have been discovered to mediate the lignin degradation by laccases, such as quinone reduction by glucose oxidase [126], the laccase-MnP system in *Rigidoporus lignosus* [127], and the laccase-veratryl alcohol oxidase interaction in *P. ostreatus* [128]. Bourbonnais and Paice [129] discovered the small-molecule redox mediator system, such as [2,2'-azino-bis(3-ethylbenzothiazoline-6-sulfonic acid); ABTS], that can overcome the barrier of redox potential for oxidizing non-phenolic lignin structures. Some white rot fungi also produce metabolites that act as a mediator like ABTS, such as 3-hydroxyanthranilate produced by *Pycnoporus cinnabarinus* [130]. The use of this mediator system (see Figure 6.14 for the chemical structures of common mediators) has become a very popular technique for the degradation of xenobotics, including PAHs, pesticides, and synthetic dyes, by fungal laccases [15,131–135].

6.6   Enzymatic Treatment of Synthetic Dyes

Virtually all of the fungal decolorization studies reviewed above clearly indicate the importance of ligninolytic enzymes, especially laccase and MnP, in decolorization of synthetic dyes. Thus, it is natural to consider

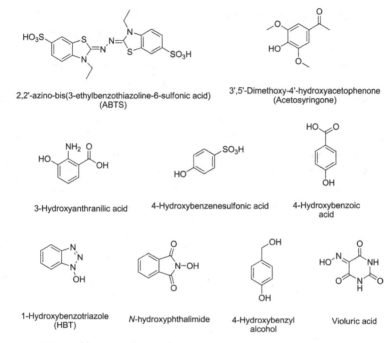

Figure 6.14 Typical laccase redox mediators.

using purified enzymes to investigate the decolorization process further. The use of individually isolated enzymes instead of live fungal biomass is attractive because of the following potential advantages:

- Less likely to be inhibited by toxic dyes
    - Some dyes are toxic to white rot fungi [89]
- Less susceptible to change in effluent quality such as pH, salinity, and temperature
- Can be added at a controlled rate
    - Does not require start-up time
- A faster kinetics
- Can be produced in more controlled, optimized cultivation conditions
    - Textile industry effluent may not be the best culture media for ligninolytic fungi
    - May be more energy efficient
    - Genetically-engineered strains may be used to produce a large quantity of enzymes [114]
- Little to no sludge production

In addition, by using isolated enzymes in a more defined mixture, the reaction mechanisms of synthetic dye decolorization and possible degradation pathways may be explored in a more controlled fashion.

In fact, the use of purified and crude phenol oxidizing enzymes for industrial wastewater treatment has been investigated extensively since the early 1980s using various plant and fungal enzymes, including horseradish peroxidase (HRP), soybean peroxidase (SBP), LiP, MnP, chloroperoxidase, non-ligninolytic fungal peroxidases (such as *Coprinopsis cinerea* peroxidase), laccase, and tyrosinase [15–17,136–140]. These phenol-oxidizing enzymes have broad substrate specificity and are capable of oxidizing many substituted phenols and aromatic amines. Among these phenol oxidizing enzymes, those enzymes secreted by lignin-degrading, white rot fungi such as *P. chrysosporium* are characterized by their exceptional ability to degrade non-phenolic, more complex aromatic compounds of natural or xenobiotic origins as mentioned earlier. Many research papers have been published on the enzymatic treatment of synthetic dyes over the last 15 years. This section reviews and discusses these research works.

6.6.1   Lignin Peroxidases

The literature concerning the use of purified LiP in dye decolorization is limited to several earlier works reported in the early 2000s or before. The requirement of veratryl alcohol for the substantial decolorization activity by LiP might have discouraged the researchers. Ollikka *et al.* [141] reported the decolorization of various azo, triarylmethane, heterocyclic and polymeric dyes by three purified LiP isozymes (0.1 U/mL) from *P. chrysosporium*. Veratryl alcohol (2 mM) was essential to decolorization of most of the dyes tested, but one of the LiP isozymes could decolorize certain dyes such as Acid Orange 52, Basic Blue 9, and Basic Blue 17 without the addition of veratryl alcohol. Young and Yu [142] investigated a LiP-catalyzed oxidation of azo (Acid Violet 7, Acid Orange 74, Reactive Black 5, and Acid Black 24), indigoid (Acid Blue 74), phthalocyanine (Reactive Blue 15), anthraquinone (Acid Green 27 and Acid Blue 25) dyes in the presence of 1 mM of veratryl alcohol. Among the dyes tested, the triazo dye Acid Black 23 was the most resistant to decolorization by the purified LiP, although fungal treatment with *P. chrysosporium* and *T. versicolor* could decolorize this dye by >96%. Ferreira *et al.* [143] reported *N*-demethylation of a heterocyclic dye, Basic Blue 9, that yielded azure B and azure A (see Figure 6.8) using LiP from *P. chrysosporium*. *N*-Demethylation of a triarylmethane dye, Basic Violet 3, with purified LiP from *P. chrysosporium* was also demonstrated by Bumpus and Brock [22] earlier. Verma and Madamwar [144] produced *P. chryso-sporium* LiP using neem hull waste under solid-substrate fermentation conditions and evaluated the use of produced enzyme for decolorization of various dyes, including Reactive Blue 4, Acid Red 119, Acid Black 194, and Ranocid Fast Blue. The highest decolorization was observed at pH 5.0 in all cases. Verma and Madamwar [145] also attempted co-cultivation of *P. chrysosporium* and *P. ostreatus* to improve the enzyme production using the lignocellulosic wastes. In this case, laccase and MnP were also produced and crude enzyme mixtures were prepared and used to decolorize several different dyes. Significant decolorization (73% to 91%) occurred in the absence of hydrogen peroxide using the crude enzyme mixtures.

6.6.2   Manganese-Dependent Peroxidases

Isolated and purified MnP from the culture of white rot fungi has been used to decolorize a number of synthetic dyes. It appears that the use of isolated MnP attracted more attention in the recent years, although the number of reports is still relatively small. Also, the identities of produced enzymes were often ambiguous in these papers at the time of publication. There might be some confusion between MnP and versatile peroxidase and/or DyP.

Moldes *et al.* [146] reported the use of *P. chrysosporium* and its MnP in a fixed-bed bioreactor for the decolorization of an anthraquinone dye, Poly R-478, and a triarylmethane dye, Basic Violet 3. They used crude MnP preparation (approximately 1,000 U/L) produced by ultrafiltration of extra-cellular liquid. By using the continuous addition of MnP, hydrogen peroxide, sodium malonate, and manganese sulfate, they could achieve up to 70% and 30% of Poly R-478 and Basic Violet 3 decolorization, respectively, in two hours. Cheng *et al.* [147] isolated a new MnP from *Schizophyllum* sp. and used the enzyme to decolorize several azo/diazo dyes, including Direct Red 28, Acid Orange 5, and Acid Orange 10. They noticed rapid decolorization of 1 mM of Acid Orange 5 with 20 U/L of purified MnP, 0.1 mM of H_2O_2, and 1 mM of manganese(II) sulfate.

Heinfling *et al.* [148] investigated the decolorization of four azo/diazo dyes (Reactive Violet 5, Reactive Black 5, Reactive Orange 96, and Reactive Red 198) and two phthalocyanine dyes (Reactive Blue 38 and Reactive Blue 15) using MnP from the culture fluids of *B. adusta* and *P. eryngii*. They found that the MnP from *B. adusta* oxidizes those dyes in a manganese independent manner, which was known with MnP isozymes from *Pleurotus* spp. (such as *P. ostreatus* and *P. eryngii*). Shin and Kim [149] described a peroxidase isolated from *P. ostreatus* that can decolorize a variety of triarylmethane and azo dyes, including Basic Violet 3, Basic Green 4, bromophenol blue, Acid Orange 52, Direct Red 28, as well as Poly R-478. Shrivastava *et al.* [150] produced MnP and manganese-independent peroxidase (MIP), which is presumably versatile peroxidase, by *P. ostreatus* in solid-state fermentation and compared their activities on the decolorization of seven related triarylmethane dyes. They found that MnP preferred the dyes that have free phenolic group on the chromophore, such as phenol red, *o*-cresol red and *m*-cresol purple, while MIP could catalyze the oxidation of brominated dyes, such as bromocresol green (see Figure 6.8). Mohorčič *et al.* [151] conducted a fungal strain screening for decolorizing diazo dye Reactive Black 5. Among the 25 strains tested, only two, namely *B. adusta* and *Geotrichum candidum*, could effectively decolorize this diazo dye. Subsequently, they used purified MnP from *B. adusta* to decolorize various azo and anthraquinone dyes, including Reactive Black 5, Acid Red 111, Acid Orange 7, Disperse Yellow 211, Reactive Blue 19, and Disperse Blue 354. Two disperse dyes were apparently more resistant to decolorization.

6.6.3   Laccases

Laccases are by far the most studied enzymes for decolorization of synthetic dyes. This subsection is organized into three parts: treatability studies, kinetics and reaction mechanisms, and toxicity. Many laccase preparations from various white rot fungi, such as *Pleurotus* spp. (*P. osteatus* and *P. sajor-caju*), *Trametes* spp. (*T. versicolor*, *T. hirsuta*, *T. modesta*, and *T. trogii*), *Sclerotium rolfsii*, *L. edodes*, and *Pycnoporus sanguineus*, as well as a laccase produced by genetically modified *Aspergillus niger*, have been evaluated.

6.6.3.1   Treatability Studies

Rodríguez *et al.* [152] performed the screening of 16 strains of white rot fungi for the production of ligninolytic enzymes and decolorization of 23 synthetic dyes. They selected laccases from *P. ostreatus* and *Trametes hispida* for enzymatic treatment. The laccase from *T. hispida* was able to decolorize a wider variety of dyes than that from *P. ostreatus*. Abadulla *et al.* [153] also conducted a screening study to compare laccase preparations from *P. ostreatus*, *Schizophyllum commune*, *N. crassa*, *Polyporus* sp., *S. rolfsii*, *Trametes villosa*, and *Myceliophtora thermophile* for the decolorization of 11 structurally different dyes.

Palmieri *et al.* [154] used *P. ostreatus* and its two laccase isozymes to study the decolorization of Reactive Blue 19. They found that one of the two isozymes (POXA3) was six-fold more efficient than the other isozyme (POXC) in Reactive Blue 19 oxidation and decolorization. Contrary to other researchers' reports (e.g., [149,155]), Palmieri *et al.* noted lesser importance of peroxidases in the decolorization of Reactive Blue 19 by *P. ostreatus*. Hou *et al.* [156] also found laccase as the only ligninolytic enzyme in the culture of *P. ostreatus* (strain 32) and used the purified enzyme for the decolorization of Reactive Blue 19. They used 0.16% ABTS (Figure 6.14) as a redox mediator. Faraco *et al.* [157] also used *P. ostreatus* laccase for the treatment of three model dye-containing wastewaters containing direct dyes, reactive dyes, and acid dyes. The *P. ostreatus* laccase could not decolorize direct or reactive dye mixtures, but acid dye model wastewater was decolorized up to 35%. They found that an azo dye, Acid Yellow 49, was resistant to laccase decolorization, while an anthraquinone dye, Acid Blue 62, was rapidly decolorized. The absence of phenolic hydroxyl or aniline amino group in Acid Yellow 49 (see Figure 6.2) was suspected to be the reason for its resistance toward the laccase-catalyzed decolorization.

Murugesan *et al.* [158] isolated and purified laccase from *P. sajor-caju* and used the purified enzyme for decolorization of three azo dyes, including Acid Red 18 (mono-azo), Acid Black 1 (diazo), and Direct Blue 71 (triazo). The laccase form *P. sajor-caju* could effectively decolorize all the dyes without any addition of redox mediators. Acid Red 18 and Acid Black 1 could be decolorized faster than Direct Blue 71, suggesting that the structural complexity of Direct Blue 71 makes it more resistant to the enzymatic treatment. Involvement of the phenolic group in the laccase-catalyzed reactions is also suggested. Murugestan *et al.* [159] employed the response surface methodology to investigate the decolorization of another diazo dye Reactive Black 5 using purified *P. sajor-caju* laccase. Addition of 1-hydroxybenzotriazole (HBT; see Figure 6.14) as a redox mediator significantly improved the decolorization of this diazo dye.

Wong and Yu [160] studied the decolorization of Acid Green 27 (anthraquinone), Acid Violet 7 (azo), and Acid Blue 74 (indigoid) using purified *T. versicolor* laccase using ABTS as a redox mediator. They found the importance of this mediator for the degradation of non-substrate dyes. The low molecular weight fraction of the *T. versicolor* culture filtrate could also mediate the dye decolorization. Nyanhongo *et al.* [161] screened four ligninolytic fungi, *T. modesta*, *T. versicolor*, *T. hirsuta*, and *S. rolfsii*, for the production of laccase and found that *Trametes modesta* was the promising laccase producer. Two anthraquinone (Reactive Blue 19 and Acid Blue 225), two triarylmethane (Acid Violet 17 and Basic Red 9), two azo (Direct Blue 71 and Reactive Black 5), indigoid (Acid Blue 74), and bifunctional (Reactive Blue 221) dyes were tested. They noted a significant difference in substrate specificity of laccase preparations from these four fungi. For example, laccases from *T. versicolor* and *S. rolfsii* could decolorize triarylmethane dyes better than the ones from *T. modesta* and *T. hirsuta*. The effect of redox mediators was also confirmed. Abadulla *et al.* [54] used purified *T. hirsuta* laccase in an ultrafiltration reactor and immobilized enzyme on alumina in a packed column reactor to decolorize seven dyes, including Reactive Blue 221, Reactive Black 5 (diazo), Direct Blue 71 (triazo), Basic Red 9 (triarylmethane), Reactive Blue 19, Acid Blue 225 (anthraquinone), and Acid Blue 71 (indigoid). They found that anthraquinone dyes and Acid Blue 71 could be decolorized twice as fast as azo dyes. They also found that the laccase immobilization improved its thermal stability and reduced the susceptibility to known laccase inhibitors such as sodium chloride, diethyldithiocarbamate, and sodium fluoride.

Zouari-Mechichi *et al.* [162] purified two isozymes of laccase from a *T. trogii* strain isolated in Tunisia and tested them for decolorizing six dyes (chemical structures unknown, likely acid metal complex dyes). They found that crude and purified enzymes performed similarly in the decolorization test and that certain dyes were resistant to decolorization even in the presence of HBT. Trupkin *et al.* [163] also used a strain of *T. trogii* to evaluate the extracellular enzymes for decolorization of various dyes, including Acid Red 26, Basic Green 4, Acid Blue 74, Reactive Blue 19, Poly R-478, and azure B. They found MnP and laccase in the liquid culture filtrate, but laccase was more predominant. The dye decolorization efficiency was higher when a whole broth with mycelium was used, suggesting that biosorption and other mechanisms are involved. Crude laccase from *T. trogii* was also used to evaluate the decolorization of three anthraquinone (anthraquinone blue, Reactive Blue 19, and Poly R-478), three azo (Direct Red 28, Acid Red 26, and Janus Green), two triarylmethane (Basic Violet 3 and Basic Green 4), one heterocyclic (azure B), and one indigoid (Acid Blue 74) dyes [84]. Poly R-478 and azure B were most resistant to decolorization by the crude *T. troggi* laccase, followed by Janus Green and Direct Red 28. A majority (>80%) of Acid Blue 74 and two anthraquinone dyes could be decolorized within 30 min.

Nagai *et al.* [164] reported the use of purified laccase from *L. edodes* for the decolorization of an azo (Acid Red 2 and Reactive Orange 16), a diazo (Acid Black 1), a triarylmethane (bromophenol blue), and two anthraquinone (Reactive Blue 19 and Poly R-478) dyes. Reactive Orange 16 and Poly R-478 were resistant to the decolorization by *L. edodes* laccase alone, although these dyes could be decolorized in the presence of a

redox mediator. They found that violuric acid was a better redox mediator than HBT for the *L. edodes* laccase. Crude and purified laccase from a
white rot fungus *Ganoderma lucidum* was investigated for the decolorization of Reactive Black 5 and an anthraquinone dye, Reactive Blue 19 [165].
The crude enzyme was able to decolorize Reactive Blue 19 without a redox
mediator, while Reactive Black 5 required a mediator HBT. With 25 U/
mL of crude *G. lucidum* laccase, 50 mg/L of Reactive Black 5 and Reactive
Blue 19 could be decolorized by 77% and 92% in two hours, respectively.
Similarly, purified laccases from *Daedalea quercina* [166] and *Perenniporia
tephropora* [167] were evaluated for the decolorization of various synthetic dyes. Lu *et al.* [168] isolated and purified laccase from the culture of
P. sanguineus and evaluated its use in decolorization of Reactive Blue 19.
Very rapid decolorization of this anthraquinone dye (50 to 200 mg/L) was
observed using 5 U/mL of *P. sanguineus* laccase without a redox mediator.

Soares *et al.* [169] performed a study on Reactive Blue 19 decolorization by a commercial laccase from genetically modified *A. niger*. They
found that in the presence of HBT as a redox mediator, the oxidation of
Reactive Blue 19 could be accelerated significantly. Within 60 min, 100%
of 20 mg/L of Reactive Blue 19 could be decolorized by 10 U/mL of laccase and 0.15% HBT. They also found that certain surfactants (e.g., Triton
X-100 and Tween 20) showed an inhibitory effect on the dye decolorization. In another report, they found that another kind of redox mediator,
violuric acid (Figure 6.14), showed improved efficiency [170].

6.6.3.2   Kinetics and Reaction Mechanisms

Chivukula and Renganathan [171] studied the mechanisms of mono-azo
compound oxidation by laccase from *Pyricularia oryzae* by using synthesized 4-(4'-sulfophenylazo)-phenol derivatives as model azo compounds.
They found that the dimethyl derivatives of the model compound were
the best substrate of *P. oryzae* laccase. Corresponding benzoquinones
and sulfophenylhydroperoxides were identified as degradation products
(Figure 6.15), suggesting the one electron oxidation of phenolic hydroxyl
group, followed by the liberation of benzoquinone and dinitrogen, and the
oxidation of benzensulfonate to peroxide. Similarly, Soares *et al.* [172] synthesized four diazo-model dyes and used them to investigate the impact of
substituents on one of the aromatic rings in the laccase-catalyzed decolorization. They used a commercial laccase formulation produced by genetically modified *A. niger*. Interestingly, without a sulfonate group ($R-SO_3^-$)
this laccase could not decolorize the synthesized dyes with or without a
redox mediator HBT. Those dyes were water insoluble. Violuric acid was

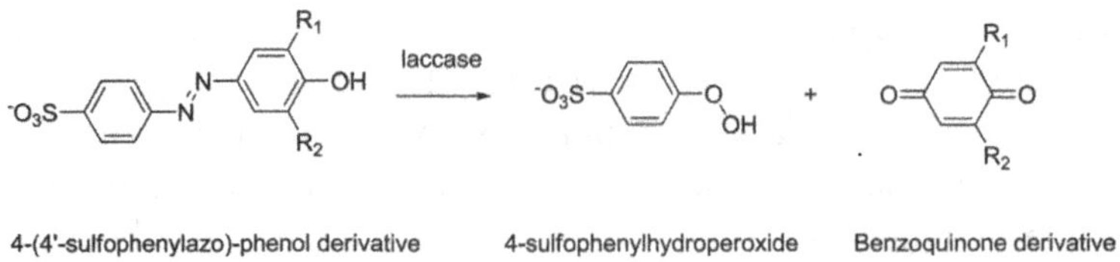

4-(4'-sulfophenylazo)-phenol derivative   4-sulfophenylhydroperoxide   Benzoquinone derivative

Figure 6.15 Degradation of 4-(4'-sulfophenylazo)-phenol derivatives by *Pyricularia oryzae* laccase [171].

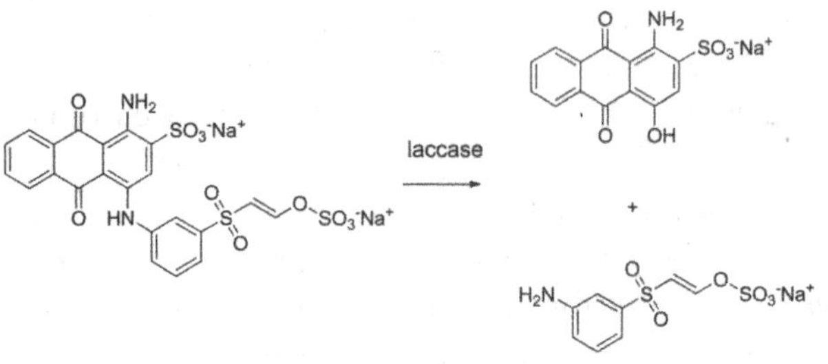

Figure 6.16 Degradation of Reactive Blue 19 by laccase from *Polyporus* sp. [174].

inhibitory to the reaction. One electron oxidation of phenolic group to phenoxy radical followed by the cleavage of C-N bond was suggested, which was proposed by Chivukula and Renganathan [171], but not confirmed experimentally.

Michniewicz *et al.* [173] studied the kinetics of laccase-catalyzed decolorization of two anthraquinone (Acid Blue 40 and Acid Blue 62) and two azo (Reactive Blue 81 and Acid Red 27) dyes and one poly-azo (Direct Black 22) dye using purified enzyme from a white rot fungus *Cerrena unicolor* without redox mediators. Among the dyes tested, Acid Blue 40 was the most preferred toward crude *C. unicolor* laccase based on the Michaelis-Menten constant (K_m). They also found that the decolorization products were more polar than the parent dyes. More recently, Hadibarata *et al.* [174] investigated the degradation pathway of an anthraquinone dye, Reactive Blue 19, by laccase from *Polyporus* sp. and detected two degradation products, which confirmed the cleavage of C-N-C bonds connecting the anthraquinone ring and the benzene ring (Figure 6.16).

Campos *et al.* [175] used purified *T. hirsuta* and *S. rolfsii* laccases for the degradation of water insoluble Vat Blue 1 (indigo), which was dispersed in water. Vat Blue 1 was oxidized by laccase and transformed to isatin (indole-2,3-dione), which was hydrolyzed and further decomposed into anthranilic acid (2-aminobenzoic acid) (Figure 6.17). The indigo particle size was reduced by the laccase treatment as well. They found that the dye could be removed from indigo-stained fabrics using the laccases. The enzyme from

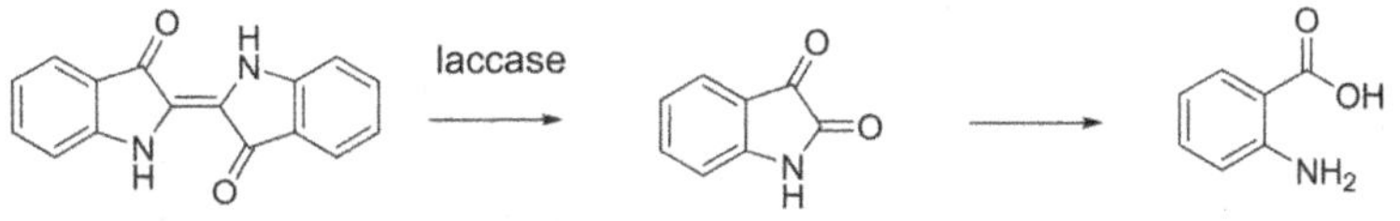

Figure 6.17 Degradation of Vat Blue 1 (indigo) by laccase from *T. hirsuta* and *S. rolfsii* [175].

T. hirsuta was superior to the one from *S. rolfsii* in dye decolorization in both water and from the fabrics. Redox mediators could slightly improve the decolorization of Vat Blue 1 by *T. hirsuta* and *S. rolfsii* laccase.

6.6.3.3 Toxicity

Abadulla *et al.* [54] reported a significant reduction in *Pseudomonas putida* toxicity of synthetic dyes by the enzymatic treatment using *T. hirsuta* laccase. Anthraquinone dyes Reactive Blue 19 and Acid Blue 225 could be detoxified by 84% and 78%, respectively, by the enzymatic treatment, while copper-chelated Reactive Blue 221 (see Figure 6.8) and diazo dye Reactive Black 5 could be detoxified by 4% and 11%, respectively. Faraco *et al.* [157] reported that enzymatic treatment of an acid dye mixture with *P. ostreatus* laccase was less effective in toxicity reduction as compared with fungal treatment with the same fungus. They used Lumistox 300 assay using *Vibrio fisheri* as a toxicity assay.

6.7 Concluding Remarks

Many research papers reviewed in this chapter clearly indicate the strong potential of enzymatic treatment for degradation and decolorization of a wide range of synthetic dyes using fungal enzymes such as laccases and peroxidases. A significant amount of knowledge has been accumulated over the years, such as sources of useful enzymes, treatability of various synthetic dyes with different chemical structures, characteristics of enzymes, and basic kinetics parameters of enzymatic dye decolorization reactions, which is very encouraging. In particular, fungal laccases have been extensively studied in the last 15 years (Table 6.3). However, fungal peroxidases, including LiP, MnP, versatile peroxidase, and DyP, have been tested to a lesser extent. The strongest advantage of this enzymatic treatment over conventional biological processes is its rapid kinetics. The enzymatic treatment can decolorize certain recalcitrant synthetic dyes in less than 1 h [84,170]. A wide substrate specificity of those ligninolytic

Table 6.3 Summary of the enzymatic treatment of synthetic dyes.

Enzyme	Organism	Dye (s)	Note	Reference
Lignin peroxidase (LiP)	*Phanerochaete chrysosporium*	Anthraquinone (Poly R-478, Reactive Blue 19), azo (Acid Orange 52), diazo (Direct Red 28), triarylmethane (bromophenol blue), heterocyclic (Basic Blue 9, Basic Blue 17, Basic Green 5), polymeric (Poly S-119, Poly T-128)	2 mM of veratryl alcohol added, Direct Red 28, Poly R-478, and Poly T-128 was decolorized <54%. Isozyme LiP 3.85 could decolorize certain dyes (e.g., Acid Orange 52) without veratryl alcohol	[141]
		Anthraquinone (Acid Green 27, Acid Blue 25), azo (Acid Violet 7, Acid Orange 74), diazo (Reactive Black 5), triazo (Acid Black 24), indigoid (Acid Blue 74), phthalocyanine (Reactive Blue 15),	1 mM of veratryl alcohol added, pH 3.5–5, Acid Black 24 was difficult to decolorize by LiP, MnP was not effective	[142]
		Heterocyclic (Basic Blue 9, azure B)	*N*-Demethylation of Basic Blue 9 yielded azure B and azure A	[143]
		Triarylmethane (Basic Violet 9)	*N*-Demethylation observed	[22]
		Anthraquinone (Reactive Blue 4), azo (Acid Black 194), diazo (Acid Red 119), Ranocid Fast Blue	2.5 mM of veratryl alcohol added	[144]
		Anthraquinone (Reactive Blue 4), azo (Acid Black 194), diazo (Acid Red 119), Ranocid Fast Blue, and several others	Co-cultivated with *P. ostreatus*, high laccase and MnP activity, decolorization occurred without H_2O_2	[145]

	Schizophyllum commune, S. rolfsii, Neurospora crassa, Polyporus sp.	Anthraquinone (Disperse Red 9, Reactive Blue 19, Acid Blue 225), azo (Acid Orange 5), diazo (Reactive Black 5), triazo (Direct Blue 71), indigoid (Acid Blue 74), triarylmethane (Acid Violet 17, Basic Violet 3), Reactive Blue 221	LiP was a minor constituent of the enzyme preparations, Laccase was predominant. MnP was also present.	[153]
Manganese-dependent peroxidase (MnP)	*Bjerkandera adusta*	Azo (Reactive Violet 5, Reactive Orange 96, Reactive Red 198), diazo (Reactive Black 5), Phthalocyanine (Reactive Blue 15, Reactive Blue 38)	MnP from *B. adusta* oxidizes dyes in a manganese-independent manner similar to *P. eryngii* MnP, maybe DyP or versatile peroxidase	[148]
		Anthraquinone (Reactive Blue 19), azo (Acid Orange 7, Disperse Yellow 211), diazo (Reactive Black 5, Acid Red 111), Disperse Blue 354	*Geotrichum candidum* also decolorized Reactive Black 5 albeit incompletely; maybe DyP? Disperse dyes were more resistant to enzymatic treatment	[151]
	P. chrysosporium	Anthraquinone (Poly R-478), triarylmethane (Basic Violet 3)	70% and 30% decolorization in two hours (continuous enzyme addition)	[146]
	Pleurotus ostreatus	Anthraquinone (Poly R-478), azo (Acid Orange 52), diazo (Direct Red 28), heterocyclic (Basic Blue 9, Basic Blue 17), triarylmethane (Basic Violet 3, Basic Green 4, bromophenol blue)	Probably versatile peroxidase, heterocyclic dyes were more resistant	[149]

(Continued)

Table 6.3 (Cont.)

Enzyme	Organism	Dye (s)	Note	Reference
		Triarylmethane (bromocresol green, bromocresol purple, bromophenol blue, bromophenol red, m-cresol purple, o-cresol red, phenol red)	Manganese-independent peroxidase (versatile peroxidase) prefer brominated dyes (e.g., phenol red)	[150]
	Schizophyllum sp.	Azo (Acid Orange 5, Acid Orange 10), diazo (Direct Red 28),	30% decolorization of 1 mM Acid Orange 5 in 60 min with 20 U/L of purified MnP	[147]
Laccase	Aspergillus niger (genetically modified)	Anthraquinone (Reactive Blue 19)	Violuric acid and HBT accelerated the decolorization, surfactants showed inhibitory effect	[169,170]
		Custom-synthesized diazo dyes	Water insoluble dyes (without sulfonate group) could not be decolorized. HBT used	[172]
	Cerrena unicolor	Anthraquinone (Acid Blue 40, Acid Blue 62), azo (Reactive Blue 81, Acid Red 27), poly-azo (Reactive Black 22)	Enzyme kinetics studied, Acid Blue 40 was the most preferred substrate, no redox mediators used	[173]
	Ganoderma lucidum	Anthraquinone (Acid Black 5), diazo (Acid Blue 19)	Crude enzyme more thermostable, 77% and 92% of Acid Black 5 and Acid Blue 19 decolorized in two hours ([dye]$_0$ = 50 mg/L, laccase dose = 25 U/mL), HBT used	[165]

Daedalea quercina	Anthraquinone (Reactive Blue 19, Reactive Blue 2), diazo dyes (Reactive Black 5, Direct Blue 1), Poly B-411	Reactive Black 5 was more resistant	[166]
Lentinula edodes	Anthraquinone (Reactive Blue 19, Poly R-478), azo (Acid Red 2, Reactive Orange 16), diazo (Acid Black 1), triarylmethane (bromophenol blue)	Poly R-478 and Reactive Orange 16 were resistant, but decolorized in the presence of a mediator, violuric acid was better than HBT	[164]
Perenniporia tephropora	Anthraquinone (Reactive Blue 19), azo (Janus green), triarylmethane (Basic Violet 3, Basic Green 4, bromocresol green, bromophenol red), among others	Reactive Blue 19 was the best substrate, HBT used	[167]
Pleurotus ostreatus	Anthraquinone (Reactive Blue 19)	One of two laccase isozymes was better, little contribution of peroxidases to decolorization	[154]
	Anthraquinone (Reactive Blue 19)	Laccase was the only ligninolytic enzyme detected, 0.16% ABTS was used	[156]

(Continued)

Table 6.3 (Cont.)

Enzyme	Organism	Dye (s)	Note	Reference
		Mixtures of anthraquinone (Acid Blue 62), azo (Reactive Red 195, Reactive Yellow 145, Acid Yello 49, Acid Red 266), diazo (Reactive Blue 222, Reactive Black 5), triazo (Direct Blue 71) poly-azo (Direct Red 80), and Direct Yellow 106	Up to 35% decolorization of acid dye model wastewater in 24 h, Acid Yellow 49 was more resistant, Acid Blue 62 was a good substrate, Acid Red 266 was more resistant, reactive dye and direct dye model wastewaters could not be decolorized, toxicity reduction was less with enzyme alone	[157]
	P. sajor-caju	Azo (Acid Red 18), diazo (Acid Black 1), triazo (Direct Blue 71)	40 μM of all dyes completely decolorized in 24 with 16 U/mL of laccase, no redox mediators used	[158]
	Polyporus sp.	Diazo (Reactive Black 5)	HBT improved the decolorization	[159]
	Polyporus sp.	Anthraquinone (Reactive Blue 19)	Two degradation products detected	[174]
	Pycnoporus sanguineus	Anthraquinone (Reactive Blue 19)	Rapid decolorization (90% within 10 min) without redox mediator	[168]
	Pyricularia oryzae	Custom synthesized seven mono-azo derivatives	Dimethyl derivatives were the best substrates, degradation products identified, degradation pathway proposed	[171]

Sclerotium rolfsii	Indigoid (Vat Blue 1)	Degradation products identified, HBT, acetosyringone, and 4-hydroxy-benzenesulfonic acid were used as mediators	[175]
	Anthraquinone (Disperse Red 60, Disperse Blue 19, Acid Blue 225), azo (Acid Orange 5), diazo (Reactive Black 5), triazo (Direct Blue 71), indigoid (Acid Blue 74), triarylmethane (Acid Violet 17, Basic Violet 3, Basic Red 9), Reactive Blue 221	Reactive Blue 221 was the most effectively decolorized, Acid Blue 225, Acid Orange 5, Reactive Black 5 were more resistant, chelating agent for copper and iron inhibit the laccase activity, laccases from six more fungi (*P. ostreatus, Schizophyllum commune, Neurospora crassa, Polyporus* sp., *Trametes villosa*, and *Myceliophtora thermophila*) were tested	[153]
Trametes versicolor	Anthraquinone (Acid Green 27), azo (Acid Violet 7), indigoid (Acid Blue74)	Acid Green 27 was readily decolorized, others required mediator (low-molecular weight fraction of culture filtrate or ABTS	[160]
T. hirsuta	Anthraquinone (Reactive Blue 19, Acid Blue 225), diazo (Reactive Black 5), triazo (Direct Blue 71), triarylmethane (Basic Red 9), indigoid (Acid Blue 74), Reactive Blue 221	Anthraquinone dyes detoxified, azo dyes were less so, Reactive Blue 221 was still toxic, immobilization improved the stability and less susceptible to inhibitors, reuse of treated water was attempted	[54]

(Continued)

Table 6.3 (Cont.)

Enzyme	Organism	Dye (s)	Note	Reference
	T. hirsuta	Vat Blue 1 (Indigo)	Degradation products identified, HBT, acetosyringone, and 4-hydroxy-benzenesulfonic acid were used as mediators	[175]
	T. hispida	Anthraquinone (Reactive Blue 19), azo (Acid Black 194), Acid Blue 185, among others	Reactive Blue 19 could be decolorized very efficiently by T. hispida laccase, 15 other white rot fungi (B. adusta, P. ostreatus, P. chrysosporium, Sporotichum pulverulentum, and T. versicolor) were also screened	[152]
	T. modesta	Anthraquinone (Reactive Blue 19, Acid Blue 225), azo (Direct Blue 71, Reactive Black 5), and indigoid (Acid Blue 74), triarylmethane (Acid Violet 17, Basic Red 9), Reactive Blue 221	Faster decolorization at 50–60°C, Laccases from T. versicolor, T. hirsuta, and Sclerotium rolfsii were also tested, six small molecules were tested for redox mediators	[161]
	T. trogii	Neolane yellow, Maxilon blue, Neolane pink, Basacryl yellow, Neolane blue, Bezaktiv yellow	Basacryl yellow and Maxilon blue were resistant to decolorization, HBT used	[162]

Anthraquinone (Reactive Blue 19, Poly R-478), azo (Acid Red 26), indigoid (Acid Blue 74), heterocyclic (azure B), triarylmethane (Basic Green 4)	Whole broth was better than culture filtrate, HBT improved the decolorization using the filtrate, MnP may be present	[163]
Anthraquinone (Reactive Blue 19, anthraquinone blue, Poly R-478), azo (Acid Red 26, Janus Green), diazo (Direct Red 28), heterocyclic (azure B), indigoid (Acid Blue 74), triarylmethane (Basic Green 4, Basic Violet 3),	Rapid decolorization of Reactive Blue 19, anthraquinone blue, and Acid Blue 74, Poly R-478 and azure B the most resistant, Janus Green and Direct Red 28 also recalcitrant	[84]

enzymes (including DyP) is particularly attractive for the treatment of textile industry wastewater where a mixture of different dyes and other constituents are present.

Many researchers noted the significant difference in substrate specificity, catalytic activity and efficiency, and stability among the enzymes from different sources (i.e., different species of white rot fungi, different strains, cultivation conditions, etc.) even if they are classified under the same class of enzymes. It is well known that those white rot fungi secrete multiple isozymes of laccases and peroxidases and those isozymes have fairly diverse catalytic properties at different levels, depending on various factors. Although this is certainly very interesting from the researchers' perspective, the inconsistency and irreproducibility in specific isozyme productivity and decolorization activity can be dreadful from the perspective of process engineering and process control. To the best of my knowledge, no industrial-scale wastewater treatment system using this technology is known, probably because of the difficulty of producing consistent quality of usable enzyme to scale up the treatment system.

The high cost associated with enzyme production is also a major drawback of this technology. In addition, many enzymes are not very stable at room temperature, especially in the aqueous phase. Although the use of crude enzyme, such as concentrated culture filtrate, would greatly reduce the cost of enzyme preparation [16,140], the shelf life of such crude enzyme preparations could be very short because of the higher susceptibility to contaminations. In addition, more research is needed to protect the enzymes from inactivation and inhibition by other wastewater constituents and to retain the catalytic activity longer. So far, immobilization of enzymes has been attempted with limited success [54]. Some membrane technologies such as ultrafiltration and nanofiltration may be used [176]. Significant research and development efforts would be required in this area.

Among the five dye-decolorizing enzymes discussed in this chapter, the use of laccases is probably the most attractive as they do not require the addition of hydrogen peroxide as an oxidizing agent. Instead, laccases use dissolved oxygen as an oxidant, which may still need to be supplied by aeration if the wastewater to be treated is anoxic or anaerobic. However, laccases often require a redox mediator such as ABTS and HBT to oxidize more structurally complex dyes, as demonstrated by many researchers [156,159,160]. Those redox mediators are also synthetic organics and are very expensive. In addition, their impact on treated water, receiving water body, and the ecosystem are unclear. It is desirable to develop a laccase-based treatment system that does not require a large dosage of such synthetic mediators.

The utility of LiP in synthetic dye decolorization appears to be limited because of the lack of reliable sources of this enzyme and the requirement of veratryl alcohol as a cofactor and redox mediator for its catalytic activity [110]. It is well known that the secretion of LiP is tightly regulated by the physiological state of white rot fungi such as *P. chrysosporium* and requires nitrogen-limited cultivation conditions for its production [177,178]. In comparison, MnP is more promising because this enzyme can be produced by many white rot fungi without strict physiological regulations. The requirement of manganese ions and organic acids is a disadvantage of MnP, although these are less expensive than veratryl alcohol and laccase-redox mediators. Manganese-independent fungal peroxidases, including versatile peroxidase from *P. eryngii* and *P. ostreatus* and DyP from *B. adusta,* should be investigated more in order to explore the potential applications of these seemingly very useful enzymes for dye decolorization.

It should be noted that information about the characteristics of the degradation byproducts and degradation mechanisms/pathways of enzyme-catalyzed dye decolorization is still largely missing. The identity, toxicity, and biodegradability of decolorized and degraded dye molecules are very important for downstream treatment processes and effluent discharge. Since the enzymatic treatment alone is not capable of mineralizing the dye molecules, an additional treatment process would be needed if the effluent from the enzymatic treatment has high organic matter content. Combinations with different unit processes, such as aerobic/anaerobic processes, ozonation, advanced oxidation processes, membrane processes, should be evaluated to achieve optimum treatment. More research needs to be done to explore the possibility of reusing the treated water for dyeing processes. In the literature reviewed in this chapter, only Abadulla *et al.* [54] investigated the reuse potential of treated wastewater with a limited success.

It is evident that certain synthetic dyes are more susceptible to enzymatic decolorization than others. It is desirable to perform more comprehensive structure activity relationship studies to elucidate the degradation mechanisms and to optimize the treatment. In addition, more research is needed to investigate the applicability of this technology to real textile wastewater treatment. One thing to note is that it appears that most of the researchers, if not all, have used raw dye molecules in their decolorization feasibility studies. However, some dye molecules, especially reactive dyes, undergo a series of chemical reactions with other constituents during the dyeing process as mentioned in Section 6.2.2. Therefore, hydrolyzed reactive dyes should be used and tested, instead of the raw dye products.

Acknowledgements

The author would like to thank Ms. Yao Jin, Ms. Dianne M. Burns, and Mr. Xiaoyan Qu at Pacific Advanced Civil Engineering for their assistance during the preparation of this chapter.

References

1. C. Zaharia, D. Suteu. Textile organic dyes – Characteristics, polluting effects and separation/elimination procedures from industrial effluents – A critical overview. In: T. Puzyn ed. *Organic Pollutants Ten Years after the Stockholm Convention - Environmental and Analytical Update.* Rijeka, Croatia, InTech, pp. 85, 2012.
2. Y. Anjaneyulu, N. Sreedhara Chary, D. S. Suman Raji. Decolourization of industrial effluents – Available methods and emerging technologies – A review. *Rev. Environ. Sci. Biotechnol.,* Vol. 4, pp. 245–273, 2005.
3. I. A. E. Bisschops, H. Sanjers. Literature reivew on textile wastewater characterization. *Environ. Technol.,* Vol. 24, pp. 1399–1411, 2003.
4. Q. Zhou. Chemical pollution and transport of organic dyes in water-soil-crop systems of the Chinese coast. *Bull. Environ. Contam. Toxicol.,* Vol. 66, pp. 784–793, 2001.
5. N. Mathur, P. Bhatnagar, P. Nagar, M. K. Bijarnia. Mutagenicity assessment of effluents from textile/dye industries of Sanganer, Jaipur (India): A case study. *Ecotoxicol. Environ. Safety,* Vol. 61, pp. 105–113, 2005.
6. N. Mathur, P. Bhatnagar, P. Bakre. Assessing mutagenicity of textile dyes from Paji (Rajasthan) using Ames bioassay. *Appl. Ecol. Environ. Res.,* Vol. 4, pp. 111–118, 2005.
7. A. M. Yousafzai, A. R. Khan, A. R. Shakoori. An assessment of chemical pollution in River Kabul and its possible impacts on fisheries. *Pakistan J. Zool.,* Vol. 40, pp. 199–210, 2008.
8. A. Agarwal. Small-scale industries drive India's economy but pollute heavily: What can be done? *Stockholm Waterfront,* Vol. 3, pp. 10–11, 2001.
9. K. Taketoshi. *Environmental Pollution and Policies in China's Township and Village Industrial Enterprises,* Bonn, Germany: Zentrum für Entwicklungsforschung (ZEF), 2001, p. 37.
10. E. J. Weber, N. L. Wolfe. Kinetic studies of the reduction of aromatic azo compounds in anaerobic sediment/water systems. *Environ. Toxicol. Chem.,* Vol. 6, pp. 911–919, 1987.
11. E. A. Clarke, R. Anliker. Organic dyes and pigments. In: I. Hutzinger ed. *The Handbook of Environmental Chemistry, Vol. 3, Part A. Anthropogenic Compounds.* Heidelberg, Germany, Springer-Verlag GmbH, pp. 181–215, 1980.

12. T. Robinson, G. McMullan, R. Marchant, P. Nigam. Remediation of dyes in textile effluent: A critical review on current treatment technologies with a proposed alternative. *Biores. Technol.*, Vol. 77, pp. 247–255, 2001.

13. G. McMullan, C. Meehan, A. Conneely, N. Kirby, T. Robinson, P. Nigam, I. M. Banat, R. Marchant, W. F. Smyth. Microbial decolourisation and degradation of textile dyes. *Appl. Microbiol. Biotechnol.*, Vol. 56, pp. 81–87, 2001.

14. V. Shah, F. Nerud. Lignin degrading system of white-rot fungi and its exploitation for dye decolorization. *Can. J. Microbiol.*, Vol. 48, pp. 857–870, 2002.

15. N. Duran, E. Esposito. Potential applications of oxidative enzymes and phenoloxidase-like compounds in wastewater and soil treatment: A review. *Appl. Catal. B: Environ.*, Vol. 28, pp. 83–99, 2000.

16. K. Ikehata, I. D. Buchanan, D. W. Smith. Recent developments in the production of extracellular fungal peroxidases and laccases for waste treatment. *J. Environ. Eng. Sci.*, Vol. 3, pp. 1–19, 2004.

17. J. Karam, J. A. Nicell. Potential application of enzymes in waste treatment. *J. Chem. Tech. Biotechnol.*, Vol. 69, pp. 141–153, 1997.

18. Society of Dyers and Colourists. *Introduction to the Colour Index: Classification System and Terminology.* West Yorkshire, United Kingdom, 2013, p. 8.

19. A. B. dos Santos, F. J. Cervantes, J. B. van Lier. Azo dye reduction by thermophilic anaerobic granular sludge, and the impact of the redox mediator anthraquinone-2,6-disulfonate (AQDS) on the reductive biochemical transformation. *Appl. Microbiol. Biotechnol.*, Vol. 64, pp. 62–69, 2004.

20. EWA. *Efficient Use of Water in the Textile Finishing Industry.* Brussels, Belgium: European Water Association (EWA), 2005.

21. P. Cooper. *Color in Dyehouse Effluent.* West Yorkshire, United Kingdom, Society of Dyers and Colourists, 1995.

22. J. A. Bumpus, B. J. Brock. Biodegradation of crystal violet by the white rot fungus *Phanerochaete chrysosporium. Appl. Environ. Microbiol.*, Vol. 54, pp. 1143–1150, 1988.

23. J. T. Spadaro, M. H. Gold, V. Renganathan. Degradation of azo dyes by the lignin-degrading fungus *Phanerochaete chrysosporium. Appl. Environ. Microbiol.*, Vol. 58, pp. 2397–2401, 1992.

24. J. K. Glenn, M. H. Gold. Decolorization of several polymeric dyes by the lignin-degrading basidiomycete *Phanerochaete chrysosporium. Appl. Environ. Microbiol.*, Vol. 45, pp. 1741–1747, 1983.

25. A. Heinfling, M. Bergbauer, U. Szewzyk. Biodegradation of azo and phthalocyanine dyes by *Trametes versicolor* and *Bjerkandera adusta. Appl. Microbiol. Biotechnol.*, Vol. 48, pp. 261–266, 1997.

26. C. F. Thurston. The structure and function of fungal laccases. *Microbiology*, Vol. 140, pp. 19–26, 1994.

27. A. T. Martinez. Molecular biology and structure-function of lignin-degrading heme peroxidases. *Enz. Microb. Technol.*, Vol. 30, pp. 425–444, 2002.

28. D. S. Arora, R. K. Sharma. Ligninolytic fungal laccases and their biotechnological applications. *Appl. Biochem. Biotechnol.*, Vol. 160, pp. 1760–1788, 2010.
29. A. Conesa, P. J. Punt, C. A. van den Hondel. Fungal peroxidases: Molecular aspects and applications. *J. Biotechnol.*, Vol. 93, pp. 143–158, 2002.
30. R. Vanholme, B. Demedts, K. Morreel, J. Ralph, Boerjan, W. Lignin biosynthesis and structure. *Plant Physiol.*, Vol. 153, pp. 895–905, 2010.
31. E. Adler. Lignin – Past, present and future. *Wood Sci. Technol.*, Vol. 11, pp. 169–218, 1977.
32. M. H. Gold, J. K. Glenn, M. Alic. Use of polymeric dyes in lignin biodegradation assays. *Methods Enzymol.*, Vol. 161, pp. 74–78, 1988.
33. L. J. Cookson. Reliability of poly B-411, a polymeric anthraquinone-based dye, in determining the rot type caused by wood-inhabiting fungi. *Appl. Environ. Microbiol.*, Vol. 61, pp. 801–803, 1995.
34. Č. Novotný, B. Rawal, M. Bhatt, M. Patel, V. Šašek, H. P. Molitoris. Capacity of *Irpex lacteus* and *Pleurotus ostreatus* for decolorization of chemically different dyes. *J. Biotechnol.*, Vol. 89, pp. 113–122, 2001.
35. S. Pointing. Feasibility of bioremediation by white-rot fungi. *Appl. Microbiol. Biotechnol.*, Vol. 57, pp. 20–33, 2001.
36. G. Davila-Vazquez, R. Tinoco, M. A. Pickard, R. Vazquez-Duhalt. Transformation of halogenated pesticides by versatile peroxidase from *Bjerkandera adusta*. *Enz. Microb. Technol.*, Vol. 36, pp. 223–231, 2005.
37. H. Hirai, S. Nakanishi, T. Nishida. Oxidative dechlorination of methoxychlor by ligninolytic enzymes from white-rot fungi. *Chemosphere*, Vol. 55, pp. 641–645, 2004.
38. M. D. Cameron, S. Timofeevski, S. D. Aust. Enzymology of *Phanerochaete chrysosporium* with respect to the degradation of recalcitrant compounds and xenobiotics. *Appl. Microbiol. Biotechnol.*, Vol. 54, pp. 751–758, 2000.
39. Č. Novotný, B. R. M. Vyas, P. Erbanová, A. Kubátová, V. Šašek. Removal of PCBs by various white rot fungi in liquid cultures. *Folia Biol.*, Vol. 42, pp. 136–140, 1997.
40. J. A. Field, E. De Jong, G. F. Costa, J. A. M. De Bont. Biodegradation of polycyclic aromatic hydrocarbons by new isolates of white rot fungi. *Appl. Environ. Microbiol.*, Vol. 58, pp. 2219, 1992.
41. H. Cabana, J. P. Jones, S. N. Agathos. Elimination of endocrine disrupting chemicals using white rot fungi and their lignin modifying enzymes: A review. *Eng. Life Sci.*, Vol. 7, pp. 429–456, 2007.
42. D. Wesenberg, I. Kyriadides, S. N. Agathos. White-rot fungi and their enzymes for the treatment of industrial dye effluents. *Biotechnol. Adv.*, Vol. 22, pp. 161–187, 2003.
43. Y. Fu, T. Viraraghavan. Fungal decolorization of dye wastewaters: A review. *Biores. Technol.*, Vol. 79, pp. 251–262, 2001.
44. M. Asgher, H. N. Bhatti, M. Ashraf, R. L. Legge. Recent developments in biodegradation of industrial pollutants by white rot fungi and their enzyme system. *Biodegradation*, Vol. 19, pp. 771–783, 2008.

45. P. Kaushik, A. Malik. Fungal dye decolourization: Recent advances and future potential. *Environ. Int.*, Vol. 35, pp. 127–141, 2009.

46. S. Rodríguez-Couto. Treatment of textile wastewater by white-rot fungi: Still a far away reality? Text. *Light Ind. Sci. Technol.*, Vol. 2, pp. 113–119, 2013.

47. C. Cripps, J. A. Bumpus, S. D. Aust. Biodegradation of azo and heterocyclic dyes by *Phanerochaete chrysosporium. Appl. Environ. Microbiol.*, Vol. 56, pp. 1114–1118, 1990.

48. P. Blánquez, N. Casas, X. Font, X. Gabarrell, M. Sarrà, G. Caminal, T. Vincent. Mechanism of textile metal dye biotransformation by *Trametes versicolor. Wat. Res.*, Vol. 38, pp. 2166–2172, 2004.

49. I. Kapdan, F. Kargi, G. McMullan, R. Marchant. Comparison of white-rot fungi cultures for decolorization of textile dyestuffs. *Bioproc. Eng.*, Vol. 22, pp. 347–351, 2000.

50. T.-H. Kim, Y. Lee, J. Yang, B. Lee, C. Park, S. Kim. Decolorization of dye solutions by a membrane bioreactor (MBR) using white-rot fungi. *Desalination*, Vol. 168, pp. 287–293, 2004.

51. Y.-C. Toh, J. J. L. Yen, J. P. Obbard, Y.-P. Ting. Decolourisation of azo dyes by white-rot fungi (WRF) isolated in Singapore. *Enz. Microb. Technol.*, Vol. 33, pp. 569–575, 2003.

52. M. Sam, O. Yeşilada. Decolorization of orange II dye by white-rot fungi. *Folia Biol.*, Vol. 46, pp. 143–145, 2001.

53. J. S. Knapp, P. S. Newby. The decolourisation of a chemical industry effluent by white rot fungi. *Wat. Res.*, Vol. 33, pp. 575–577, 1999.

54. E. Abadulla, T. Tzanov, S. Costa, K.-H. Robra, A. Cavaco-Paulo, G. M. Gübitz. Decolorization and Detoxification of Textile Dyes with a Laccase from *Trametes hirsuta. Appl. Environ. Microbiol.*, Vol. 66, pp. 3357–3362, 2000.

55. M. Sathiya, S. Periyar, A. Sasikalaveni, K. Murugesan, P. T. Kalaichelvan. Decolorization of textile dyes and their effluents using white rot fungi. *Afr. J. Biotechnol.*, Vol. 6, pp. 424–429, 2007.

56. S. Rodríguez-Couto, M. A. Sanromán, D. Hofer, G. M. Gübitz. Stainless steel sponge: A novel carrier for the immobilisation of the white-rot fungus *Trametes hirsuta* for decolourization of textile dyes. *Biores. Technol.*, Vol. 95, pp. 67–72, 2004.

57. T. Robinson, B. Chandran, P. Nigam. Studies on the production of enzymes by white-rot fungi for the decolourisation of textile dyes. *Enz. Microb. Technol.*, Vol. 29, pp. 575–579, 2001.

58. T. Korniłłowicz-Kowalska, K. Rybczyńska. Anthraquinone dyes decolorization capacity of anamorphic *Bjerkandera adusta* CCBAS 930 strain and its HRP-like negative mutants. *World J. Microbiol. Biotechnol.*, Vol. 30, pp. 1725–1736, 2014.

59. I. Eichlerová, L. Homolka, F. Nerud. Decolorization of high concentrations of synthetic dyes by the white rot fungus *Bjerkandera adusta* strain CCBAS 232. *Dyes Pigments*, Vol. 75, pp. 38–44, 2007.

60. I. Eichlerová, L. Homolka, O. Benada, O. Kofroňová, T. Hubálek, F. Nerud. Decolorization of Orange G and Remazol Brilliant Blue R by the white rot fungus *Dichomitus squalens:* Toxicological evaluation and morphological study. *Chemosphere*, Vol. 69, pp. 795–802, 2007.

61. P. R. Moreira, E. Almeida-Vara, G. Sena-Martins, I. Polónia, F. X. Malcata, J. C. Duarte. Decolourisation of Remazol Brilliant Blue R via a novel *Bjerkandera* sp. *strain*. *J. Biotechnol.*, Vol. 89, pp. 107–111, 2001.

62. K. Harazono, K. Nakamura. Decolorization of mixtures of different reactive textile dyes by the white-rot basidiomycete *Phanerochaete sordida* and inhibitory effect of polyvinyl alcohol. *Chemosphere*, Vol. 59, pp. 63–68, 2005.

63. B. R. M. Vyas, H. P. Molitoris. Involvement of an extracellular H_2O_2-dependent ligninolytic activity of the white rot fungus *Pleurotus ostreatus* in the decolorization of Remazol brilliant blue R. *Appl. Environ. Microbiol.*, Vol. 61, pp. 3919–3927, 1995.

64. X. Zhao, I. R. Hardin. HPLC and spectrophotometric analysis of biodegradation of azo dyes by *Pleurotus ostreatus*. *Dyes Pigments*, Vol. 73, pp. 322–325, 2007.

65. X. Zhao, I. R. Hardin, H.-M. Hwang. Biodegradation of a model azo disperse dye by the white rot fungus *Pleurotus ostreatus*. *Int. Biodeter. Biodegr.*, Vol. 57, pp. 1–6, 2006.

66. M. Asgher, S. A. H. Shah, M. Ali, R. L. Legge. Decolorization of some reactive textile dyes by white rot fungi isolated in Pakistan. *World J. Microbiol. Biotechnol.*, Vol. 22, pp. 89–93, 2006.

67. M. Krishnaveni, R. Kowsalya. Characterization and decolorization of dye and textile effluent by laccase from *Pleurotus florida* – A white-rot fungi. *Int. J. Pharm. Bio Sci.*, Vol. 2, pp. B117-B123, 2011.

68. E. P. Chagas, L. R. Durrant. Decolorization of azo dyes by *Phanerochaete chrysosporium* and *Pleurotus sajorcaju*. *Enz. Microb. Technol.*, Vol. 29, pp. 473–477, 2001.

69. D. S. L. Balan, R. T. R. Monteiro. Decolorization of textile indigo dye by ligninolytic fungi. *J. Biotechnol.*, Vol. 89, pp. 141–145, 2001.

70. I. Nilsson, A. Mölle, B. Mattiasson, M. S. T. Rubindamayugi, U. Welander. Decolorization of synthetic and real textile wastewater by the use of white-rot fungi. *Enz. Microb. Technol.*, Vol. 38, pp. 94–100, 2006.

71. S. Pointing, L. L. P. Vrijmoed. Decolorization of azo and triphenylmethane dyes by *Pycnoporus sanguineus* producing laccase as the sole phenoloxidase. *World J. Microbiol. Biotechnol.*, Vol. 16, pp. 317–318, 2000.

72. K.-S. Shin. The role of enzymes produced by white-rot fungus *Irpex lacteus* in the decolorization of the textile industry effluent. *J. Microbiol.*, Vol. 42, pp. 37–41, 2004.

73. M. Chander, D. S. Arora, H. K. Bath. Biodecolourisation of some industrial dyes by white-rot fungi. *J. Ind. Microbiol. Biotechnol.*, Vol. 31, pp. 94–97, 2004.

74. C. G. Boer, L. Obici, C. G. M. de Souza, R. M. Peralta. Decolorization of synthetic dyes by solid state cultures of *Lentinula (Lentinus) edodes* producing

manganese peroxidase as the main ligninolytic enzyme. *Biores. Technol.*, Vol. 94, pp. 107–112, 2004.

75. N. Hatvani, I. Mécs. Effects of certain heavy metals on the growth, dye decolorization, and enzyme activity of *Lentinula edodes*. *Ecotoxicol. Environ. Safety*, Vol. 55, pp. 199–203, 2003.

76. M. S. Revankar, S. S. Lele. Synthetic dye decolorization by white rot fungus, *Ganoderma* sp. *WR-1*. *Biores. Technol.*, Vol. 98, pp. 775–780, 2007.

77. T. Yoshida, H. Tsuge, H. Konno, T. Hisabori, Y. Sugano. The catalytic mechanism of dye-decolorizing peroxidase DyP may require the swinging movement of an aspartic acid residue. *FEBS J.*, Vol. 278, pp. 2387–2394, 2011.

78. Y. Wang, R. Vazquez-Duhalt, M. A. Pickard. Purification, characterization, and chemical modification of manganese peroxidase from *Bjerkandera adusta* UAMH 8258. *Curr. Microbiol.*, Vol. 45, pp. 77–87, 2002.

79. I. Eichlerová, L. Homolka, L. Lisá, F. Nerud. Orange G and Remazol Brilliant Blue R decolorization by white rot fungi *Dichomitus squalens, Ischnoderma resinosum* and *Pleurotus calyptratus*. *Chemosphere*, Vol. 60, pp. 398–404, 2005.

80. K. Selvam, K. Swaminathan, K.-S. Chae. Decolourization of azo dyes and a dye industry effluent by a white rot fungus *Thelephora* sp. *Biores. Technol.*, Vol. 88, pp. 115–119, 2003.

81. J. S. Knapp, P. S. Newby, L. P. Reece. Decolorization of dyes by wood-rotting basidiomycete fungi. *Enz. Microb. Technol.*, Vol. 17, pp. 664–668, 1995.

82. P. Sivasakthivelan. Decolorization of textile dyes and their effluents using white rot fungi. *Int. J. ChemTech Res.*, Vol. 5, pp. 1309–1312, 2013.

83. M. Tekere, A. Y. Mswaka, R. Zvauya, J. S. Read. Growth, dye degradation and ligninolytic activity studies on Zimbabwean white rot fungi. *Enz. Microb. Technol.*, Vol. 28, pp. 420–426, 2001.

84. L. Levin, E. Melignani, A. M. Ramos. Effect of nitrogen sources and vitamins on ligninolytic enzyme production by some white-rot fungi. Dye decolorization by selected culture filtrates. *Biores. Technol.*, Vol. 101, pp. 4554–4563, 2010.

85. F.-M. Zhang, J. S. Knapp, K. N. Tapley. Development of bioreactor systems for decolorization of Orange II using white rot fungus. *Enz. Microb. Technol.*, Vol. 24, pp. 48–53, 1999.

86. I. Mielgo, M. T. Moreira, G. Feijoo, J. M. Lema. A packed-bed fungal bioreactor for the continuous decolourisation of azo-dyes (Orange II). *J. Biotechnol.*, Vol. 89, pp. 99–106, 2001.

87. D. Wesenberg, F. Buchon, S. N. Agathos. Degradation of dye-containing textile effluent by the agaric white-rot fungus *Clitocybula dusenii*. *Biotechnol. Lett.*, Vol. 24, pp. 989–993, 2002.

88. Q. Yang, M. Yang, K. Pritsch, A. Yediler, A. Hagn, M. Schloter, A. Kettrup. Decolorization of synthetic dyes and production of manganese-dependent peroxidase by new fungal isolates. *Biotechnol. Lett.*, Vol. 25, pp. 709–713, 2003.

89. E. A. Erkurt, A. Ünyayar, H. Kumbur. Decolorization of synthetic dyes by white rot fungi, involving laccase enzyme in the process. *Proc. Biochem.*, Vol. 42, pp. 1429–1435, 2007.

90. C. Máximo, M. T. P. Amorim, M. Costa-Ferreira. Biotransformation of industrial reactive azo dyes by *Geotrichum* sp. *CCMI 1019*. *Enz. Microb. Technol.*, Vol. 32, pp. 145–151, 2003.

91. C. Máximo, M. Costa-Ferreira. Decolourisation of reactive textile dyes by *Irpex lacteus* and lignin modifying enzymes. *Proc. Biochem.*, Vol. 39, pp. 1475–1479, 2004.

92. P. Baldrian, J. Šnajdr. Production of ligninolytic enzymes by litter-decomposing fungi and their ability to decolorize synthetic dyes. *Enz. Microb. Technol.*, Vol. 39, pp. 1023–1029, 2006.

93. S. Singh, K. Pakshirajan. Enzyme activities and decolourization of single and mixed azo dyes by the white-rot fungus *Phanerochaete chrysosporium*. *Int. Biodeter. Biodegr.*, Vol. 64, pp. 146–150, 2010.

94. W. Liu, Y. Chao, X. Yang, H. Bao, S. Qian. Biodecolorization of azo, anthraquinonic and triphenylmethane dyes by white-rot fungi and a laccase-secreting engineered strain. *J. Ind. Microbiol. Biotechnol.*, Vol. 31, pp. 127–132, 2004.

95. J. A. Libra, M. Borchert, S. Banit. Competition strategies for the decolorization of a textile-reactive dye with the white-rot fungi *Trametes versicolor* under non-sterile conditions. *Biotechnol. Bioeng.*, Vol. 82, pp. 736–744, 2003.

96. H.-R. Kariminiaae-Hamedaani, A. Sakurai, M. Sakakibara. Decolorization of synthetic dyes by a new manganese peroxidase-producing white rot fungus. *Dyes Pigments*, Vol. 72, pp. 157–162, 2007.

97. R. ten Have, P. J. M. Teunissen. Oxidative mechanisms involved in lignin degradation by white-rot fungi. *Chem. Rev.*, Vol. 101, pp. 3397–3414, 2001.

98. D. Koua, L. Cerutti, L. Falquet, C. J. A. Sigrist, G. Theiler, N. Hulo, C. Dunand. PeroxiBase: A database with new tools for peroxidase family classification. *Nucleic Acid Res.*, Vol. 37, pp. D261-D266, 2009.

99. H. B. Dunford. *Heme Peroxidases.* New York, NY, John Wiley & Sons, Inc., 1999.

100. T. L. Poulos. Peroxidases. *Curr. Opin. Biotechnol.*, Vol. 4, pp. 484–489, 1993.

101. M. Kuwahara, J. K. Glenn, M. A. Morgan, M. H. Gold. Separation and characterization of two extracellular H_2O_2-dependent oxidases from ligninolytic cultures of *Phanerochaete chrysosporium*. *FEBS Lett.*, Vol. 169, pp. 247–250, 1984.

102. M. Tien, T. K. Kirk. Lignin-degrading enzyme from the hymenomycete *Phanerochaete chrysosporium* Burds. *Science*, Vol. 221, pp. 661–663, 1983.

103. M. Tien, T. K. Kirk. Lignin-degrading enzyme from *Phanerochaete chrysosporium*: Purification, characterization, and catalytic properties of a unique H_2O_2-requiring oxygenase. *Proc. Natl. Acad. Sci. USA*, Vol. 81, pp. 2280–2284, 1984.

104. T. L. Poulos, S. L. Edwards, H. Wariishi, M. H. Gold. Crystallographic refinement of lignin peroxidase at 2 A. *J. Biol. Chem.*, Vol. 268, pp. 4429–4440, 1993.

105. M. Sundaramoorthy, K. Kishi, M. H. Gold, T. L. Poulos. The crystal structure of manganese peroxidase from Phanerochaete chrysosporium at 2.06-A resolution. *J. Biol. Chem.*, Vol. 269, pp. 32759–32767, 1994.

106. M. Sundaramoorthy, M. H. Gold, T. L. Poulos. Ultrahigh (0.93 A) resolution structure of manganese peroxidase from *Phanerochaete chrysosporium*: Implications for the catalytic mechanism. *J. Inorg. Biochem.*, Vol. 104, pp. 683–690, 2010.

107. T. K. Kirk, M. Tien, P. J. Kersten, M. D. Mozuch, B. Kalyanaraman. Ligninase of *Phanerochaete chrysosporium* mechanism of its degradation of the nonphenolic arylglycerol b-aryl ether substrate of lignin. *Biochem. J.*, Vol. 236, pp. 279–287, 1986.

108. T. Umezawa, T. Higuchi. Cleavages of aromatic ring and β-O-4 bond of synthetic lignin (DHP) by lignin peroxidase. *FEBS Lett.*, Vol. 242, pp. 325–329, 1989.

109. P. J. Harvey, J. M. Palmer, H. E. Schoemaker, H. L. Dekker, R. Wever. Presteady-state kinetic study on the formation of compound I and II of ligninase. *Biochim. Biophys. Acta*, Vol. 994, pp. 59–63, 1989.

110. P. J. Harvey, H. E. Schoemaker, J. M. Palmer. Veratryl alcohol as a mediator and the role of radical cations in lignin biodegradation by *Phanerochaete chrysosporium*. *FEBS Lett.*, Vol. 195, pp. 242–246, 1986.

111. J. K. Glenn, L. A. Aran, M. H. Gold. Mn (II) oxidation is the principal function of the extracellular Mn-peroxidase from *Phanerochaete chrysosporium*. *Arch. Biochem. Biophys.*, Vol. 251, pp. 688–696, 1986.

112. A. Paszczynski, V.-B. Huynh, R. Crawford. Enzymatic activities of an extracellular, manganese-dependent peroxidase from *Phanerochaete chrysosporium*. *FEMS Microbiol. Lett.*, Vol. 29, pp. 37–41, 1985.

113. A. Paszczynski, V.-B. Huynh, R. Crawford. Comparison of ligninase-I and peroxidase-M2 from the white-rot fungus *Phanerochaete chrysosporium*. *Arch. Biochem. Biophys.*, Vol. 244, pp. 750–765, 1986.

114. K. Ikehata, I. D. Buchanan, D. W. Smith. Production of enzymes for environmental applications. A review. In: A. Sakurai ed. *Wastewater Treatment Using Enzymes*. Kerala, India, Research Signpost, pp. 1–40, 2003.

115. S. Camarero, S. Sarkar, F. J. Ruiz-Dueñas, M. J. Martínez, A. T. Martínez. Description of a versatile peroxidase involved in the natural degradation of lignin that has both manganese peroxidase and lignin peroxidase substrate interaction sites. *J. Biol. Chem.*, Vol. 274, pp. 10324–10330, 1999.

116. A. Heinfling, F. J. Ruiz-Dueñas, M. J. Martínez, M. Bergbauer, U. Szewzyk, A. T. Martínez. A study on reducing substrates of manganese-oxidizing peroxidases from Pleurotus eryngii and Bjerkandera adusta. *FEBS Lett.*, Vol. 428, pp. 141–146, 1998.

117. Y. Sugano. DyP-type peroxidases comprise a novel heme peroxidase family. *Cell. Mol. Life Sci.*, Vol. 66, pp. 1387–1403, 2009.

118. Y. Sugano, Y. Matsushima, K. Tsuchiya, H. Aoki, M. Hirai, M. Shoda. Degradation pathway of an anthraquinone dye catalyzed by a unique

peroxidase DyP from *Thanatephorus cucumeris* Dec 1. *Biodegradation*, Vol. 20, pp. 433–440, 2009.

119. Y. Sugano, R. Muramatsu, A. Ichiyanagi, T. Sato, M. Shoda. DyP, a unique dye-decolorizing peroxidase, represents a novel heme peroxidase family: ASP171 replaces the distal histidine of classical peroxidases. *J. Biol. Chem.*, Vol. 282, pp. 36652–36658, 2007.

120. R. Pogni, M. C. Baratto, S. Giansanti, C. Teutloff, J. Verdin, B. Valderrama, F. Lendzian, W. Lubitz, R. Vazquez-Duhalt, R. Basosi. Tryptophan-based radical in the catalytic mechanism of versatile peroxidase from *Bjerkandera adusta*. *Biochemistry*, Vol. 44, pp. 4267–4274, 2005.

121. M. Pérez-Boada, F. J. Ruiz-Dueñas, R. Pogni, R. Basosi, T. Choinowski, M. J. Martínez, K. Piontek, A. T. Martínez. Versatile peroxidase oxidation of high redox potential aromatic compounds: Site-directed mutagenesis, spectroscopic and crystallographic investigation of three long-range electron transfer pathways. *J. Mol. Biol.*, Vol. 354, pp. 385–402, 2005.

122. E. I. Solomon, U. M. Sundaram, T. E. Machonkin. Multicopper oxidases and oxygenases. *Chem. Rev.*, Vol. 96, pp. 2563–2605, 1996.

123. J.-M. Bollag, A. Leonowicz. Comparative studies of extracellular fungal laccases. *Appl. Environ. Microbiol.*, Vol. 48, pp. 849–854, 1984.

124. K. Piontek, M. Antorini, T. Choinowski. Crystal structure of a laccase from the fungus *Trametes versicolor* at 1.90-A resolution containing a full complement of coppers. *J. Biol. Chem.*, Vol. 277, pp. 37663–37669, 2002.

125. C. Eggert, U. Temp, K.-E. L. Eriksson. Laccase is essential for lignin degradation by the white-rot fungus *Pycnoporus cinnabarinus*. *FEBS Lett.*, Vol. 407, pp. 89–92, 1997.

126. G. Szklarz, A. Leonowicz. Cooperation between fungal laccase and glucose oxidase in the degradation of lignin derivatives. *Phytochemistry*, Vol. 25, pp. 2537–2539, 1986.

127. H. Galliano, G. Gas, J. L. Seris, A. M. Boudet. Lignin degradation by *Rigidoporus lignosus* involves synergistic action of two oxidizing enzymes: Mn peroxidase and laccase. *Enz. Microb. Technol.*, Vol. 13, pp. 478–482, 1991.

128. L. Marzullo, R. Cannio, P. Giardina, M. T. Santini, G. Sannia. Veratryl alcohol oxidase from *Pleurotus ostreatus* participates in lignin biodegradation and prevents polymerization of laccase-oxidized substrates. *J. Biol. Chem.*, Vol. 270, pp. 3823–3827, 1995.

129. R. Bourbonnais, M. G. Paice. Oxidation of non-phenolic substrates. An expanded role for laccase in lignin biodegradation. *FEBS Lett.*, Vol. 267, pp. 99–102, 1990.

130. C. Eggert, U. Temp, J. F. D. Dean, K.-E. L. Eriksson. A fungal metabolite mediates degradation of non-phenolic lignin structures and synthetic lignin by laccase. *FEBS Lett.*, Vol. 391, pp. 144–148, 1996.

131. R. Bourbonnais, D. Leech, M. G. Paice. Electrochemical analysis of the interactions of laccase mediators with lignin model compounds. *Biochim. Biophys. Acta*, Vol. 1379, pp. 381–390, 1998.

132. C. Johannes, A. Majcherczyk. Natural mediators in the oxidation of poly-cyclinc aromatic hydrocarbons by laccase mediator systems. *Appl. Environ. Microbiol.*, Vol. 66, pp. 524–528, 2000.

133. E. Srebotnik, K. E. Hammel. Degradation of nonphenolic lignin by the laccase/1-hydroxybenzotriazole system. *J. Biotechnol.*, Vol. 81, pp. 179–188, 2000.

134. P. J. Collins, M. J. J. Kotterman, J. A. Field, A. D. W. Dobson. Oxidation of anthracene and benzo[a]pyrene by laccases from *Trametes versicolor. Appl. Environ. Microbiol.*, Vol. 62, pp. 4563–4567, 1996.

135. M. A. Pickard, R. Roman, R. Tinoco, R. Vazquez-Duhalt. Polycyclic aromatic hydrocarbon metabolism by white rot fungi and oxidation by *Coriolopsis gallica* UAMH 8260 laccase. *Appl. Environ. Microbiol.*, Vol. 65, pp. 3805–3809, 1999.

136. K. Ikehata, J. A. Nicell. Characterization of tyrosinase for the treatment of aqueous phenols. *Biores. Technol.*, Vol. 74, pp. 191–199, 2000.

137. M. D. Aitken, R. Venkatadri, R. L. Irvine. Oxidation of phenolic pollutant by a lignin degrading enzyme from the white-rot fungus *Phanerochaete chrysosporium. Wat. Res.*, Vol. 23, pp. 443–450, 1989.

138. S. C. Atlow, L. Bonadonna-Aparo, A. M. Klibanov. Dephenolization of industrial wastewaters catalyzed by polyphenol oxidase. *Biotechnol. Bioeng.*, Vol. 26, pp. 599–603, 1984.

139. A. M. Klibanov, B. N. Alberti, E. D. Morris, L. M. Felshin. Enzymatic removal of toxic phenols and anilines from waste waters. *J. Appl. Biochem.*, Vol. 2, pp. 414–421, 1980.

140. K. Ikehata, I. D. Buchanan, D. W. Smith. Treatment of oil refinery wastewater using crude *Coprinus cinereus* peroxidase and hydrogen peroxide. *J. Environ. Eng. Sci.*, Vol. 2, pp. 463–472, 2003.

141. P. Ollikka, K. Alhonmäki, V.-M. Leppänen, T. Glumoff, T. Raijola, I. Suominen. Decolorization of azo, triphenyl methane, heterocyclic, and poly-meric dyes by lignin peroxidase isoenzymes from *Phanerochaete chrysosporium. Appl. Environ. Microbiol.*, Vol. 59, pp. 4010–4016, 1993.

142. L. Young, J. Yu. Ligninase-catalysed decolorization of synthetic dyes. *Wat. Res.*, Vol. 31, pp. 1187–1193, 1997.

143. V. S. Ferreira, D. B. Magalhães, S. H. Kling, J. G. da Silva Júnior, E. P. S. Bon. N-Demethylation of methylene blue by lignin peroxidase from *Phanerochaete chrysosporium*: Stoichiometric relation for H_2O_2 consumption. *Appl. Biochem. Biotechnol.*, Vol. 84–86, pp. 255–265, 2000.

144. P. Verma, D. Madamwar. Decolorization of synthetic textile dyes by lignin peroxidase of *Phanerochaete chrysosporium. Folia Biol.*, Vol. 47, pp. 283–286, 2002.

145. P. Verma, D. Madamwar. Production of ligninolytic enzymes for dye decolorization by cocultivation of white-rot fungi *Pleurotus ostreatus* and *Phanerochaete chrysosporium* under solid-state fermentation. *Appl. Biochem. Biotechnol.*, Vol. 102, pp. 109–118, 2002.

146. D. Moldes, S. Rodríguez-Couto, C. Cameselle, M. A. Sanroman. Study of the degradation of dyes by MnP of *Phanerochaete chrysosporium* produced in a fixed-bed bioreactor. *Chemosphere*, Vol. 51, pp. 295–303, 2003.

147. X. Cheng, R. Jia, P. Li, S. Tu, Q. Zhu, W. Tang, X. Li. Purification of a new manganese peroxidase of the white-rot fungus *Schizophyllum* sp. F17, and decolorization of azo dyes by the enzyme. *Enz. Microb. Technol.*, Vol. 41, pp. 258–264, 2007.

148. A. Heinfling, M. J. Martínez, A. T. Martínez, M. Bergbauer, U. Szewzyk. Transformation of industrial dyes by manganese peroxidases from *Bjerkandera adusta* and *Pleurotus eryngii* in a manganese-independent reaction Appl. *Environ. Microbiol.*, Vol. 64, pp. 2788–2793, 1998.

149. K.-S. Shin, C.-J. Kim. Decolorisation of artificial dyes by peroxidase from the white-rot fungus, *Pleurotus ostreatus*. *Biotechnol. Lett.*, Vol. 20, pp. 569–572, 1998.

150. R. Shrivastava, V. Christian, B. R. M. Vyas. Enzymatic decolorization of sulfonphthalein dyes. *Enz. Microb. Technol.*, Vol. 36, pp. 333–337, 2005.

151. M. Mohorčič, S. Teodorovič, V. Golob, J. Friedrich. Fungal and enzymatic decolourisation of artificial textile dye baths. *Chemosphere*, Vol. 63, pp. 1709–1717, 2006.

152. E. Rodríguez, M. A. Pickard, R. Vazquez-Duhalt. Industrial dye decolorization by laccases from ligninolytic fungi. *Curr. Microbiol.*, Vol. 38, pp. 27–32, 1999.

153. E. Abadulla, K.-H. Robra, G. M. Gübitz, L. M. Silva, A. Cavaco-Paulo. Enzymatic decolorization of textile dyeing effluents. *Text. Res. J.*, Vol. 70, pp. 409–414, 2000.

154. G. Palmieri, G. Cennamo, G. Sannia. Remazol Brilliant Blue R decolourisation by the fungus *Pleurotus ostreatus* and its oxidative enzymatic system. *Enz. Microb. Technol.*, Vol. 36, pp. 17–24, 2005.

155. K.-S. Shin, I. K. Oh, C.-J. Kim. Production and purification of Ramazol Brilliant Blue R decolorizing peroxidase from the culture filtrate of *Pleurotus ostreatus*. *Appl. Environ. Microbiol.*, Vol. 63, pp. 1744–1748, 1997.

156. H. Hou, J. Zhou, J. Wang, C. Du, B. Yan. Enhancement of laccase production by *Pleurotus ostreatus* and its use for the decolorization of anthraquinone dye. *Proc. Biochem.*, Vol. 39, pp. 1415–1419, 2004.

157. V. Faraco, C. Pezzella, A. Miele, P. Giardina, G. Sannia. Bio-remediation of colored industrial wastewaters by the white-rot fungi *Phanerochaete chrysosporium* and *Pleurotus ostreatus* and their enzymes. *Biodegradation*, Vol. 20, pp. 209–220, 2009.

158. K. Murugesan, M. Arulmani, I.-H. Nam, Y.-M. Kim, Y.-S. Chang, P. T. Kalaichelvan. Purification and characterization of laccase produced by a white rot fungus *Pleurotus sajor-caju* under submerged culture condition and its potential in decolorization of azo dyes. *Appl. Microbiol. Biotechnol.*, Vol. 72, pp. 939–946, 2006.

159. K. Murugesan, A. Dhamija, I.-H. Nam, Y.-M. Kim, Y.-S. Chang. Decolourization of reactive black 5 by laccase: Optimization by response surface methodology. *Dyes Pigments*, Vol. 75, pp. 176–184, 2007.

160. Y. Wong, J. Yu. Laccase-catalyzed decolorization of synthetic dyes. *Wat. Res.*, Vol. 33, pp. 3512–3520, 1999.

161. G. S. Nyanhongo, J. Gomes, G. M. Gubitz, R. Zvauya, J. Read, W. Steiner. Decolorization of textile dyes by laccases from a newly isolated strain of *Trametes modesta*. *Wat. Res.*, Vol. 36, pp. 1449–1456, 2002.

162. H. Zouari-Mechichi, T. Mechichi, A. Dhouib, S. Sayadi, A. T. Martínez, M. J. Martínez. Laccase purification and characterization from *Trametes trogii* isolated in Tunisia: Decolorization of textile dyes by the purified enzyme. *Enz. Microb. Technol.*, Vol. 39, pp. 141–148, 2006.

163. S. Trupkin, L. Levin, F. Forchiassin, A. Viale. Optimization of a culture medium for ligninolytic enzyme production and synthetic dye decolorization using response surface methodology. *J. Ind. Microbiol. Biotechnol.*, Vol. 30, pp. 682–690, 2003.

164. M. Nagai, T. Sato, H. Watanabe, K. Saito, M. Kawata, H. Enai. Purification and characterization of an extracellular laccase from the edible mushroom *Lentinula edodes*, and decolorization of chemically different dyes. *Appl. Microbiol. Biotechnol.*, Vol. 60, pp. 327–335, 2002.

165. K. Murugesan, I.-H. Nam, Y.-M. Kim, Y.-S. Chang. Decolorization of reactive dyes by a thermostable laccase produced by *Ganoderma lucidum* in solid state culture. *Enz. Microb. Technol.*, Vol. 40, pp. 1662–1672, 2007.

166. P. Baldrian. Purification and characterization of laccase from the whiterot fungus *Daedalea quercina* and decolorization of synthetic dyes by the enzyme. *Appl. Microbiol. Biotechnol.*, Vol. 63, pp. 560–563, 2004.

167. S. B. Younes, T. Mechichi, S. Sayadi. Purification and characterization of the laccase secreted by the white rot fungus *Perenniporia tephropora* and its role in the decolourization of synthetic dyes. *J. Appl. Microbiol.*, Vol. 102, pp. 1033–1042, 2007.

168. L. Lu, M. Zhao, B.-B. Zhang, S.-Y. Yu, X.-J. Bian, W. Wang, Y. Wang. Purification and characterization of laccase from *Pycnoporus sanguineus* and decolorization of an anthraquinone dye by the enzyme. *Appl. Microbiol. Biotechnol.*, Vol. 74, pp. 1232–1239, 2007.

169. G. M. B. Soares, M. Costa-Ferreira, M. T. P. de Amorim. Decolorization of an anthraquinone-type dye using a laccase formulation. *Biores. Technol.*, Vol. 79, pp. 171–177, 2001.

170. G. M. B. Soares, M. T. P. De Amorim, M. Costa-Ferreira. Use of laccase together with redox mediators to decolourize Remazol Brilliant Blue R. *J. Biotechnol.*, Vol. 89, pp. 123–129, 2001.

171. M. Chivukula, V. Renganathan. Phenolic azo dye oxidation by laccase from *Pyricularia oryzae*. *Appl. Environ. Microbiol.*, Vol. 61, pp. 4374–4377, 1995.

172. G. M. B. Soares, M. T. P. Amorim, R. Hrdina, M. Costa-Ferreira. Studies on the biotransformation of novel disazo dyes by laccase. *Proc. Biochem.*, Vol. 37, pp. 581–587, 2002.

173. A. Michniewicz, S. Ledakowicz, R. Ulrich, M. Hofrichter. Kinetics of the enzymatic decolorization of textile dyes by laccase from *Cerrena unicolor*. *Dyes Pigments*, Vol. 77, pp. 295–302, 2008.

174. T. Hadibarata, A. R. M. Yusoff, R. A. Kristianti. Decolorization and metabolism of anthraquionone-type dye by laccase of white-rot fungi *Polyporus* sp. *S133*. *Water Air Soil Poll.*, Vol. 223, pp. 933–941, 2012.

175. R. Campos, A. Kandelbauer, K.-H. Robra, A. Cavaco-Paulo, G. M. Gübitz. Indigo degradation with purified laccases from *Trametes hirsuta* and *Sclerotium rolfsii*. *J. Biotechnol.*, Vol. 89, pp. 131–139, 2001.

176. C. López, I. Mielgo, M. T. Moreira, G. Feijoo, J. M. Lema. Enzymatic membrane reactors for biodegradation of recalcitrant compounds. Application to dye decolourisation. *J. Biotechnol.*, Vol. 99, pp. 249–257, 2002.

177. J. A. Buswell, B. Mollet, E. Odier. Ligninolytic enzyme production by *Phanerochaete chrysosporium* under conditions of nitrogen sufficiency. *FEMS Microbiol. Lett.*, Vol. 25, pp. 295–299, 1984.

178. B. D. Faison, T. K. Kirk. Factors involved in the regulation of a ligninase activity in *Phanerochaete chrysosporium*. *Appl. Environ. Microbiol.*, Vol. 49, pp. 299–304, 1985.

7

Single and Hybrid Applications of Ultrasound for Decolorization and Degradation of Textile Dye Residuals in Water

Nilsun H. Ince* and Asu Ziylan

Institute of Environmental Sciences, Boğaziçi University, Istanbul, Turkey

Abstract

The discharge of textile dyeing wastewater into sewage treatment facilities is of major concern due to the complexity and intense color of the effluent, the strict regulations on color, the low biodegradability of synthetic dyes and the political dimensions of the problem. As such, it has been a common practice in most parts of the world to pretreat dyeing process effluent before discharging it into sewage treatment works. Emerging public interest on environmentally friendly methodologies for treating industrial effluents has lately led to the search and development of "green processes" that are recognized with low or no chemical and energy consumption, and with no hazardous sludge generation. Ultrasound has lately emerged as a suitable option for serving this objective via the unique means of generating reactive oxidizing and reducing species that are highly capable of destroying the color components in dyeing mill process effluents. The aim of this chapter is to give an overview of the nature of textile dyeing mill wastewater and its treatability by ultrasonic irradiation and/or a variety of hybrid processes that utilize ultrasound, focusing particularly on the degradation and mineralization of reactive azo dyes, which are vastly consumed throughout the globe, although they exert toxic effects in water via their ability to transform into carcinogenic amines under anaerobic conditions.

Keywords: Reactive azo dyes, ultrasound, bulk solution, bubble-liquid interface, hybrid processes, sonocatalysis, advanced oxidation

---

Corresponding author: ince@boun.edu.tr

Sanjay K. Sharma (ed.) Green Chemistry for Dyes Removal from Wastewater, (261–293)
© 2015 Scrivener Publishing LLC

7.1   Overview of the Textile Industry, Dyestuff and Dyeing Mill Effluents

Global annual production of textiles requires 0.7 million tons of dyestuff per year and a well-equipped dyeing/spinning mill consumes 3 tons of dye and 600 tons of water per 100 tons of fiber [1,2]. An old-fashioned mill with unmodified equipment may use more than 1000 tons of water to process the same quantity of fiber. Hence, textile dyeing industries expend large volumes of water, which is ultimately discharged with intense color, chemical oxygen demand (COD), suspended/dissolved solids and recalcitrant material as unfixed dye residuals and spent auxiliaries. A typical reactive dyebath effluent contains 20–30% of the input dye mass (1500–2200 mgL^{-1}) and traces of heavy metals (i.e., cobalt, chromium and copper) that arise from the use of metal-complex azo dyes [2–4]. As such, proper treatment of textile process effluent is of utmost significance to prevent transmission of dye residuals and other recalcitrant matter to sewage treatment facilities and the receiving waters, where they impose aesthetic, environmental and health problems via hindering of oxygen and light transmission and production of toxic end products by hydrolysis [5].

Textile dyestuff is classified either by reference to the dye properties [6] or by chemical constitution and usage [7]. Accordingly, classification of dyestuff based on Color Index chemical constitutions is presented in Table 7.1(a); the associated chemical structures and classification based on application are given in Table 7.1(b). As stated previously, spent dyebaths of dyeing operations contain not only the residuals of dyestuff, but also the spent auxiliaries (added to enhance the process efficiency), some of which are wasted with the dyebaths, thus further increasing the complexity of the effluent. A summary of the most commonly used auxiliary chemicals and their expected benefits is given in Table 7.2 [8].

The expanded use of reactive dyes in textile dyeing operations (due to their high color fastness and wide spectrum) has significantly impacted the quality of the process effluent, which has lately become particularly identified with the dyeing of cotton products, i.e., the use of reactive azo dyes [10]. However, a common feature of these dyes is that they have poor fixing properties (70–90%) and 10–30% of the input dye mass is wasted with the spent dyebaths. This is why dyeing mill effluents are recognized not only by their intense color, but also by their low biochemical oxygen demand (BOD), variable pH, high chemical oxygen demand (COD), high temperature and salinity, as presented in Table 7.3.

Reactive azo dyes constitute 50% of synthetic dye production globally (7×10^5 tons) [12] and are non-biodegradable in conventional biotreatment

Table 7.1(a) Dye classification with respect to chemical structure.

Classification by structure	Representative moiety
Azo	
Anthraquinone	
Indigoid	
Nitro	
Phthalocyanine	
Stilbene	HC = CH
Sulphur	
Triphenylmethane	

Table 7.1(b) Dye classification with respect to application [9].

Classification by application	Applied fiber
Acid	silk, wool, nylon, modified acrylic, leather
Complexed Acid	nylon, wool
Azoic	cotton
Direct	cellulose, wool, silk, nylon

(Continued)

Table 7.1(b) (*Cont.*)

Classification by application	Applied fiber
Disperse	polyester, acetate
Basic	wool, silk, acrylic/modified acrylic
Mordant	wool, wool blends, silk, cotton, modified cellulose
Pigment	cellulose
Reactive	cellulose, cotton, flax, wool, nylon fibers
Sulfur	cotton
Vat	cellulose, cotton, wool

Table 7.2 The auxiliary chemicals and their functions in textile dyeing operations.

Auxiliaries used	Composition	Function
Salts	Sodium chloride, sodium sulfate	Neutralize electrical charge on fiber; retard
Acids	Acetic, sulfuric	pH control
Bases	Sodium hydroxide, sodium carbonate	pH control
Buffers	Phosphate	pH control
Sequestering agents	EDTA	Complex, retard
Dispersers/surface active reagents	Anionic, cationic, non-ionic	Regulate, soften
Oxidizing agents	Hydrogen peroxide, sodium nitrite	Insolubilize
Reducing agents	Sodium hydrosulfite, sodium sulfide	Solubilize, remove unfixed dyes
Carriers	Phenyl phenols, chlorinated benzenes	Enhance absorption

operations due to the stability of the azo structure. On the other hand, the azo bond is disintegrated both anaerobically and by hydrolysis to produce carcinogenic amines [13–18]. Additionally, the presence of one or more sulfonic groups in nearly all azo dye structures makes them highly polar and water soluble and thus very difficult to separate by conventional physical-chemical operations, as well [19].

Current experience on the treatment of dyeing mill wastewater is focused on processes such as adsorption, coagulation, precipitation and

Table 7.3 Characteristics of textile dying mill wastewater [8,11].

pH	7.0–12
BOD, Biochemical Oxygen Demand (mgL^{-1})	80–6000
COD, Chemical Oxygen Demand (mgL^{-1})	150–12000
TOC, Total Organic Carbon (mgL^{-1})	130–1120
TSS, Total Suspended Solids (mgL^{-1})	15–8000
TDS, Total Dissolved Solids (mgL^{-1})	2900–12500
Chloride (mgL^{-1})	1000–1600
TKN, Total Kjeldahl Nitrogen (mgL^{-1})	70–80
Color (Pt-Co)	50–12500

nanofiltration, all of which are indeed phase transfer operations that remove dissolved and suspended solutes from the aqueous to the solid phase, thus generating large volumes of hazardous sludge. As such, these processes need to be modified to satisfy the strict regulations on sludge treatment/disposal and/or adsorbent regeneration/replacement, which are all costly operations [20,21]. The above challenges to destroy dye residuals in biotreated wastewater effluents seem to be resolved by the introduction of advanced oxidation processes (AOPs), whereby highly reactive hydroxyl radicals are generated chemically, photochemically and/or by radiolytic/sonolytic means. Hence, AOPs not only offer complete decolorization of aqueous solutions without the production of huge volumes of sludge, but also promise considerable degrees of mineralization and detoxification of the dyes and their oxidation/hydrolysis byproducts.

7.2 Sonication: A Viable AOP for Decolorizing/Detoxifying Dying Process Effluents

Advanced oxidation processes (AOPs) are powerful methods of destroying recalcitrant compounds in water, and ultrasound is a unique AOP that is particularly effective in decolorizing intensely colored solutions [5,17,22–29]. The potential of ultrasound as an AOP is based on cavitation phenomenon, i.e., the formation, growth and implosive collapse of acoustic cavity bubbles in water and the generation of local hot spots with very extreme temperatures and pressures. Hence, cavitation is a source of high-energy microreactors, where water vapor and volatile gases such as N_2 and O_2 are thermally fragmented to produce highly reactive oxidizing species as hydroxyl ($HO\bullet$), peroxyl ($HO_2^-\bullet$) and superoxide ($O_2^-\bullet$) radicals.

Despite the instability of all these species and their tendency to recombine (to form water and hydrogen peroxide), some of them rapidly react with organic molecules, while some are ejected into the bulk solution to initiate aque ous phase oxidation reactions. A summary of chemical reactions taking place in sonicated water (saturated with air) is given in the following sequence, where the symbol ")))" represents ultrasonic irradiation [30]:

$$H_2O\ (g) +))) \rightarrow HO\bullet + H\bullet \tag{7.1}$$

$$HO\bullet + H\bullet \rightarrow H_2O \tag{7.2}$$

$$HO\bullet + HO\bullet \rightarrow H_2O_2 \tag{7.3}$$

$$O_2 +))) \rightarrow O_2^-\bullet \tag{7.4}$$

$$2N_2 + O_2 \rightarrow 2N_2O \tag{7.5}$$

$$2N_2O + H_2O \rightarrow 2HNO_2 + 2N_2 \tag{7.6}$$

$$\bullet H + N_2O \rightarrow N_2 + \bullet OH \tag{7.7}$$

$$\bullet OH + N_2O \rightarrow 2NO + \bullet H \tag{7.8}$$

$$HNO_2 + H_2O_2 \rightarrow HNO_3 + H_2O \tag{7.9}$$

$$HNO_3 +))) \rightarrow \bullet OH + NO_2\bullet \tag{7.10}$$

$$HNO_3 +))) \rightarrow \bullet H + \bullet NO_3\bullet \tag{7.11}$$

It is these properties of ultrasound that offer a unique medium for the oxidative and/or thermal destruction of organic compounds without the addition of chemical reagents and generation of hazardous sludge that makes it a promising means of "green technology." On the other hand, mineralization or conversion of organic carbon to CO_2 by ultrasound alone is time-consuming and this is why recent studies with ultrasound generally focus on the addition of small quantities of soluble or insoluble catalysts to generate additional nucleation and reaction sites and/or to provide excess radical species. The method provides an excellent synergy based on the presence of excess nucleation/reaction sites, and the ubiquitous properties of ultrasound for accelerating rates of mass transfer and chemical reactions and for improving surface properties of solid particles [31–33].

7.2.1   Sonochemical Degradation of Azo Dyes

The majority of published literature on sonochemical destruction of color and dyestuff is carried out with reactive azo dyes due to their globally high rate of consumption and undesired impacts on aquatic life [5,17,22,24,25,34–41]. The mechanism of azo dye destruction by sonolysis is based on HO• addition to the chromophore, i.e., the N=N or C-N bonds of the molecule, as depicted in Figure 7.1. Generally, it has been shown by density functional theory (DFT) calculations or other means that the priority of HO• attack is the azo bond [23,42], while the role of electron transfer reactions is also of significance [43].

The high water solubility and polar nature of azo dyes hinder their diffusion to the gaseous cavitation bubbles, where very extreme conditions are available for high-energy chemistry. As such, azo dyes are not expected to undergo thermal decomposition under ultrasonic irradiation. This hypothesis was validated by Singla *et al.* [3], who carried out sonoluminescence (SL) analysis in the present of an azo dye and reported that no SL quenching relative to that of water occurred, thus neither the dye nor its degradation byproducts were thermally decomposed at the cavitational bubble or the interfacial area [3]. On the other hand, other researchers demonstrated that depending on the applied frequency and the experimental conditions, some of the dye molecules may diffuse away from the bulk solution to reach the bubble-liquid interface, where they are exposed to considerably larger concentrations of HO• than in the bulk liquid [17,44]. As such, the chemical reactions and the potential reaction sites of an azo dye exposed to ultrasonic irradiation are as presented in Equations 7.12–7.13 and Figure 7.2, respectively.

$$Dye + HO• \rightarrow [Dye\text{-}OH\ adduct]^* \rightarrow Oxidized\ dye + CO_2 + H_2O \quad (7.12)$$

$$Dye +))) \rightarrow R• + Dissociated\ dye\ fragments + C_2H_4 \quad (7.13)$$

where [Dye-OH adduct]* is an excited intermediate product resulting from the addition of an HO• onto –C–N– or –N=N– bonds, and R• is an organic radical dissociated from the dye.

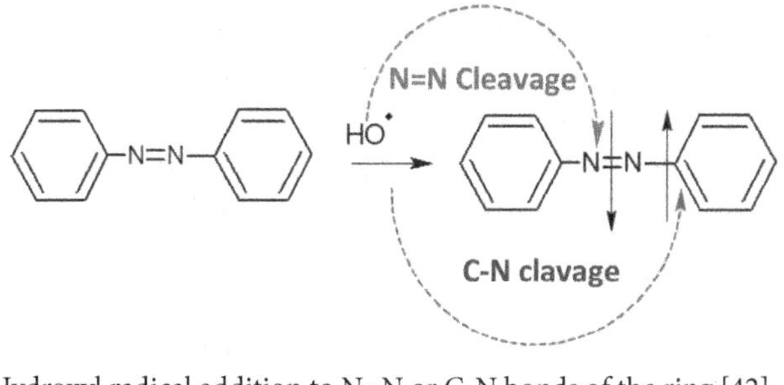

Figure 7.1 Hydroxyl radical addition to N=N or C-N bonds of the ring [42].

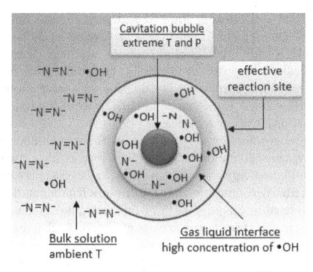

Figure 7.2 Potential reaction sites during sonolysis of an azo dye [35].

The chemical structure and nature of o-substituents also impact the rate of decolorization by ultrasound. For example, a simple molecule with a low molecular weight and no more than one or two substituents has a higher decolorization rate constant than a complex one with several aromatic rings and a larger number of o-substituents. Moreover, a high pKa, low SO_3^- content, and high preference of the hydrazone isomer over the azo form are factors that facilitate the decay of color by ultrasound [41,42,45]. High pKa and low SO_3^- are believed to decrease the anionicity of the molecule, which in turn facilitates its approach to the negatively charged cavity bubbles; while the hydrazone form is associated with internal hydrogen bonding and reactivity with H_2O_2/HO_2^- pair.

Figure 7.3 shows chemical structures and tautomerization of two dissimilarly structured azo-naphtol dyes, both with an o-hydroxyl substituent about the azo bond to indicate the preference of the hydrazone tautomer. The apparent decolorization rate constants as determined under equivalent ambient/ultrasonic conditions were 4.90×10^{-4} and 2.35×10^{-4} min^{-1}, respectively [40]. The main reason for the lower reactivity of the second dye is its more complex structure (618 g mol^{-1}) and the presence of a second o-substituent (SO_3^-), which weakens the internal H-bonding via steric effects and competition for H-abstraction from the HO•-adduct. Moreover, delocalization of the C-O bond electrons of the carboxyl amine group (attached to the naphthol-ring) stabilizes the molecule, thus further decelerating the rate of bleaching.

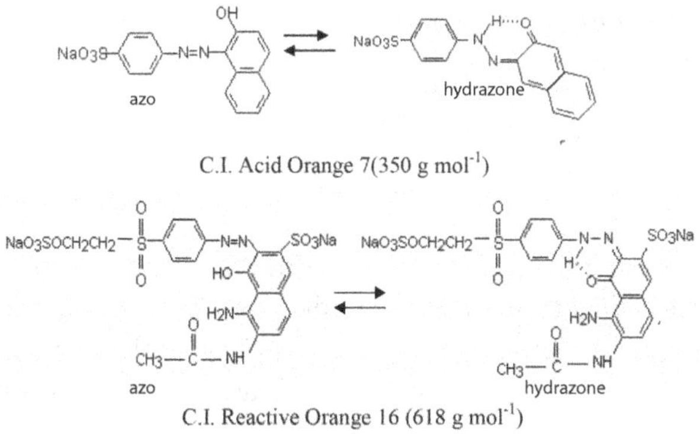

C.I. Acid Orange 7(350 g mol^{-1})

C.I. Reactive Orange 16 (618 g mol^{-1})

Figure 7.3 Tautomerization of two azo dyes with dissimilar structures.

The rate of azo dye degradation by ultrasound follows second-order reaction kinetics as r=k[Dye][HO•], where r is the rate of decolorization and k is the second order reaction rate constant. Accordingly, in many of the published literature the rate of decolorization has been fit to the simple pseudo-first-order kinetic model r=k′[Dye], where k′ is the apparent rate constant as estimated by regression analysis. Some other researchers agreeing on pseudo-first-order kinetics have claimed that the rate must follow a Langmuirian type of adsorption model that considers gas-liquid heterogeneity [35,46]:

$$r = \frac{k' K[Dye]_0}{1 + K[Dye]_0} \qquad (7.14)$$

where: k′ and K are the apparent rate and adsorption equilibrium constants, respectively and $[Dye]_0$ is the initial dye concentration.

7.2.2 Operation Parameters in Decolorization/Degradation of Textile Dyes by Ultrasound

It has been much reported that the relation between cavitational yield and the applied power is parabolic, i.e., the yield increases with an increase in the applied power up to a definite optimum or a threshold, beyond which it begins declining. The mutual agreement in the literature on reduced efficiency at too high powers is based on the "cushioning" effect, i.e., the barrier exerted by the dense cloud of cavity bubbles to the transfer of acoustic energy through the liquid [47]. Among a number of mathematical formulas that describe the correlation between the efficiency of cavitational yield and the applied acoustic power, that proposed by Sivakumar *et al.* [24]

specifically provides an expression that relates the efficacy of decolorization to the applied power:

$$E = 75.9\left(\frac{P}{V}\right)^{0.64} \tag{7.15}$$

where E is the decolorization efficiency and P/V is the specific power.

Another significant operation parameter is the frequency, but there is such a wide range of application that it is not possible to define a specific frequency that provides maximum decolorization or mineralization of textile dyes [29,34,35,48]. In practice, three ranges of frequencies are available for three distinct uses of ultrasound: (i) high frequency or diagnostic ultrasound (2–10 MHz), (ii) low frequency or conventional "power ultrasound" (20–100 kHz), and medium frequency, or "sonochemical-effects" ultrasound (300–1000 kHz) [33]. It is this latter range, where chemical reactions of particularly hydrophilic compounds are uniquely catalyzed through unstable and short-lived cavitation recognized with effective oscilations, that leads to more frequent collapse events, allowing the ejection of some radical species to the bulk solution.

Hence, selection of the right frqency is of utmost significance to obtain good reaction yields, and this is possible by considering the solubility and/or volatility of the target contaminants. For example, if the target is hydrophobic with a high vapor pressure, it easily diffuses to the gas-liquid interface and the gaseous bubble interior to undergo oxidative and thermal decomposition, respectively. The most suitable frequency range for the reaction of such compounds is 20–100 kHz, which is characterized by long-lived "stable" cavities that generate very extreme temperatures during their violent collapse. In contrast, hydrophilic compounds, particularly those at very low concentrations, tend to remain in the bulk liquid, where they undergo sonochemical degradation only if there is sufficient •OH and/or other reactive species in solution. The condition is satisfied at the 200–800 kHz range or under "unstable" and short-lived cavitation that allows the ejection of some radicals to the bulk liquid [49].

Owing to the high solubility of textile dyes, power ultrasound (20–100 kHz) has been found ineffective for their degradation unless applied in the pulse mode or in the presence of a volatile reagent that produced additional radicals via pyrolysis [28,29,50]. The inefficiency is due to the nature of long-lived and large-resonate cavitation bubbles (1.0×10^{-5} s; 170 μm), which contain massive quantities of water vapor that reduce the theoretical values of their collapse temperatures (4000–5000 K) [51–53].

Consequently, the degradation of textile dyes by high-frequency ultrasound is much more effective, however, with the hindrance caused by the "threshold frequency," above which sonochemical effects fade away. As such, the effect of frequency on the rate of decolorization of thiazine dyes is reported as $k'_{22.8kHz} < k'_{490kHz} < k'_{127kHz}$, highlighting the impact of "cavitation threshold" [34].

On the other hand, at equivalent power densities, the collapse temperature under short frequency ultrasound is always higher than that of high frequency (4558 K at 20 kHz, 2458 K at 500 kHz) [54] implying that the degradation of hydrophilic solutes such as textile dyes is not directly related to the temperature. As such, a more significant parameter affecting the efficiency of hydrophilic substrate degradation by ultrasound is the incidence of efficient bubble collapse and its duration that dictate the probability of radical combination reactions. This explains why these solutes react faster at high frequency ultrasound despite the milder collapse conditions, and why the bubbles are very small and short-lived (transient cavitation), undergoing a larger number of oscillations with a more frequent occurrence of efficient bubble collapse (25-fold) that promotes free radical transfer to the bulk solution. Moreover, our experience has shown that the effective frequency is not only a matter of bubble dynamics and the solubility of the dye, but also that of the solution matrix, which controls the rate of production/depletion of OH radicals.

The range of frequencies tested on the degradation of azo and a variety of other dyes is listed in Table 7.4 with additional information on the operation parameters such as pH, the sparge gas and the acoustic power. Note that the applied pH in general (except for C.I. Basic Violet) lies within a highly acidic to closely neutral range, because most of the investigated dyes at acidic conditions are protonated with a slight enrichment in hydrophobicity, which lowers their resistance to diffuse to the negatively-charged cavity bubbles, or the interfacial region. Hence, acidification is a positive encouragement to the mass transfer of highly soluble dye molecules from the bulk solution to the gas-liquid interface. Note that acidity is not only maintained by the addition of strong acids to the solution, but also by air bubbling to form reactive nitrogen species and HNO_3. In fact, the injection of air or another gas is also essential to enhance the number of nucleation sites and the violence of collapse. The use of inert gases for this purpose is very popular due their high pyrolytic gas ratios γ (Cp/Cv), while bubbling of the solution with a mixture of two gases (e.g., air/O_2, Ar/O_2) in a well-selected molar ratio is also very common due to a variety of additional benefits [55].

Table 7.4   A summary of published research and some of the critical experimental conditions in decolorization of dye solutions by ultrasound.

C.I Generic Name/ Structure	Conditions	Reference
Acid Orange 7/mono-azo	300 kHz (18W), 520 kHz (88 W), air/Ar, pH 3–7 24 kHz-80 kHz/pulse	[27,42,50]
Acid Orange 8/mono-azo	300 kHz (18 W), air/Ar, pH 3–6	[42] [28]
Acid Orange 24/mono-azo	355 kHz (18W), 520 kHz (88W), pH 3.5–6.5	[3]
Acid Orange 52/mono-azo	200 kHz (86W), pH 2 45 kHz (40W), pH 6, CCl_4 addition	[56]
Basic Blue 3/oxazine	520 kHz (88W), air/Ar, pH 5.5–6.5	[5]
Basic Blue 24/thiazine	22.8–490 kHz (2–8W), N_2	[34]
Basic Brown 4/di-azo	520 kHz (88W), air, ph 5.5–6.5	[5]
Basic Green 4/ triphenylmethane	300 kHz (20–100W), air/Ar, pH 5.5–6.5	[57]
Basic Violet 3/ triarylmethane	300 kHz (20–80W), pH 2–7	[48]
Basic Violet 10/xanthene	300 kHz (60W), pH 8.3	[58]
Direct Yellow 9/mono-azo	20 kHz (37W), CCl_4 addition, air, pH 6.9 577–1145 kHz (40–49 W), air, pH 6.9	[29]
Reactive Black 5/di-azo	520 kHz (88W), air/Ar, pH 5.5–6.5	[5]
Reactive Orange 16/ mono-azo	300 kHz (18W), air, pH 3–9.5	[41]
Reactive Blue 19/ triarylmethane	20 kHz, pH 2–10	[59]
Reactive Red 22/mono-azo	200 kHz (86W), air/Ar, pH 2–6.5	[35]
Reactive Red 24/mono-azo	28 kHz, pH 1–10	[60]
Reactive Red 141/mono-azo	520 kHz (88W), air/Ar, pH 5.5–6.5 20 kHz (37W), 577, 861, 1145 kHz (40–49W) air, pH 5.5–6.9	[27] [29]
Reactive Red 195/mono-azo	20 kHz (20–60W), pH 7.0	[61]

7.2.3   Addition of Chemical Reagents

The most commonly added chemical reagents to enhance sonochemical degradation of textile effluents are hydrogen peroxide (H_2O_2), ferrous/ferric ions (Fe^{2+}/Fe^{3+}), carbon tetrachloride (CCl_4) and surface active compounds. The major contribution of H_2O_2 was specified in the previous section as dissociation to produce $\overline{HOO}$ radicals [62], but it has a more vital role in the presence of Fe^{2+}/Fe^{3+} via initiation of Fenton reactions (to be discussed in the next section on hybrid processes). The effect of CCl_4 arises from its high volatility (p_i = 91 mm Hg/20°C), i.e., the ability to diffuse to the gaseous cavity bubbles to undergo molecular fragmentation and release reactive chlorine radicals, hypochlorous acid and hydrogen chloride. Moreover, CCl_4 reacts strongly with hydrogen in the interface and the bubble interior, reducing the probability of its combination with HO• (k_{CCl4-H} = 3.8 × 10^7 M s^{-1}, $k_{H-HO•}$ = 4 × 10^{10}M s^{-1}) [63]. This is why it is considerably effective under low-frequency ultrasound, at which the rate of HO• ejection to the bulk solution is low [28,56]. Surfactants such as linear alkyl sulfonates (LAS) are also unique with their potential to migrate to the cavity bubble, where they undergo pyrolysis during the collapse phase to release reactive organic radicals as •CH [64].

Despite their reactivity with •OH, the presence of anions such as carbonate, nitrate, bromide and iodide may also accelerate decolorization reactions, provided that they are maintained at a proper dose. For example, it was found that rates of degradation of C.I. Acid Blue 40 and C.I. Reactive Red 141 gradually increased with increasing concentrations of CO_3^{-2} up to a plateau above which they started to decline [28,44,58]. Acceleration at moderate doses of carbonate ions is due to the formation of carbonate radicals ($CO_3•^-$) and also the fact that although these radicals are less reactive with organic compounds than HO•, their recombination is less likely. This and their larger stability than HO• make them more mobile and transferrable to the bulk solution, where they react with many hydrophilic compounds including synthetic dyes [46,48].

Finally, the addition of peroxymonosulfate (HSO_5^-) and persulfate ($S_2O_8^{2-}$) has been found to be remarkably effective not only for decolorization but also for the mineralization of synthetic dyebaths containing radical scavenging reagents [65]. This effect was attributed to ultrasonic activation of the additives to generate $SO_4^-•$ and HO•, which at pH 8.6 were equally powerful as oxidizing agents.

7.2.4   Reactors

There are two essential components in sonochemistry: the liquid and a source of high-energy vibrations [66]. The former is necessary, because it is the only medium for the formation of acoustic cavity bubbles; while the latter is the transducer, which is a device capable of converting electrical to sound energy [67]. The most common equipment for generating acoustic cavitation in the laboratory are the bath, the probe and the plate-type reactors; but the formation and ejection of radicals to the bulk solution are 20–25-fold higher in plate-type reactors than in probe or bath systems at equivalent acoustic powers [37]. Although in general all lab-scale systems are operated on continuous mode, pulse mode of irradiation is also common with the advantage of more effective utilization of hydroxyl radicals via temperature control, i.e., the availability of a longer time for heat dissipation [50,68].

A selection of most commonly used lab-scale reactors for sonochemical decolorization and mineralization of textile dyes is presented in Figure 7.4. Note that the common feature of all reactors is that they are cooled by circulating water to prevent excessive heating. The "horn" and "bath" are associated with low frequency emission and thus mechanical effects (e.g., cleaning, dispersing) and thermal decomposition of volatile solutes, while the "plate" is recognized as a source of high-frequency irradiation and thus aqueous-phase sonochemistry. However, there is plenty of literature on the use of short-frequency or "power" ultrasound with very effective decolorization.

7.3   Hybrid Processes with Ultrasound: A Synergy of Combinations

In the previous sections, we have tried to demonstrate the power of ultrasound for decolorizing and decomposing a wide range of synthetic dyes, pointing out, however, that ultrasound alone is ineffective for sufficient mineralization of these compounds. The rest of this chapter, therefore, is devoted to the description of the most commonly used hybrid processes that employ combined application of ultrasound with an AOP or heterogeneous catalysis. A list of hybrid processes used in decolorization and mineralization of dyestuff and dying effluents and their selected operation parameters is presented in Table 7.5.

7.3.1   Sono-Ozonolysis (US/O$_3$)

Decolorization/mineralization of synthetic dye solutions and simulated dyebaths by US/O$_3$ process has been found considerably more effective

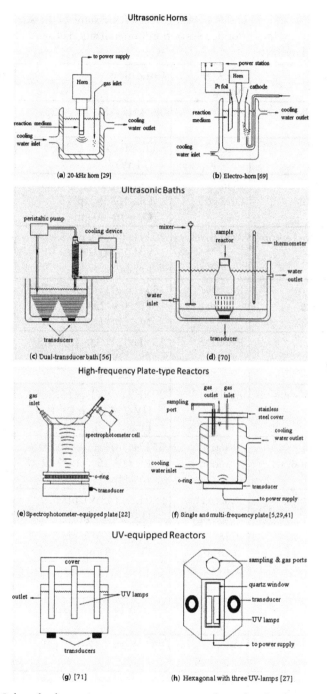

Figure 7.4 Lab-scale ultrasonic reactor systems commonly used in decolorization and degradation of textile dyes.

Table 7.5 A list of US-assisted hybrid processes and some of the selected experimental conditions for decolorization/mineralization of synthetic dye solutions and dyebaths.

Process	C.I. Generic Name	Conditions	Reference
Sonophotolysis	Acid Orange 7	520 kHz (88W), UV-254 (36–108 W) UV, pH 5.5	[27]
	Direct Yellow 68	Bath, pH 3–11, UV-254 (11 W)	[71]
Sonophoto-ozonolysis	Acid Orange 7	520 kHz (88 W), UV-254 (36–108 W), pH 5.5, O_3 = 10–60 gm^{-3}	[27]
Sono-photo-Fenton	Acid Black 2	20 kHz (49W), pH 3, UV-254 ($I = 6.4 \times 10^{-5}$ $Einsm^{-2}s^{-2}$)	[77,80]
Sono-photo-persulfate	Reactive Violet 2	35 kHz, pH = 3.0–10, PMS/PS[1], UV-254, ($I = 2.12 \times 10^{-4}$ $Eins\ L^{-1}s^{-1}$)	[65]
	Reactive Blue 7	35 kHz, pH 3.0–10, PMS/PS[1], UV-254, ($I = 2.12 \times 10^{-4}$ $Eins\ L^{-1}s^{-1}$)	[65]
Sono-photo-catalysis	Acid Black 2	20 kHz (49W), pH 3, UV-254 ($I = 6.4 \times 10^{-5}$ $Einsm^{-2}s^{-2}$), ZnO	[80]
Sono-photo-electrocatalysis	Reactive Violet 2	35 kHz, pH = 3.0–10, PMS/PS, UV-254, ($I = 2.12 \times 10^{-4}$ $Eins\ L^{-1}s^{-1}$)	[65]
Sono-ozonolysis	Reactive Orange7	520 kHz (88W), pH 5.5, O_3 = 10–60 gm^{-3}	[27]
	Acid Orange 8	300 kHz (18W), pH 5, O_3 = 2–8 gm^{-3}	[28]
	Acid Orange 52	20 kHz, pH 5.5, O_3 = 25–113 gm^{-3}	[81]
	Reactive Blue 19	20 kHz (44–88 W), pH 8,	[59]
	Direct Red 23	20 kHz (44–176W), pH 4–12,	[82]
	Reactive Black 5	520 kHz (25–100W), pH 7, O_3 = 3.36 gm^{-3} 20 kHz (44W), pH 11, O_3	[25] [74]

Table 7.5 (*Cont.*)

Process	C.I. Generic Name	Conditions	Reference
	Reactive Yellow 84	20 kHz (22–88W), pH 2–12, O_3	[83]
Sono-Fenton	Acid Red 14	20 kHz, pH 3, $FeSO_4$	[64]
	Acid Black 1	40 kHz (5W), pH 2–6, $FeSO_4$	[84]
	Reactive Violet 2	20 kHz, pH 3, $FeSO_4$	[64]
Sono-Fenton-like	Acid Orange 7	20 kHz (10–20W), pH 2–5, Fe^0	[85]
	Acid Red 14	59 kHz, pH 2–8, Fe^0	[86]
	Reactive Red 141	20 kHz (37W), 577–1145 kHz (40–49W), pH 6.9, air, Cu^0	[29]
Sono-Fenton-persulfate	Reactive Violet 2	40 kHz, pH 3, Fe^{2+}/S_2O_8	[87]
	Reactive Yellow3	40 kHz, pH 3, Fe^{2+}/S_2O_8	[87]
Sonocatalysis	Direct Yellow 9	20 kHz (37W), 577–1145 kHz (40–49W), pH 6.9, air, TiO_2	[29]
	Basic Blue 24	39 kHz, pH 7, TiO_2, H_2O_2	[88]
	Acid Red 14	40 kHz, pH 7, TiO_2, ZnO, ZnO/TiO_2	[89]
	Acid Red 14	40 kHz, pH 5.5, Er^{+3}:$YAlO_3$/TiO_2-ZnO	[90]
	Reactive Red 141	220 kHz (37W), 577,861,1145 kHz (40–49W), pH 6.9, air, TiO_2	[29]
	Acid Red 14	50 kHz, pH 3, Ar/O_2, MnO_2	[26]
Sono-sorption	Acid Black 210	25 kHz (1.5W), pH 1–11, **exfo-graphite**[4] + **GAC**[5], H_2O_2	[60]
	Direct Black 168	40 kHz, pH 1–5, H_2O_2, **kaol**[6]	[91]

(*Continued*)

Table 7.5 (*Cont.*)

Process	C.I. Generic Name	Conditions	Reference
	Basic Blue 24	20 kHz, **waste paper**	[92]
Sono-persulfate	Basic Green 4	22.5 kHz, pH 4–12, **poly-nano-clay**[7]	[93]
	Direct Black 168	40 kHz, pH 1–5, fly ash	[91]
	Acid Orange 7	40 kHz, pH 4, **Fe^0/GAC**[5]	[94]
	Acid Orange2	35 kHz, pH 2–11, Co/S$_2$O$_8$	[95]
	Acid Red 18	35 kHz, pH 2–11, **Co/S$_2$O$_8$**	[95]
Sonoelectrocatalysis	Reactive Red 195	20 kHz, **MMO**[2] pH 7, NaCl, KCl, Na$_2$SO$_4$, Na$_2$CO$_3$	[61]
	Reactive Brilliant Red X-3B, Acid Orange 52, Basic Blue 24, Basic violet10, Reactive	22 kHz, **Pt-ACF**[3], pH 2.8–10.8, Na$_2$SO$_4$ 20–80 kHz, PbO$_2$-mesh steel,	[69]
	Blue 19	pH 3.0–9.0	[96]
	Acid Yellow 75	20, 40 kHz, NaCl, **Pt**, 40°C,	[97]
	Acid Black, Methyl orange, Reactive Black 5	20 kHz, **Pt-graphite**, Na$_2$SO$_4$,	[98]

[1]peroxymono/persulfate
[2]mixed metal oxide electrode
[3]active carbon fiber
[4]exfoliated graphite
[5]granular activated carbon
[6]kaoline
[7]polyacrylic nano-clay

than that by sonolysis alone, particularly for mineralization. The synergy is due to: (i) increased mass transfer, dispersion and solubility of ozone and the formation of excess radical species via its thermal decomposition in the gas phase; and (ii) the additional pathways of oxidation upon the formation of secondary radicals as $O_2^-\bullet$ and $HO_2\bullet$. The depiction of these phenomena by chemical reactions is as follows [72–75]:

$$O_3 +))) \rightarrow O_2 (g) + O (^3P)(g) \qquad (7.16)$$

$$O (^3P) (g) + H_2O (g) +))) \rightarrow 2HO\bullet \qquad (7.17)$$

$$HO\bullet + O_3 \rightarrow HO_2\bullet + O_2 \qquad (7.18)$$

The solution temperature is a crucial parameter in US/O_3 hybrid processes, as the rate of bleaching/mineralization gradually increases by heating the solution and rapidly declines when the temperature exceeds 65°C [74]. The acceleration above the ambient temperature is due to the reduction in viscosity and/or surface tension, which in turn lowers the threshold intensity required for the formation of cavitation bubbles [74]. On the other hand, in some cases temperatures higher than 40°C were found to decelerate the degradation of some dyes due to the reduction in the solubility and rate of decomposition of ozone [22,76–77].

7.3.2 Sonophotolysis (US/UV) and Sonophoto-Ozonolysis (US/UV/O_3)

Integration of ultrasonic reactors with UV-irradiation results in the formation of excess $HO\bullet$ (upon photolysis of US-generated H_2O_2) that is responsible for the observed enhancement in the rate of oxidation. The addition of moderate doses of peroxide may further improve the efficiency provided that the light source emits at around 200 nm, where H_2O_2 absorbs strongly. A more effective additive than peroxide is persulfate, which in the presence of UV and ultrasonic irradiation is converted to $HO\bullet$ and $SO_4^-\bullet$, the latter being highly reactive with the intermediate oxidation byproducts of synthetic dyes [65]. The most commonly used additive to enhance mineralization is gaseous ozone, the catalytic effect of which was discussed in the previous subsection. In the combined presence of US/UV, ozone undergoes sonolytic and photolytic decomposition at the same time to generate singlet state oxygen and additional radical species such as $HO\bullet$ and $HO_2\bullet$:

$$O_3 + h\upsilon \rightarrow O_2 + O(^1D) \qquad (7.19)$$

$$O(^1D) + H_2O \rightarrow HO\bullet + HO\bullet \rightarrow H_2O_2 \qquad (7.20)$$

$$H_2O_2 + h\upsilon \rightarrow 2HO\bullet \qquad (7.21)$$

Relative performance of singly applied ultrasound and ozone, and combinations thereof, for the mineralization of C.I. Acid Orange 7 in a hexagonal UV-equipped sonoreactor (Figure 7.4h) is presented in Figure 7.5. The insets on top show the geometry of the reactor and the position/distribution of UV-irradiation relative to the cavitation clouds. The efficiencies of each process for bleaching and mineralization of the dye under the same conditions were as follows: US: 70%, 0%; O_3: 72%, 24%; US/UV: 84%, 10%; US/UV/O_3: 100%, 41%. The outstanding synergy of the trio is the result of enhanced rate of ozone mass transfer and decomposition (sonolytic and photolytic), excess HO• and other reactive species formation including the singlet oxygen, which is not only more stable than HO•, but also reacts strongly with a wide range of organic chemicals via hydrogen abstraction [78,79].

7.3.3   Sono-Fenton (US/Fe²⁺) and Sonophoto-Fenton (US/UV/Fe²⁺)

The sono-Fenton process, which involves sonolysis in the presence of peroxide and a ferrous salt to generate ferro-hydroxyl complexes and reactive oxygen species has received considerable attention for destroying textile dyestuff and other synthetic dyes [99,100,84,44]. The process is quite sensitive to the operation parameters, specifically to pH, temperature, concentrations of ferrous/ferric ions and peroxide, and to the power density.

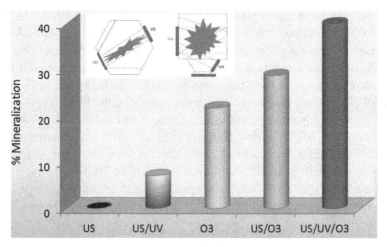

Figure 7.5 Relative performance of single and combined US, UV, O_3 for the mineralization of C.I. Acid Orange 7 during 1-h reaction (C_0 = 57 µM, T = 25°C, I_0 = 2.4x10⁻⁴ Einstein m⁻²s⁻¹, 520 kHz, O_3 flow = 0.25 L min⁻¹).

For example, the efficiency of sono-Fenton process for the decolorization and degradation of dyestuff at short frequency ultrasound was found to increase sharply with increasing solution temperature (T) and power density (P) up to some threshold (T = 40°C, P = 50 WL^{-1} pH = 3.0), above which it declined [84]. The result was attributed to enhanced solubility and mass transfer of ferrous ions at the given pH, and to the excess of energy that provided additional $Fe-O_2H^{2+}$ and HOO radicals. The deceleration of the reactions at more extreme conditions of T and P is the result of an increase of the equilibrium vapor pressure, which led to a higher vapor concentration in the bubbles or a lower collapse temperature. A simplified reaction scheme that describes the sono-Fenton process in the presence of a dye molecule is outlined in Equations 7.13–7.16.

$$Fe^{2+} + H_2O_2 + H^+ \rightarrow Fe^{3+} + HO\bullet + H_2O \qquad (7.22)$$

$$Fe^{3+} + H_2O_2 \rightarrow Fe-O_2H^{2+} + H^+ \qquad (7.23)$$

$$Fe-O_2H^{2+} +))) \rightarrow Fe^{2+} + HOO\bullet \qquad (7.24)$$

$$Dye + HO\bullet \rightarrow [Dye-OH\ adduct]^* \rightarrow Oxidized\ dye + CO_2 + H_2O \quad (7.25)$$

A costly means of enhancing the efficiency of sono-Fenton reaction is the immersion of a light source into the reactor (sonophoto-Fenton) to provide excess OH radicals via photolytic decomposition of Fe-OOH complex and H_2O_2:

$$[Fe\ (OOH)^{2+}] + h\upsilon \rightarrow Fe^{2+} + HO\bullet \qquad (7.26)$$

$$H_2O_2 + h\upsilon \rightarrow 2HO\bullet \qquad (7.27)$$

As such, the degradation of C.I. Reactive Violet 2 was found to be 10-times faster than that by the classical photo-Fenton reaction under equivalent conditions, and the addition of a surfactant rendered an initial rapid mineralization via the reductive cleavage of the surfactant to provide excess radical species [64].

7.3.4 Sonocatalysis

The basic principle of sonocatalysis is diffusion and sorption of soluble solutes onto a solid surface or its interface followed by a sequence of heterogeneous chemical reactions on the active sites. It is important at this point to distinguish three kinds of solid media that are of value in sonocatalytic processes: (i) those which originally have no reactive sites, but are partially covered with them upon the formation and collapse of cavity bubbles on their surface; (ii) those which originally contain reactive metals on their surface, or furnish metals from their cores upon exposure to ultrasound; and (iii) those which generate electron-whole pairs and/or reactive oxygen species on their surface upon sonication.

The first group of solids is made up of a wide range of non-reactive adsorbents (e.g., activated carbon, fly ash) and the governing mechanism in the degradation of dyestuff in their presence is sono-sorption of the solutes on the adsorbent surface with the advantages of ultrasound for continuous cleaning and disaggregation of the particles and for reducing the micro-vortex motion of the adsorbate on the surface of the adsorbent [92,101,102]. The most commonly tested adsorbents for the elimination of dye residuals in process effluents are natural zeolite, clay/nanoclay, chitosan, sepiolite, limestone, and cellulosic materials [92,95,103–108]. Note that sono-sorption is considerably more cost-effective for the elimination of dyestuff than conventional adsorption processes, owing to a lower sorbent requirement and a lower cost of adsorbent regeneration.

The second group of solids consists of zero-valent metals, metal oxides and metallic nanoparticles, all of which are excellent adsorbents with active heterogeneous reaction sites. Some of the unique benefits that ultrasound provides to such catalytic systems are perfect mixing, cleaning, surface improving, disintegrating and disaggregating of the solid particles. The activities of two such solids, namely zero-valent iron and manganese dioxide, will be discussed later in this section within the scope of heterogeneous catalysis. The third group of solids is made up of semiconductors and their composites, which owing to their unique properties will also be given special emphasis in the rest of this chapter.

Heterogeneous sonocatalysis has attracted considerable attention over the last few years for decomposing and mineralizing synthetic dyes [15,29,60,85,89,91]. The advantages of the process rise not only from the presence of surplus nuclei for excess cavity formation, but also from the turbulent flow conditions that accelerate rates of mass transfer and chemical reactions at solid surfaces, which are remarkably enlarged via disaggregation of the particles. Among a wide range of catalysts tested in various laboratories, zero-valent metals, nanopowders, composites of semiconductors, graphite and cast iron have been found considerably effective.

Nanoparticles, particularly of TiO_2 and its composites, play a special role in catalyzing US-mediated reactions. Figure 7.6 shows relative efficacies of nano-TiO_2 and zero-valent copper on the degradation of C.I. Direct Yellow 9 by short-frequency US [29]. Some of the possible reasons for the higher performance of TiO_2 than copper are: (i) the slightly positive character of the semiconductor surface at pH < 6.8 (zero point charge) and the anionic character of the dye that enables attractive forces between the

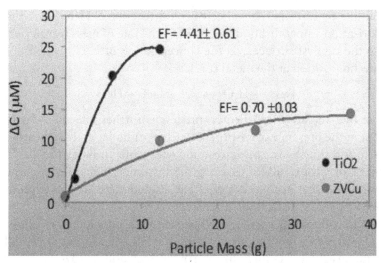

Figure 7.6 Estimated efficiencies of TiO_2 and zero-valent Cu for catalyzing sonochemical (20 kHz) degradation of C.I. Direct Yellow 9 at pH 6.9. The solid lines display the quadratic relations between moles of dye removed per mass of particle added, and "EF" is the efficiency factor defined as the linear slope of the fitted curve.

negatively charged moieties and the solid surface (note that the solution pH originally at 6.9 decreased readily to around 5.0 via oxidation); (ii) the larger surface area of the nanoparticles and the likely formation of e-h pairs on their massive surfaces upon the extreme conditions of non-symmetrical collapse; (iii) the wide wavelength range of single-bubble sonolumines-cence (200–500 nm), at which the semiconductor surface is likely excited, as in photocatalytic processes [88].

The effectiveness of US/TiO_2 for dye degradation may further be improved by the use of composite nano-semiconductors such as TiO_2/ZnO to enrich the active sorption sites and the number of reactive species [89]. For example, a unique composite as Er^{3+}: $YAlO_3$/TiO_2–ZnO was found to significantly improve the degradation of organic dyes under ultrasound via direct decomposition of dye molecules on the surface of the catalyst, separation of the electron-hole pair and the up-conversion luminescence of the Er^{3+}: $YAlO_3$ composite by ultrasonic cavitation [90,101].

The use of zero-valent iron (ZVI) together with ultrasonic irradiation also has a special role owing to the unique activity of the catalyst surface and its vicinity, where dye molecules are exposed to reductive and oxidative decomposition while the metal surface is continuously cleaned by hydro-dynamic sheer forces of ultrasound [94,100]. The reactivity of the surface arising from corrosion of the metal (at acidic pH) leads to heterogeneous

Fenton-like or so-called "advanced Fenton reactions." Hence, the rate of decolorization/mineralization is limited by the rate of mass transfer of solutes to the catalyst surface and the concentration of H_2O_2, which undergoes decomposition at the metal surface:

$$Fe\ (s) + H_2O_2 \rightarrow Fe^+ + OH^- + HO\bullet \tag{7.28}$$

Note that the advanced Fenton process is further effective when carried out in the presence of adsorbents such as granular activated carbon to generate additional adsorption/reaction sites while utilizing the mechanical effects of ultrasound for cleaning the surface of the adsorbent and the catalyst to prevent accumulation of the oxidation/reduction byproducts [85,86].

Another important catalyst to integrate with ultrasound is exfoliated graphite, which in the presence of peroxide has been found to exhibit an outstanding activity in mineralization of fourteen different azo dyes at short frequency ultrasound [60]. The synergy was explained by the likely rupture of the graphite shield via hydrodynamic sheer forces and the subsequent decreases and increases in its particle size and surface area, respectively.

The role of monosulfate and persulfate anions in the presence of ultrasound has already been discussed, but will once again be highlighted in heterogeneous media. It has been shown that persulfate is easily activated by heat, cavitation, hydrogen peroxide or catalysts such as Co (II), Fe (II), Ag (I) to generate sulfate radicals and HO•, both of which are reactive with textile dyes to produce dye-OH adducts and oxidized dye intermediates as:

$$S_2O_8^{2-} +))) \rightarrow SO_4^{2-} + SO_4^-\bullet \tag{7.29}$$

$$SO_4^-\bullet + H_2O \rightarrow SO_4^{2-} + HO\bullet + H\bullet \tag{7.30}$$

$$S_2O_8^{2-} + M^{n+} \rightarrow SO_4^{2-} + SO_4^-\bullet \tag{7.31}$$

$$Dye + SO_4^-\bullet + HO\bullet \rightarrow Dye^-\bullet + Dye\text{-}OH\ adduct + SO_4^{2-} \tag{7.32}$$

Accordingly, 90–97% decolorization and more than 60% mineralization of reactive azo dyes have been reported for Co/US-activated persulfate system, and complete decolorization with 73% mineralization reported for Fe/US-activated persulfate [87].

Last but not least, manganese dioxide at acidic pH was found to exhibit considerable activity in decolorization and mineralization of synthetic azo dyes by ultrasound via the presence of very active reaction sites on the mineral surface and the solid-liquid interface for redox reactions [26]. The results were attributed to: (i) reduced diameter and increased surface area of the mineral via cavitation effects, (ii) enhanced heterogeneous reactions via passivation of the outer oxide layer by the micro-jets, (iii) enhanced dissolution of the catalyst and replenishment of its surface with fresh mineral via the presence of US-generated H_2O_2.

7.3.5   Sonoelectrocatalysis

The superiority of US-assisted electrocatalysis over standard electrochemical oxidation processes is based on the presence of high energy microjets, which render intense agitation of the liquid and continuous degassing on the electrode surface [109,110]. Agitation facilitates ion transport across the electrode double layer and reduces the rate of ion depletion in the diffusion layer, while degassing at the electrode surface prevents the accumulation of gas bubbles or the formation of a current barrier [96]. The efficacy of the method has been tested on a variety of synthetic dyes (e.g., methylene blue, reactive brilliant red X-3B, Rhodamine B, methyl orange) and dyebaths and it has been reported that the process is highly sensitive to the applied potential, the acoustic power and frequency, and to the solution pH [69,96–98]. As such, decolorization was appreciable only at $E \geq 4V$ and within a power range of 20–60 W [61]. Note also that the process was found more effective at short frequency irradiation, which allows more intense microjets and thus a higher degree of agitation.

As in most hybrid processes, sonoelectrocatalysis has been found more cost-effective than simple electrochemical oxidation, as demonstrated for C.I. Reactive Blue 19, whose decomposition was enhanced by 50% (to reach 81% mineralization) and energy consumption reduced by 38–50% via integrating a conventional electrochemical oxidation process with low-frequency ultrasound [96].

7.4   Conclusions

Treatment of textile dyeing wastewater and process streams has long been a major concern of environmental engineers, scientists and entrepreneurs not only due to the complexity of the effluent and the strict regulations on its discharge, but also to the socioeconomic and political dimensions of the problem. The aim of this chapter was to present an overview of the nature of textile dyeing mill wastewater and its treatability by ultrasonic irradiation and/or a variety of hybrid processes that utilize ultrasound. The discussion has focused particularly on the degradation and mineralization of reactive azo dyes, which due to their high color fastness are vastly consumed throughout the globe, but exert toxic effects in water via their low biodegradability and transformation to carcinogenic amines under anaerobic conditions.

The first part of the chapter has highlighted some fundamental concepts of sonochemistry that are essential in understanding the mechanism of dye degradation under ultrasonic irradiation, with sufficient amount of data to support the viability of sonochemistry as a "green process" for decolorizing and degrading dyeing process effluents and/or dye residuals. The second part was devoted to hybrid processes such as US/O$_3$, US/UV, US/Fenton, US/persulfate and US/solids, the latter with nanoparticles such as titanium dioxide, or reactive metals such as zero-valent iron and manganese dioxide. The efficiency of all hybrid processes for enhancing decolorization and mineralization of the test dyes was attributed to the synergy arising from the turbulent flow conditions of ultrasonication that accelerate rates of mass transfer (of solutes and chemical reagents) and radical formation reactions while enlarging and cleaning the solid surfaces enriched with reactive species and active sites, where dye molecules and their reaction intermediates undergo oxidative and reductive decomposition. Finally, it was noted that the power of hybrid processes also lied on ultrasonic activation of the additives or the solid surfaces to generate excess reactive species such as HO•, Fe^{2+}/Fe^{3+}, SO$_4^-$•, and the sonoluminescence of bubble collapse that provided sufficient energy for the formation and separation of electron-hole pairs in semiconductors and their composites, respectively.

References

1. A. M. Talarposhti, T. Donnelly, G. K. Anderson, Color removal from a simulated dye wastewater using a two-phase anaerobic packed bed reactor. *Water Res.* 35 (2001) 425–432.
2. Kasar-Dual, Personal communications. Çorlu, Tekirdag, Turkey, 2014.
3. R. Singla, F. Grieser, M. Ashokkumar. Sonochemical degradation of Martius Yellow dye in aqueous solution. *Ultrason. Sonochem.* 16 (2009) 28–34.
4. I. Arslan-Alaton, G. Tureli, T. Olmez-Hanci, Treatment of azo dye production wastewaters using Photo-Fenton-like advanced oxidation processes: Optimization by response surface methodology. *J. Photoc. Photobio. A.* 202(2–3) (2009) 142–153.
5. G. Tezcanli-Güyer, N. H. Ince, Degradation and toxicity reduction of textile dyestuff by ultrasound. *Ultrason. Sonochem.* 10 (2003) 235–240.
6. S. V. Kulkarni, C. D. Blackwell, A. L. Blackard, C. W. Stackhouse, M. W. Alexander, Textile dyes and dyeing equipment: Classification, properties, and environmental aspects air and energy engineering research laboratory. *EPA* EPA/600/S2–85/010 Apr. 1985.

7. A. Abel, The new color index international of pigments and solvent Dyes. *Surf. Coat. Int.* 81 (2) (1998) 77–85.

8. V. M. Correia, T. Stephenson, S. J. Judd, Characterization of textile wastewaters – A review. *Environ. Technol.* 15 (1994) 917–929.

9. G. I. Alaton, Electrocoagulation and advanced oxidation as intermediate effluent treatment steps for water reuse in a textile dyeing plant. Ph.D Thesis. Institute of Environmental Sciences, Bogazici University, 2005.

10. I. Arslan, I. Akmehmet Balcioglu, D. W. Bahnemann, Heterogeneous photocatalytic treatment of simulated dyehouse effluents using novel TiO2-photocatalysts. *Appl. Catal. B: Environ.* 26 (2000) 193–206.

11. A. Al-Kdasi, A. Idris, K. Saed, C. T. Guan, Treatment of textile wastewater by advanced oxidation processes – A review. *Global Nest: The Int. J.* 6(3) (2004) 222–230.

12. C. Long, Z. Lu, A. Li, W. Liu, Z. M. Jiang, J. L. Chen, Q. X. Zhang, Adsorption of reactive dyes onto polymeric adsorbents: Effect of pore structure and surface chemistry group of adsorbent on adsorptive properties. *Sep. Purif. Technol.* 44 (2005) 91.

13. M. F. Boeniger, Carcinogenity of azo dyes derived from benzidine. Department of Health and Human Services (NIOSH). Publ. No. 80–119. Cincinnati, OH, 1980.

14. N. H. Ince, "Critical" effect of hydrogen peroxide in photochemical dye degradation. *Water Res.* 33(4) (1999) 1080–1084.

15. H. N. Liu, G. T. Li, J. H. Qu, H. J. Li, Degradation of azo dye Acid Orange 7 in water by Fe^{0}/granular activated carbon system in the presence of ultrasound. *J. Hazard. Mater.* 144 (2007) 180–186.

16. I. Arslan-Alaton, Advanced oxidation of textile industry dyes in advanced oxidation proccesses for water and wastewater treatment. S. Parsons, ed., IWA Publishing, Alliance House, London, UK, 2004.

17. I. Gultekin, G. Tezcanli-Guyer, N. H. Ince, Sonochemical decay of C.I. Acid Orange 8: Effects of CCl_4 and t-butyl alcohol. *Ultrason. Sonochem.* 16 (2009) 577–581.

18. A. Rehorek, M. Tauber, G. Gübitz, Application of power ultrasound for azo dye degradation. *Ultrason. Sonochem.* 11 (2004) 177–182.

19. M. Stylidi, D. I. Kondarides, X. E. Verykios, Visible light induced photocatalytic degradation of Acid Orange 7 in aqueous TiO_2 suspensions. *Appl. Catal. B: Environ.* 47 (2004) 189–201.

20. N. H. Ince, I. G. Apikyan, Combination of activated carbon adsorption with light-enhanced chemical oxidation via hydrogen peroxide. *Water Res.* 34(17) (2000) 167–176.

21. I. Arslan, Treatability of a simulated disperse dye-bath by ferrous iron coagulation, ozonation, and ferrous iron-catalyzed ozonation. *J. Hazard. Mater.* 85(3) (2001) 229–241.

22. K. Vinodgopal, J. Peller, O. Makogon, P. V. Kamata, Ultrasonic mineralization of a reactive textile azo dye, remazol black B. *Water Res.* 32(12) (1998) 3646–3650.

23. J. M. Joseph, H. Destaillats, H. M. Hung, M. R. Hoffmann, The sonochemical degradation of azobenzene and related azo dyes: Rate enhancements via Fenton's reactions, *J. Phys. Chem. A* 104 (2000) 301–307.
24. M. Sivakumar. A. B. Pandit, Ultrasound enhanced degradation of Rhodamine B: optimization with power density. *Ultrason. Sonochem.* 8 (2001) 233–240.
25. N. H. Ince, G. Tezcanli-Güyer, Reactive dyestuff degradation by combined sonolysis and ozonation. *Dyes and Pigments* 49(3) (2001) 145–153.
26. J. Ge, J. Qu, Degradation of azo dye acid red B on manganese dioxide in the absence and presence of ultrasonic irradiation. *J. Hazard. Mater. B* 100 (2003) 197–207.
27. G. Tezcanli-Güyer, N. H. Ince, Individual and combined effects of ultrasound, ozone and UV irradiation: A case study with textile dyes. *Ultrasonics* 42(19) (2004) 591–596.
28. I. Gultekin, N. H. Ince, Degradation of aryl-azo-naphthol dyes by ultrasound, ozone and their combination: Effect of alpha substituents. *Ultrason. Sonochem.* 13 (2006) 208.
29. Z. Eren, N. H. Ince, Sonolytic and sonocatalytic degradation of azo dyes by low and high frequency ultrasound. *J. Hazard. Mater.* 177 (2010) 1019–1024.
30. R. Kidak, N. H. Ince, Effects of operating parameters on sonochemical decomposition of phenol. *J. Hazard. Mater. B* 137 (2006) 1453–1457.
31. K. S. Suslick. Sonochemistry. *Science* 247 (1990) 1439–1445.
32. T. J. Mason. *Sonochemistry.* Oxford University Press, New York, 1998.
33. N. H. Ince, G. Tezcanli, R. K. Belen, Ultrasound as a catalyzer of aqueous reaction systems: The state of the art and environmental applications. *Appl. Catal. B-Environ.* 29(3) (2001) 167–176.
34. D. Koyabashi, C. Honma, A. Suzuki, T. Takahashi, H. Matsumoto, D. Kuroda, K. Otake, A. Shono, Comparison of ultrasonic degradation rates constants of methylene blue at 22.8 kHz, 127 kHz, and 490 kHz. *Ultrason. Sonochem.* 19 (2012) 745–749.
35. K. Okitsu, K. Iwasaki, Y. Yobiko, H. Bandow, R. Nishimura, Y. Maeda, Sonochemical degradation of azo dyes in aqueous solution: A new heterogeneous kinetics model taking into account the local concentration of OH radicals and azo dyes. *Ultrason. Sonochem.* 12 (2005) 255–262.
36. A. Rehorek, P. Hoffmann, A. Kandelbauer, G. M. Gübitz, Sonochemical substrate selectivity and reaction pathway of systematically substituted azo compounds. *Chemosphere* 67 (2007) 1526–1532.
37. S. Vajnhandl, A. M. L. Marechal, Case study of the sonochemical decolouration of textile azo dye Reactive Black 5. *J. Hazard. Mater.* 141 (2007) 329–335.
38. L. Wang, L. Zhu, W. Luo, Y. Wu, H. Tang, Drastically enhanced ultrasonic decolorization of methyl orange by adding CCl_4. *Ultrason. Sonochem.* 14 (2007) 253.
39. C. H. Wu, Effects of sonication on decolorization of C.I. Reactive Red 198 in UV/ZnO system. *J. Hazard. Mater.* 153 (2008) 1254–1261.

40. P. Chowdhury, T. Viraraghavan, Sonochemical degradation of chlorinated organic compounds, phenolic compounds and organic dyes – A review. *Sci. Total Environ.* 407 (2009) 2474–2492.

41. N. H. Ince, G. Tezcanli-Guyer, Impacts of pH and molecular structure on ultrasonic degradation of azo dyes. *Ultrasonics* 42 (2004) 591–596.

42. A. S. Özen, V. Aviyente, R. A. Klein, Modeling the oxidative degradation of azo dyes: A density functional theory study. *J. Phys. Chem. A* 107 (2003) 4898–4907.

43. S. Das, P. V. Kamat, S. Padmaja, V. Au, S. A. Madison, Free radical induced oxidation of the azo dye Acid Yellow 9. *J. Chem. Soc., Perkin Trans.* 2 (1999) 1219–1223.

44. C. Minero, P. Pellizzari, V. Maurino, E. Pelizzetti, D. Vione, Enhancement of dye sonochemical degradation by some inorganic anions present in natural waters. *Appl. Catal. B: Environ.* 77 (2008) 308–316.

45. J. Oakes, P. Gratton, Kinetic investigations of azo dye oxidation in aqueous media. *J. Chem. Soc., Perkin Trans.* 2 (1998) 1857–1864.

46. S. Dalhatou, C. Petrier, S. Laminsi, S. Baup, Sonochemical removal of napthol blue black azo dye: Influence of paramaters and effect of mineral ions. *Int. J. Environ. Sci. Technol.* (2013) DOI 10.1007/s13762–013–0432–8.

47. T. J. Mason. *Sonochemistry: The Use of Ultrasound in Chemistry*. Roy. Soc. Ch. Cambridge, U. K., 1990.

48. F. Guzman-Duque, C. Pétrier, C. Pulgarin, G. Peñuela, R. A. Torres-Palma, Effects of sonochemical parameters and inorganic ions during the sonochemical degradation of crystal violet in water. *Ultrason. Sonochem.* 18 (2011) 440–446.

49. C. Petrier, A. Francony, Ultrasonic waste-water treatment: Incidence of ultrasonic frequency on the rate of phenol and carbon tetrachloride degradation. *Ultrason. Sonochem.* 4 (1997) 295–300.

50. T. Velegraki, I. Poulios, M. Charalabak, Photocatalytic and sonolytic oxidation of acid orange 7 in aqueous solution. *Appl. Catal. B: Environmental* 62 (2006) 159–168.

51. M. Guiterrez, A. Henglein, F. Ibanez, Radical scavenging in the sonolysis of aqueous solutions of iodide, bromide, and azide. *J. Phys. Chem.* 95(15) (1991) 6044–6047.

52. C. Petrier, M. F. Lamy, A. Francony, A. Benahcene, B. David, V. Renaudin, N. Gondrexon, Sonochemical degradation of phenol in dilute aqueous solutions: Comparison of the reaction rates at 20 and 487 kHz. *J. Phys. Chem.* 98(41) (1994) 10514–10520.

53. E. Ciawi, M. Ashokkumar, F. Grieser, Limitations of the methyl radical recombination method for acoustic cavitation bubble temperature measurements in aqueous solutions. *J. Phys. Chem. B* 110(20) (2006) 9779–9781.

54. S. Sochard, A. M. Wilhelm, H. Delmas, Modelling of free radicals production in a collapsing gas-vapour bubble. *Ultrason. Sonochem.* 4(82) (1997) 77–84.

55. C. Petrier, M. Micolle, G. Merlin, J. L. Luche, G. Reverdy, Characteristics of pentachlorophenate degradation in in aqueous solution by means of ultra-sound. *Environ. Sci. Technol.* 26 (1992) 1639–1642.

56. K. Okitsu, K. Kawasaki, B. Nanzai, N. Takenaka, H. Bandow, Effect of carbon tetrachloride on sonochemical decomposition of methyl orange in water. *Chemosphere* 71 (2008) 36–42.

57. O. Moumeni, O. Hamdaoui, C. Pétrier, Sonochemical degradation of mala-chite green in water. *Chem. Eng. Process.* 62 (2012) 47–53.

58. S. Merouani, O. Hamdaoui, F. Saoudi, M. Chiha, C. Pétrier, Influence of bicarbonate and carbonate ions on sonochemical degradation of Rhodamine B in aqueous phase. *J. Hazard. Mater.*175 (2010) 593–599.

59. Z. He, L. Lin, S. Song, M. Xia, L. Xu, H. Ying, J. Chen, Mineralization of C.I. Reactive Blue 19 by ozonation combined with sonolysis: Performance optimization and degradation mechanism. *Sep. Purif. Technol.* 62 (2008) 376.

60. M. Li, J. T. Li, H. W. Sun, Decolorizing of azo dye Reactive red 24 aqueous solution using exfoliated graphite and H2O2 under ultrasound irradiation. *Ultrason. Sonochem.* 15 (2008) 717–723.

61. A. Somayajula, P. Asaithambi, M. Susree, M. Matheswaran, Sonoelectrochemical oxidation for decolorization of Reactive Red 195. *Ultrason. Sonochem.* 19 (2012) 803–811.

62. Y. Jiang, C. Petrier, T. D. Waite, Sonolysis of 4-chlorophenol in aqueous solu-tion: Effects of substrate concentration, aqueous temperature and ultrasonic frequency. *Ultrason. Sonochem.* 13(5) (2006) 415–422.

63. G. V. Buxton, C. L. Greenstock, W. P. Helman, A. B. Ross, Critical-review of rate constants for reactions of hydrated electrons, hydrogen-atoms and hydroxyl radicals (•OH/•O-) in aqueous solution. *J. Phys. Chem.* 17 (1988) 513–886.

64. I. Grcic, M. Maljkovic, S. Papic, N. Koprivanac, Low frequency US and UV-A assisted Fenton oxidation of simulated dyehouse wastewater. *J. Hazard. Mater.* 197 (2011) 272–284.

65. I. Grcic, S. Papi, N. Koprivanac, I. Kovacic, Kinetic modeling and synergy quantification in sono- and photo-oxidative treatment of simulated dye-house effluent. *Water Res.* 46 (2012) 5683–5695.

66. S. Parsons. Advanced Oxidation Processes for Water and Wastewater Treatment, S. Parsons, ed., IWA Publishing, Alliance House, London, UK, 2004.

67. T. Mason, D. Peters. *Practical Sonochemistry, 2nd ed.: Power Ultrasound Uses and Applications.* Horwood Chemical Science Series, Cambridge, 2002.

68. D. J. Casadonte Jr., M. Flores, C. Petrier, Enhancing sonochemical activity in aqueous media using power-modulated pulsed ultrasound: An initial study. *Ultrason. Sonochem.* 12 (2005) 147.

69. Z. Ai, J. Li, L. Zhang, S. Lee, Rapid decolorization of azo dyes in aqueous solu-tion by an ultrasound-assisted electrocatalytic oxidation process. *Ultrason. Sonochem.* 17 (2010) 370–375.

70. J. Wang, Y. Lv, L. Zhang, B. Liu, R. Jiang, G. Han, R. Xu, X. Zhang, Sonocatalytic degradation of organic dyes and comparison of catalytic activities of CeO_2/ TiO_2, SnO_2/TiO_2 and ZrO_2/TiO_2 composites under ultrasonic irradiation. *Ultrason. Sonochem.* 17 (2010) 642–648.

71. C. S. Poon, Q. Huang, P. C. Huang, Degradation kinetics of cuprophenyl yellow RL by UV/H2O2/ultrasonication (US) process in aqueous solution. *Chemosphere* 38(5) (1999) 1005–1014.

72. T. M. Olson, P. F. Barbier, Oxidation kinetics of natural organic matter by sonolysis and ozone. *Water Res.* 28 (1994) 1383.

73. J. W. Kang, M. R. Hoffmann, Kinetics and mechanism of the sonolytic destruction of Methyl tert-Butyl Ether by ultrasonic irradiation in the presence of ozone. *Environ. Sci. Technol.* 32 (1998) 3194.

74. Z. He, S. Song, H. Zhou, H. Ying, J. Chen, C.I. Reactive Black 5 decolorization by combined sonolysis and ozonation. *Ultrason. Sonochem.* 14 (2007) 298–304.

75. H. Zhang, L. Duan, D. Zhang, Absorption kinetics of ozone in water with ultrasonic radiation. *Ultrason. Sonochem.* 14 (2007) 552–556.

76. A. O. Martins, V. M. Canalli, C. M. N. Azevedo, M. Pires, Degradation of pararosaniline (C.I. Basic Red 9 monohydrochloride) dye by ozonation and sonolysis. *Dyes Pigments* 68 (2006) 227–234.

77. L. K. Weavers, M. R. Hoffmann, Sonolytic decomposition of ozone in aqueous solution: Mass transfer effects. *Environ. Sci. Technol.* 32 (1998) 3941.

78. A. Ziylan, N. H. Ince, Ozonation-based advanced oxidation for pre-treatment of water with residuals of anti-inflammatory medication. *Chem. Eng. J.* 220 (2013) 151–160.

79. E. Reisz, W. Schmidt, H. P. Schuchmann, Von C. Sonntag, Photolysis of ozone in aqueous solutions in presence of tertiary butanol: Considerations regarding the usefulness of the O3/UV process in water purification. *Environ. Sci. Technol.* 37 (2003) 1941–1948.

80. Z. Eren, N. H. Ince, F. N. Acar, Degradation of textile dyes, dyebaths and dyeing wastewater by homogeneous and heterogeneous sonophotolysis. *J. Adv. Oxid. Technol.* 13(2) (2010) 206–211.

81. H. Zhang, L. Duan, D. Zhang, Decolorization of methyl orange by ozonation in combination with ultrasonic irradiation. *J. Hazard. Mater.* 138 (2006) 53–59.

82. S. Song, H. Ying, Z. He, J. Chen, Mechanism of decolorization and degradation of CI Direct Red 23 by ozonation combined with sonolysis. *Chemosphere* 66 (2007) 1782–1788.

83. Z. He, S. Song, M. Xia, J. Qiu, H. Ying, B. Lu, Y. Jiang, J. Chen, Mineralization of C.I. Reactive Yellow 84 in aqueous solution by sonolytic ozonation. *Chemosphere* 69 (2007) 191–199.

84. J. H. Sun, S. P. Sun, J. Y. Sun, R. X. Sun, L. P. Qiao, H. Q. Guo, M. H. Fan, Degradation of azo dye Acid black 1 using low concentration iron of Fenton process facilitated by ultrasonic irradiation. *Ultrason. Sonochem.* 14 (2007) 761–766.

85. H. Zhang, H. Fu, D. Zhang, Degradation of C.I. Acid Orange 7 by ultrasound enhanced heterogeneous Fenton-like process. *J. Hazard. Mater.* 172 (2009) 654–660.

86. J. J. Lin, X. Zhao, D. Liu, Z. Yu, Y. Zhang, H. Xua, The decolorization and mineralization of azo dye C.I. Acid Red 14 by sonochemical process: Rate improvement via Fenton's reactions. *J. Hazard. Mater.* 157 (2008) 541–546.

87. I. Grcic, D. Vujevic, N. Koprivanac. Modelling the mineralization and discoloration in colored systems by $(US)Fe^{2+}/H_2O_2/S_2O_8^{2-}$ processes: A proposed degradation pathway. *Chem. Eng. J.* 157 (2010) 35–44.

88. N. Shimizu, C. Ogino, M. F. Dadjour, T. Murata, Sonocatalytic degradation of methylene blue with TiO_2 pellets in water. *Ultrason. Sonochem.* 14 (2007) 184–190.

89. J. Wang, Z. Jiang, L. Zhang, P. Kang, Y. Xie, Y. Lv, R. Xu, X. Zhang, Sonocatalytic degradation of some dyestuffs and comparison of catalytic activities of nano-sized TiO_2, nano-sized ZnO and composite TiO_2/ZnO powders under ultrasonic irradiation. *Ultrason. Sonochem.* 16 (2009) 225–23.

90. J. Gao, R. Jiang, J. Wang, P. Kang, B. Wang, Y. Li, K. Li, X. Zhang, The investigation of sonocatalytic activity of $Er^{3+}:YAlO_3/TiO_2$-ZnO composite in azo dyes degradation. *Ultrason. Sonochem.* 18 (2011) 541–548.

91. Y. L. Song, J. T. Li, Degradation of C.I. Direct Black 168 from aqueous solution by fly ash/H_2O_2 combining ultrasound. *Ultrason. Sonochem.* 16 (2009) 440–444.

92. M. H. Entezari, Z. Sharif Al-Hoseini, Sono-sorption as a new method for the removal of methylene blue from aqueous solution. *Ultrason. Sonochem.* 14 (2007) 599–604.

93. S. Sonawane, P. Chaudhari, S. Ghodke, S. Phadtare, S. Meshram, Ultrasound assisted adsorption of basic dye onto organically modified bentonite (nanoclays). *J. Sci. Ind. Res. India* 68 (2009) 162–167.

94. H. N. Liu, G. T. Li, J. H. Qu, H. J. Li, Degradation of azo dye Acid Orange 7 in water by Fe^0/granular activated carbon system in the presence of ultrasound. *J. Hazard. Mater.* 144 (2007) 180–186.

95. P. Gayathri, R. Praveena Juliya Dorathi, K. Palanivelu, Sonochemical degradation of textile dyes in aqueous solution using sulphate radicals activated by immobilized cobalt ions. *Ultrason. Sonochem.* 17 (2010) 566–571.

96. M. Siddique, R. Farooq, Z. M. Khan, Z. Khan, S. F. Shaukat, Enhanced decomposition of reactive blue 19 dye in ultrasound assisted electrochemical reactor. *Ultrason. Sonochem.* 18 (2011) 190–196.

97. M. Rivera, M. Pazos, M. A. Sanroman, Improvement of dye electrochemical treatment by combination with ultrasound technique. *J. Chem. Technol. Biotechnol.* 84 (2009) 1118–1124.

98. J. P. Lorimer, T. J. Mason, M. Plattes, S. S. Phull, Dye effluent decolourisation using ultrasonically assisted electro-oxidation. *Ultrason. Sonochem.* 7 (2000) 237–242.

99. J. Liang, S. Komarov, N. Hayashi, E. Kasai, Recent trends in the decomposition of chlorinated aromatic hydrocarbons by ultrasound irradiation and Fenton's reagent. *J. Mater. Cycles Waste Manage.* 9 (2007) 47–55.

100. H. Zhang, J. Zhang, C. Zhang, F. Liu, D. Zhang, Degradation of C.I. Acid Orange 7 by the advanced Fenton process in combination with ultrasonic irradiation. *Ultrason. Sonochem.* 16 (2009) 325–330.

101. M. H. Entezari, A. Heshmari, A. Sarafraz-Yazdi, A combination of ultrasound and inorganic catalyst removal of 2-chlrophenol from aqueous solution. *Ultrason. Sonochem.* 12 (2005) 137–141.

102. A. B. Pandit, M. Sivakumar, Ultrasound enhanced degradation of Rhodamine B: Optimization with power density. *Ultrason. Sonochem.* 8 (2001) 233–240.

103. Z. Al-Qodah, Adsorption of dyes using shale oil ash. *Water Res.* 34 (2000) 4295–4303.

104. V. Meshko, L. Markovska, M. Mincheva, A. E. Rodrigues, Adsorption of basic dyes on granular activated carbon and natural zeolite. *Water Res.* 35 (2001) 3357–3366.

105. R. S. Juang, R. L. Tseng, F. C. Wu, S. H. Lee, Adsorption behavior of reactive dyes from aqueous solutions on chitosan. *J. Chem. Technol. Biotechnol.* 70 (1997) 391–399.

106. M. H. Entezari, Z. Sharif Al-Hoseini, N. Ashraf, Fast and efficient removal of Reactive Black 5 from aqueous solution by a combined method of ultrasound and sorption process. *Ultrasonics Sonochemistry* 15 (2008) 433–437.

107. M. S. El-Geundi, Homogeneous surface diffusion model of basic dyestuffs onto natural clay in batch adsorbers. *Adsorpt. Sci. Technol.* 8 (1991) 217–225.

108. G. Rytwo, S. Nir, M. Crespin, L. Margulies, Adsorption and interactions of Methyl Green with Montmorillonite and Sepiolite. *J. Colloid Interf. Sci.* 222 (2000) 12–19.

109. R. Pelegrini, P. Peralta, A. R. de Andrade, J. Reyes, N. Duran, Electrochemically assisted photocatalytic degradation of reactive dyes. *Appl. Catal. B. Environ.* 22 (1999) 83–90.

110. J. P. Lorimer, T. J. Mason, M. Plattes, S. S. Phull, D. J. Walton, Degradation of dye effluent. *Pure Appl. Chem.* 73 (2001) 1957.

8

Biosorption of Organic Dyes: Research Opportunities and Challenges

Guilherme L. Dotto[1], Sanjay K. Sharma[2] and Luiz A. A. Pinto*[,3]

[1]*Chemical Engineering Department, Federal University of Santa Maria-UFSM,
Santa Maria, Brazil*
[2]*Green Chemistry & Sustainability Research Group, Department of Chemistry,
JECRC University, Jaipur, India*
[3]*School of Chemistry and Food, Federal University of Rio Grande-FURG,
Rio Grande-RS, Brazil*

This chapter is dedicated to the research group of the Unit Operation Laboratory-Federal University of Rio Grande (FURG), Brazil.

Abstract

Biosorption is an emerging green technology to remove organic dyes from effluents. However, many efforts are still necessary to make biosorption an attractive option in relation to the conventional treatment processes. This chapter presents some important aspects regarding biosorption technology. Firstly, the general aspects of biosorption are presented. Secondly, the classification, advantages and properties of the biosorbents are discussed. Factors affecting biosorption are explained, after which studies on equilibrium, thermodynamics, kinetics and mechanisms are addressed. Finally, future perspectives and challenges are presented. It is expected that this information will assist future research in the biosorption field.

Keywords: Biosorption, biosorbents, dyes, effluents, equilibrium, mechanism

Acronyms

DCEFs: Dye-containing effluents
EDS: Energy dispersive X-ray spectroscopy
FT-IR: Fourier transform infrared spectroscopy

Corresponding author: dqmpinto@furg.br

Sanjay K. Sharma (ed.) Green Chemistry for Dyes Removal from Wastewater, (295–329)
© 2015 Scrivener Publishing LLC

GlcN: Glucosamine
GlcNAc: N-acetyl-glucosamine
pH_{zpc}: point of zero charge
SEM: Scanning electron microscopy
SODs: Synthetic organic dyes
T: Temperature

8.1   General Considerations

8.1.1   Dye-Containing Effluents

Mankind has used dyes for thousands of years, and the first known use of an organic colorant was nearly 4000 years ago, when the blue dye indigo was found in the wrappings of mummies in Egyptian tombs. In 1856, William Henry Perkin accidentally discovered the world's first commercially successful synthetic dye. Starting that year, dyes were manufactured synthetically and on a large scale [1]. Synthetic organic dyes (SODs) are extensively used in many fields such as the textile industry, leather tanning industry, paper production, food technology, agricultural research, light-harvesting arrays, photoelectrochemical cells and in hair colorings [2]. The SODs are indentified by their color index (C.I.) and classified as: acid dyes, basic dyes, disperse dyes, direct dyes, reactive dyes, solvent dyes, sulfur dyes and vat dyes [3]. Overall, at present there are more than 100,000 commercial dyes with a rough estimated production of $7 \times 10^5 - 1 \times 10^6$ tons per year [4]. A portion of these dyes is lost in the manufacturing and processing units, and are destined for industrial effluents [5].

The dye-containing effluents (DCEFs) are characterized by high alkalinity, biological oxidation demand, chemical oxidation demand, and total dissolved solids with dye concentrations generally below 1 g L^{-1} [6]. The inadequate disposal of untreated DCEFs in water bodies causes serious direct and indirect impacts on the environment and human health. Some direct impacts are: color change, poor sunlight penetration, damage on flora and fauna, ground water pollution, depletion of dissolved oxygen and suppression in the reoxygenation capacity. The indirect impacts are: killing of aquatic life, eutrophication, genotoxicity, microtoxicity and damage to the immune system of human beings [7]. Thus, several governments have established environmental restrictions with regard to the quality of DCEFs and this has obligated the industries to remove dyes from their effluents before discharge [8]. However, DCEFs are very difficult to treat, since the dyes are recalcitrant molecules with a complex aromatic structure, resistant to aerobic digestion, and stable to oxidation agents [3–7]. This is why many efforts have been made by environmental researchers in order to develop effective and low-cost technologies to remove dyes from industrial effluents.

8.1.2 Technologies for Dye Removal

As reported in recent review articles [2–7,9–11], several technologies have been developed to treat DCEFs such as coagulation-flocculation, filtration, sedimentation, precipitation-flocculation, electrocoagulation-electroflotation, biodegradation, photocatalysis, oxidation, electrochemical treatment, membrane separation, ion-exchange, incineration, irradiation, advanced oxidation, bacterial decolorization, electrokinetic coagulation and adsorption on activated carbon. It is evident that each technique has its own advantages and drawbacks. From the industrial viewpoint, no single process provides adequate treatment, being that significant reduction of expenses and enhancement of dye removal can be achieved by the combination of different methods in hybrid treatments [9,11]. So, there is a need to develop alternative decolorization methods that are effective and acceptable in industrial use [9]. Among the above-mentioned technologies, it is recognized that adsorption on activated carbon is one of the most powerful to treat DCEFs, providing benefits such as ease of operation and high efficiency [5,10]. However, its use is limited by the obtention and regeneration costs of activated carbon [4,6,9]. Alternatively, biosorption can be employed to treat DCEFs, because it combines the advantages of adsorption with the use of natural, low-cost, eco-friendly and renewable biosorbents [6,12,13].

8.1.3 General Aspects of Biosorption

Probably, the work of Adams and Holmes [14] represented an early attempt at biosorption. They described the removal of Ca and Mg ions by tannin resin, black wattle bark (*Acacia mollissima*) [14]. However, biosorption has gained special attention from 1970 on, due to the increase in environmental awareness, which has led to a search for new techniques capable of the inexpensive treatment of polluted effluents [15]. Two definitions are commonly accepted in literature for the term "biosorption": 1) Biosorption is the removal of materials (compounds, metal ions, organic dyes, etc.) by inactive, non-living biomass (materials of biological origin) due to high attractive forces present between the two [15]; 2) Biosorption is the passive uptake of pollutants from aqueous solutions by the use of non-growing or non-living microbial mass, thus allowing the recovery and/or environmentally acceptable disposal of the pollutants. The term is used to indicate a number of metabolism-independent processes (physical and chemical adsorption, electrostatic interaction, ion exchange, complexation, chelation, and microprecipitation) taking place essentially in the cell wall [12].

It is common sense in the literature that the biosorption process involves a solid phase (sorbent, biosorbent or biological material) and a liquid phase (solvent, normally water) containing a dissolved species to be sorbed (adsorbate, organic dyes for example). Due to the high affinity between the biosorbent and adsorbate, the latter is attracted and bound there by different interaction mechanisms. The process continues till equilibrium is established between the amount of solid-bound adsorbate species and its portion remaining in the solution. The degree of biosorbent affinity for the adsorbate determines its distribution between the solid and liquid phases [12,13,15–17].

In the same way as other techniques for the treatment of DCEFs, biosorption has its own advantages and drawbacks. The main advantages of biosorption are: biosorbent materials can be found easily as wastes or byproducts and at almost no cost; the biosorbents are cheap, eco-friendly and renewable; the biosorbents have several functional groups on the surface, capable of binding with many classes of dyes; selectivity for many dyes; competitive performance (biosorption is capable of a performance comparable to the most similar technique, ion exchange treatment) [12,16,18,19]; there is no need of costly growth media; the process is independent of physiological constraints of living cells; the process is very rapid, as non-living material behaves as an ion-exchange resin; the conditions of the process are not limited by the living biomass; no aseptic conditions are required; the process is reversible and the dye can be desorbed easily, thus recycling of the materials is quite possible; chemical or biological sludge is minimized [15–17,20]. On the other hand, the drawbacks are: the characteristics of the biosorbents cannot be biologically controlled; centrifugation or filtration operations are necessary for the solid-liquid separation after the biosorption process; mass loss after regeneration; difficulties for scale-up [12,20].

From the operational viewpoint, biosorption is generally based on two types of investigations: batch systems (discontinuous operation) and dynamic systems (continuous operation) [9,12]. The first and fundamental investigations regarding biosorption are made in batch systems. These operations are cheap, simple to operate and, consequently, are adequate for small- and medium-size applications, using simple and readily available mixing tank equipment. Other advantages are, simplicity, well-established experimental methods, controlled conditions and easily interpretable results [4,9,13]. For dynamic systems (continuous operation), the most common configuration is the packed column, but others, such as fluidized bed and moving bed are possible. This method has a number of process engineering advantages including high efficiency operations and relatively easy scaling up from a laboratory scale procedure. The stages in the separation protocol can also be automated and a high degrees of purification can be attained in a single-step process. A large volume of DCEFs can be continuously treated using a defined quantity of biosorbent in the column [12,21]. This chapter is mainly focused on the batch system, since this operation is more extensively studied in the literature and its investigation is fundamental.

8.2 Biosorbents

Many alternative biological materials are capable of removing pollutants (SODs for example) from aqueous solutions by biosorption. These biological materials are known as "biosorbents" [6,12,15–19]. According to the literature, the main biosorbents can be classified as follows: agricultural wastes, algae biomass, bacterial biomass, chitosan and fungal biomass [2,4,6,9,12]. In this section, these classes are discussed in detail.

8.2.1 Agricultural Wastes

Agricultural wastes are lignocellulosic materials that consist of three main structural components, which are lignin, cellulose and hemicelluloses. These organic compounds are useful for binding SODs through different mechanisms [22,23]. Agricultural wastes are low-cost, renewable and available in large amounts. Furthermore, they are usually used without or with a minimum of processing (washing, drying, grinding) and thus reduce production costs by using a cheap raw material and eliminating energy costs associated with thermal treatment [3,24–30]. Some examples of these biosorbents are sawdust, bark, roots, hulls, leaves, stalks, bagasse, peels, seeds and others [23–35]. Table 8.1 summarizes the biosorption capacities of some important agricultural wastes.

Alencar *et al.* [23] studied the application of *Mangifera indica* (mango) seeds as biosorbent for removal of Victazol Orange-3R (VO-3R) dye from aqueous solutions. They obtained the best results (51.2 mg g^{-1}) at a pH

Table 8.1 Biosorption capacities of agricultural wastes.

Agricultural waste	Dye	pH	T (K)	Biosorption capacity (mg g⁻¹)	Reference
Mangifera indica seeds	Victazol Orange-3R	2.0	323	51.2	[23]
Coffee residues	Remazol Blue RN	2.0	298	179.0	[24]
Coffee residues	Basic Blue 3G	10.0	298	295.0	[24]
Wood apple shell	Methylene Blue	10.0	305	95.2	[26]
Wood apple shell	Crystal Violet	10.0	305	130.0	[26]
Grapefruit peel	Crystal Violet	6.0–10.0	318	254.2	[27]
Peanut husk	Methylene Blue	7.0	293	72.1	[28]
Princess tree leaf	Basic Red 46	8.0	298	43.1	[29]
Pine cone	Acid Black 26	2.0	338	62.9	[30]
Pine cone	Acid Green 25	2.0	338	43.3	[30]
Pine cone	Acid Blue 7	2.0	338	37.4	[30]
Pineapple leaf powder	Basic Green 4	9.0–10.0	298	54.6	[31]
Aqai palm stalk	Reactive Black 5	2.0	298	52.3	[32]
Aqai palm stalk	Reactive Orange 16	2.0	298	61.3	[32]
Cupuassu shell	Reactive Red 194	2.0	298	64.1	[33]
Cupuassu shell	Direct Blue 53	2.0	298	37.5	[33]
Capsicum annuum seeds	Reactive Blue 49	2.0	298	96.3	[34]
Jujuba seeds	Congo Red	2.0	333	55.6	[35]

of 2 and temperature of 323 K. They proposed the following interaction mechanism: H- bonds were formed between the VO-3R and lingocellulose through water molecules, which were essential for the interaction. Kyzas *et al.* [24] proposed the use of untreated coffee residues for the removal of Remazol Blue RN (RB) and Basic Blue 3G (BB) from aqueous solutions.

The maximum adsorption capacities for RB and BB were 179 mg g^{-1} (at pH of 2) and 295 mg g^{-1} (at pH of 10), respectively. Furthermore, they found that after 10 cycles of biosorption-desorption, the reduction in biosorption percentages from the 1st to 10th cycle was approximately 7% for both dyes. Çelekli et al. [25] verified the potential of walnut husk to remove Lanaset Red G from aqueous solutions. Based on the artificial neural network studies, they verified that the pH was the most important parameter, followed by the initial dye concentration. More information about the use of agricultural solid wastes as biosorbents can be obtained in a nice review recently published by Salleh et al. [3].

8.2.2   Algae Biomass

The term algae refers to a large and diverse assemblage of organisms that contain chlorophyll and carry out oxygenic photosynthesis [36]. However, in this section, blue green algae (which are cyanobacteria) are also included because they have similar characteristics from a biosorption viewpoint. Algae biomass has been found to be potential biosorbents because of their availability in both fresh and saltwater [6]. This type of biomass is composed mainly of proteins, carbohydrates and lipids, which contain many functional groups such as carboxyl, hydroxyl, sulfate, phosphate, amines, aldehydes, ketones and others [37–40]. These functional groups have high binding affinity for SODs [12,18,19,37–42]. The dye removal by algae can be attributed to the accumulation of dye ions on the surface of algal biopolymers and further to the diffusion of the dye molecules from aqueous phase onto the solid phase of the biopolymer [6,12]. Fast and easy growth in simple medium, low cost, availability in large quantities, high binding affinity and renewability are some of the advantages for the use of algae biomass as biosorbents [38,39,41,42]. Table 8.2 summarizes the biosorption capacities of algae biomass.

Dogar et al. [39] studied the biosorption of methylene blue (MB) from aqueous solutions using green algae Ulothrix sp. They concluded that electrostatic interactions occurred between Ulothrix sp. and MB. The brown macroalga, Stoechospermum marginatum, was tested by Daneshvar et al. [42] to remove Acid Blue 25 (AB25), Acid Orange 7 (AO7) and Acid Black 1 (AB1) from aqueous solutions. They found that the equilibrium curves were successfully described by the Freundlich model, and the estimated biosorption capacities were 22.2, 6.73 and 6.57 mg g^{-1} for AB25, AO7 and AB1, respectively. Dotto et al. [43] investigated the equilibrium and thermodynamics of the biosorption of azo dyes onto Spirulina platensis. They found that Tartrazine biosorption occurred by formation

Table 8.2 Biosorption capacities of algae biomass.

Algae biomass	Dye	pH	T (K)	Biosorption capacity (mg g⁻¹)	Reference
Spirulina platensis	Acid Blue 9	2.0	298	1653.0	[37]
Spirulina platensis	FD&C Red 40	2.0	298	400.3	[37]
Azolla filiculoides	Basic Orange	7.0	303	833.3	[38]
Ulothrix sp.	Methylene Blue	7.9	293	86.1	[39]
Stoechospermum marginatum	Acid Blue 25	2.0	300	22.2	[42]
Stoechospermum marginatum	Acid Orange 7	2.0	300	6.7	[42]
Stoechospermum marginatum	Acid Black 1	2.0	300	6.6	[42]
Chlorella vulgaris	Remazol Black B	2.0	308	555.6	[45]
Chlorella vulgaris	Remazol Red RR	2.0	298	196.1	[45]
Chlorella vulgaris	Remazol Golden	2.0	298	71.9	[45]
Azolla rongpong	Acid Red 88	2.5	303	81.3	[46]
Azolla rongpong	Acid Green 3	2.5	303	83.3	[46]
Azolla rongpong	Acid Orange 7	2.5	303	76.9	[46]
Azolla rongpong	Acid Blue 15	2.5	303	76.3	[46]
Caulerpa scalpelliformis	Sandocryl Yellow	8.0	293	27.0	[47]
Spirogyra sp.	Acid Orange 7	4.0	318	6.2	[48]
Spirogyra sp.	Basic Red 46	10.0	318	13.2	[48]
Spirogyra sp.	Basic Blue 3	10.0	318	12.2	[48]

of a multilayer, but for Allura red biosorption occurred by formation of a monolayer on the S. platensis surface. The maximum biosorption capacities were 363.2 mg g⁻¹ and 468.7 mg g⁻¹ for Tartrazine and Allura red, respectively. The thermodynamic evaluation showed that the biosorption of azo dyes onto S. platensis was a spontaneous, favorable and exothermic process. In another study, Cardoso *et al.* [44] verified that *Spirulina platensis* was more efficient than commercial activated carbon for the removal of Reactive Red 120 dye.

8.2.3   Bacterial Biomass

Bacteria constitute a large domain or kingdom of prokaryotic microorganisms [49]. In the last years, bacterial biomass has been tested as biosorbent to remove SODs from aqueous solutions [50–54]. Two general types of bacteria exist, Gram-positive and Gram-negative. Gram-positive bacteria are comprised of a thick peptidoglycan layer connected by amino acid bridges. Imbedded in the Gram-positive cell wall are polyalcohols, some of which are lipid linked to form lipoteichoic acids. The cell wall of Gram-negative bacteria is thinner, and composed of only 10–20% peptidoglycan. Furthermore, the cell wall contains an additional outer membrane composed of phospholipids and lipopolysaccharides [6,55]. These chemical functional groups on the cell walls are potential biosorption sites to interact with SODs [56,57]. The biosorption of SODs by bacterial biomass is an eco-friendly and cost-effective process, since this biomass is generally obtained in large quantities as residue of full-scale fermentation processes [52–54]. Table 8.3 summarizes the biosorption capacities of bacterial biomass.

In the study of Won *et al.* [52], the binding mechanisms involved in the biosorption of reactive dyes onto *Corynebacterium glutamicum* were

Table 8.3 Biosorption capacities of bacterial biomass.

Bacterial biomass	Dye	pH	T (K)	Biosorption capacity (mg g^{-1})	Reference
Corynebacterium glutamicum	Basic Blue 3	6.0–9.0	298	50.4	[50]
Corynebacterium glutamicum	Reactive Blue 4	2.0	298	184.9	[52]
Corynebacterium glutamicum	Reactive Orange 16	2.0	298	156.6	[52]
Corynebacterium glutamicum	Reactive Yellow 2	2.0	298	155.0	[52]
Bacillus subtilis	Reactive Blue 4	2.0	303	36.3	[53]
Pseudomonas sp.	Acid Black 172	3.0	298	2961.2	[54]

elucidated. They concluded that at pH < 7, electrostatic interactions occurred between the functional groups on the surface of the biomass and the dye molecules. However, at pH > 7, one different mechanism for each dye was observed: For the Reactive Blue 4 (RB4), they concluded that a chemical reaction between the hydroxyl group of the biomass and two chlorine groups occurred. For the Reactive Orange 16 (RO16), the hydroxyl groups of the biomass reacted with the vinyl sulfone group of RO 16 by nucleophilic addition. For the Reactive Yellow 2 (RY2), a nucleophilic substitution reaction occurred. Du *et al.* [54] studied the biosorption of the metal-complex dye Acid Black 172 by live and heat-treated biomass of *Pseudomonas* sp. They found that the heat-treated biomass was more efficient than live biomass. The maximum amount of the dye adsorbed by the heat-treated biomass could reach up to 2961.2 mg g^{-1}. They also concluded that the amine groups played a major role in the biosorption of Acid Black 172 and that the heating of the biomass significantly increased the permeability of the cell wall so that the dye could enter into the cells and be adsorbed into intracellular proteins. Arunarani *et al.* [51] evaluated the potential of *Pseudomonas putida* for the removal of Basic Violet 3 and Acid Blue 93 from aqueous solutions. They verified that this microbe is an efficient way to treat textile effluents.

8.2.4   Chitosan

Chitosan, a de-N-acetylated analog of chitin, is a heteropolysaccharide consisting of linear b-1,4-linked GlcN and GlcNAc units [58]. The chemical structure of chitosan is presented in Figure 8.1. It has been found by many researchers that the OH and NH$_2$ on the chitosan structure (Figure 8.1) are the main functional groups responsible for dye binding [59–65].

The use of chitosan for SODs removal from aqueous solutions is based mainly on three factors: First, due to the fact that the chitosan-based

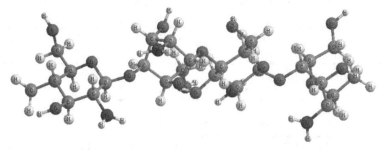

Figure 8.1 Chemical structure of chitosan (Haworth projection, based on references [9] and [58]).

polymers are low-cost materials obtained from natural resources and their use as biosorbents is extremely cost-effective. Generally, chitosan is obtained from seafood residues (shrimp, crab, lobster...) which are abundant and zero-cost materials [6,9]. Second, the biosorption capacities are high and higher biosorption rates are found. As a consequence, the amount of biosorbent used is reduced when compared to conventional adsorbents and also the process is fast [66–68]. The third factor is the development of complex materials by chitosan [69,70]. Since chitosan is versatile, it can be manufactured into films, membranes, fibers, sponges, gels, composites, beads and nanoparticles, or supported on inert materials [66,70–73]. Table 8.4 summarizes the biosorption capacities of chitosan and chitosan-based materials.

Dotto and Pinto [67] studied the biosorption of food dyes Acid Blue 9 and Food Yellow 3 onto chitosan. They found that the chitosan surface was covered by food dyes and biosorption occurred by chemisorption.

Table 8.4 Biosorption capacities of chitosan and chitosan-based materials.

Chitosan based material	Dye	pH	T (K)	Biosorption capacity (mg g^{-1})	Reference
Powder	FD&C Red 40	6.6	308	529.0	[59]
Powder	FD&C Yellow 5	3.0	298	350.0	[60]
Hydrogel beads	Congo Red	5.0	308	209.3	[61]
Crosslinked beads	Reactive Black 5	3.0	308	2043.1	[62]
Hydrogel composites	Acid Orange 7	2.0	298	221.1	[63]
Hydrogel composites	Methyl Orange	2.0	298	185.2	[63]
Hydrogel composites	Acid Red 18	2.0	298	342.5	[63]
Nanoparticles	Acid Orange 10	4.0	298	800.7	[64]
Nanoparticles	Acid Orange 12	4.0	298	1516.9	[64]
Nanoparticles	Acid Red 18	4.0	298	828.2	[64]
Nanoparticles	Acid Red 73	4.0	298	1185.3	[64]
Films	Acid Red 18	7.0	298	194.6	[69]
Films	FD&C Blue 2	7.0	298	154.8	[69]
Hollow fibers	Reactive Blue 19	3.5	298	454.5	[70]
Powder	Acid Blue 9	3.0	298	1134.0	[74]
Powder	FD&C Yellow 5	3.0	298	1977.0	[74]
Powder	FD&C Yellow 6	3.0	298	1684.0	[74]

Furthermore, SEM and EDS analysis indicated the chitosan surface coverage and the interaction between the sulphonate groups of the dyes and the amino groups of the chitosan. In the work of Mirmohseni *et al.* [70], chitosan hollow fibers were prepared by dry-wet spinning process, and applied for the biosorption of Reactive Blue 19 (RB 19). They obtained chitosan hollow fibers with high mechanical strength (47.57 MPa) which were suitable to remove RB19 from aqueous solutions. Wong *et al.* [75] evaluated the effects of temperature, particle size and percentage deacetylation on the biosorption of acid dyes on chitosan. They found that the biosorption was favored by the temperature increase, particle size decrease and deacetylation decrease. These results were explained on the basis of the swelling effect and changes in crystallinity. On the other hand, Piccin *et al.* [59], in the biosorption of FD&C Red 40 by chitosan, verified that the process was favored by the temperature decrease and deacetylation degree increase. Their results were explained based on the dye solubility and increase in chitosan protonate amino groups. Detailed information about the use of chitosan as biosorbent can be obtained in Crini and Badot [9] and Wan Ngah *et al.* [66].

8.2.5   Fungal Biomass

Fungal biomass is another class of biosorbents used for the removal of SODs from aqueous solutions [76–85]. A fungus is a member of a large group of eukaryotic organisms that includes microorganisms such as yeasts and molds. The fungal cell wall is composed of glucans and chitin. Furthermore, in its internal cellular structure, there are proteins, lipids, disaccharides, polysaccharides, alcohols and other compounds [86]. The above compounds contain a series of functional groups such as amino, carboxylic acid, phosphate and others [6,80–85]. These functional groups are responsible for dye binding [80–85]. Fungi are easy to grow, produce high yields of biomass and at the same time can be manipulated genetically and morphologically. The fungal organisms are widely used in a variety of large-scale industrial fermentation processes. The biomass can be cheaply and easily procured in substantial quantities as a byproduct of established industrial fermentation processes [87]. Table 8.5 summarizes the biosorption capacities of some fungi.

Khambhaty *et al.* [81] studied the biosorption of Brilliant Blue G (BBG) from aqueous solutions by marine *Aspergillus wentii*. Some parameters such as contact time (0–80 min), initial dye concentration (119.3–544.8 mg L^{-1}) and pH (2–10) were evaluated. They verified that the contact time necessary to reach equilibrium was 180 min. The BBG biosorption was strictly pH dependent. The biosorption isotherm data fitted well to the

Table 8.5 Biosorption capacities of some fungi.

Fungal biomass	Dye	pH	T (K)	Biosorption capacity (mg g^{-1})	Reference
Aspergillus foetidus	Reactive Black 5	2.0–3.0	323	76.0	[76]
Cunninghamella elegans	Acid Blue 62	5.0	298	300.0	[77]
Cunninghamella elegans	Acid Red 266	5.0	298	610.0	[77]
Aspergillus lentulus	Acid Blue 120	6.5	303	97.5	[78]
Cephalosporium aphidicola	Acid Red 57	1.0	293	109.4	[79]
Cunninghamella elegans	Orange II	5.6	301	6.9	[80]
Cunninghamella elegans	Reactive Black 5	5.6	301	4.2	[80]
Cunninghamella elegans	Reactive Red198	5.6	301	23.5	[80]
Aspergillus wentii	Brilliant Blue G	2.0	298	312.5	[81]
Aspergillus parasiticus	Reactive Red198	2.0	323	101.4	[82]
Saccharomyces cerevisiae	Remazol Blue	3.0	298	84.6	[83]
Saccharomyces cerevisiae	Remazol Black B	3.0	298	88.5	[83]
Saccharomyces cerevisiae	Remazol Red RB	3.0	298	48.8	[83]
Thamnidium elegans	Reactive Red198	2.0	298	234.2	[84]
Thamnidium elegans	Methyl Violet	4.0–10.0	288–318	579.4	[85]

Langmuir model. The pseudo-second-order kinetic model described the biosorption kinetics accurately and the biosorption process was found to be controlled by pore and surface diffusion. In the work of Akar *et al.* [84], a filamentous fungus, *Thamnidium elegans*, was tested as biosorbent for the removal of Reactive Red 198 (RR198) from aqueous solutions.

They concluded that the Langmuir and the pseudo-second-order models appropriately described the equilibrium and kinetics of the dye biosorption process, respectively. Kinetic studies showed that the suggested biomaterial offered fast decolorization kinetics. Biosorption equilibrium was established within 40 min. The possible interaction of *Thamnidium elegans*-RR198 was clarified by the FTIR and zeta potential analysis. It was found that –OH and –NH groups on the *Thamnidium elegans* surface contributed to dye binding.

8.3   Factors Affecting Biosorption

The biosorption process is affected by various factors, including pH, temperature, biosorbent dosage, particle size, contact time, initial dye concentration and stirring rate [2–13]. The selection of the more adequate operation ranges for these factors is a crucial study in the biosorption field [6,9,12,13]. From this study, it is possible to improve the biosorbent potential and also the performance of the biosorption system. Thus it is very important to know how these variables can affect the biosorption process. In this section, the effects of pH, temperature, biosorbent dosage, particle size, contact time, initial dye concentration and stirring rate are discussed in details.

8.3.1   pH

One of the most important factors that affects the biosorption process is the pH, which is a measure of acidity (pH < 7) or basicity (pH > 7) of an aqueous solution [9,11,12]. This factor influences the surface charge of the biosorbent, the degree of ionization of the material present in the solution, the dissociation of functional groups on the active sites of the biosorbent and the chemistry of the dyes [9]. In relation to the biosorbent, the point of zero charge (pH_{zpc}) is indicative of the surface charge as a function of pH [19]. When the pH of the solution is lower than pH_{zpc}, the surface of the biosorbent surface is positively charged and the biosorbent surface is negatively charged at pH values higher than pH_{zpc} [19,25,31,44]. As a result, the biosorption of cationic dyes is increased at high pH values and the anionic dyes biosorption is favored at low pH values [3]. Regarding the dyes, pH affects the structural stability of dye molecules (in particular the dissociation of their ionizable sites), and therefore their color intensity. Hydrolysis, protonation, deprotonation, complexation by organic and/or inorganic ligands, redox reactions and precipitation of the dyes are strongly influenced by pH [3,4,9].

In a relevant study, Chatterjee et al. [88] analyzed the chitosan-Congo Red interactions as a function of pH. The main findings of their research were: i) the high biosorption of the dye at lower pH (pH < 6.4) was probably due to the increase in electrostatic attraction between negatively charged dye molecule (–SO3– Na+) and positively charged amine group of chitosan (–NH$_3$$^+$); ii) at pH 6.4, where surface charge of chitosan was neutral, biosorption of the dye was attributed to hydrogen bond formation between some of the molecular components of Congo Red such as N, S, O, benzene ring and CH$_2$OH groups of the chitosan molecule; iii) at pH above 6.4, surface charge of chitosan was negative and the chitosan-Congo Red interactions were due to hydrogen bonding and van der Waals forces. In another investigation, Weber et al. [89] studied the effect of pH (2.5–8.5) in the biosorption of Direct Blue 38 onto Papaya seeds. They found that at low pH values, high electrostatic attraction occurred between the positively charged surface of papaya seeds and the negatively charged anionic dye. As a consequence, pH 2.5 was selected as the more adequate. On the contrary, Khataee et al. [48] obtained best biosorption capacities at pH 10.0 in the biosorption of Basic Red 46 onto *Spirogyra* sp.

8.3.2 Temperature

Generally, the discharge temperature of DCEFs is in the range of 298–303 K [8]. However, various textile dye effluents are discharged at relatively high temperatures (323–333 K) [12]. Thus, temperature is an important design parameter, which should be evaluated in biosorption studies. The temperature can affect the biosorbent and the dyes from a negative or positive viewpoint. Dotto et al. [19] studied the temperature effect on the biosorption of food dyes by *Spirulina platensis*. They found that the temperature increase from 298 to 328 K caused a strong decrease in biosorption capacity. They assumed that the temperature increase causes an increase in the solubility of the dyes, so, the interaction forces between the dyes and the solvent become stronger than those between dyes and biosorbent. Furthermore, they suggested that at temperatures above 318 K the damage of sites can occur on the surface of biosorbent and, consequently, a decrease in the surface activity. This behavior was also observed by Piccin et al. [59] and Dotto et al. [74] in the biosorption of SODs onto chitosan, Barka et al. [90] in the biosorption of synthetic dyes onto dried prickly pear cactus and Cengiz et al. [91] in the biosorption of Astrazon Red onto

Posidonia oceanica. On the other hand, in a study on the biosorption of Acid Orange 10, Acid Red 18, and Acid Red 73 onto chitosan, Wong *et al.* [75] verified that the biosorption capacities increased when the temperature increased from 298 to 333 K. They attributed this phenomenon to the fact that the temperature increase produces a swelling effect within the internal structure of the chitosan, facilitating the biosorption. Also, they affirmed that the dyes mobility was increased due to the faster rate of diffusion of adsorbate molecules from the solution to the biosorbent. The same dependence in relation to the temperature was observed by Reddy *et al.* [35] in the biosorption of Congo Red onto Indian Jujuba Seeds and Mahmoud *et al.* [92] in the biosorption of Methylene Blue onto *Hibiscus cannabinus.*

8.3.3   Biosorbent Dosage

The biosorbent dosage is particularly important because it determines the extent of decolorization and may also be used to predict the cost of biosorbent per unit of solution to be treated [9]. The study of the effect of biosorbent dosage gives an idea of the effectiveness of a biosorbent and the ability of a dye to be adsorbed with a minimum dosage [3]. Generally, the percentage of dye removal increases with an increase in biosorbent dosage, since the number of biosorption sites is higher. However, the biosorption capacity increases significantly as biosorbent dosage decreases, since it is expressed in mg of dye adsorbed per gram of biosorbent. This can be attributed to overlapping or aggregation of biosorption sites resulting in a decrease in total biosorbent surface area available to the dye and an increase in diffusion path length. Zhang *et al.* [93] verified the biosorbent dosage effect in the biosorption of Rhodamine B (RhB) by sugarcane bagasse. They found that the percentage of RhB removal increased from 28.0 to 99.0% when the biosorbent dosage was increased from 1 to 20 g L^{-1}. As expected, the biosorption capacity decreased from 65.7 to 11.0 mg g^{-1}. Dawood and Sen [94] verified that in the biosorption of Congo Red by pine cone powder, an increase in biosorbent dosage from 0.01 to 0.03 g resulted in a decrease of the amount of adsorbed dye from 13.44 to 6.28 mg g^{-1}. Elkady *et al.* [95] verified that in the biosorption of Remazol Reactive Red 198 onto eggshells, an increase in biosorbent dosage from 5 to 20 g L^{-1} caused an increase in percentage of dye removal from 38 to 65%.

8.3.4   Particle Size

The biosorbent particle size is another important effect which should be studied. As biosorption is a surface phenomenon and can be attributed to the relationship between the effective specific surface area of the biosorbent particles and their sizes [9,12,96,97], the surface area usually increases when the particle size is decreased [12,98]. As a consequence, more biosorption sites are exposed to the dyes for the same amount of biosorbent, increasing the biosorption capacity [99]. Furthermore, McKay *et al.* [100,101] indicated that the equilibrium is attained faster with small particles, since the intraparticle diffusion resistance is lower, and the accessibility of dye molecules to internal sites is facilitated. Piccin *et al.* [98] evaluated the particle size effect on the biosorption of FD&C Red 40 by chitosan. They used particles of 100, 180 and 260 μm, and obtained specific surface areas of 4.2, 3.4 and 1.6 m^2 g^{-1}, respectively. As a consequence, the biosorption capacities were 191.6, 157.5 and 104.9, respectively. Daneshvar *et al.* [42] studied the particle size effect in the biosorption of Acid Black 1 onto *Sargassum glaucescens* and *Stoechospermum marginatum*. Their results showed that the biosorption capacity of the biosorbents increased with the decrease in particle size. The same effect was observed by Prola *et al.* [102] in the biosorption of Reactive Red 120 onto *Jatropha curcas* shells.

8.3.5   Contact Time

In a biosorption process, the contact time is one of the more important parameters from the industrial viewpoint. A good biosorbent should not only provide high biosorption capacities, but also furnish a fast process. In this way, contact time is fundamental to select an adequate biosorbent. Generally, during the process, the biosorbent surface is progressively blocked by the adsorbate molecules, becoming covered after some time. When this happens, the biosorbent cannot adsorb any more dye molecules [9,37,67]. This reflects that the biosorption capacity increases with time and, at some point in time, reaches a constant value where no more dye is removed from the solution [99]. At this point, the amount of dye being adsorbed onto the material is in a state of dynamic equilibrium with the amount of dye desorbed from the biosorbent [9,96,97]. Deniz and Saygideger [29], using princess tree leaf as biosorbent, verified that the removal of Basic Red 46 increased with time and attained saturation in about 70 min. Furthermore, they found that, initially, the biosorption rate of BR 46 was rapid, but it gradually decreased with time until it reached equilibrium. This behavior was attributed to the aggregation of the dye molecules around the biosorbent particles. Yang *et al.* [103] observed that the biosorption rate of Acid Black 172 and Congo Red onto *Penicillium* YW 01 was initially fast and then gradually decreased to reach equilibrium. In the same way, Kumar and Ahmad [104] found that in the biosorption of Crystal Violet onto ginger waste, the dye was rapidly adsorbed in the first 45–60 min, and then the biosorption rate decreased gradually and reached equilibrium in about 150 min. They assumed that at the beginning, the biosorption rate was very fast because the dye was adsorbed on the exterior surface, but, when the exterior surface was saturated, the dye entered into the pores of the biosorbent particles and was adsorbed on the interior surface.

8.3.6   Initial Dye Concentration

The initial dye concentration also affects the biosorption process. Initial concentration provides an important driving force to overcome all mass transfer resistances of the dye between the aqueous and solid phases [12]. The effect of the initial dye concentration factor depends on the immediate relation between the concentration of the dye and the available binding sites on a biosorbent surface [3]. The amount of the dye adsorbed onto the biosorbents increases with an increase in the initial concentration, if the amount of biosorbent was kept unchanged. This is due to the increase in the driving force of the concentration gradient with the higher initial dye concentrations [105]. On the other hand, the percentage of dye removal decreases with an increase in the initial dye concentration, which may be due to the saturation of biosorption sites on the biosorbent surface [106]. At a low concentration there will be unoccupied active sites on the biosorbent surface, and when the initial dye concentration increases, the active sites required for biosorption of the dye molecules will be lacking [107]. Zhang *et al.* [93] showed that in the biosorption of Rhodamine B (RhB) and Basic Blue 9 onto sugarcane bagasse, the removal of both dyes by bagasse decreased with increasing initial dye concentration. However, the biosorption capacity increased with the initial dye concentration. In the study of Vucurovic *et al.* [108], the equilibrium biosorption increased from 1.9 to 4.8 mg g^{-1} when the initial Methylene Blue (MB) concentration increased from 20 to 50 mg L^{-1}. Fan *et al.* [109] indicated that in the biosorption of Azure Blue onto *Cladosporium* sp., the uptake of dye increased up to 55 mg g^{-1} with the increasing initial dye concentration, but the percentage of dye removal decreased from 98% to 56% with the increase in the initial dye concentration from 100 to 500 mg L^{-1}.

8.3.7   Stirring Rate

The stirring rate is an important parameter in the biosorption process, since it influences the distribution of solute in the bulk solution and the formation of the external boundary film [96,97]. This parameter influences the initial stages of biosorption, where the external mass transfer is the controlling mechanism. However, the intraparticle diffusion is not influenced [110]. As a consequence, the equilibrium time is affected, but the equilibrium biosorption capacity is not a function of stirring rate [111]. The stirring rate increase causes an increase in impeller Reynolds number [112], so, the energy dissipation and turbulence in the mixing zone is increased. As a consequence, a decrease in film thickness occurs and the boundary layer resistance is lower [68]. The above facts cause an increase in biosorption capacity at the initial stages.

Dotto and Pinto [68] studied the stirring rate effect on the biosorption of Acid Blue 9 and Food Yellow 3 onto chitosan. They observed that, at 40 min, the increase in stirring rate from 15 to 400 rpm caused a large increase in Acid Blue 9 biosorption capacity, from 110 to 220 mg g^{-1}. In the same way, for Food Yellow 3, an increase in stirring rate from 15 to 400 rpm increased biosorption capacity from 220 to 350 mg g^{-1}. Furthermore, they proved that the stirring rate increase caused a decrease in film diffusion effect. Hanafiah *et al.* [113] studied the biosorption of Acid Blue 25 onto *Shorea dasyphylla* sawdust. They verified that the stirring rate increase from 100 to 200 rpm improved the biosorption. They explained that at a higher stirring rate, a good degree of mixing is achieved, and the boundary layer thickness around the biosorbent particles is reduced. Similar behavior was found by Uzun and Guzel [114] in the biosorption of Orange II and Crystal Violet onto chitosan. Dotto and Pinto [111] also corroborate this effect in the biosorption of Acid Blue 9 and FD&C Red 40 by *Spirulina platensis* nanoparticles.

8.4 Biosorption Isotherms, Thermodynamics and Kinetics

8.4.1 Equilibrium Isotherms

The biosorption equilibrium is normally described by the biosorption isotherm curves. These curves relate the amount of dye adsorbed on the biosorbent at equilibrium (q_e, mg g^{-1}) and the equilibrium dye concentration remaining in liquid phase (C_e, mg L^{-1}). From the isotherm curves it is possible to infer how the SODs-biosorbent interactions occur, and so these curves are crucial to improve the biosorbent performance [12,13]. Furthermore, the isotherm curves are a key tool for obtaining the maximum biosorption capacity of the biosorbents in a specific condition, as well as for elucidating the possible biosorption mechanisms [115,116].

Blázquez *et al.* [117] presented a classification with five isotherm shapes for biosorption in liquid phase. Figure 8.2 shows these five isotherm shapes.

Several different types of isotherms have been presented in the literature; the isotherm shape depends on the type of biosorbent, the type of adsorbate, and intermolecular interactions between the adsorbate and biosorbent [116]. The biosorption isotherms of agricultural wastes generally belong to type I (which has a convex shape). This shape is associated with monomolecular layer biosorption of nonporous or microporous biosorbents. Types II and III depict multimolecular adsorption layer formation and strong and weak adsorbate-biosorbent interaction on macroporous adsorbents, respectively. Type IV describes multimolecular layer

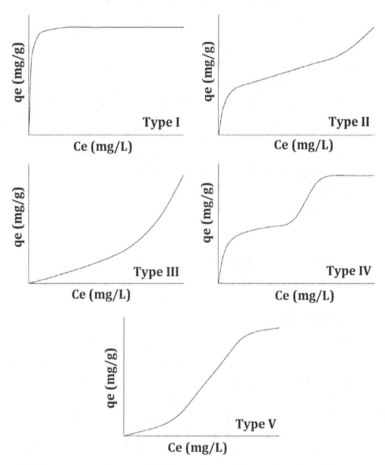

Figure 8.2 Characteristic shapes of the biosorption isotherms (Based on Blázquez *et al.* [117]).

formation via condensation in mesopores, whereas type V describes a similar process, but with both strong and weak adsorbate-adsorbent interactions [96,97,112,115,117].

In order to improve the design of a biosorption process, it is important to establish the more adequate model to represent the equilibrium curve. There are many mathematical descriptions of biosorption isotherms, some of which are based on a simplified physical picture of sorption and desorption, whereas others are purely empirical and intended to correlate the experimental data in simple equations with two or more empirical parameters [116–119]. Among all of these models, the more usual for the biosorption of SODs are: Langmuir [120], Freundlich [121], Redlich-Peterson [122], Dubinin-Radushkevich [123], Sips [124] and Tóth [125].

A basic assumption of the Langmuir model is that biosorption takes place at specific homogeneous sites within the biosorbent, and once a dye molecule occupies a site, no further biosorption can take place at that site [120]. The Langmuir model is presented in Equation 8.1:

$$q_e = \frac{q_m k_L C_e}{1 + k_L C_e} \tag{8.1}$$

where q_m is the maximum biosorption capacity (mg g^{-1}) and k_L is the Langmuir constant (L mg^{-1}). The Langmuir model was found to be adequate for the following biosorption systems: wood apple shell basic dyes [26], grapefruit peel-Crystal Violet [27], princess tree leaf-Basic Red 46 [29] and chitosan-acid dyes [64].

The Freundlich isotherm model is the earliest known relationship describing the biosorption process. The model applies to biosorption on heterogeneous surfaces with interaction between adsorbed molecules. The application of the Freundlich equation also suggests that biosorption energy exponentially decreases on completion of the sorptional centers of a biosorbent. This isotherm is an empirical equation which can be employed to describe heterogeneous systems and is expressed in Equation 8.2 [121]:

$$q_e = k_F C_e^{1/n_F} \tag{8.2}$$

where k_F is the Freundlich constant ((mg g^{-1})(mg L^{-1})$^{-1/n_F}$) and $1/n_F$ the heterogeneity factor. The following biosorption systems were well described by the Freundlich model: pine cone-Acid Blue 7 [30], rice husk-Malachite Green [126], palm shell powder-Reactive Blue 21 [127] and chitosan-Reactive Red 141 [127].

The Redlich-Peterson isotherm is used to represent biosorption equilibrium over a wide concentration range, and can be applied either in homogeneous or in heterogeneous systems due to its versatility. The Redlich-Peterson isotherm is given by Equation 8.3 [122]:

$$q_e = \frac{k_{RP}C_e}{1+\left(a_{RP}C_e^{\beta}\right)} \tag{8.3}$$

where k_{RP} and a_{RP} are the Redlich–Peterson constants (L g^{-1}) and (L mg^{-1})$^{\beta}$, respectively, and β is the heterogeneity coefficient, which varies between 0 and 1. The Redlich-Peterson model was adequate to represent the biosorption of FD&C Red 40 onto chitosan [59] and food dyes onto chitosan films [69].

Another model usually employed to represent the biosorption processes is the Dubinin-Radushkevich model [123], as demonstrated in Equation 8.4:

$$q_e = q_S \exp(-B\varepsilon^2) \tag{8.4}$$

where q_s is the D-R constant (mg g^{-1}) and ε can be related as Equation 8.5:

$$\varepsilon = RT\ln(1+\frac{1}{C_e}) \tag{8.5}$$

where R is the universal gas constant (8.314 J mol^{-1} K^{-1}) and T the temperature (K). The B constant is relative to the biosorption free energy E (kJ mol^{-1}), as demonstrated in Equation 8.6 [123]:

$$E = \frac{1}{\sqrt{2B}} \tag{8.6}$$

The Dubinin-Radushkevich model was adequate to describe the biosorption of Basic Blue 9 onto nut shells and Eichhornia [128].

The Sips isotherm is a combination of the Langmuir and Freundlich isotherms, and is given by Equation 8.7 [124]:

$$q_e = \frac{q_{mS}(k_S C_e)^m}{1+(k_S C_e)^m} \tag{8.7}$$

where q_{mS} is the maximum biosorption capacity from the Sips model (mg g^{-1}), k_S is the Sips Constant (L mg^{-1}) and m the fractionary exponent related with the biosorption mechanism [124]. The Sips model was adequate to describe the biosorption of food dyes onto *Spirulina platensis* [19], Reactive Black 5 onto aqai stalks [32] and Remazol Black B onto Brazilian pine-fruit shells [129].

The Tóth model was found from a modification of the Langmuir model and suggests that the biosorption occurs via multilayer formation. This model is given by Equation 8.8 [125]:

$$q_e = \frac{q_{mT}C_e}{\left(\dfrac{1}{k_T} + C_e^{\,mT}\right)^{mT}} \qquad (8.8)$$

where q_{mT} is the maximum biosorption capacity from the Tóth model (mg g^{-1}), k_T is the Tóth constant (mg L^{-1})^{mT} and m_T is Tóth exponent. This model was adequate to describe the biosorption of Methylene Blue and Eriochrome Black onto *Scolymus hispanicus* L. [130].

The fit of these mathematical models on the biosorption isotherm curves and the consequent determination of the adjustable isotherm parameters are fundamental to know a biosorption process, since these can furnish the following information: i) a equilibrium relation which can be used in kinetic studies; ii) an idea about the biosorbent-dye interactions; iii) the affinity between biosorbent and dye; iv) the maximum biosorption capacity of the biosorbent in a specific condition; v) the amount of biosorbent which should be used for the removal of a certain quantity of dye from an effluent; vi) the influence of the experimental conditions on the equilibrium.

8.4.2   Thermodynamic Parameters

In the biosorption field, the thermodynamic is generally studied thought the estimation of Gibbs free energy change (ΔG^0, kJ mol^{-1}), enthalpy change (ΔH^0, kJ mol^{-1}) and entropy change (ΔS^0, kJ mol^{-1}K^{-1}) [6,16,19,43,44]. From a thermodynamic viewpoint, the biosorption is considered as a uniphasic reaction [12,13], and so the Gibbs free energy change can be estimated by Equation 8.9 [131]:

$$\Delta G^0 = -RT\ln(K) \qquad (8.9)$$

where R is the universal gas constant (8.314 J mol^{-1} K^{-1}), T the temperature (K) and K thermodynamic equilibrium constant (dimensionless) [132]. The K values are obtained in different ways in the literature [23,35,43,90,92,102,113,132–135]. Detailed and consistent forms to obtain K can be found in the works of Liu [132], Milonjic [133] and Zhou *et al.* [134].

The relationship of ΔG^0 to enthalpy change (ΔH^0) and entropy change (ΔS^0) of biosorption is expressed in the form of Equation 8.10 [131]:

$$\Delta G^0 = \Delta H^0 - T\Delta S^0 \qquad (8.10)$$

By combining Equations 8.9 and 8.10, the known Equation 8.11 can be obtained [132–135]:

$$\ln(K) = -\left(\frac{\Delta H^0}{RT}\right) + \frac{\Delta S^0}{R} \qquad (8.11)$$

The plot of ln K against 1/T (van't Hoff plot) theoretically yields a straight line that allows calculation of ΔH^0 and ΔS^0 from the respective slope and interception of Equation 8.11 [132–135].

In biosorption systems, the values of ΔG^0, ΔH^0 and ΔS^0 are employed to obtain information about the process. The Gibbs energy change (ΔG^0) indicates the degree of spontaneity of a biosorption process, and a higher negative value reflects a more energetically favorable biosorption [132]. In relation to ΔH^0, negative values reflect exothermic processes and positive values reflect endothermic processes [19,23,35,90,135]. From the magnitude of ΔH^0 it is possible to infer the biosorbent-dye interactions. Physisorption, such as van der Waals interactions, are usually lower than 20 kJ mol^{-1}, and electrostatic interaction ranges from 20 to 80 kJ mol^{-1}. Chemisorption bond strengths can be from 80–450 kJmol^{-1} [23]. Regarding ΔS^0, the negative valuesshow that the randomness decreases at the solid-solution interface during the biosorption [9,12,13,43]. Positive values suggest the possibility of some structural changes or readjustments in the dye-biosorbent complex [136]. Furthermore, by comparing the values of ΔH^0 and $T\Delta S^0$ it can be verified if the biosorption is an enthalpy-controlled or an entropy-controlled process [69]. Table 8.6 shows typical values for Gibbs free energy change (ΔG^0), enthalpy change (ΔH^0) and entropy change (ΔS^0) in some biosorption systems.

8.4.3   Kinetic Models

Kinetic study is fundamental in biosorption systems. From kinetic analysis, the solute uptake rate, which determines the residence time required for completion of biosorption reaction, may be established [137]. This study explains how fast the biosorption occurs and also provides information on the factors affecting the process [13]. Furthermore, it is possible to investigate the rate-controlling steps [12]. Several biosorption reaction models and diffusional models have been used to study the biosorption of SODs [9–13,23–35,37,44,45,60,67–69]. Among these models, the most commonly used ones are the pseudo-first-order [138], pseudo-second-order [139], general order [13], Elovich [140], Avrami [141] and Weber-Morris [142].

Table 8.6 Thermodynamic parameters in biosorption systems.

Biosorbent	Dye	ΔG^0 (kJ mol^{-1})	ΔH^0 (kJ mol^{-1})	ΔS^0 (kJ mol^{-1} K^{-1})	Reference
Mango seeds	Victazol Orange 3R	−26.6 to −23.6	12.3	0.120	[23]
Jujuba seeds	Congo Red	−6.4 to −3.5	12.9	0.058	[35]
Spirulina platensis	Tartrazine	−18.3 to −17.5	−10.9	0.020	[43]
Spirulina platensis	Allura Red	−21.3 to −19.7	−3.3	0.050	[43]
Dried prickly pear cactus	Methylene Blue	−2.6 to 0.7	−31.0	0.095	[90]
Dried prickly pear cactus	Eriochrome Black	1.1 to 2.6	−10.2	−0.038	[90]
Dried prickly pear cactus	Alizarin S	0.6 to 2.1	−11.7	−0.042	[90]
Sugarcane bagasse	Rhodamine B	−7.4 to −7.1	−2.2	0.016	[93]
Sugarcane bagasse	Basic Blue 9	−6.9 to −5.8	−23.5	−0.055	[93]
Jatropha curcas shell	Reactive Red 120	−30.5 to −26.6	20.4	0.157	[102]
Shorea dasyphylla sawdust	Acid Blue 25	−32.2 to −29.3	−11.5	0.073	[113]
Citrus waste biomass	Reactive Red 45	−0.6 to 2.6	−25.5	−0.080	[135]
Citrus waste biomass	Reactive Yellow 42	0.4 to 5.2	−36.0	−0.120	[135]
Citrus waste biomass	Reactive Blue 19	3.1 to 4.7	−9.0	−0.040	[135]
Citrus waste biomass	Reactive Blue 49	−2.0 to −0.45	−14.2	−0.040	[135]

The kinetic models of pseudo-first-order [138] and pseudo-second-order [139] are based on the biosorption capacity. The pseudo-first-order model (Equation 8.12) is generally applicable over the initial 20–30 min of the sorption process, while the pseudo-second-order model (Equation 8.13) is suitable for the whole range of contact time.

$$q_t = q_1(1 - \exp(-k_1 t)) \tag{8.12}$$

$$q_t = \frac{t}{(1/k_2 q_2{}^2) + (t/q_2)} \tag{8.13}$$

where k_1 and k_2 are the rate constants of pseudo-first-order and pseudo-second-order models, respectively, in (min^{-1}) and ($g\ mg^{-1}\ min^{-1}$), q_1 and q_2 are the theoretical values for the biosorption capacity ($mg\ g^{-1}$), and q_t is the biosorption capacity at any time t. The pseudo-first-order model was suitable to represent the biosorption of Reactive Black 5 onto chitosan beads [143] and Reactive Blue 4 onto *Bacillus subtilis* [53]. The pseudo-second-order model was suitable to represent the biosorption of basic dyes onto wood apple shell [26], Methylene Blue onto peanut husk [27] and Congo red onto cashew nut shell [144].

An alternative to the pseudo-first-order and pseudo-second-order models is the general order model [13]. For this model, the biosorption process on the surface of biosorbent is assumed to be a rate-controlling step. The general order model can be described by Equation 8.14 [13,44]:

$$q_t = q_n - \frac{q_n}{\left[k_n \left(q_n\right)^{n-1} t(n-1) + 1\right]^{1/(n-1)}} \tag{8.14}$$

where q_n is the biosorption capacity at the equilibrium ($mg\ g^{-1}$), k_n the rate constant ($min^{-1} (g\ mg^{-1})^{n-1}$) and n is the biosorption reaction order with regard to the effective concentration of the biosorption sites available on the surface of biosorbent. This model was adequate to represent the biosorption of Reactive Red 120 onto *Spirulina platensis* [44].

When the biosorption processes occur through chemisorption on solid surface, and the biosorption velocity decreases with time due to covering of the superficial layer, the Elovich model is most used. The Elovich kinetic model is described according to Equation 8.15 [140]:

$$q_t = \frac{1}{a} \ln(1 + abt) \tag{8.15}$$

where a is the initial velocity due to dq/dt with $q_t = 0$ (mg g^{-1} min^{-1}) and b is the desorption constant of the Elovich model (g mg^{-1}). The Elovich model was adequate to represent the biosorption of food dyes onto chitosan [67,68].

Another model used in biosorption systems is the fractionary Avrami kinetic equation, which was proposed based on thermal decomposition modeling [141]. The Avrami model is show in Equation 8.16 [141]:

$$q_t = q_{AV}(1 - \exp(-k_{AV}t)^n)$$ (8.16)

where k_{AV} is the Avrami kinetic constant (min^{-1}), q_{AV} is the Avrami theoretical values for the biosorption capacity (mg g^{-1}) and n is a fractionary reaction order which can be related to the biosorption mechanism. The Avrami model was found to be appropriate for the biosorption of Reactive Black 5 and Reactive Orange 16 onto aqai stalks [32], Acid Blue 9 and FD&C Red 40 onto *Spirulina platensis* [37] and Remazol black B onto Brazilian pine-fruit shells [129].

The Weber-Morris intraparticle diffusion model [142] is one of the most used models in biosorption systems for identifying the possible mass transfer steps [9,12,13,32,44,60,65,68,102,111]. According to Weber and Morris [142], the plot q_t versus $t^{1/2}$ shows multilinearity, and each linear portion represents a distinct mass transfer mechanism. The first linear portion can be attributed to the external mass transfer mechanism. The second portion is relative to the intraparticle diffusion, and the third is the final equilibrium. Furthermore, if the regression passes through the origin, then the rate-limiting process is only due to the intraparticle diffusion. Otherwise, the intraparticle diffusion is not the only rate-controlling step and some degrees of boundary layer diffusion also control the biosorption [44,65,68,145]. The Weber-Morris intraparticle diffusion model is given by Equation 8.17:

$$q_t = k_{di}t^{1/2} + C_i$$ (8.17)

where k_{di} is the intraparticle diffusion rate constant (mg g^{-1} min$^{-1/2}$), C is the constant in the intraparticle diffusion model (mg g^{-1}), which is proportional to the boundary layer thickness, and i is the number of each biosorption stage.

Tan *et al.* [38] applied the Weber-Morris model to represent the biosorption of Basic Orange onto *Azolla filiculoides* biomass. They found that the plot q_t versus $t^{1/2}$ was not linear over the full time range; however, it exhibited a tri-linearity, suggesting three successive steps of the biosorption. The first linearity was attributed to the boundary layer diffusion, the second part was assigned to the intraparticle diffusion step, and the third linearity was related to the equilibrium. Dotto and Pinto [68] observed multilinearity with two distinct phases in the biosorption of Acid Blue 9 and Food Yellow 3 onto chitosan. The initial portion was related to the boundary layer diffusion (film diffusion). The second portion was assigned to the gradual biosorption step, where intraparticle diffusion control is rate limiting. So, they assumed that film diffusion and intraparticle diffusion were simultaneously operating during the biosorption process.

8.5 Future Perspectives and Challenges

The biosorption of organic dyes is a crescent research area and many efforts have been made for this technology to become an alternative to the conventional treatment processes. Several biomasses have been successfully employed and the factors affecting biosorption are known. Furthermore, equilibrium, thermodynamics and the kinetic aspects have also been well studied. Nevertheless, these efforts have not been sufficient to apply biosorption on a full industrial scale. Thus, to assist future research in the biosorption field, some of the studies listed below are still necessary.

   i.   Despite the fact that several biomasses were successfully employed for the biosorption of organic dyes, the search for new, low-cost, efficient, eco-friendly and alternative biosorbents is always necessary.
   ii. It is know that specific compounds and functional groups on biomasses are responsible for dyes binding. Thus, in order to apply as biosorbents, the identification and extraction of these specific compounds from biomasses are another perspective in the biosorption field.
   iii. A relevant way of investigation is relative to fungi, bacteria and algae. These biomasses can be obtained from cultivations, and their characteristics and properties are directly related with the culture conditions. To this end, it is necessary to control the culture conditions (temperature, pH, lightness, stirring rate, cell concentration, nutrients) in order to obtain biosorbents with specific functionalities.
   iv. The majority of biosorption studies use synthetic aqueous solutions, with only one dye, in batch systems. However, from a realistic viewpoint, the dye-containing effluents are composed of more than one dye. In addition, other

contaminants such as metals and surfactants are present. So, future works regarding multicomponent biosorption (two, three or several dyes), treatment of simulated effluents and real industrial effluents are necessary.

v. About 85% of the biosorption studies are made in batch systems. These systems are good for lab-scale, but, impracticable for industrial applications. It is evident that studies regarding biosorption in continuous mode are necessary. Some alternatives are fixed bed, fluidized bed, spouted bed and others.

vi. Several biosorption studies were made in lab-scale. Thus, another point which should be highlighted is the scale-up of the biosorption processes.

vii. A current problem in the biosorption field is the phase separation after the process. Generally, filtration and centrifugation are employed, but these unit operations are expensive from an industrial viewpoint. The development of biosorbents and operations to facilitate the phase separation after the biosorption process is desirable.

viii. The development and application of predictive mathematical models, which explain in detail the biosorption process, are crucial at this actual stage.

References

1. K. Hunger, *Industrial Dyes: Chemistry, Properties, Applications*, Wiley-VCH, Weinheim, 2003.
2. E. Forgacs, T. Cserháti, G. Oros, *Environment International*, Vol. 30, p. 953, 2004.
3. M.A.M. Salleh, D.K. Mahmoud, W.A.W.A. Karim, A. Idris, *Desalination*, Vol. 280, p. 1, 2011.
4. V.K. Gupta, Suhas, *Journal of Environmental Management*, Vol. 40, p. 2313, 2009.
5. G. Mezohegyi, F.P. Van der Zee, J. Font, A. Fortuny, A. Fabregat, *Journal of Environmental Management*, Vol. 102, p. 148, 2012.
6. A. Srinivasan, T. Viraraghavan, *Journal of Environmental Management*, Vol. 91, p. 1915, 2010.
7. A.K. Verma, R.R. Dash, P. Bhunia, *Journal of Environmental Management*, Vol. 93, p. 154, 2012.
8. C. Hessel, C. Allegre, M. Maisseu, F. Charbit, P. Moulin, *Journal of Environmental Management*, Vol. 83 p. 171, 2007.
9. G. Crini, P.M. Badot, *Progress in Polymer Science*, Vol. 33, p. 399, 2008.

10. A. Demirbas, *Journal of Hazardous Materials*, Vol. 167, p. 1, 2009.
11. R.G. Saratale, G.D. Saratale, J.S. Chang, S.P. Govindwar, *Journal of the Taiwan Institute of Chemical Engineers*, Vol. 42, p. 138, 2011.
12. Z. Aksu, *Process Biochemistry*, Vol. 40, p. 997, 2005.
13. Y. Liu, Y.J. Liu, *Separation and Purification Technology*, Vol. 61, p. 229, 2008.
14. B.A. Adams, E.L. Holmes, *Journal of the Society of Chemical Industry*, Vol. 54, p. 1, 1935.
15. B. Volesky, Z.R. Holan, *Biotechnology Progress*, Vol. 11, p. 235, 1995.
16. U. Farooq, J.A. Kozinski, M.A. Khan, M. Athar, *Bioresource Technology*, Vol. 101, p. 5043, 2010.
17. N. Das, R. Vimala, P. Karthika, *Indian Journal of Biotechnology*, Vol. 7, p. 159, 2008.
18. G.L. Dotto, T.R.S. Cadaval Jr., L.A.A. Pinto, *Process Biochemistry*, Vol. 47, p. 1335, 2012.
19. G.L. Dotto, E.C. Lima, L.A.A. Pinto, *Bioresource Technology*, Vol. 103, p. 123, 2012.
20. B. Volesky, *Water Research*, Vol. 41, p. 4017, 2007.
21. E. Worch, *Journal of Water Supply: Research and Technology*, Vol. 57, p. 171, 2008.
22. R. Sanghi, B. Bhattacharya, *Coloration Technology*, Vol. 118, p. 256, 2002.
23. W.S. Alencar, E. Acayanka, E.C. Lima, B. Royer, F.E. de Souza, J. Lameira, C.N. Alves, *Chemical Engineering Journal*, Vol. 209, p. 577, 2012.
24. G.Z. Kyzas, N.K. Lazaridis, A.C. Mitropoulos, *Chemical Engineering Journal*, Vol. 189–190, p. 148, 2012.
25. A. Çelekli, S.S. Birecikligil, F. Geyik, H. Bozkurt, *Bioresource Technology*, Vol. 103, p. 64, 2012.
26. S. Jain, R.V. Jayaram, *Desalination*, Vol. 250, p. 921, 2010.
27. A. Saeed, M. Sharif, M. Iqbal, *Journal of Hazardous Materials*, Vol. 179, p. 564, 2010.
28. J. Song, W. Zou, Y. Bian, F. Su, R. Han, *Desalination*, Vol. 265, p. 119, 2011.
29. F. Deniz, S.D. Saygideger, *Desalination*, Vol. 268, p. 6, 2011.
30. N.M. Mahmoodi, B. Hayati, M. Arami, C. Lan, *Desalination*, Vol. 268, p. 117, 2011.
31. S. Chowdhury, S. Chakraborty, P. Saha, *Colloids and Surfaces B: Biointerfaces*, Vol. 84, p. 520, 2011.
32. N.F. Cardoso, E.C. Lima, T. Calvete, I.S. Pinto, C.V. Amavisca, T.H.M. Fernandes, R.B. Pinto, W.S. Alencar, *Journal of Chemical and Engineering Data*, Vol. 56, p. 1857, 2011.
33. N.F. Cardoso, E.C. Lima, I.S. Pinto, C.V. Amavisca, B. Royer, R.B. Pinto, W.S. Alencar, S.F.P. Pereira, *Journal of Environmental Management*, Vol. 92, p. 1237, 2011.
34. S.T. Akar, A. Gorgulu, T. Akar, S. Celik. *Chemical Engineering Journal*, Vol. 168, p. 125, 2011.

35. M.C.S. Reddy, L. Sivaramakrishna, A.V. Reddy, *Journal of Hazardous Materials*, Vol. 203–204, p. 118, 2012.
36. T.A. Davis, B. Volesky, A. Mucci, *Water Research*, Vol. 37, p. 4311, 2003.
37. G.L. Dotto, V.M. Esquerdo, M.L.G. Vieira, L.A.A. Pinto, *Colloids and Surfaces B: Biointerfaces*, Vol. 91, p. 234, 2012.
38. C. Tan, M. Li, Y.M. Lin, X.Q. Lu, Z. Chen, *Desalination*, Vol. 266, p. 56, 2011.
39. Ç. Doğar, A. Gürses, M. Açıkyildiz, and E. Özkan, *Colloids and Surfaces B: Biointerfaces*, Vol. 76, p. 279, 2010.
40. S.V. Mohan, S.V. Ramanaiah, P.N. Sarma, *Biochemical Engineering Journal*, Vol. 38, p. 61, 2008.
41. M. Kousha, E. Daneshvar, M.S. Sohrabi, M. Jokar, A. Bhatnagar, *Chemical Engineering Journal*, Vol. 192, p. 67, 2012.
42. E. Daneshvar, M. Kousha, M. Jokar, N. Koutahzadeh, E. Guibal, *Chemical Engineering Journal*, Vol. 204–206, p. 225, 2012.
43. G.L. Dotto, M.L.G. Vieira, V.M. Esquerdo, L.A.A. Pinto, *Brazilian Journal of Chemical Engineering*, Vol. 30, p. 13, 2013.
44. N.F. Cardoso, E.C. Lima, B. Royer, M.V. Bach, G.L. Dotto, L.A.A. Pinto, T. Calvete, *Journal of Hazardous Materials*, Vol. 241–242, p. 146, 2012.
45. Z. Aksu, S. Tezer, *Process Biochemistry*, Vol. 40, p. 1347, 2005.
46. T.V.N. Padmesh, K. Vijayaraghavan, G. Sekaran, M. Velan, *Chemical Engineering Journal*, Vol. 122, p. 55, 2006.
47. R. Aravindhan, J.R. Rao, B.U. Nair, *Journal of Hazardous Materials*, Vol. 142, p. 68, 2007.
48. A.R. Khataee, F. Vafaei, M. Jannatkhah, *International Biodeterioration & Biodegradation*, Vol. 83, p. 33, 2013.
49. C. Michael Hogan, *Encyclopedia of Earth: Bacteria,* National Council for Science and the Environment, Washington, 2010.
50. J. Mao, S.W. Won, S.B. Choi, M.W. Lee, Y.S. Yun, *Biochemical Engineering Journal*, Vol. 46, p. 1, 2009.
51. A. Arunarani, P. Chandran, B.V. Ranganathan, N.S. Vasanthi, S.S. Khan, *Colloids and Surfaces B: Biointerfaces*, Vol. 102, p. 379, 2013.
52. S.W. Won, M.H. Han, Y.S. Yun, *Water Research*, Vol. 42, p. 4847, 2008.
53. A.R. Binupriya, M. Sathishkumar, C.S. Ku, S.I. Yun, *Colloids and Surfaces B: Biointerfaces*, Vol. 76, p. 179, 2010.
54. L.N. Du, B. Wang, G. Li, S. Wang, D.E. Crowley, Y.H. Zhao, *Journal of Hazardous Materials*, Vol. 205–206, p. 47, 2012.
55. I. Ali, M. Asim, T.A. Khan, *Journal of Environmental Management*, Vol. 113, p. 170, 2012.
56. S.W. Won, S.B. Choi, Y.S. Yun, *Colloids and Surfaces A: Physicochemical Engineering Aspects*, Vol. 262, p. 175, 2005.
57. K. Vijayaraghavan, Y.S. Yun, *Journal of Hazardous Materials*, Vol. 141, p. 45, 2007.
58. K.V.H. Prashanth, R.N. Tharanathan, *Trends in Food Science & Technology*, Vol. 18, p. 117, 2007.

59. J.S. Piccin, M.L.G. Vieira, J.O. Gonçalves, G.L. Dotto, L.A.A. Pinto, *Journal of Food Engineering*, Vol. 95, p. 16, 2009.
60. G.L. Dotto, M.L.G. Vieira, L.A.A. Pinto, *Industrial Engineering Chemistry Research*, Vol. 51, p. 6862, 2012.
61. S. Chatterjee, T. Chatterjee, S.H. Woo, *Separation Science and Technology*, Vol. 46, p. 986, 2011.
62. T.Y. Kim, S.S. Park, S.Y. Cho, *Journal of Industrial and Engineering Chemistry*, Vol. 18, p. 1458, 2012.
63. S. Zhao, F. Zhou, L. Li, M. Cao, D. Zuo, H. Liu, *Composites: Part B*, Vol. 43, p. 1570, 2012.
64. W.H. Cheung, Y.S. Szeto, G. McKay, *Bioresource Technology*, Vol. 100, p. 1143, 2009.
65. W.H. Cheung, Y.S. Szeto, G. McKay, *Bioresource Technology*, Vol. 98, p. 2897, 2007.
66. W.S. Wan Ngah, L.C. Teong, M.A.K.M. Hanafiah, *Carbohydrate Polymers*, Vol. 83, p. 1446, 2011.
67. G.L. Dotto, L.A.A. Pinto, *Carbohydrate Polymers*, Vol. 84, p. 231, 2011.
68. G.L. Dotto, L.A.A. Pinto, *Journal of Hazardous Materials*, Vol. 187, p. 164, 2011.
69. G.L. Dotto, J.M. Moura, T.R.S. Cadaval, L.A.A. Pinto, *Chemical Engineering Journal*, Vol. 214, p. 8, 2013.
70. A. Mirmohseni, M.S. Seyed Dorraji, A. Figoli, F. Tasselli, *Bioresource Technology*, Vol. 121, p. 212, 2012.
71. A.R. Fajardo, L.C. Lopes, A.F. Rubira, E.C. Muniz, *Chemical Engineering Journal*, Vol. 183, p. 253, 2012.
72. H.Y. Zhu, R. Jiang, L. Xiao, *Applied Clay Science*, Vol. 48, p. 522, 2010.
73. B.H. Hameed, M. Hasan, A.L. Ahmad, *Chemical Engineering Journal*, Vol. 136, p. 164, 2008.
74. G.L. Dotto, M.L.G. Vieira, J.O. Gonçalves, L.A.A. Pinto, *Quimica Nova (In Portuguese)*, Vol. 34, p. 1193, 2011.
75. Y.C. Wong, Y.S. Szeto, W.H. Cheung, G. McKay, *Adsorption*, Vol. 14, p. 11, 2008.
76. R. Patel, S. Suresh, *Bioresource Technology*, Vol. 99, p. 51, 2008.
77. M.E. Russo, F. Di Natale, V. Prigione, V. Tigini, A. Marzocchella, G.C. Varese, *Chemical Engineering Journal*, Vol. 162, p. 537, 2010.
78. P. Kaushik, A. Malik, *Journal of Hazardous Materials*, Vol. 185, p. 837, 2011.
79. I. Kiran, T. Akar, A.S. Ozcan, A. Ozcan, S.T. Akar, *Biochemical Engineering Journal*, Vol. 31, p. 197, 2006.
80. S.T. Ambrósio, J.C. Vilar Jr., C.A.A. da Silva, K. Okada, A.E. Nascimento, R.L. Longo, G.M. Campos Takaki, *Molecules*, Vol. 17, p. 452, 2012.
81. Y. Khambhaty, K. Mody, S. Basha, *Ecological Engineering*, Vol. 41, p. 74, 2012.
82. S.T. Akar, T. Akar, A. Çabuk, *Brazilian Journal of Chemical Engineering*, Vol. 26, p. 399, 2009.
83. Z. Aksu, *Process Biochemistry*, Vol. 38, p. 1437, 2003.

84. T. Akar, S. Arslan, S.T. Akar, *Ecological Engineering*, Vol. 58, p. 363, 2013.
85. T. Akar, A. Kulcu, S.T. Akar, *Chemical Engineering Journal*, Vol. 221, p. 461, 2013.
86. M.J. Pelczar, F.C.S. Chan, N.R. Krieg, *Microbiology*, McGraw-Hill, Columbus, 1998.
87. A. Kapoor, T. Viraraghavan, *Bioresource Technology*, Vol. 53, p. 195, 1995.
88. S. Chatterjee, S. Chatterjee, B.P. Chatterjee, A.K. Guha, *Colloids and Surfaces A: Physicochemical Engineering Aspects*, Vol. 299, p. 146, 2007.
89. C.T. Weber, E.L. Foletto, L. Meili, *Water Air Soil Pollution*, Vol. 224, p. 1427, 2013.
90. N. Barka, K. Ouzaouit, M. Abdennouri, M. El Makhfouk, *Journal of the Taiwan Institute of Chemical Engineers*, Vol. 44, p. 52, 2013.
91. S. Cengiz, F. Tanrikulu, S. Aksu, *Chemical Engineering Journal*, Vol. 189–190, p. 32, 2012.
92. D.K. Mahmoud, M.A.M. Salleh, W.A.W. Abdul Karim, A. Idris, Z.Z. Abidin, *Chemical Engineering Journal*, Vol. 181–182, p. 449, 2012.
93. Z. Zhang, I.M. O'Hara, G.A. Kent, W.O.S. Doherty, *Industrial Crops and Products*, Vol. 42, p. 41, 2013.
94. S. Dawood, T.K. Sen, *Water Research*, Vol. 46, p. 1933, 2012.
95. M.F. Elkady, A.M. Ibrahim, M.M.A El-Latif, *Desalination*, Vol. 278, p. 412, 2011.
96. D.M. Ruthven, *Principles of Adsorption and Adsorption Processes*, John Wiley & Sons, New York, 1984.
97. M. Suzuki, *Adsorption Engineering*, Kodansha, Tokyo, 1990.
98. J.S. Piccin, G.L. Dotto, M.L.G. Vieira, L.A.A. Pinto, *Journal of Chemical and Engineering Data*, Vol. 56, p. 3759, 2011.
99. S.V. Yadla, V. Sridevi, M.V.V.C. Lakshmi, *Journal of Chemical, Biological and Physical Sciences*, Vol. 2, p. 1585, 2012.
100. G. McKay, H.S. Blair, J.R. Gardner, *Journal of Applied Polymer Science*, Vol. 27, p. 3043, 1982.
101. G. McKay, H.S. Blair, J.R. Gardner, *Journal of Applied Polymer Science*, Vol. 28, p. 1767, 1983.
102. L.D.T. Prola, E. Acayanka, E.C. Lima, C.S. Umpierres, J.C.P. Vaghetti, W.O. Santos, S. Laminsi, P.T. Djifon, *Industrial Crops and Products*, Vol. 46, p. 328, 2013.
103. Y. Yang, G. Wang, B. Wang, Z. Li, X. Jia, Q. Zhou, Y. Zhao, *Bioresource Technology*, Vol. 102, p. 828, 2011.
104. R. Kumar, R. Ahmad, *Desalination*, Vol. 265, p. 112, 2011.
105. N. Kannan, M.M. Sundaram, *Dyes and Pigments*, Vol. 51, p. 25, 2001.
106. Z. Eren, F.N. Acar, *Desalination*, Vol. 194, p. 1, 2006.
107. Y. Bulut, H. AydIn, *Desalination*, Vol. 194, p. 259, 2006.
108. V.M. Vucurovic, R.N. Razmovski, M.N. Tekic, *Journal of the Taiwan Institute of Chemical Engineers*, Vol. 43, p. 108, 2012.

109. H. Fan, J.S. Yang, T.G. Gao, H.L. Yuan, *Journal of the Taiwan Institute of Chemical Engineers*, Vol. 43, p. 386, 2012.

110. P. Li, G. Xiu, A.E. Rodrigues, *Chemical Engineering Science*, Vol. 58, p. 3361, 2003.

111. G.L. Dotto, L.A.A. Pinto, *Biochemical Engineering Journal*, Vol. 68, p. 85, 2012.

112. V.J. Inglezakis, S.G. Poulopoulos, *Adsorption, ion exchange and catalysis: Design of operations and environmental applications*, Elsevier, Amsterdam, 2006.

113. M.A.K.M. Hanafiah, W.S. Wan Ngah, S.H. Zolkafly, L.C. Teong, Z.A. Abdul Majid, *Journal of Environmental Sciences*, Vol. 24, p. 261, 2012.

114. I. Uzun, F. Guzel, *Journal of Hazardous Materials*, Vol. 118, p. 141, 2005.

115. C.H. Giles, D. Smith, A.A. Huitson, *Journal of Colloid and Interface Science*, Vol. 47, p. 755, 1974.

116. J.F. Gao, Q. Zhang, J.H. Wang, X.L. Wu, S.Y. Wang, Y.Z. Peng, *Bioresource Technology*, Vol. 102, p. 805, 2011.

117. G. Blázquez, M. Calero, F. Hernáinz, G. Tenorio, M.A. Martín-Lara, *Chemical Engineering Journal*, Vol. 160, p. 615, 2010.

118. O. Hamdaoui, E. Naffrechoux, *Journal of Hazardous Materials*, Vol. 147, p. 381, 2007.

119. O. Hamdaoui, E. Naffrechoux, *Journal of Hazardous Materials*, Vol. 147, p. 401, 2007.

120. I. Langmuir, *Journal of the American Chemical Society*, Vol. 40, p. 1361, 1918.

121. H.M.F. Freundlich, *Journal Physical Chemistry*, Vol. 57, p. 385, 1906.

122. O. Redlich, D.L. Peterson, *Journal of Chemical Physics*, Vol. 63, p. 1024, 1959.

123. M.M. Dubinin, L.V. Radushkevich, *Chemical Zentrum*, Vol. 1, p. 875, 1947.

124. R. Sips, *Journal of Chemical Physics*, Vol. 16, p. 490, 1948.

125. J. Tóth, *Journal of Colloid Interface Science*, Vol. 225, p. 378, 2000.

126. S. Chowdhury, R. Mishra, P. Saha, P. Kushwaha, *Desalination*, Vol. 265, p. 159, 2011.

127. G. Sreelatha, V. Ageetha, J. Parmar, P. Padmaja, *Journal of Chemical and Engineering Data*, Vol. 56, p. 35, 2011.

128. Sumanjit, S. Rani, R.K. Mahajan. *Arabian Journal of Chemistry, proof* 2012

129. N.F. Cardoso, R.B. Pinto, E.C. Lima, T. Calvete, C.V. Amavisca, B. Royer, M.L. Cunha, T.H.M. Fernandes, I.S. Pinto, *Desalination*, Vol. 269, p. 92, 2011.

130. N. Barka, M. Abdennouri, M. EL Makhfouk, *Journal of the Taiwan Institute of Chemical Engineers*, Vol. 42, p. 320, 2011.

131. J.M. Smith, H.C. Van Ness, M.M. Abbott, *Introduction to Chemical Engineering Thermodynamics*, LTC, Rio de Janeiro, 2000.

132. Y. Liu, *Journal of Chemical and Engineering Data*, Vol. 54, p. 1981, 2009.

133. S.K. Milonjić, *Journal of the Serbian Chemical Society*, Vol. 72, p. 1363, 2007.

134. X. Zhou, H. Liu, J. Hao, *Adsorption Science & Technology*, Vol. 30, p. 647, 2012.

135. M. Asgher, H.N. Bhatti, *Ecological Engineering*, Vol. 38, p. 79, 2012.

136. T. Calvete, E.C. Lima, N.F. Cardoso, S.L.P. Dias, F.A. Pavan, *Chemical Engineering Journal*, Vol. 155, p. 627, 2009.
137. H. Qiu, L.L. Pan, Q.J. Zhang, W. Zhang, Q. Zhang, *Journal of Zhejiang University Science A*, Vol. 10, p. 716, 2009.
138. S. Lagergren, *Kungliga Svenska Vetenskapsakademien*, Vol. 24, p. 1, 1898.
139. Y.S. Ho, G. McKay, *Process Safety and Environmental Protection*, Vol. 76, p. 332, 1998.
140. F.C. Wu, R.L. Tseng, R.S. Juang, *Chemical Engineering Journal*, Vol. 150, p. 366, 2009.
141. M. Avrami, *Journal of Chemical Physics*, Vol. 7, p. 1103, 1939.
142. W.J. Weber, J.C. Morris, *Journal of Sanitary Engineering Division of American Society of Civil Engineering*, Vol. 89, p. 31, 1963.
143. P. Senthil Kumar, S. Ramalingam, C. Senthamarai, M. Niranjana, P. Vijayalakshmi, S. Sivanesan, *Desalination*, Vol. 261, p. 52, 2010.
144. S. Chatterjee, T. Chatterjee, S.H. Woo, *Chemical Engineering Journal*, Vol. 166, p. 168, 2011.
145. M.I. El-Khaiary, G.F. Malash, *Hydrometallurgy*, Vol. 105, p. 314, 2011.

9

Dye Adsorption on Expanding Three-Layer Clays

Tolga Depci[1] and Mehmet S. Çelik[*,2]

[1]Faculty of Engineering, Department of Mining Engineering, Inonu University, Malatya, Turkey
[2]Faculty of Mining, Mineral Processing Department, Istanbul, Turkey

Abstract

Many dyes that are used in the textile industry are toxic and tend to accumulate in living organisms directly or indirectly, causing various diseases and disorders. Clay minerals have recently been shown to be good candidates as dye adsorbents for wastewater treatment. This review discusses the use of both natural and modified three-layer clays such as bentonite, montmorillonite, and vermiculite to remove both cationic and anionic dyes. The most important parameters governing adsorption capacities are pH, ionic strength, and modification processes. Specific surface area is somewhat less important. Modification processes include thermal activation, acid activation, surfactant addition, and combinations of these, all of which increase adsorption capacities. Adjusting the pH changes both adsorbent surface properties and adsorbate dye ionization.

Keywords: Bentonite, montmorillonite, cationic dyes, anionic dyes, adsorption, dye removal

9.1   Introduction

Dye-containing wastewaters from the textile industry are one of the main pollutant sources worldwide. There are over 100,000 different textile dyes with an estimated annual production of 7.10^5 metric tons. Of these dyes, 30% are used in excess of 1,000 tons per annum and 90% are used at the level of 100 tons per annum [1–5]. In addition to the textile industry, the

---

Corresponding author: mcelik@itu.edu.tr

Sanjay K. Sharma (ed.) Green Chemistry for Dyes Removal from Wastewater, (331–358)
© 2015 Scrivener Publishing LLC

consumption of dyes has increased in other industries like cosmetics, pulp and paper, paint, pharmaceutical, food, carpet, and printing industries. In addition to coloring the wastewater, dyes and their breakdown products are toxic, carcinogenic, and mutagenic to life forms, and they can cause allergies and skin diseases [6–9]. Removal of dyes from wastewater is generally difficult. They are recalcitrant organic molecules that have complex aromatic structures. They are usually biologically non-degradable; are resistant to aerobic digestion; and are stable to light, heat and oxidizing agents [10,11]. The treatment and decolorization processes that have emerged during the past three decades remain plagued by cost and disposal problems. Their specific advantages and disadvantages are compared in Table 9.1 [5,12–15].

Adsorption is one of the best treatment methods due to its flexibility, simplicity of design, and insensitivity to toxic pollutants. Activated carbon is the most popular adsorbent, and has been cited by the US Environmental Protection Agency as one of the best available control technologies [15]. However, because activated carbon production and regeneration is expensive, considerable attention has been given to other low cost and easily available adsorbents for dye removal. Recently, clay and its modified forms have been used as adsorbents, and there has been an upsurge of interest in the interactions between dyes and clay particles.

Clay may serve as an ideal adsorbent because of its low cost. It has relatively large specific surface area, excellent physical and chemical stability, and other advantageous structural and surface properties. For example,

Table 9.1 Principal, existing and emerging processes with their advantages and disadvantages for removal of dyes [5,12].

Methods	Advantages	Disadvantages
Oxidation	Rapid and efficient process	(H_2O_2) agent needs to activate by some means; high energy cost; chemicals required; sludge generation
Advanced oxidation process	No sludge production; little or no consumption of chemicals; efficiency for recalcitrant dyes	Economically unfeasible; formation of byproducts; technical constraints
Ozonation	Ozone can be applied in its gaseous state and does not increase the volume of wastewater and sludge	Short half-life (20 min)

(Continued)

Table 9.1 (*Cont.*)

Methods	Advantages	Disadvantages
Biodegradation	Economically attractive, publicly acceptable treatment	Slow process; necessary to create an optimal favorable environment; maintenance and nutrition requirements
Biomass	Low operating cost; good efficiency and selectivity; no toxic effect on microorganisms	Slow process; performance depends on some external factors (pH, salts); not effective for all dyes
Anaerobic textile-dye bioremediation	Allows azo and other water-soluble dyes to be decolorized	Anaerobic breakdown yields methane and hydrogen sulfide
Coagulation Flocculation	Simple; economically feasible	High sludge production; handling and disposal problems
Membrane separations	Remove all dye types; produce a high-quality treated effluent	High pressures; expensive; incapable of treating large volumes
Ion exchange	No loss of sorbent on regeneration; effective	Not effective for all dyes, especially disperse dye; economic constraints
Irradiation	Effective oxidation at lab scale	Requires a lot of dissolved O_2
Adsorption on activated carbons	The most effective adsorbent; great capacity; produce a high-quality treated effluent	Very expensive; ineffective against disperse and vat dyes; the regeneration is expensive and results in loss of the adsorbent; nondestructive process

montmorillonite clay is cheaper than activated carbon, and it has a large specific surface area and a high cation exchange capacity [14]. In addition, clay minerals exhibit strong affinity for both cationic and anionic dyes. However, the sorption capacity for basic dye is usually higher than for acid dye because of the ionic charges on the dyes and clay. The adsorption of dyes on clay minerals is mainly dominated by ion-exchange processes. This means that the sorption capacity can vary strongly with pH.

This review presents (i) brief information on textile dyes; (ii) a critical analysis of natural and modified expanding three-layer clay minerals; (iii) their characteristics, advantages and limitations; and (iv) discussion of adsorption mechanisms involved. The adsorption mechanisms, capacities, thermodynamics, and kinetics are given as text or tables. It should be remembered that comparisons in these tables are necessarily limited because of differences in surface area, pore structure, functional groups, modification processes, and experimental conditions. Therefore, the reader is encouraged to refer to original articles for more detailed information.

9.2   Classification of Dyes

Textile dyes represent a category of organic compounds and are mainly classified in two different ways: (1) based on their application characteristics (i.e., C.I. Generic Name such as acidic, basic, direct, disperse, mordant, reactive, sulphur dye, pigment, vat, azo insoluble), and (2) based on their chemical structure (i.e., C.I. Constitution Number such as nitro, azo, carotenoid, diphenylmethane, xanthene, acridine, quinoline, indamine, sulphur, amino and hydroxy ketone, anthraquinone, indigoid, phthalocyanine, inorganic pigment, etc.). Table 9.2 shows the characteristics of textile dyes. They are also classified as anionic, cationic, and nonionic dyes [5,7,15]. Cationic dyes are basic dyes, while the anionic dyes include direct, acid, and reactive dyes [16,17]. The literature on adsorption of cationic and anionic dyes by natural and modified clays has recently been expanding.

9.2.1   Anionic Dye

Anionic dyes are used with silk, wool, polyamide, modified acrylic and polypropylene fibers, and are the most varied class of dyes. Examples are azoic, anthraquinone, triphenylmethan and nitro dyes. They possess common features of being water-soluble, having ionic substituents, possessing a negative charge, and having harmful effects on humans [21,22]. Some of the adsorption studies by clay minerals include: the removal of acid fuchsin (AF) by sodium montmorillonite [23], acid green B and direct pink 3B by organophilic montmorillonite [24]; Congo red (CR) by natural, thermal activated, and acid activated bentonite [25]; and Acid scarlet, Acid turquoise blue, and Indigo carmine by organo-bentonite [26].

9.2.2   Cationic Dyes

Cationic dyes are used in acrylic, wool, nylon and silk dyeing, have a positive charge, and are known as basic dyes. Examples are azo and methane dyes, anthraquinone, di- and tri-arylcarbenium and phthalocyanine dyes, and various polycarbocyclic and solvent dyes [21]. Depending on aromatic

Table 9.2 Characteristics of typical dyes used in textile dyeing operations [17–20].

Type of Dye	Description	Example	Typical Pollutants Associated with Various Dyes
Acid Dyes	Water-soluble anionic compounds	Methyl Orange, Congo Red	color; organic acids; unfixed dyes
Basic Dyes	Water-soluble, applied in weakly acidic dyebaths; very bright dyes	Methylene Blue, Malachite Green	Degradation by physical, chemical or biological treatments may produce small amount of toxic and carcinogenic products
Direct Dyes	Water-soluble, anionic compounds; can be applied directly to cellulosics without mordants (or metals like chromium and copper)	Congo Red and Matius Yellow	color; salt; unfixed dye; cationic fixing agents; surfactant; defoamer; leveling and retarding agents; finish; diluents
Disperse Dyes	Not water-soluble	Pink B and Blue B	color; organic acids; carriers; leveling agents; phosphates; defoamers; lubricants; dispersants; delustrants; diluents
Reactive	Water-soluble, anionic compounds; largest dye class	Procion dyes	color; salt; alkali; unfixed dye; surfactants; defoamer; diluents; finish
Sulfur	Organic compounds containing sulfur or sodium sulfide	Alizarin	color; alkali; oxidizing agent; reducing agent; unfixed dye
Vat	Water-insoluble; oldest dyes; more chemically complex	Indigo, Tyrian Purple	color; alkali; oxidizing agents; reducing agents

groups, they are resistant to breakdown by chemical, physical and biological treatments [27]. They cause allergic reactions, dermatitis, skin irritation, mutations, and cancer [28]. Some of the adsorption studies by clay minerals have been carried out on crystal violet [28], methylene blue [29], and basic yellow 28 [30].

9.3 The Expanding Three-Layer Clay Minerals and Dye Adsorption

Clay minerals are hydrous aluminum silicates with large interlayer spaces. These spaces can hold significant amounts of water and other substances and can allow swelling and shrinking [31]. Clay minerals are classified into four types by Grim and Güven [32]: two-layer types; three layer types; regular mixed-layer types; and the chain structure types.

Three-layer type clay minerals have a sheet structure with two layers of silica tetrahedrons and one central dioctahedral or trioctahedral alumina layer. These types of clay minerals are divided into two categories: the nonexpanding illites; and the expanding montmorillonites, smectites and vermiculites. It is the expanding clays that are most used for dye adsorption [31].

Montmorillonite is a 2:1 layered silicate with negative charge due to ionic substitution in its structure. This charge is balanced by exchangeable cations such as Na^+ or Ca^{2+} present in the interlayer space [23,33]. The Na, Ca, Mg, Fe, and Li-Al silicates in the smectite group include Na-montmorillonite, Ca-montmorillonite, saponite (Mg), nontronite (Fe), and hectorite [34]. Smectites are the dominant clay mineral in bentonite, and montmorillonite is the most common smectite [35]. These clay minerals are altered or metamorphic forms of glassy igneous material, usually a tuff or volcanic ash [36]. Montmorillonite has received considerable recognition as an adsorbent because of its high adsorption capacity [29,37–44]. Vermiculite is another expanding, three-layer type, clay mineral that is derived by alteration or weathering from black mica, chlorite, illite, and is a secondary metamorphic mineral containing magnesium, iron, aluminum and silicate [45,46].

Montmorillonite, vermiculite and bentonite are the clay minerals most often mentioned in studies to remove dyes from wasterwater [29,30,37–40,47,48]. Modified or activated forms enhance dye adsorption capacities [37,49–55]. Activation methods include acid activation [56], treatment with surfactants [57], thermal treatment [58], polymer addition, pillaring by different types of poly (hydroxo metal) cations [59], and grafting of organic compounds [60].

Acid activation enhances surface area and average pore volume due to the decomposition of the crystalline structure [56,61,62]. In addition, acid activation can also change the chemical properties such as cation exchange capacity and the surface acidity [63], but excessive acid decreases surface area [64]. Another favorable method, thermal activation, increases surface area and removes surface impurities [65,66]. Sennour *et al.* [67] have mentioned that thermal activated clays are used in the textile, oil, and sugar industries. Excess heating collapses structure and reduces surface area [39,65]. There have been many wastewater applications using acid and thermal modified three-layer type clays [28,68–70].

Pillared clay is another modified clay which is prepared by intercalating natural clays with bulky polyoxycations such as Al (aliminium) or Zr (zirconium). Calcination at high temperatures results in transforming the intercalated polyoxycations into rigid oxide pillars with a highly porous structure. Although the use of pillared clay is limited due to lack of thermal stability [71], the removal of cationic and anionic dyes from aqueous solutions by pillared bentonite and montmorillonite has been reported [72,73].

Anionic and cationic surfactants can be used to increase dye adsorption and create so-called organoclay [74–76]. Anionic surfactant enhances adsorption of basic dyes, and cationic surfactant enhances adsorption of acid dyes. Excess surfactant may cause aggregation or dye solubilization. In addition, large-scale production may not be practical due to the complexity of the process [77] and high cost of surfactants [78,79].

9.3.1 Removal of Anionic Dyes by Expanding Three-Layer Clays

Expanding three-layer clays, with their high cation exchange capacity, are readily available, reusable and offer a low-cost alternative for dye removal [80–85]. The three most important dye adsorption parameters are surface charge, surface area and ion exchange.

However, as clays are generally hydrophilic, they are not effective adsorbents for the nonpolar organic compounds in water if not modified by surfactants. The pH of the solution determines surface charge and ionization/dissociation of the adsorbate molecule [86,87], and H^+ and OH^- are the potential determining ions for clay minerals. Zeta (ζ) potential for natural bentonite increases in the negative direction with increasing pH and the isoelectrical point (IEP) occurs in the pH range of 2–3 [49,50,88,121]. Above the IEP pH point, the surface charge is negative, and anionic dye is repelled. For example, Özcan *et al.* [70] used Na-bentonite for adsorption of anionic Acid Blue 193 while varying the pH from 1 to 11. Maximum removal occurred at pH 1.5 and decreased at higher pH values as surface charges became more negative [89,90].

Baskaralingam *et al.* [83] failed to remove very much Acid Red 151 using Na-bentonite, and the maximum 5% uptake decreased with increasing pH. Similar trends for bentonite were observed for adsorption of Procion Navy Hexl [38], Congo red [39], Supranol Yellow 4GL [73], and Sulfacid brilliant pink [91]. Similarly, Zohra *et al.* [80] mentioned that the anionic dye (Benzopurpurin 4B) did not adsorb onto natural Na-bentonite because of the similar surface charge.

The ability of Ca-bentonite to remove anionic dye from aqueous solutions has also been studied. Ca-Bentonite has a permanent negative charge due to the isomorphous substitution of Al^{3+} for Si^{4+} in the tetrahedral layer and Mg^{2+} for Al^{3+} in the octahedral layer [73,83]. The high surface calcium content also serves to neutralize any OH^- and decreases the effect of pH on adsorption. Lian *et al.* [92] selected natural Ca-bentonite as a cheaper bentonite adsorbent for the removal of Congo red from aqueous solutions. Their results showed that adsorption is dependent on concentration, but changed only slightly (10% decrease) in the pH range of 5–11.

Natural bentonite is sometimes reasonably efficient as an adsorbent for anionic dye in water [53,54]. However, surface-modified bentonite is much more efficient and has a greater potential to provide an alternative to activated carbon [84,95,96]. Surface modification includes heat treatment [58], acid activation [63], treating with cationic surfactants [78] and polymer modification [60,93]. Introducing quaternary ammonium salts imparts hydrophobicity and makes an "organoclay" with an affinity for organic compounds [94], and organoclays have been widely used in wastewater treatment processes [84,95,96]. Table 9.3 lists adsorption capacities of organo-bentonites for selected anionic dyes.

Literature data show that the adsorption capacities of the expanding three-layer clay increases upon modification [28,38,50,51,63,82,83]. For example, natural and modified bentonites using dodecyl trimethylammonium bromide (DTMA) for the removal of Acid Blue 193 were compared by Özcan *et al.* [70]. The adsorption capacity of DTMA-bentonite (740.5 mg/g) is about 11-fold higher than that of raw Na-bentonite (67.1 mg/g) at 20°C. Recently, Akl *et al.* [37] also showed similar results for hexadecyl trimethyl ammonium bromide (CTAB)-modified bentonite, where 37.05 mg/g and 210.104 mg/g of Congo red were removed. In the modification process, the alkyl chain length of surfactants affected the adsorption capacity of bentonite. Ma *et al.* [50] further emphasized that higher adsorption capacity is obtained using long-chain surfactants. Decreasing the surfactant carbon chain length from C16 to C8 during modification decreases the adsorption capacities of Orange II from 298.39 mg/g for C16-bentonite to 44.08 mg/g for C8-bentonite.

Table 9.3 Organo-bentonites used for removal of anionic dyes from aqueous solutions together with their adsorption capacities.

Clay Type	Surfactant	Dye Name	Specific Surface Area	Adsorption Capacity (mg/g)	Reference
Bentonite	Cetyltrimethylammonium bromide	Congo red	18.41	210.104	[37]
Bentonite	Hexadecyltrimethylammonium bromide; Octyltrimethyl ammonium bromide; Myristyltrimethylammonium bromide	Orange II and Orange G	17.46 4.86 0.21	228.74 170.46	[50]
Bentonite	Polyepicholorohydrin-dimethylamine	Violet K-3R and Acid Dark Blue 2G	Not Provided	44.64 40.77	[51]
Bentonite	Dodecyltrimethylammonium bromide	Acid Blue 193 (AB193)	Not Provided	740.5	[70]
Montmor.	KSF protonated dodecylamine	Remazol brilliant blue R	128	38.99	[72]
Bentonite	Cetyltrimethyl ammonium bromide	Benzopurpurin 4B	Not Provided	153.84	[80]
Bentonite	Dodecyltrimethyl ammonium bromide	Reactive blue (RB19)	32.86	206.58	[82]
Bentonite	Cetyldimethylbenzylammonium chloride and Cetylpyridinium chloride	Acid Red 151	8.92 2.41	357.14 416.60	[83]
Montmor.	Calcium alginateorganophilic	Acid Green B Direct Pink 3B	Not Provided	26 16	[97]

Surface charge is also important for organoclays. Acid and reactive dyes are water-soluble anionic dyes with negative charge, so the surface of natural clays has to be modified by a cationic surfactant for dye removal. The adsorption of anionic dyes on natural bentonite is difficult. Adsorption depends on pH due to the increasing negative charge with increasing pH. The zeta potential variation of bentonite versus pH is given in Figure 9.1 [50]. Although the zeta potential trend is the same for hexadecyltrimethylammonium bromide (HTAB)-modified bentonite [38] and cationic-polymer/bentonite, these modified bentonites remain positive for a much higher pH range [51].

Figure 9.1 illustrates an IEP of pH = 3 for natural bentonite and an IEP of pH = 7.5 for HTAB-modified bentonite. Below the IEP the positively charged surface attracts negative dye molecules. Baskaralingam *et al.* [83] have shown that natural bentonite had virtually zero adsorption of Acid Red 1515, while cetyldimethylbenzylammonium- and cetylpyridinium-modified bentonite yield good removal of the dye in the acidic pH range and decrease with increasing pH. The same trend was observed by Socias-Viciana *et al.* [98], who used Na-bentonite and dodecyltrimethylammonium bromide-modified bentonite (DTMA-bentonite) to remove Acid Blue dye. This trend was also observed by Khenifi *et al.* [99], who studied the removal of Supranol Yellow 4GL using organo bentonite. They described how surfactants adsorb to the external surfaces of bentonite via cation exchange. Then, due to extensive hydrophobic bonding, tail-tail interactions between surfactant and dye cause positive charge development on the surfaces that ultimately leads to clay dispersion [51,80,100,101]. Baskaralingam *et al.* [83] also showed that dye uptake decreases as pH approaches 10, and the

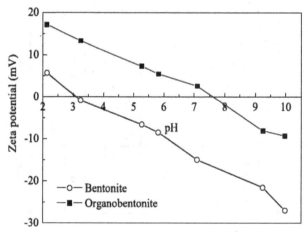

Figure 9.1 Effect of pH on zeta potential of bentonite and HTAB-modified bentonite [50].

same trend was also observed elsewhere [51,70,79,80,81]. Even though the electrostatic mechanism is very important, it is not the only mechanism for dye adsorption in the system. Adsorbents can also interact with dye molecules via other mechanisms such as hydrophobic interaction and chemisorption [83]. Talep *et al.* [97] observed this by using alginate-organophilic montmorillonite composite to remove two acidic dyes (Acid Green B and Direct Pink 3 B).

In addition, Ma *et al.* [50] maintain the idea that the main driving mechanism for the adsorption of acid dye is anion exchange. The counter-ion bromide in the organo-bentonite is replaced by the dye anion and the adsorption capacity of organo-bentonite is affected by the surfactant alkyl chain length. When the longer alkyl chain surfactant is modified, bentonite gives higher adsorption capacity because inorganic cations in bentonite exchange with surfactants to increase interlamellar spacing and thus expose new sorption sites [83]. The exchange capacity of bentonite increases with the amount of surfactant [50,102]. These results support the study done by Li *et al.* [79]. They modified bentonite using epicholorohyrin-dimethylamine polyamine (EPI-DMA) and showed that the layer space is expanded due to the intercalation of EPI-DMA into the clay layers, leading to a more hydrophobic and more positive surface. Their results also showed that modified clay is suitable for adsorption of Disperse Yellow SE-6GRL, Disperse Red S-R, Reactive Reddish Violet K2-BP and Reactive Jade Blue K-GL. However, their study also indicates that relatively large amounts of polymer are required to make negative clay surface become positive.

Silva *et al.* [72] and Talep *et al.* [97] mentioned similar results for montmorillonite. Organic modification can significantly improve the clay's adsorption capability towards anionic dyes. Adsorption is ascribed to the binding between anionic groups (e.g., sulfonic groups) of the dye and the positively charged surface of organoclays [76,83].

The above-mentioned mechanisms illustrate how modification changes the surface charge of the expanding three-layer clays and increases their active sites. The ion-exchange mechanism offers an exchange of the inorganic cation with the organic surfactant cation. As a result, the adsorption capacity of organoclay increases compared to natural clay mineral [70,94,103].

In addition to pH and modification or activation processes, surface area and ion exchange mechanisms are used to explain adsorption of anionic dye on bentonite. Some researchers mentioned that the adsorption capacity of natural bentonite is hampered by its small surface area [84,104], and modification of bentonite further decreases the surface area [37,50,83].

Xu and Boyd [101] pointed out that large hexadecyltrimethylammonium (HDTMA) molecules are initially adsorbed by cation exchange in the interlayer, which causes extensive clay aggregation and loss of surface area. As loading increases, HDTMA is adsorbed on the external surfaces of aggregates. In addition, recent investigations have shown that surfactants may block the fine pores and decrease the specific surface area of bentonite [105]. However, these investigations also show that the specific surface area has no effect if maximum adsorption density of any dye is way below the monolayer coverage. Surface area is not as important as other parameters.

Contrary to some surfactant modification processes, acid and thermal modification tends to increase the surface area. Acid activation decreases overall average pore size with a concomitant increase in the surface area [68,70]. Thermal activation similarly enhances the surface area [99,106]. Recently, Toor and Jin [25] modified natural bentonite by thermal activation (TA), acid activation (AA) and combined acid and thermal activation (ATA) for the removal of anionic dye, Congo red (CR). More dye was removed because interlayer spaces collapsed, creating a more tightly bound structure and an increase in surface area (Table 9.4).

9.3.2   Removal of Cationic Dyes by Expanding Three-Layer Clays

The adsorption mechanism of cationic dyes on natural and modified three-layer clays has also been investigated by many researchers. Cationic dyes are known as basic dye and carry a positive charge in their molecule [21]. Positive cationic dyes are attracted to the negative clay surfaces [39,72], and above the IEP the surface of the expanding three-layer clays are negative due to successive deprotonation of positively charged groups and the underlying negative charge density. Adsorption increases with increasing pH. Tahir and Naseem [88] showed that bentonite removal of cationic malachite green oxalate increases from 29% to 91% with an increase in pH

Table 9.4 Characteristics of acid, thermal and acid + thermal modified bentonites [25].

Characteristics	Pore size (Å)	Surface area (m² g⁻¹)
Raw bentonite	50.28 ± 0.004	25.7 ± 0.008
Acid activated bentonite (0.5 M HCl)	53.78 ± 0.12	75.5 ± 0.10
Thermal activated bentonite	51.65 ± 0.10	34.6 ± 0.06
Acid and thermal activated bentonite	56.13 ± 0.09	84.6 ± 0.12

from 2.0 to 9.0, and then remained almost constant. Similar results were observed by Al-Khatib *et al.* [29] for methylene blue uptake on bentonite.

A contrary result by Eren and Afsin [27] showed that cationic Crystal Violet adsorption on natural and Ni-, Co-, Zn-pretreated bentonite is pH independent. They explained that adsorption occurs partly by ion exchange in the interlayer and on basal plane surfaces, and partly via noncoloumbic interactions between an adsorbed cation and a neutralized site. Turabik [30] also found that bentonite adsorption of Basic Red 46 and Basic Yellow 28 basic dyes are pH independent in the range of pH 2.0–8.0.

Kurniawan *et al.* [49] prepared rarasaponin-bentonite to adsorb basic methylene blue (MB) and malachite green (MG). They found that dye removal gradually increases in the pH range of 4–7 and reaches its maximum at around pH 8 (Figure 9.2). They described a competitive adsorption mechanism between hydrogen ions (H^+) and dye molecules. Moreover, a high concentration of H^+ ions causes protonation of silanol groups that repels cationic dye molecules. As the pH of the system increases, the concentration of H^+ ions decreases and the silanol sites become deprotonated, in turn increasing the negative charge density on the adsorbent surface and facilitating the adsorption of positive dye molecules. Even though uptake of pollutants generally increases with increasing surface area, Kurniawan *et al.* [49] also showed that in this case specific surface area has no effect on the adsorption. The surface area of bentonite decreases with these added surfactants (51.8 m²/g decreases to 45.3 m²/g), but dye recovery increases,

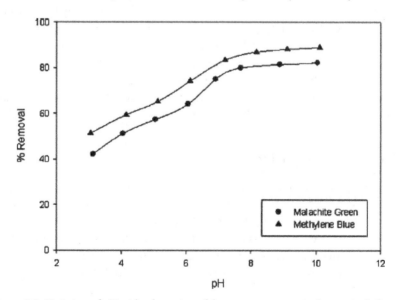

Figure 9.2 Variation of pH with adsorption of dyes onto rarasaponin-bentonite [49].

and this is much like other organo-bentonites used in anionic dye adsorption. Examples of cationic dye removal by surfactant activation are given in Table 9.5.

9.3.3   Effect of Ionic Strength on Uptake of Anionic and Cationic Dyes

Li *et al.* [51] described how actual dye wastewaters contain various types of salts and surfactants, among which NaCl and the phenol sodium dodecyl benzenesulfonate (SDBS) are most common. Salt gives high ionic strength and increases the efficiency of the adsorption process [109]. Recently, Akl *et al.* [37] found that increasing ionic strength with NaCl, KCl, and $CaCl_2$ compresses the diffuse double layer and improves the removal of acid dye from aqueous solution by both bentonite and modified bentonite (Figure 9.3). Theoretically, when the electrostatic forces are attractive, an increase in ionic strength will decrease the adsorption capacity [51,110]. To the contrary, the studies done by Akl *et al.* [37] and Li *et al.* [51] showed that the uptake of anionic dyes on positively charged modified bentonite increases with ionic strength.

Eren and Afsin [27] studied the effect of ionic strength using NaCl on bentonite adsorption of cationic Crystalline Violet. They explained that increasing ionic strength entraps dye molecules in newly generated aggregates [111]. These aggregates initially form by edge-to-edge type agglomeration, and then with a more concentrated solution, form by face-to-face type agglomeration, giving more compact, irregular, multilayered aggregates [112]. There is an overall increase in mean aggregate radius.

Li *et al.* [51] also investigated the effect of ionic strength using SDBS on epichlorhydrin-dimethylamine (Epi-DMA)-modified bentonite for four different dyes. Increasing ionic strength decreases dye removal because SDBS reacts with EPI-DMA/bentonite much stronger than the dyes [113].

9.3.4   Adsorption Kinetics

Physical or chemical characteristics of the adsorbent system affect the nature of the adsorption process. To explain and examine the controlling mechanism of adsorption process, time-dependent experimental adsorption data are used and several kinetics models are available. Literature generally shows that the equilibrium adsorption between the dye and bentonite is attained within a very short time. In the initial stage of adsorption,

Table 9.5 Examples of modified bentonite clays for the removal of cationic dyes.

Modification Process	Dye Name	Specific Surface Area	Adsorption Capacity	Reference
Co-saturated (Co–)	Crystal violet	Not provided	0.56 mmol/g	[27]
Zn-saturated (Zn–)	Crystal violet	Not provided	0.50 mmol/g	[27]
Ni-saturated (Ni–)	Crystal violet	Not provided	0.41 mmol/g	[27]
Acid Activation (for 0.2 M H_2SO_4)	Crystal violet	110 m^2/g Raw Bentonite: 36.17 m^2/g	0.51 mmol/g	[28]
Acid Activation (for 5 M HCl)	Methylene blue	Not provided	20.16 mg/g	[29]
Tetrabutyl ammonium chloride (TBAC)	Methylene blue	Not provided	34.84 mg/g	[107]
Trimethyl ammonium bromide (CTAB)	Methylene blue	Not provided	6.41 mg/g	[107]
Cetyltrimethylammonium (CTMA)	Crystal violet	Not provided	106.56 L/g	[108]
Benzyltriethylammonium (BTEA)	Crystal violet	Not provided	43.26 L/g	[108]

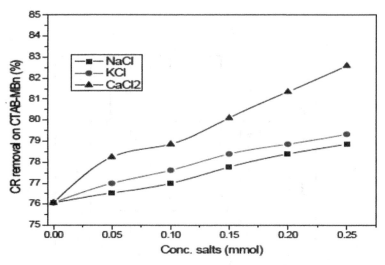

Figure 9.3 Effect of ionic strength on adsorption capacity of bentonite and CTAB-MBn for Congo red [37].

a large number of vacant surface sites are available for adsorption and then the remaining vacant surface sites are difficult to occupy due to repulsive forces between the dye molecules on the bentonite surface [27].

In the literature, the pseudo-first- and pseudo-second-order kinetic models are generally used to describe adsorption rates and capacities [17]. The pseudo-second-order model is most prominent [72,97]. Toor and Jin [25], Vimonses *et al.* [39] and Baskaralingam *et al.* [83] have mentioned that this model is more likely to fit adsorption behavior over the whole time range. This model is also in agreement with the rate-controlling step being chemical adsorption, where valence electrons are shared between dye and adsorbent.

The formula of a pseudo-second-order model [114] obtained by Toor and Jin [25] is given below (Figure 9.4).

Pseudo-second-order: $$q_t = \frac{k_2 q_e^2 t}{(1+k_2 q_e t)} \qquad (9.1)$$

where q_e and q_t are the amounts (mg/g) of solute bound at the interface at the equilibrium and after time t (min), respectively, and k_2 is the rate constant of the pseudo-second-order adsorption (g/mg min).

9.3.5   Adsorption Isotherms

The adsorption equilibrium isotherm is important for describing how the adsorbate molecules partition between the liquid and the solid

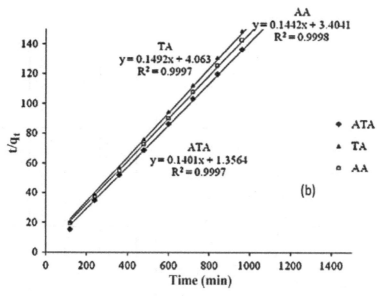

Figure 9.4 The second-order-pseudo kinetics for adsorption of Congo red by modified bentonite [25].

phases at equilibrium. Adsorption isotherms are mathematical models that describe distribution of the adsorbate species among liquid and solid phases. A literature survey indicates that Langmuir and Freundlich adsorption iosotherms have been used extensively to describe clay-dye adsorption.

Table 9.6 shows adsorption isotherm parameters for both anionic and cationic dye adsorbed onto natural and modified clays. The Langmuir iostherm generally fits the experimental data well. However, it should be noted that comparative evaluation is limited because the solid-to-liquid ratio might differ from one case to another.

Li *et al.* [51] investigated the adsorption properties and mechanisms of a cationic-polymer/bentonite complex (EPI-DMA/bentonite) for anionic dyes (Reactive Violet K-3R and Acid Dark Blue 2G). Their results on anionic dye/EPI-DMA/bentonite system showed that the Freundlich model is most suitable. Other investigations also showed that the Freundlich isotherm can be a good fit. Rarasponin-modified bentonite was used to remove basic dyes (MB and MG) and follows both Langmuir and Freundlich models [49]. Adsorption of MB is higher than MG dye because small molecules gain easier access to the internal pore network of the adsorbent. Similar results are observed for montmorillonite [23].

Table 9.6 Adsorption of cationic and anionic dyes by the expanding three-layer clays using the Langmuir and Freundlich models.

Adsorbents	Dyes name	Langmuir		Freundlich		Reference
		q_{max} mg/g	K_L L/g	K_f mol/g	n	
Sodium montmorillonite	Acid fuchsin	$2.76*10^{-4}$	$5.81*10^{-4}$	1.48	1.21	[23]
Thermal Activated	Congo red (CR)	54.64	0.061	4.84	1.68	[25]
Acid Activated		69.44	0.060	5.62	1.73	
Combined acid and thermal activation		75.75	0.068	6.99	1.79	
Natural bentonite	Basic Yellow 28 Basic Red 46	256.4 333.3	0.257 0.769			[30]
Natural Bentonite	Congo Red	35.84	0.23			[39]
Ca-bentonite	Congo Red	107.41	0.11	26.91	3.23	[63]
Acid-activated bentonite	Acid Red 57 and Acid Blue 294	641.9 117.8	2.46×10^{-3} 7.16×10^{-3}	3.18 12.2	1.27 3.16	[70]
KSF-montmorillonite	Remozal Brilliant Blue R	30.76	0.28			[72]
Surfactant-Al- bentonite	Supranol Yellow 4GL	111.11	0.2			[73]
CDBA-bentonite And CP-bentonite	Acid Red 151	357. 14 416.66	0.5714 12.04			[83]

9.3.6   Adsorption Thermodynamics

The temperature dependence of adsorption reveals whether adsorption is exothermic (decrease of adsorption capacity with increasing temperature) or endothermic (increase of adsorption capacity with increasing temperature). The rate of diffusion of the adsorbate molecules across the external boundary layer and into the internal pores of the adsorbent particles increases with increasing temperature as a result of the reduced viscosity of the solution [115]. The temperature dependence also gives valuable information about the enthalpy and entropy changes accompanying adsorption [116].

Tahir and Rauf [88] used bentonite to remove cationic dye (Malahite Green) from aqueous solution. The $\Delta H°$ is positive (endothermic), as has been found in most cases [18,83,117], and adsorption capacity increases with temperature [118]. The $\Delta S°$ is also positive, corresponding to an increased degree of freedom in the system as a result of adsorption of the dye molecules. Rytwo *et al.* [116] attributed this overall increase in entropy to the release of hydrated inorganic cations from the clay. In addition, Seki and Yurdakoç [118] explained the positive value of $\Delta S°$ by structural changes which take place as a result of interactions of dye molecules with active groups on the clay surface. Tahir and Rauf [88] also found negative $\Delta G°$ values in the −20 to 0 kJmol^{-1} range, corresponding to spontaneous physical processes. Table 9.7 summarizes studies on the temperature dependence of adsorption of cationic dyes, including endothermic effects.

Eren and Afsin [27] found negative $\Delta H°$ and $\Delta S°$ for the adsorption of cationic Crystalline Violet (CV) on bentonite pretreated with Ni, Co and Zn. They ascribed the negative value of $\Delta H°$ to an ion-exchange mechanism that also includes the release of water molecules hydrated around the exchangeable cations. They also showed large differences between natural and pretreated bentonite and conclude that pretreatment created an irregular increase of the randomness at the bentonite solution interface during adsorption.

Interestingly, cationic and anionic dye adsorption responds differently to temperature. The adsorption of anionic dye adsorption on natural and modified bentonites is illustrated in Table 9.8. Toor and Jin [25] studied the adsorption of Congo Red (CR) by thermal activated (TA), acid activated (AA), and combined acid and thermal activated (ATA) bentonites. They showed decreasing adsorption rates with increasing temperature and described weakening sorptive forces both between dye-clay sites and between adsorbed dye molecules [119]. Similar results have been reported by Vimoneses *et al.* [39]. They showed a decrease in adsorption with

Table 9.7 Effect of temperature on the adsorption of cationic dyes.

Adsorbents	Dyes name	Adsor dosage	Temp Range °K	$\Delta H°$ kJ/mol	$\Delta S°$ kJ/mol	$\Delta G°$ kJ/mol	Reference
Co-bentonite	Crystal Violet	1 g/L	295 309	41.94	0.14	−0.99 −3.02	[27]
Acid Activated Bentonite	Crystal Violet	2 g/L	290 338	45.5	0.193	−10.5 −19.8	[28]
Natural bentonite	Basic Yellow 28 Basic Red 46	1 g/L	293 333 293 333	8.30 8.80	0.042 0.057	−4.16 −5.79 −7.70 −9.92	[30]
BTEA-Bentonite	Crystal Violet	1 g/L	303 333	54.28	0.19	−4.11 −9.92	[108]

Table 9.8 Effect of temperature on the adsorption of anionic dyes.

Adsorbents	Dyes name	Adsor dosage	Temp Range °K	ΔH kJ/mol	ΔS kJ/mol	ΔG kJ/mol	Reference
Na-Montmorillonite	acid fuchsin	1 g/L	288 313	−186.02	−590.78	−10.12	[23]
Bentonite	Congo Red	2 g/L	298 328	12.37	0.11	−21.104 −23.36	[37]
CTAB-bentonite	Congo Red	2 g/L	298 328	12.71	0.13	−27.19 −29.87	[37]
Natural bentonite	Congo Red	5 g/L	298 333	−13.02	7.41	−15.25 −15.53	[39]
Ca-bentonite	Congo Red	2 g/L	293 323	5.14	0.004	−6.49 −11.17	[63]
KSF-DP Montmorillonite	Remazol Brilliant Blue R	2.5 g/L	298 323	77.02	277	−6.87 −12.5	[72]
Sodium-pillared bentonite	SupranoYellow 4GL	1 g/L	303 323	−32.32	−0.008	−14.77 −12.30	[73]
Al-pillared bentonite	SupranoYellow 4GL	1 g/L	303 323	15.92	0,009	−13.00 −14.42	[73]
DTMA-bentonite	Reactive Blue 19	2 g/L	293 333	−4.55	0.69	−25.04 −26.43	[82]
Alginate/organophilic montmorillonite	Acid Green	4 g/L	303 323	−42.6	46	−2.4	[97]
CTAB-bentonite	SupranoYellow 4GL	2 g/L	278 323	33.68	0.19	−19.14 −27.69	[99]

increasing temperature and described relatively weak hydrogen bonds and van der Waals forces. Toor and Jin [25] showed negative $\Delta S°$ for CR adsorption onto bentonite, indicative of a decrease in randomness at the solid/solution interface. However, CR adsorption did not result in any noticeable changes in the internal structure of the adsorbent. Results given in Table 9.8 also support the findings of Toor and Jin [25]. In addition, Table 9.8 shows that the adsorption of anionic dye onto montmorillonite is an exothermic process, and a negative value of the Gibbs free energy, ΔG, indicates spontaneity and thermodynamic stability. The negative change of the entropy ΔS further suggests a decrease in the degree of freedom of the adsorbed species [23,97].

On the other hand, Table 9.8 also shows how the adsorption of the same dye with different bentonites may result in endothermic or exothermic heats. For instance, Congo red uptake on Ca-bentonite increases with an increase in temperature (positive enthalpy and endothermic) and also has physical bonding [63]. Similar results have been reported for the adsorption of Congo red onto both bentonite and surfactant-modified bentonite [37], and for the adsorption of Supranol Yellow 4GL onto Al-pillared bentonite [73]. It is common that increasing temperature may create a swelling influence inside the adsorbent structure, leading to additional penetration of big dye molecules [120], so the nature of the adsorption of dyes is endothermic. Endothermic adsorption has also been reported for modified montmorillonite [72].

9.4   General Remarks

The ready availability of clay makes it a viable candidate as an alternative adsorbent. Various methods and adsorbents can potentially be used for the removal of dyes from aqueous solutions, and each treatment process has its own specific disadvantage. For instance, activated carbon and ion exchange processes are expensive and sophisticated. Clay is low cost, has great specific surface area, and has excellent physical and chemical stability. The expanding three-layer clays have recently been used as an efficient dye adsorbent. This chapter has attempted to highlight some important studies of anionic and cationic dye adsorption on both natural and modified clays. It has been shown that natural expanding three-layer clays are effective adsorbents for cationic and anionic dyes, and when they are modified by different methods and chemicals, the uptake of anionic and cationic dyes is markedly increased. The most important adsorption parameters are pH, ionic strength and modification process. Specific surface area has less of an effect. A high pH value is preferred for cationic dye adsorption, and a low pH value is preferred for anionic dye adsorption. Increased ionic strength generally enhances adsorption due to the compression of the diffuse double layer on the adsorbent. Langmuir and Freundlich isotherms have been used extensively by many researchers, with the Langmuir isotherm being the most predominant. Langmuir models also give good R_L values, and they confirm a favorable adsorption process for all systems. Kinetic data usually follows the pseudo-second-order model, and thermodynamic data usually indicates an endothermic process. Anionic dye adsorption is sometimes exothermic. From the economic point of view, clay can be considered an efficient adsorbent when compared to activated carbon.

References

1. A. Baban, A. Yediler, and N. K. Ciliz, *Clean*, Vol. 38, No. 1, p. 84–90, 2010.
2. P. A. Soloman, C. A. Basha, V. Ramamurthi, K. Koteeswaran, and N. Balasubramanian, *Clean*, Vol. 37, No. 11, p. 889–900, 2009.
3. C.I. Pearce, J. R. Lloyd, and J. T. Guthrie, *Dyes and Pigments*, Vol. 58, p. 179–196, 2003.
4. G. McMullan, C. Meehan, A. Conneely, N. Kirby, T. Robinson, P. Nigam, I. M. Banat, R. Marchant, and W. F. Smyth, *Appield Microbiological Biotechnology*, Vol. 56, p. 81–87, 2001.
5. T. Robinson, G. McMullan, R. Marchant, and P. Nigam, *Bioresource Technology*, Vol. 77, p. 247–255, 2001.
6. D. Suteu, C. Zaharia, D. Bilba, A. Muresan, R. Muresan, and A. Popescu, *Industria Textila*, Vol. 60, No. 5, p. 254–263, 2009.
7. C. Zaharia, D. Suteu, A. Muresan, R. Muresan, and A. Popescu, *Environmental Engineering and Management Journal*, Vol. 8, No. 6, p. 1359–1369, 2009.
8. S. Chatterjee, D. S. Lee, M. W. Lee, and S. H. Woo, *Bioresource Technology*, Vol. 100, p. 3862–3868, 2009.
9. A. Bhatnagar, and A. K. Jain, *Journal of Colloid Interface Science*, Vol. 281, p. 49–55, 2005.
10. Q. Sun, and L. Yang, *Water Resource*, Vol. 37, p. 1535–1544, 2003.
11. M. N. V. Ravi Kumar, T. R. Sridhari, K. D. Bhavani, and P. K. Dutta, *Colorage*, Vol. 40, p. 25–34, 1998.
12. G. Crini, *Bioresource Technology*, Vol. 97, p. 1061–1085 2006.
13. F. Derbyshire, M. Jagtoyen, R. Andrews, A. Rao, I. Martin-Gullon, and E. Grulke, Carbon materials in environmental applications, in: L. R. Radovic, ed., *Chemistry and Physics of Carbon*, Vol. 27. Marcel Dekker, New York, pp. 1–66, 2001.
14. S. Babel, and T. A. Kurniawan, *Journal of Hazardous Materials*, B97, p. 219–243, 2003.

15. D. Suteu, C. Zaharia, and T. Malutan, Biosorbents based on lignin used in biosorption processes from wastewater treatment (chap. 7). In: *Lignin: Properties and Applications in Biotechnology and Bioenergy*, R. J. Paterson, ed., Nova Science Publishers, 27 pp., ISBN 978-1-61122-907-3, New York, U. S. A., 2011.
16. G. Mishra, and M. Tripathy, *Colourage*, Vol. 40, p. 35–38, 1993.
17. M. A. M. Salleh, D. K. Mahmoud, W. A. W. A. Kari, and A. Idris, *Desalination*, Vol. 280, p. 1–13, 2011.
18. O. Demirbas, M. Alkan, and M. Dogan, *Adsorption*, Vol. 8, p. 341–349, 2002.
19. A. C. Martínez-Hutile, and E. Brillas, *Applied Cat. B: Environment*, Vol. 87, p. 105–145, 2009.
20. E. Forgacs, T. Cserhati, and G. Oras, *Environ International*, Vol. 30, p. 953–971, 2004.
21. K. Hunger, *Industrial Dyes, Chemistry, Properties, Applications*, Wiley-VCH, Weinheim, Germany, pp. 1–10, 2003.
22. O. D. Tyagi, M. S. Yadav, and M. Yadav, *A Textbook of Synthetic Dyes*, 67, Anmol-PVT. LTD., 2002.
23. A. S. Elsherbiny, *Applied Clay Science*, Vol. 83–84, p. 56–62, 2013.
24. M. F. A. Taleb, D. E. Hegazy, and S. A. Ismail, *Carbohydrate Polymers*, Vol. 87, p. 2263–2269, 2012.
25. M. Toor, and B. Jin, *Chemical Engineering Journal*, Vol. 187, p. 79–88, 2012.
26. D. Shen, J. Fan, W. Zhou, B. Gao, Q. Yue, and Q. Kang, *Journal of Hazardous Materials*, Vol. 172, p. 99–107, 2009.
27. E. Eren, and B. Afsin, *Dyes Pigments*, Vol. 76, p. 220–225, 2008.
28. E. Eren, and B. Afsin, *Journal of Hazardous Materials*, Vol. 166, p. 830–835, 2009.
29. L. Al-Khatib, F. Fraige, M. Al-Hwaiti, and O. Al-Khashman, *American Journal of Environmental Science*, Vol. 8(5), p. 510–522, 2012.
30. M. Turabik, *Journal of Hazardous Materials*, Vol. 158, p. 52–64, 2008.
31. M. Rafatullah, O. Sulaiman, R. Hashim, and A. Ahmad, *Journal of Hazardous Materials*, Vol. 177, p. 70–80, 2010.
32. E. G. Ralph, and G. Necip, *Bentonites: Geology, mineralogy, properties and uses* ISBN 0444416137, p. 256, 1978.
33. G. E. Christidis, and D. D. Eberl, *Clays and Clay Minerals*, Vol. 51, p. 644–655, 2003.
34. World Health Organization Geneva, Environmental Health Criteria 231: Bentonite, Kaolin and Selected Clay Minerals, 2005.
35. C. S. Ross, and E. V. Shannon, *American Ceramic Society Bulletien*, Vol. 9, p. 77–96, 1926.
36. H. H. Murray, *Clay Minerals*, Vol. 34, p. 39, 1999.
37. M. A. Akl, A. M. Youssef, and M. M. Al-Awadhi, *Journal of Anal Bioanal Technology*, 4:174. doi:10.4172/2155-9872.1000174, 2013.
38. I. Gulgonul, *Physicochemical Problems in Minerial Processing*, Vol. 48(2), p. 369–380, 2012.

39. V. Vimonses, B. Jin, C. W. K. Chow, and C. Saint, *Journal of Hazardous Material*. doi: 10.1016/j.jhazmat.2009.06.094, 2009.
40. V. Vimonses, S. Lei, B. Jin, C. W. K. Chow, and C. Saint, *Chemical Engineering Journal*, Vol. 148, p. 354–364, 2009.
41. E. Bulut, M. Özacar, and I. A. Şengil, *Journal of Hazardous Materials*, Vol. 154, p. 613- 622, 2008.
42. M. Ozacar, and I. A. Sengil, *Journal of Environmental Management*, Vol. 80, p. 372–379, 2006.
43. C. Bilgic, *Journal of Colloid Interface Science*, Vol. 281, p. 33–38, 2005.
44. C. C. Wang, L. C. Juang, T. C. Hsu, C. K. Lee, J. F. Lee, and F. C. Huang, *Journal of Colloid Interface Science*, Vol. 273, p. 80–86, 2004.
45. Y. Liu, Y. Zheng, and A. Wang, *Journal of Environmental Sciences*, Vol. 22 (4), p. 486–493, 2010.
46. L. Chmielarz, P. Kuśtrowski, Z. Piwowarska, B. Dudek, B. Gil, and M. Michalik, *Applied Catalysis B: Environmental*, Vol. 88 (3–4), p. 331–40, 2009.
47. J. Tang, Z. F. Yang, and Y. J. Yi, *Procedia Environmental Sciences*, Vol. 8, p. 2205–2213, 2011.
48. I. Chaari, M. Feki, M. Medhioub, J. Bouzid, E. Fakhfakha, and F. Jamoussi, *Journal of Hazardous Materials*, Vol. 172, p. 1623–1628, 2009.
49. A. Kurniawan, H. Sutiono, N. Indraswati, and S. Ismadji, *Chemical Engineering Journal*, Vol. 189–190, p. 264–274, 2012.
50. J. Ma, B. Cui, J. Dai, and D Li, *Journal of Hazardous Materials*, Vol. 186, p. 1758–1765, 2011.
51. Q. Li, Q. Y. Yue, H. J. Sun, Y. Su, and B. Y. Gao, *Journal of Environmental Management*, Vol. 91, p. 1601–1611, 2010.
52. M. Zhao, Z. Tang, and P. Liu, *Journal of Hazardous Materials*, Vol. 158, p. 43–51, 2008
53. J. H. An, and S. Dultz, *Applied Clay Science*, Vol. 36, p. 256–264, 2007.
54. Q. U. Jiuhui, *Journal of Environmental Science*, Vol. 20, p. 1–13, 2008.
55. G. E. Chistidis, P. W. Scott, and A. C. Dunham, *Applied Clay Science*, Vol. 12, p. 329–347, 1997.
56. A. Steudel, L. F. Batenburg, H. R. Fischer, P. G. Weidler, and K. Emmerich, *Applied Clay Science* Vol. 44, p. 105–115, 2009.
57. H. He, R. L. Frost, T. Bostrom, P. Yuan, L. Duong, D. Yang, Y. Xi, and T. Kloprogge, *Applied Clay Science*, Vol.31, p. 262–271, 2006.
58. S. Al-Asheh, F. Banat, L. Abu-Aitah, *Seperation and Purufication Technology*, Vol. 33, p. 1–10, 2003.
59. L. B. de Paiva, A. R. Morales, and F. R. V. Díaz, *Applied Clay Science*, Vol. 42, p. 8–24, 2008.
60. P. Liu, *Applied Clay Science*, Vol. 38, p. 64–76, 2007.
61. D. Doulia, Ch. Leodopoloud, K. Gimouhopoulos, and F. Rigas, *Journal of Colloid and Interface Science*, Vol. 340, p. 131–141, 2009.
62. I. Chaari, E. Fakhfakh, S. Chakroun, J. Bouzid, N. Boujelben, M. Feki, F. Rocha, and F. Jamoussi, *Journal of Hazardous Materials*, Vol. 156, p. 545–551, 2008.

63. L. Lian, L. Guo, and C. Guo, *Journal of Hazardous Materials*, 161, 126–131, 2009.
64. F. R. V. Díaz, and P. de Souza Santos, *Quim. Nova*, Vol. 24, p. 343–353, 2001.
65. F. Beragaya, B. K. G. Theng, and G. Lagaly, Modified clays and clay minerals, in: F. Beragaya, B. K. G. Theng, and G. Lagaly, eds., *Handbook of Clay Science: Development in Clay Science*, Vol. 1, Elsiever, The Netherlands, 2006.
66. S. Al-Asheh, F. Banat, L. Abu-Aitah, *Separation, Purification Technology*, Vol. 33, p. 1–10, 2003.
67. R. Sennour, G. Mimane, A. Benghalem, and S. Taleb, *Applied Clay Science*, Vol. 43, p. 503–506, 2009.
68. B. Benguella, and A. Yacouta-Nour, *Desalination*, Vol. 235, p. 276–292, 2009.
69. M. Hajjaji, and H. El Arfaoui, *Applied Clay Science*, Vol. 46, p. 418–421, 2009.
70. A. S. Ozcan, and A. Ozcan, *Journal of Colloid and Interface Science*, Vol. 276, p. 39–46, 2004.
71. A. P. Carvalho, A. Martin, J. M. Silva, J. Pires, H. Vasques, and M. Brotas de Carvalho, *Clays and Clay Minerals*, Vol. 51, p. 340–349, 2003.
72. M. M. F. Silva, M. M. Oliveira, M. C. Avelino, M. G. Fonseca, R. K. S. Almeida, and E. C. S. Filho, *Chemical Engineering Journal*, Vol. 203, p. 259–268, 2012.
73. Z. Bouberka, S. Kacha, M. Kameche, S. Elmaleh, and Z. Derriche, *Journal of Hazardous Materials*, B119, p. 117–124, 2005.
74. R. Devi, V. Singh, and A. Kumar, *Journal of Bioresource Technology*, Vol. 99, p. 1853–1860, 2008.
75. P. Janos, and V. Smidova, *Colloid Interface Science*, Vol. 291, p. 19–27, 2005.
76. L. Zhu, and J. Ma, *Chemical Engineering Journal*, Vol. 139, p. 503–509, 2008.
77. C. Faur-Brasquet, Z. Reddad, K. Kadirvelu, and P. Le Cloirec, *Applied Surface Science*, Vol. 196, p. 356–365, 2002.
78. L. Wang, and A. Wang, *Journal of Hazardous Materials*, Vol. 160, p. 173–180, 2008.
79. Q. Li, Q. Y. Yue, Y. Su, B. Y. Gao, and L. Fu, *Journal of Hazardous Materials*, Vol. 147, p. 370–380, 2007.
80. B. Zohra, K. Aicha, S. Fatima, B. Nourredine, and D. Zoubir, *Chemical Engineering Journal*, Vol. 136, p. 295–305, 2008.
81. A. S. Ozcan, B. Erdem, and A. Ozcan, *Colloid and Surfaces A: Physicochemical Engineering Aspects*, Vol. 266, p. 73–81, 2005.
82. A. Ozcan, C. Omeroğlu, Y. Erdoğan, and A. S. Ozcan, *Journal of Hazardous Materials*, Vol. 140, p. 173–179, 2007.
83. P. Baskaralingam, M. Pulikesi, D. Elango, V. Ramamurthi, and S. Sivanesan, *Journal of Hazardous Materials*, B128, p. 138–144, 2006.
84. C. Bertagnolli, A. L. P. Araujo, S. J. Kleinübing, and M. G. C. Silva, *Chemical Engineering Transaction*, Vol. 24, p. 1537–1542, 2011.
85. C. Bertagnolli, S. J. Kleinnübing, and M. G. C. Silva, *Applied Clay Science*, Vol. 53, p. 73–79, 2011.
86. T. Depci, A. R. Kul, and Y. Onal, *Chemical Engineering Journal*, Vol. 200–202, p. 224–236, 2012.

87. A. Mittal, J. Mittal, A. Malviya, D. Kaur, and V. K. Gupta, *Journal of Colloid and Interface Science*, Vol. 343, p. 463–473, 2010.
88. S. S. Tahir, and N. Rauf, *Chemosphere*, Vol. 63, p. 1842–1848, 2006.
89. M. Ozacar, and I. A. Sengil, *Bioresource Technology*, Vol. 96, p. 791–795, 2005.
90. C. Namasivayam, and D. J. S. E. Arasi, *Chemosphere*, Vol. 34, p. 401–471, 1997.
91. O. Bouras, T. Chami, M. Houari, H. Khalaf, J. C. Bollinger, and M. Baudi, *Environmental Technology*, Vol. 23(4), p. 405–411, 2002.
92. L. Lian, L. Guo, and C. Guo, *Journal of Hazardous Materials*, Vol. 161, p. 126–131, 2009.
93. R. Chen, F. Peng, and S. Su, *Journal of Applied Polymer Science*, Vol. 108, p. 2712–2717, 2008.
94. Y. H. Shen, *Chemosphere*, Vol. 44, p. 989–995, 2001.
95. O. Gok, A. S. Ozcan, and A. Özcan, *Applied Surface Science*, Vol. 256, p. 5439–5443, 2010.
96. A. F. Almeida Neto, M. G. A. Vieira, and M. G. C. Silva, *Materials Research*, Vol. 15, p. 114–124, 2012.
97. M. F. A. Taleb, D. E. Hegazy, and S. A. Ismail, *Carbohydrate Polymers*, Vol. 87, p. 2263–2269, 2012.
98. M. M. Socias-Viciana, M. C. Hermosin and J. Cornejo, *Chemosphere*, Vol. 37(2), p. 289–300, 1998.
99. A. Khenifi, Z. Bouberka, F. Kameche, and Z. Derriche, *Adsorption*, Vol. 13, p. 149–158, 2007.
100. B. Chen, L. Zhu, J. Zhu, and B. Xing, *Environmental Science Technology*, Vol. 39, p. 6093–6100, 2005.
101. S. Xu, and S. A. Boyd, *Environmental Science Technology*, Vol. 29, p. 3022, 1995.
102. J. Ma, B. Cui, J. Dai, and D. Li, *Journal of Hazardous Materials*, Vol. 186, p. 1758–1765, 2011.
103. G. Sheng, S. Xu, and S. A. Boyd, *Water Resource*, Vol. 30, p. 1483–1489, 1996.
104. P. Monvisade, and P. Siriphannon, *Applied Clay Science*, Vol. 42, p. 427–431, 2009.
105. K. G. Bhuttacharyya, and S. S. Gupta, *Seperation Purification Technology*, Vol. 50, p. 388–397, 2006.
106. O. Zuzana, M. Annamária, D. Silvia, and B. Jaroslav, *Arhiv za tehnicke nauke*, Vol. 7(1), p. 49–56, 2012.
107. S. H. Sonawane, P. L. Chaudhari, S. A. Ghodke, M. G. Parande, V. M. Bhandari, S. Mishra, R. D. Kulkarni, *Ultrason. Sonochem.* Vol. 16(3), p. 351–355, 2009.
108. R. Jian-min, W. Si-wei, and J. Wei, *World Academy of Science, Engineering and Technology*, Vol. 41, 2010.
109. T. S. Anirudhan, and M. Ramachandran, *Applied Clay Science*, Vol. 35, p. 276–281, 2007.
110. G. Alberghina, R. Bianchini, M. Fichera, and S. Fisichella, *Dyes Pigment*, Vol. 46, p. 129–137, 2000.

111. A. P. P. Cione, C. C. Schmitt, M. G. Neumann, and F. Gessner, *Journal of Colloid Interface Science*, Vol. 226, p. 205–209, 2000.
112. P. F. Luckham, and S. Rossi, *Advanced Colloid Interface Science*, Vol. 82, p. 43–92, 1999.
113. D. Voisin, and B. Vincent, *Advances in Colloid and Interface Science*, Vol. 106(1–3), p. 1–22, 2003.
114. Y. S. Ho, G. Mckay, *Process Biochem.* 34, p. 451–465, 1999.
115. M. Dogan, M. Alkan, A. Turkyilmaz, and Y. Özdemir, *Journal of Hazardous Materials*, B109, p. 141–148, 2004.
116. M. Alkan, O. Demirbas, S. Celikcapa, and M. Dogan, *Journal of Hazardous Materials*, Vol. 116, p. 135–145, 2004.
117. G. Rytwo, R. Huterer-Harari, S. Dultz, and Y. Gonen, *Journal of Thermal Anal. Calorimetry*, Vol. 84, p. 225, 2006.
118. Y. Seki, and K. Yurdakoc, *Adsorption*, Vol. 12, p. 89, 2006.
119. A. E. Ofomaja, and Y. S. Ho, *Dyes Pigments*, Vol. 74, p. 60–66, 2007.
120. K. G. Bhattacharyya, and A. Sarma, *Dyes Pigments*, Vol. 57, p. 211–222, 2003.
121. M. S. Çelik, Electrokinetic behavior of clay surfaces, in: *Clay Surfaces: Fundamentals and Applications*, F. Wypych, Interface Science and Technology Series, (series vol. ed., Ed. K. G. Satyanarayana), Academic Press, Ch. 2, p. 57–89, 2004.

10

Non-conventional Adsorbents for Dye Removal

Grégorio Crini

Laboratoire Chrono-environnement UMR6249 usc INRA, University of Franche-Comté, Besançon, France

Abstract

Amongst the numerous techniques of pollutant removal or recovery in commercial and industrial use, adsorption on activated carbons and ion-exchange on synthetic organic resins are the procedures of choice and give the best results, as they can be used to remove different types of dyes. However, although these commercial materials are the preferred conventional adsorbents for dye removal, their widespread industrial use is restricted due to high cost. As such, alternative non-conventional adsorbents are proposed as inexpensive and efficient materials. In this chapter, an extensive list of non-conventional adsorbent literature has been compiled and results in terms of adsorption capacities discussed. Each material is also illustrated by precise examples from among the works published in the literature.

Keywords: Adsorption, batch method, chemisorption, dyes, non-conventional adsorbent, pollutant removal, wastewater treatment, water pollution

10.1 Introduction

During the past three decades, several physical, chemical and biological decontamination/decolorization methods have been reported; few, however, have been accepted by the industrial world [1–7]. Amongst the numerous techniques of pollutant removal or recovery in commercial and industrial use, adsorption on activated carbons and ion-exchange on synthetic organic resins are the procedures of choice and give the best results,

Corresponding author: gregorio.crini@univ-fcomte.fr

Sanjay K. Sharma (ed.) Green Chemistry for Dyes Removal from Wastewater, (359–407)
© 2015 Scrivener Publishing LLC

as they can be used to remove different types of pollutants [6]. In particular, adsorption/separation techniques are widely used to remove certain classes of chemical (recalcitrant) pollutants from waters, especially those that are hardly destroyed in conventional biological wastewater treatments. Because of their great capacity to adsorb dyes, commercial activated carbon (CAC) and synthetic polymeric organic resins are the most effective adsorbents. This capacity is mainly due to their structural characteristics and their porous texture which gives them a large surface area, and their chemical nature which can be easily modified by chemical treatment in order to increase their properties. However, although these commercial materials are preferred conventional adsorbents for dye removal, their widespread industrial use is restricted due to high cost. As such, alternative non-conventional adsorbents were proposed, studied and employed as inexpensive and efficient adsorbents [8].

For many adsorption processes, the separation is caused by a mass separating agent. The mass separating agent is adsorbent. Consequently, the performance of any adsorptive separation or purification process is directly determined by the quality of the adsorbent. So, the first important step to an efficient adsorption process is the search for a solid porous material with high capacity. In principle, as adsorption is a surface phenomenon, any porous solid having a large surface area may be an adsorbent [9]. However, a suitable adsorption process of pollutants should also meet several other requirements. Indeed the selection of an adsorbent is based on the following criteria: low cost and readily available, granular type with a good particle size distribution, a well-developed (infra)structure, porous material with a large total surface area, surface charge (acidity/basicity properties), presence of adsorption sites or functional groups, high physical strength (not disintegrating) in solution, suitable mechanical properties, efficient for removal of a large range of dyes, high capacity, high rate of adsorption, high selectivity, a long life, and able to be regenerated if required [9]. By plotting solid phase concentration against liquid phase concentration graphically, it is possible to depict an equilibrium adsorption isotherm. An adsorption isotherm represents the relationship existing between the amount of pollutant adsorbed and the pollutant concentration remaining in solution. Equilibrium is established when the amount of pollutant being adsorbed onto the material is equal to the amount being desorbed. Among the numerous theories relating to adsorption equilibrium, the Langmuir adsorption isotherm is the best known of all isotherms describing adsorption [10–13]. Using the Langmuir equation, it is possible to obtain an interesting parameter widely used in the literature to promote a solid material as adsorbent, i.e., the theoretical monolayer capacity or the maximum adsorption capacity of an adsorbent (q_{max}). The equation of Langmuir is represented by the Equations 10.1 and 10.2, where x is the amount of dye adsorbed (mg); m is the amount of adsorbent used (g); C_e (mg L^{-1}) and q_e (mg g^{-1}) are the liquid phase concentration and solid phase concentration of adsorbate at equilibrium, respectively; K_L (L g^{-1}) and a_L (L mg^{-1}) are the Langmuir isotherm constants. The Langmuir isotherm constants, K_L and a_L are evaluated through linearization of Equation 10.1. Hence by plotting C_e/q_e against C_e it is possible to obtain the value of K_L from the intercept, which is $1/K_L$, and the value of a_L from the slope, which is a_L/K_L. The theoretical monolayer capacity is q_{max} and is numerically equal to K_L/a_L.

$$q_e = \frac{x}{m} = \frac{K_L C_e}{1 + a_L C_e} \tag{10.1}$$

$$\frac{C_e}{q_e} = \frac{1}{K_L} + \frac{a_L}{K_L} C_e \tag{10.2}$$

The main non-conventional adsorbents studied in the literature and employed as non-conventional adsorbent for dye removal include activated carbons (AC) from byproducts or solid wastes, clays and clay-based materials, zeolites, agricultural wastes (including byproducts from forest industries), industrial byproducts, biosorbents such as peat, chitin/chitosan and biomass, polysaccharides such as starch-based materials and alginates, and other miscellaneous materials such as cotton or substances capable of forming host-guest complexes such as cyclodextrins and calixarenes (Figure 10.1). In this chapter, an extensive list of non-conventional adsorbent literature has been compiled. Results in terms of adsorption

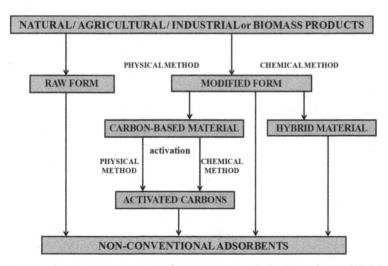

Figure 10.1 Schematic representation of non-conventional adsorbents from solids [8].

capacities using values of the monolayer capacity obtained from batch studies are compiled and discussed.

10.2   Activated Carbons from Solid Wastes

Adsorption and ion-exchange processes from aqueous solutions are important processes in water purification and wastewater decontamination. Amongst all the adsorbent materials proposed, activated carbon (AC) is the more popular adsorbent for the removal of pollutants from wastewater [2,14–16]. However, the use of carbons based on relatively expensive starting materials is unjustified for most pollution control applications [17]. This has led many workers to search for more economic carbon-based adsorbents from non-conventional resources. Indeed, certain waste products from industrial and agricultural operations, including wood byproducts, represent potentially economical alternative materials to prepare AC. These waste materials have little or no economic value and often present a disposal problem. Therefore, there is a need to valorize these low-cost byproducts. So, their conversion into AC would add economic value, help reduce the cost of waste disposal, and most importantly provide a potentially inexpensive alternative to the existing commercial activated carbons.

A wide variety of carbons have been prepared from agricultural and wood wastes such as bagasse, coir pith, banana pith, date pits, sago waste, silk cotton hull, corn cob, maize cob, straw, rice husk, rice hulls, fruit stones, nutshells, pinewood, sawdust, coconut tree sawdust, bamboo, and cassava peel. Many of them have been tested and proposed for dye removal (Table 10.1). The excellent ability and economic promise of the activated carbons prepared from agricultural byproducts have been presented and described in a comprehensive and interesting review by Oliveira and Franca [18]. There are also several reports on the production of AC from various city wastes and industrial byproducts such as waste PET bottles, waste tires, refuse-derived fuel, wastes generated during lactic acid fermentation from garbage, sewage sludges, waste newspaper, waste carbon slurries and blast furnace slag. These non-conventional ACs exhibited high adsorption properties as shown in Table 10.1. However, it should be point out that the adsorption capacities of a non-conventional carbon depend on the different sources of raw materials, the history of its preparation and treatment conditions such as pyrolysis temperature and activation time. Many other factors can also affect the adsorption capacity in the same adsorption conditions such as surface chemistry (heteroatom content), surface charge and pore structure. The adsorption mechanisms

Table 10.1 Reported adsorption capacities q_{max} (mg g^{-1}) for carbon materials made from agricultural solid wastes and industrial byproducts (selected papers).

Raw material	Dye	q_{max}	Reference
Bagasse	Basic Red 22	942	[19]
Bagasse	Acid Blue 25	674	[19]
Bagasse	Acid Blue 80	391	[20]
Bagasse	Rhodamine B	65.5	[21]
Bagasse	Basic Blue 9	30.7	[21]
Bituminous coal	Acid Red 88	26.1	[22]
Cane pith	Basic Red 22	941.7	[23]
Cane pith	Acid Blue 25	673.6	[23]
Cedar sawdust	Basic Blue 9	142.36	[24]
Cereal chaff	Basic Blue 9	20.3	[25]
Charcoal	Acid Red 114	101	[10]
Charfines	Acid Red 88	33.3	[22]
Charfines	Direct Brown 1	6.4	[22]
Cherry sawdust	Basic Blue 9	39.84	[26]
Coal	Basic Blue 9	250	[27]
Coal	Basic Red 2	120	[27]
Coir pith	Acid Violet	8.06	[28]
Coir pith	Direct Red 28	6.72	[29]
Coir pith	Acid Blue 15	2.6	[30]
Coir pith	Basic Violet 10	2.56	[28]
Corncob	Acid Blue 25	1060	[19]
Corncob	Basic Red 22	790	[19]
Date pits	Basic Blue 9	17.3	[31]
Flamboyant pods	Acid Red 18	551.799	[32]
Flamboyant pods	Acid Yellow 23	643.041	[32]
Flamboyant pods	Acid Yellow 6	673.687	[32]
Grass waste	Basic Blue 9	457.64	[33]
Hazelnut shell	Basic Blue 9	8.82	[34]
Lignite coal	Basic Blue 9	32	[35]
Lignite coal	Acid Red 88	30.8	[22]
Mahogany sawdust	Acid Yellow 36	183.8	[36]

(Continued)

Table 10.1 (*Cont.*)

Raw material	Dye	q_{max}	Reference
Papaya seeds	Basic Blue 9	555.55	[37]
Pinewood	Acid Blue 264	1176	[38]
Pinewood	Basic Blue 69	1119	[38]
Pinewood	Basic Blue 9	556	[38]
Pine-fruit shell	Reactive Red 120	275	[39]
Rice husk	Basic Green 4	511	[40]
Rice husk	Acid Yellow 36	86.9	[36]
Rice husk	Acid Blue	50	[41]
Rice husk	Basic Blue 9	40.59	[42]
Rice husk	Basic Blue 9	19.83	[43]
Rice husk	Direct Red 31	25.63	[44]
Rice husk	Direct Orange 26	19.96	[44]
Scrap tires	Rhodamine B	307.2	[45]
Sewage sludge	Basic Red 46	188	[46]
Sewage sludge	Basic Blue 9	114.94	[47]
Straw	Basic Blue 9	19.82	[43]
Sugarcane bagasse	Acid Orange 10	5.78	[48]
Waste carbon slurries	Acid Blue 113	219	[49]
Waste carbon slurries	Acid Yellow 36	211	[49]
Waste carbon slurries	Ethyl Orange	198	[49]
Waste newspaper	Basic Blue 9	390	[50]

by which the adsorption of dyes takes place on these non-conventional adsorbents are still not clear. This is because adsorption is a complicated process depending on several interactions such as electrostatic and non-electrostatic (hydrophobic) interactions.

10.3   Clays

Clays are hydrous aluminosilicates broadly defined as those minerals that make up the colloid fraction (<2 μm) of soils, sediments, rocks and water. Clays may be composed of mixtures of fine-grained clay minerals and clay-sized crystals of other minerals such as quartz, carbonate and metal oxides.

Clays invariably contain exchangeable ions on their surface and play an important role in the environment by acting as a natural scavenger of molecules by taking up cations and/or anions either through ion-exchange or adsorption, or both. The prominent ions found on the clay surface are Ca^{2+}, Mg^{2+}, H^+, K^+, NH_4^+, Na^+, and SO_4^{2-}, Cl^-, PO_4^{3-}, NO_3^-. These ions can be exchanged with other ions easily without affecting the structure of clay mineral. Clays are classified by the differences in their layered structures and there are several classes such as smectites (montmorillonite, saponite), mica (illite), kaolinite, serpentine, pylophyllite (talc), vermiculite and sepiolite [51]. Because of their low cost, abundance in most continents of the world, high adsorption properties and potential for ion-exchange, clay materials are strong candidates as adsorbents [52]. Indeed natural clay materials possess a layered structure and are considered as host materials. From the literature data, the adsorption capabilities mainly result from a net negative charge on the structure of minerals. This negative charge gives clay the capability to adsorb positively charged species. Their adsorption properties also come from their high surface area and high porosity. Among the clays, montmorillonite has the largest surface area and the highest cation exchange capacity. Its current market price is considered to be 20 times cheaper than that of activated carbon [14]. In recent years, there has been an increasing interest in utilizing clay minerals like montmorillonite, bentonite, kaolinite, diatomite and Fuller's earth for their capacity to adsorb not only inorganic but also organic molecules. Clay minerals exhibit a strong affinity for both heteroatomic cationic and anionic dyes as reported in Table 10.2. However, the adsorption capacity for basic dye is much higher than for acid dye because of the ionic charges on the dyes and character of the clay. The adsorption process is mainly dominated by ion-exchange. This means that the adsorption capacity can strongly vary with the pH. In addition, some clays such as bentonite require activation by acid washing before they exhibit adsorptive properties.

Good removal capability of clay materials to uptake dye was previously demonstrated by Bagane and Guiza [53], Harris et al. [54], Ghosh and Bhattacharyya [55], Espantaleon et al. [56], Al-Ghouti et al. [57], Shawabkeh and Tutunji [58], Atun and Hisarli [59], Ozdemir et al. [60], Özcan et al. [61], and Tsai et al. [62], and more recently by Demirbas and Alkan [63], Feddal et al. [64], and Vanaamudan et al. [65]. For instance, the removal performance of Fuller's earth and CAC for Basic Blue 9 was compared by Atun and Hisarli [59]. They showed that the adsorption capacity is greater on Fuller's earth than on CAC. Fuller's earth is efficient because it contains variable amounts of dioctahedral smectites, natural zeolites and other sepiolites. Shawabkeh and Tutunji [58], studying the adsorption of Basic Blue 9

Table 10.2 Reported adsorption capacities q_{max} (mg g^{-1}) for clays (selected papers).

Adsorbent	Dye	S[a]	q_{max}	Reference
Activated bentonite (Turkey)	Acid Blue 193	767	740.5	[61]
Activated bentonite (Spain)	Sella Fast Brown H		360.5	[56]
Activated bentonite (Algeria)	Gentian Violet		108.56	[66]
Acid treated clay (Morocco)	Basic Blue 9		500	[67]
Activated clay (Singapore)	Basic Red 18		157	[68]
Activated clay (Singapore)	Acid Blue 9		57.8	[68]
Bentonite (Australia)	Acid Yellow 23		75.4	[69]
Bentonite (Australia)	Acid Red 18		69.8	[69]
Charred dolomite (Ireland)	Reactive Dye E-4BA	36	950	[70]
Clay (Morocco)	Basic Blue 9	125	350	[67]
Clay (Tunisia)	Basic Blue 9	71	300	[53]
Clay (Turkey)	Basic Blue 9	30	58.2	[71]
Clay (Morocco)	Basic Red 46	42.43	54	[72]
Clay (Morocco)	Astrazon Red		54	[73]
Clay (Turkey)	Brilliant Cresyl Blue	80	42	[74]
Clay (Turkey)	Nile Blue	80	25	[74]
Clay (Turkey)	Basic Blue 9	30	6.3	[75]
Clay/Carbons mixture	Acid Blue 9		64.7	[76]
Diatomite (Jordan)	Basic Blue 9	27.8	198	[57]
Diatomite (Jordan)	Basic Blue 9	33	156.6	[58]
Hydrotalcite	Reactive Yellow 208	100	47.8	[77]
Kaolin (India)	Brilliant Green	13.69	65.42	[78]
Kaolin (Algeria)	Basic Blue 9	61.13	45	[79]
Kaolin (Australia)	Congo Red	20.28	5.44	[80]
Kaolinite (India)	Rhodamine B		46.08	[81]
Montmorillonite clay (Brazil)	Basic Blue 9	62	289.12	[82]
Sepiolite (Spain)	Direct Blue 85	107	202	[83]
Sepiolite (Turkey)	Reactive Yellow 176	50.5	169.1	[60]
Sepiolite (Turkey)	Reactive Black 5	50.5	120.5	[60]

Table 10.2 (*Cont.*)

Adsorbent	Dye	S^a	q_{max}	Reference
Sepiolite (Turkey)	Reactive Red 239	50.5	108.8	[60]
Sepiolite (Brazil)	Astrazon Red		108	[83]
Sepiolite (Spain)	Basic Red 46	107	106	[83]
Sepiolite (Turkey)	Basic Blue 9	322	79	[84]
Sonicated sepiolite (Turkey)	Basic Blue 9	487	128	[84]
Treated diatomite (Jordan)	Basic Blue 9	572.9	126.6	[85]

^{a}surface area (in $m^2 g^{-1}$)

onto diatomite, showed that this naturally occurring material could sub-stitute the use of CAC as an adsorbent due to its availability and low cost, and its good adsorption properties. Al-Ghouti *et al.* [57] also investigated the feasibility of using diatomite for the removal of the problematic reactive dyes and concluded that clay materials were efficient adsorbents. Tsai *et al.* [62] demonstrated that the beer brewery waste mostly consisting of diatomite could be directly used as a porous adsorbent mainly based on its pore properties. Clay materials can be modified to improve their adsorption capacity. Ozdemir *et al.* [60] investigated the modification of sepiolite to adsorb a variety of azo-reactive dyes, and showed that the adsorption capacities are substantially improved upon modifying its surface with quaternary amines. The adsorption capacity of kaolinite can also be improved by purification and by treatment with NaOH solution [55]. The acid-treated bentonite showed a higher adsorption capacity than non-modified bentonite [61,56]. It is evident from the literature survey that clay materials may be promising adsorbents for environmental and purification purposes [3].

10.4 Siliceous Materials

The use of natural siliceous materials [86] such as dolomite, perlite, alunite, and glasses for wastewater is increasing because of their abundance, availability and low price, and interesting adsorption properties (Table 10.3). Walker *et al.* [70] previously proposed dolomite, both a mineral and a rock, as adsorbent for dye removal. The structure can be visualized as alternative layers of calcite and magnesite [87]. Outstanding removal capability of dolomite to uptake dye was obtained [70]. A comparison was made

Table 10.3 Reported adsorption capacities q_{max} (mg g^{-1}) for siliceous materials and silica gel (selected papers).

Adsorbent	Dye	q_{max}	Reference
Alunite (Turkey)	Reactive yellow 64	5	[94]
Alunite (Turkey)	Reactive Blue 114	2.92	[94]
Alunite (Turkey)	Reactive Red 124	2.85	[94]
Calcined alunite (Turkey)	Acid Red 88	832.81	[102]
Calcined alunite (Turkey)	Reactive Red 124	153	[94]
Calcined alunite (Turkey)	Acid Yellow 17	151.5	[93]
Glass powder	Acid Red 4	4.03	[59]
Modified silica	Acid Blue 25	45.8	[99]
Perlite (Turkey)	Basic Blue 9	162.3	[89]
Silica (Taiwan)	Basic Blue 9	11.21	[98]

with uptake using a commercial carbon indicating the adsorption by the dolomite char was greater than that of carbon and untreated dolomite. Indeed charred dolomite has a higher equilibrium capacity for reactive dye removal than CAC, with a capacity of 950 mg g^{-1} of adsorbent for dolomite compared to 650 mg dye adsorbed per g of adsorbent for carbon. However, the mechanism was not clear (probably a combination of precipitation and adsorption).

The use of perlite as a low-cost adsorbent for the removal of dyes has been investigated for the first time by Alkan and co-workers [88–92]. Perlite is a glassy volcanic rock with a rhyolithic composition which can be processed into an expanded form for cellular structure formation. The expansion takes place due to the presence of water in perlite when it is heated to high temperature. Perlite has high silica content, usually greater than 70% and this material is inexpensive and easily available in many countries. Interesting adsorption performances were obtained and it was suggested that dye molecules were physically adsorbed onto the surface material. However, perlite of different types (expanded and unexpanded) and of different origin have different properties because of the differences in composition. Alunite is one of the minerals of the jarosite group and contains approximately 50% SiO$_2$. This material is so cheap that regeneration was not necessary. Its surface charge and the pH play a significant role in influencing the capacity of alunite towards dyes. However, Özacar and Sengil [93,94] showed that untreated alunite does not have good adsorbent properties, but after a suitable process, alunite-type layered compounds are useful as adsorbents for removing color, and in particular to interact efficiently with acid dyes from wastewater. The authors obtained adsorption capacities of 57.47 mg and 212.8 mg of Acid Blue 40/g commercial activated carbon and treated alunite, respectively. Among other inorganic materials, silica beads also deserve particular attention [95–98,54,99], considering the chemical reactivity of their hydrophilic surface resulting from the presence of silanol groups. Their porous texture, high surface area and mechanical stability also make them attractive as adsorbents for decontamination applications. However, due to their low resistance toward alkaline solutions their usage is limited to media of pH less than 8 [100]. Moreover, the surface of siliceous materials contains acidic silanol (among other surface groups) which causes a strong and often irreversible non-specific adsorption. For that reason, it is necessary to eliminate the negative features of these adsorbents. In order to promote their interaction with dyes, the silica surface can be modified using silane coupling agents with the amino functional group [96]. Phan *et al.* [99] also showed that modified silica beads have a better potential for the removal of acid dyes from colored effluents. Recent strategies involve silica monoliths either as composites or biocomposites when they include immobilized living cells or biopolymers (enzymes, alginates, etc.). Recently, Rodrigues *et al.* [101] reviewed some of the most promising materials and pointed out their advantages.

10.5   Zeolites

Adsorption provides an attractive alternative for the treatment of polluted waters, especially if the adsorbent is inexpensive and does not require an additional pretreatment step before its use. From these two points of view, zeolites represent an interesting alternative. Natural zeolites are highly porous aluminosilicates with different cavity structures. These abundant and low-cost resources are becoming widely used as alternative materials in areas where adsorptive applications are required due to their high cation-exchange ability as well as their molecular sieve properties. The general chemical formula of zeolites is $M_{x/n}[Al_xSi_yO_{2(x+y)}]pH_2O$, where M is (Na, K, Li) and/or (Ca, Mg, Ba, Sr), n is cation charge; x/y = 1–6, p/x = 1–4. The characteristics of zeolites and their applications in water and wastewater treatment have been reviewed by Ghobarkar *et al.* [103], and more recently by Wang and Peng [104]. Zeolites consist of a wide variety of species, more than 40 natural species such as clinoptilolite, mordenite, chabazite, stilbite, and launmontite. However, the most abundant and frequently studied zeolite in the world is clinoptilolite, a mineral of the heulandite group (chemical formula: $(K_2, Na_2, Ca)_3Al_6Si_{30}O_{72}.21H_2O$). In its structure, three independent components are found: the aluminosilicate framework, exchangeable cations, and zeolitic water. It has been intensively studied recently because of its applicability in removing trace quantities of pollutants such as cations including heavy metals, and organics including phenols and dyes, thanks to its cage-like structures suitable for ion exchange.

High ion exchange capacity and relatively high specific surface areas, and more importantly their relatively cheap prices, make zeolites attractive adsorbents for dye removal [104,103]. Another advantage of zeolites over resins is their ion selectivities generated by their rigid porous structures. Several studies have been conducted on the adsorbent behavior of natural zeolites (Table 10.4). These literature data shows natural zeolites were effective for cationic dye adsorption mainly due to their cation-exchange characteristic, while they exhibited low capacity in anionic dyes. Another problem of zeolites is their low permeability which requires an artificial support when used in column operations. However, in their excellent review, Wang and Peng [104] concluded that application of natural or modified zeolites for water and wastewater treatment is a promising technique in environmental cleaning processes. The adsorption mechanism on zeolite particles is also complex because of their porous structure, inner and outer charged surfaces, mineralogical heterogeneity, and other

Table 10.4 Reported adsorption capacities q_{max} (mg g^{-1}) for zeolites (selected papers).

Adsorbent	Dye	q_{max}	Reference
Modified zeolite	Congo Red	69.94	[107]
Modified zeolite	Basic Blue 9	42.7	[108]
Modified zeolite	Basic Blue 9	15.68	[109]
Zeolite (Turkey)	Reactive Red 239	111.1	[60]
Zeolite (Turkey)	Reactive Yellow 176	88.5	[60]
Zeolite (Turkey)	Methyl Violet	75.25	[110]
Zeolite (Turkey)	Basic Blue 9	70.42	[89]
Zeolite (Turkey)	Basic Blue 9	28.6	[108]
Zeolite (Algeria)	Basic Blue 9	22	[79]
Zeolite (China)	Basic Blue 9	8.67	[109]
Zeolite (Turkey)	Reactive Black 5	60.5	[60]
Zeolite (Macedonia)	Basic Dye	55.86	[111]

imperfections on the surface [105,106]. However, it is recognized that like clay, the adsorption properties of zeolites result mainly from their ion-exchange capabilities.

10.6 Agricultural Solid Wastes

The byproducts from the agricultural and forest industries could be assumed as low-cost adsorbents since they are abundant in nature, inexpensive, require little processing and are effective materials. Agricultural solid wastes from cheap and readily available resources such as date pits [31], pith, corncob, barley husk, wheat straw, wood chips and orange peel have been successfully employed for the removal of dyes from an aqueous solution (Table 10.5). There have been numerous studies on the use of these agricultural solid wastes for dye removal and recent information about the most important features of these wastes can be found in the recent review by Rangabhashiyam *et al.* [112]. In this excellent review, the authors comprehensively discussed the sequestration of dye from textile industry wastewater using agricultural waste products as non-conventional adsorbents. Sawdust and bark have also been the subject of numerous studies. These low-cost materials are available in large quantities and may have potential as adsorbents due to their physicochemical characteristics and adsorption properties. Sawdust contains various organic compounds (lignin, cellulose, and hemicellulose) with polyphenolic groups that might be useful for binding dyes through different mechanisms. Shukla *et al.* [113] have reviewed the role of sawdust materials in the removal of pollutants from aqueous solutions. Baouab *et al.* [114], Garg *et al.* [115,116] and Özacar and Sengil [117] demonstrated that sawdust is a promising effective material for the removal of dyes from wastewaters. Examples of sawdust adsorption capacities are reported in Table 10.5. The adsorption mechanisms can be explained by the presence of several interactions such as complexation, ion-exchange due to a surface ionization, and hydrogen bonds. One problem with sawdust materials is that the adsorption results are strongly pH-dependent and in general, the adsorption capacity of basic dye is much higher than acid dye because of the ionic charges on the dyes and the ionic character of sawdust. Garg *et al.* [115,116,118] and Batzias and Sidiras [119] proposed chemical pretreatment of sawdust in order to improve the adsorption capacity and to enhance the efficiency of sawdust adsorption. Another waste product from the timber industry is bark, a polyphenol-rich material. Bark is an abundant forest residue which has been found to be effective in removing dyes from water solutions. Because

Table 10.5 Reported adsorption capacities q_{max} (mg g^{-1}) for waste materials from agriculture (selected papers).

Adsorbent	Dye	q_{max}	Reference
Banana peel	Basic Blue 9	20.8	[124]
Banana peel	Basic Violet 10	20.6	[124]
Banana peel	Methyl orange	21	[124]
Banana pith	Direct Red	5.92	[126]
Bark	Basic Red 2	1119	[27]
Bark	Basic Blue 9	914	[27]
Broad bean peels	Basic Blue 9	192.7	[126]
Coconut bunch waste	Basic Blue 9	70.92	[127]
Coir pith	Acid Yellow 99	442.13	[128]
Coir pith	Basic Blue 9	120.43	[129]
Coir pith	Basic Violet 10	94.73	[129]
Coir pith	Basic Violet 10	14.9	[130]
Coir pith	Direct Red 31	76.3	[130]
Coir pith	Acid Violet	7.34	[28]
Eucalyptus bark	Remazol BB	90	[122]
Egyptian bagasse pith	Basic Blue 69	168	[131]
Egyptian bagasse pith	Basic Blue 69	152	[132]
Egyptian bagasse pith	Basic Red 22	75	[132]
Egyptian bagasse pith	Acid Red 114	20	[132]
Egyptian bagasse pith	Acid Blue 25	17.5	[132]
Egyptian bagasse pith	Acid Blue 25	14.4	[131]
Neem sawdust	Basic Violet 3	3.78	[133]
Neem sawdust	Basic Green 4	3.42	[133]
Oil palm trunk fibre	Basic Green 4	149.35	[134]
Orange peel	Methyl Orange	20.5	[124]
Orange peel	Acid Violet	19.88	[135]
Orange peel	Basic Blue 9	18.6	[124]
Orange peel	Basic Violet 10	14.3	[124]
Palm-fruit bunch	Basic Yellow	320	[136]
Pine tree leaves	Astrazon Red	71.94	[137]
Pine sawdust	Acid Yellow 132	398.8	[117]
Pine sawdust	Acid Blue 256	280.3	[117]

Table 10.5 (*Cont.*)

Adsorbent	Dye	q_{max}	Reference
Pumpkin seed hull	Basic Blue 9	141.92	[138]
Raw date pits	Basic Blue 9	80.3	[31]
Rice hull ash	Direct Red 28	171	[139]
Rice husk	Basic Red 2	838	[27]
Rice husk	Basic Blue 9	312	[27]
Sugar beet pulp	Gemazol Turquoise Blue-G	234.8	[140]
Sugar cane dust	Basic Green 4	4.88	[141]
Sugar-industry-mud	Basic Red 22	519	[142]
Tree fern	Basic Red 13	408	[12]
Treated Parthenium	Basic Blue 9	88.49	[143]
Treated sawdust	Basic Green 4	74.5	[118]
Treated sawdust	Basic Green 4	26.9	[118]
Vine	Basic Red 22	210	[144]
Vine	Basic Yellow 21	160	[144]
Wood sawdust	Basic Blue 69	74.4	[145]
Wood sawdust	Acid Blue 25	5.99	[145]

of its low cost and high availability, bark is very attractive as an adsorbent. Palma *et al.* [120] reported that, like sawdust, the cost of forest wastes is only associated with the transport cost from the storing place to a site where they will be utilized. Bark is an effective adsorbent because of its high tannin content [121,122]. The polyhydroxy polyphenol groups of tannin are thought to be the active species in the adsorption process. There are promising perspectives for the utilization of bark as adsorbent on an industrial scale. However, the adsorption mechanisms are not clearly identified. Tree fern, an agricultural byproduct, was also proposed to remove pollutants from aqueous solutions [123,12]. This is a complex material containing lignin and cellulose as major constituents, giving interesting adsorption capacity values. The adsorption mechanism is due to chemisorption (mainly ion-exchange).

10.7 Industrial Byproducts

Industrial solid wastes such as metal hydroxide sludge, fly ash and red mud can be used as low-cost and locally available adsorbents for dye removal, although the adsorption capacities reported were low (Table 10.6). Netpradit *et al.* [146–148] studied the capacity and mechanisms of metal hydroxide sludge in removing azo reactive dyes. The sludge is a dried waste from the electroplating industry, which is produced by precipitation of metal ions in wastewater with calcium hydroxide. It contains insoluble metal hydroxides and other salts. The authors demonstrated that metal hydroxide sludge was an effective positively charged adsorbent with maximum adsorption capacities for azo reactive dyes. The charge of the dyes is an important factor for the adsorption due to the ion exchange mechanism. Similar interesting results were reported by Santos *et al.* [83],

Table 10.6 Reported adsorption capacities q_{max} (mg g^{-1}) for industrial byproducts (selected papers).

Adsorbent	Dye	q_{max}	Reference
Activated red mud	Reactive Blue 19	454.54	[160]
Activated red mud	Acid Blue 113	83.33	[161]
Activated red mud	Reactive Black 5	35.58	[161]
Boron waste	Astrazon Red	74.73	[162]
Chrome sludge	Basic Blue 9	0.51	[163]
Fe (III)/Cr (III) hydroxide	Basic Blue 9	22.8	[164]
Fly ash	Alizarin Sulfonic	11.21	[98]
Fly ash	Rhodamine B	10	[165]
Fly ash	Basic Blue 9	5.57	[166]
Fly ash	Basic Blue 9	5.52	[149]
Fly ash	Basic Blue 9	4.6	[167]
Leather waste	Reactive Red	163	[158]
Leather waste	Basic Blue 9	80	[158]
Metal hydroxide sludge	Reactive Blue 19	91	[168]
Metal hydroxide sludge	Reactive Red 2	62.5	[148]
Metal hydroxide sludge	Reactive Red 141	56.18	[148]
Metal hydroxide sludge	Reactive Red 120	48.31	[148]
Palm oil mill sludge	Basic Blue 9	50.7	[169]
Red mud	Basic Blue 9	2.49	[155]
Red mud	Direct Red 28	4.05	[156]
Sewage sludge	Basic Blue 9	114.9	[47]
Sludge waste	Bomaplex Red CR-L	192.31	[170]

studying the adsorption of Reactive Blue 19 onto a waste metal hydroxide sludge. Another industrial byproduct shown to adsorb dyes is fly ash, the major solid waste byproduct from coal-fired power plants. Although it may contain some hazardous substances, such as heavy metals, it is widely utilized in industry in many countries [149]. Gupta *et al.* [150] showed that bagasse fly ash generated in the sugar industry does not contain large amounts of toxic metals and could be used efficiently for adsorption of dyes. Numerous other studies reported that fly ash has interesting adsorption properties [151,152,16,153]. However, although fly ash is a waste material originating in great amounts in combustion processes, its properties are extremely variable and depend strongly on their origin [151,149,154]. Another abundant industrial byproduct is red mud [155–157]. Red mud emerges as a waste byproduct during the alkaline leaching of bauxite in the Bayer process, and causes serious environmental problems due to its high alkalinity and large amount. Owing to its high aluminum, iron and calcium content, red mud has been able to remove many types of pollutants. Waste red mud is a bauxite processing residue discarded in alumina production. Namasivayam and Arasi [156] proposed red mud as adsorbent for the removal of Congo red. The maximum capacity was 4.05 mg g⁻¹. Wang *et al.* [155] showed that physical and chemical treatment can significantly change the adsorption capacity. The industrial tanning of leather produces considerable amounts of chromium-containing solid wastes and liquid effluents [158,159]. Oliveira *et al.* [158] proposed the use of these solid wastes as complexing materials for textile dye removal. Their results showed that no chromium lixiviation from chromium-containing leather waste occurs during the adsorption experiments. However, the adsorption mechanisms are not clearly identified [158].

10.8   Peat

Peat is porous and rather complex soil material with organic matter in various stages of decomposition. This natural material is a plentiful, relatively inexpensive and widely available biosorbent, which has adsorption capabilities for a variety of pollutants [8,171]. It is a heterogeneous material with some hydrophobic fractions (such as aromatic and aliphatic moieties contained in humic substances) having natural affinity with hydrophobic molecules in aqueous solution. Raw peat contains lignin, cellulose, fulvic and humic acid as major constituents. These constituents, especially lignin and humic acid, bear polar functional groups such as alcohols, aldehydes, ketones, carboxylic acids, phenolic hydroxides and ethers that can be involved in chemical bonding. Brown *et al.* [172] reported that the precise composition of peat formed depends on the nature of the vegetation, regional climate, acidity of the water and the degree of metamorphosis.

Peat is known to have excellent ion exchange properties similar to natural zeolites. Indeed, because of its polar character andcellular structure, peat can effectively remove dyes from solution [173–177,131,178,179,16]. Peat is shown to be a particularly effective adsorbent for basic dyes but has a lower capacity for acid dyes. Peat adsorption capacities are reported in Table 10.7. Adsorption of a cationic dye generally increases with an increase in pH consistent with a mechanism of adsorption by cation exchange with acidic functional groups. There are then more adsorption sites for cation uptake from solution. Similarly a decrease in pH produces more competition for adsorption between cations and the increased H^+ presence, and as a result of this, cations will be desorbed. For the acid and basic dye, the removal performance was comparable with that of activated carbon, while for the disperse dyes, the performance was much better. Peat tends to have a high cation exchange capacity, and is a very effective adsorbent for decolorization purposes. The adsorption was strongly dependent on electrical charge density on the surface because the interactions, for example, between the dye molecules and carboxylic groups of the peat, could be either electrostatic or non-electrostatic. This was dependent on each dye structure.

When raw peat is directly used as an adsorbent, there are possible limitations because natural peat has a low mechanical strength, a high affinity for water, a poor chemical stability, a tendency to shrink and/or swell, and to leach fulvic acid [180,181]. Chemical pretreatment and the development of immobilized biomass beads can produce a more robust media. As with

Table 10.7 Reported adsorption capacities q_m (mg g^{-1}) for peat (selected papers).

Biosorbent	Dye	q_{max}	Reference
Magellanic peat	Acid Black 1	33.7	[183]
Magellanic peat	Basic Blue 3	33.1	[183]
Peat	Basic Blue 69	195	[184]
Peat	Basic Blue 9	190	[173]
Peat	Astrazone Blue	26.32	[185]
Peat	Acid Blue 25	12.7	[184]
Peat	Reactive Black 5	7	[186]
Treated peat	Basic Violet 14	400	[179]
Treated peat	Basic Green 4	350	[179]

other adsorbents, chemical processes are also used for improving adsorption properties and selectivity. For example, Sun and Yang [179] prepared modified peat-resin by mixing oxidizing peat with polyvinyl alcohol and formaldehyde. These materials possess a macroreticular porous structure with high physical characteristics and interesting adsorption performances. Their studies demonstrated that modified peat can be used for the removal of a variety of basic dyes. For example, the maximum adsorption capacities for Basic Violet 14 and Basic Green 4 were 400 and 350 mg g^{-1} treated peat, respectively. Vecino *et al.* [182] also proposed a peat-based hybrid material for effective dye removal: peat was entrapped in calcium alginate beads. The mechanism by which dyes are adsorbed onto peat has been a matter of considerable debate. Different studies have reached different conclusions. Various pollutant-binding mechanisms are thought to be involved in the biosorption process including physical adsorption, ion-exchange, complexation, adsorption-complexation and chemisorption [172]. Variations in peat type and adsorbent preparation also make the comparison of results difficult [183]. However, it is now recognized that ion-exchange is the most prevalent mechanism.

10.9   Chitin and Chitosan

Abundant information on chitin and chitosan for environmental purposes can be found in the literature [187–193]. Crini and Badot [190] reviewed over 100 papers (1998–2008) on the application of chitin and chitosan as non-conventional efficient biosorbents for the removal of dyes from aqueous solutions even in low concentration. They concluded that the adsorption of dyes using these biopolymers is one of the reported emerging and efficient biosorption methods for dye removal. Chitin and its derivative, chitosan, are abundant, renewable and biodegradable resources. These biopolymers are commercially extracted from crustaceans (crab, krill, crayfish) primarily because a large amount of the crustacean's exoskeleton is available as a byproduct of food processing [191]. Utilization of industrial solid wastes for the treatment of wastewater from another industry could be helpful not only to the environment in solving the solid waste disposal problem, but also the economy [194]. Chitin contains 2-acetamido-2-deoxy-β-D-glucose through a β (1→4) linkage. This waste product is second only to cellulose in terms of abundance in nature. Chitosan contains 2-acetamido-2-deoxy-β-D-glucopyranose and 2-amino-2-deoxy-β-D-glucopyranose residues. Chitosan has drawn particular attention as a complexing agent due to its low cost compared to activated carbon and its high contents of amino and hydroxy functional groups showing high potentials of a wide range of molecules, including phenolic compounds, dyes and metal ions. This biopolymer represents an attractive alternative as non-conventional materials because of its physicochemical characteristics, chemical stability, high reactivity, excellent chelation behavior and high selectivity toward pollutants, including dye molecules. Indeed, numerous studies demonstrated that chitosan-based biosorbents are efficient materials and have an extremely high affinity for many classes of dyes (Table 10.8). They are also versatile materials, allowing the adsorbent to be used under different forms, from flake-types, to gels, bead-types, or fibers. Wong et al. [199] found that the maximum adsorption capacities of chitosan for Acid Orange 12, Acid Orange 10, Acid Red 73 and Acid Red 18 were 973.3, 922.9, 728.2, and 693.2 mg g^{-1}, respectively. Wu et al. [195] also reported the usefulness of chitosan for the removal of reactive dyes. Chiou and Li [196] found that 1 g of chitosan adsorbed 2498 mg of Reactive Blue 2. In general, chitosan-based biosorbents have demonstrated outstanding removal capabilities for direct dyes [190]. In comparison with commercial activated carbons, the beads exhibited excellent performance for adsorption of anionic dyes: the adsorption values were 3–15 times higher at the same pH [197]. Hence chitosan chelation is a procedure of choice for dye removal from aqueous solution. However, performances depend on the different sources of chitin, the degree of N-acetylation, molecular weight

Table 10.8 Reported adsorption capacities q$_{max}$ (mg g^{-1}) for chitosan and chitosan-based biosorbents (selected papers).

Biosorbent	Dye	q$_{max}$	Reference
Chitosan	Acid Orange 12	973.3	[198]
Chitosan	Acid Orange 12	973.3	[199]
Chitosan	Acid Orange 10	922.9	[199]
Chitosan	Acid Orange 10	922.9	[198]
Chitosan	Acid Red 73	728.2	[199]
Chitosan	Acid Red 73	728.2	[198]
Chitosan	Acid Red 18	693.2	[199]
Chitosan	Acid Red 18	693.2	[198]
Chitosan	Acid Green 25	645.1	[199]
Chitosan	Acid Green 25	645.1	[198]
Chitosan	Remazol Black 13	91.47	[200]
Chitosan	Congo Red	78.9	[201]

Table 10.8 (*Cont.*)

Biosorbent	Dye	q_{max}	Reference
Chitosan beads	Reactive Red 189	1189	[202]
Chitosan beads	Acid Blue 25	263.15	[203]
Chitosan beads	Acid Red 37	128.21	[203]
Chitosan beads	Basic Green 4	93.55	[204]
Chitosan beads	Congo Red	92.59	[205]
Chitosan beads (crab)	Reactive Red 222	1106	[195]
Chitosan beads (lobster)	Reactive Red 222	1037	[195]
Chitosan flakes (lobster)	Reactive Red 222	398	[195]
Chitosan flakes (crab)	Reactive Red 222	293	[195]
Chitosan (powder)	FD&C Red 40	529	[2006]
Chitosan (powder)	FD&C Yellow 5	350	[2007]
Chitosan (powder)	FD&C Blue 1	219	[2007]
Crosslinked chitosan beads	Reactive Blue 2	2498	[197]
Crosslinked chitosan beads	Reactive Red 2	2422	[197]
Crosslinked chitosan beads	Direct Red 81	2383	[197]
Crosslinked chitosan beads	Reactive Yellow 86	1911	[197]
Crosslinked chitosan beads	Reactive Red 189	1936	[202]
Crosslinked chitosan beads	Reactive Blue 15	1334	[208]
Crosslinked chitosan beads	Acid Blue 25	142.86	[203]
Crosslinked chitosan beads	Metanil Yellow	722	[208]
Crosslinked chitosan beads	Reactive Red 120	361.9	[209]
Crosslinked chitosan beads	Acid Red 37	59.52	[203]
Crosslinked chitosan composite beads	Reactive Blue 19	1060	[210]
Crosslinked quaternary chitosan	Reactive Orange 16	1060	[211]
Grafted chitosan	Remazol Yellow	1211	[212]
Grafted chitosan	Basic Yellow 37	595	[212]
Grafted chitosan	Congo Red	330.62	[201]
Magnetic chitosan nanoparticles	DF&C Blue 1	475.61	[213]
Magnetic chitosan nanoparticles	D&C Yellow 5	292.07	[213]
Modified chitosan beads	Congo Red	433.12	[214]

and solution properties and vary with crystallinity, affinity for water, percent deacetylation and amino group content [191–193]. Adsorption properties are also dependent on the type of material used and the uptake is strongly pH-dependent. These problems can explain why it is difficult to develop chitosan-based materials as adsorbents at an industrial-scale. In their review, Crini and Badot [190] concluded chitosan-based materials may be promising biosorbents for adsorption processes since they demonstrated outstanding removal capabilities for dyes.

10.10   Biomass

The literature clearly shows a greater number of studies on the adsorption of metal ions by biomass as compared with organic pollutants [215,216]. Indeed, biomass has received considerable interest in metal adsorption due its excellent metal-binding capacities through various mechanisms and interesting selectivity. However, attention has also been focused on the interaction between dyes and biomass. Biosorbents derived from suitable microbial biomass can be used for the effective removal of dyes from solutions since certain dye molecules have a particular affinity for binding with microbial species [217–223]. The use of biomass for wastewater is increasing not only because of its high potential as a complexing material due to its specific physicochemical characteristics, but also because of its availability in large quantities and at low prices. A wide variety of microorganisms including algae, yeasts, bacteria and fungi are capable of decolorizing a wide range of dyes with a high efficiency. Fungi can be classified into two kinds according to their life state: living cells to biodegrade and biosorb dyes, and dead cells (fungal biomass) to adsorb dyes. Most of the studies have concentrated on living fungi for biosorption of the dyes. There are few studies on dye removal using dead fungal biomass, except in recent years. Table 10.9 shows some of the adsorption capacities reported in the literature. Fu and Viraraghavan [224–227] demonstrated that, compared with commercial activated carbons, dead fungal biomass of *Aspergillus niger* is a promising biosorbent for dye removal. Aksu and Tezer [228] demonstrated uptake of 588.2 mg of Reactive Black 5 per g using *Rhizopus arrhizus* biomass. Waranusantigul *et al.* [229] and Chu and Chen [230,231] also reported the usefulness of biomass for the removal of basic dyes. The biosorption capacity of fungal biomass could be increased by some pretreatment (by autoclaving or by reacting with chemicals). Other types of biomass such as yeasts have been studied for their dye uptake capacities [232,233]. Yeasts are extensively used in a variety of large-scale industrial

Table 10.9 Reported adsorption capacities q_{max} (mg g^{-1}) for biomass, including living and dead materials, yeasts, and alga (selected papers).

Adsorbent	Dye	q_{max}	Reference
Activated sludge biomass	Reactive Yellow 2	333.3	[238]
Activated sludge biomass	Basic Red 18	285.71	[239]
Activated sludge biomass	Basic Blue 9	256.41	[239]
Activated sludge biomass	Reactive Blue 2	250	[238]
Activated sludge biomass	Basic Blue 47	157.5	[230]
Activated sludge biomass	Basic Red 18	133.9	[230]
Activated sludge biomass	Reactive Black 5	116	[240]
Activated sludge biomass	Basic Red 29	113.2	[230]
Activated sludge biomass	Direct Yellow 12	98	[235]
Activated sludge biomass	Basic Yellow 24	56.98	[231]
Algae *Gelidium*	Basic Blue 9	171	[241]
Algae *Sargassum muticum*	Basic Blue 9	279.2	[242]
Algal waste	Basic Blue 9	104	[241]
Baker's yeast	Astrazone Blue	70	[243]
Brown alga *S. marginatum*	Acid Orange 7	35.62	[244]
Chlorella vulgaris	Reactive Red 5	555.6	[245]
Dead fungus *Aspergillus niger*	Basic Blue 9	18.54	[227]
Dead fungus *Aspergillus niger*	Direct Red 28	14.72	[224]
Dead fungus *Aspergillus niger*	Acid Blue 29	13.82	[226]
Immobilized *Phanerochaete chrysosporium* biomass	Reactive Blue 19	98.8	[246]
Immobilized Aspergillus fumigatus biomass	Reactive Blue 19	86.7	[247]
Immobilized Aspergillus fumigatus biomass	Reactive Red 24	78	[247]
Lentinus sajor-caju biomass	Reactive Red 120	117.8	[248]
Living biomass	Acid Blue 29	6.63	[226]
Living biomass	Basic Blue 9	1.17	[227]
Macroalga *C. lentillifera*	Red GTLN	113.64	[249]
Macroalga *C. lentillifera*	Blue FGRL	94.34	[249]
Macroalga *C. lentillifera*	Basic Yellow 28	35.46	[249]

(Continued)

Table 10.9 *(Cont.)*

Adsorbent	Dye	q_{max}	Reference
Modified fungal biomass	Disperse Red 1	5.59	[225]
Moss	Basic Blue 9	185	[250]
Paenibacillus macerans	Acid Blue 062	95.08	[251]
Paenibacillus macerans	Acid Blue 225	94.98	[251]
Penicillium sp.	Acid Violet	4.32	[252]
Phanerocheate chrysosporium	Reactive Blue 4	211.6	[253]
Phanerochaete chrysosporium	Reactive Blue 19	80.91	[246]
Posidonia oceanica	Astrazon Red	68.97	[254]
Posidonia oceanica	Alpacide Yellow	15.11	[255]
Rhizopus arrhizus	Reactive Black 5	588.2	[228]
Rhizopus arrhizus	Reactive Orange 16	190	[236]
Rhizopus arrhizus	Reactive Red 4	150	[236]
Rhizopus arrhizus	Reactive Blue 19	90	[236]
Rhizopus nigricans	Reactive Black 8	122	[256]
Rhizopus nigricans	Reactive Brown 9	112	[256]
Spirodela polyrrhiza	Basic Blue 9	144.93	[229]
Spirodela polyrrhiza (duckweed)	Basic Blue 9	144.9	[229]
Trametes versicolor	Direct Red 128	152.3	[257]
Trametes versicolor	Direct Blue 1	101.1	[257]
Treated *Lentinus sajor-caju* biomass	Reactive Red 120	182.9	[248]
Treated brown alga *S. marginatum*	Acid Orange 7	71.05	[244]
Yeasts	Remazol Blue	173.1	[232]
Yeasts	Reactive Black 5	88.5	[233]

fermentation processes and waste biomass from these processes is a potential source of cheap adsorbent material.

The major advantages of biosorption technology are its effectiveness in reducing the concentration of dyes to very low levels and the use of inexpensive biosorbent material. Fungal biomass can be produced cheaply using relatively simple fermentation techniques and inexpensive growth media [224,234]. The use of biomass is especially interesting when the dye-containing effluent is very toxic. Biosorption is also an emerging technology that attempts to overcome the selectivity disadvantage of conventional adsorption processes. The use of dead rather than live biomass eliminates the problems of waste toxicity and nutrient requirements. Biomass adsorption is effective when conditions are not always favorable for the growth and maintenance of the microbial population. In spite of good adsorption properties and high selectivity some problems can occur. The adsorption process is slow: in the case of biomass of *Aspergillus niger* equilibrium was reached in 42 h. Another problem is that the initial pH of the dye solution strongly influenced the biosorption [232]. Biosorption was also influenced by the functional groups in the fungal biomass and its specific surface properties [235]. Biosorption performance depends on some external factors such as salts and ions in solution which may be in competition. Another limitation of the technology include the fact that the method has only been tested for limited practical applications since biomass is not appropriate for the treatment of effluents using column systems, due to clogging effect. Because of major limitations for their efficient utilization in a column reactor, there is the need for their immobilization. This step forms a major cost factor of the process. In general, decolorization by living and dead cells involves several complex mechanisms such as surface adsorption, ion-exchange, complexation (coordination), complexation-chelation and microprecipitation. Cell walls consisting mainly of polysaccharides, proteins and lipids, offer many functional groups. The dyes can interact with these active groups on the cell surface in a different manner. The accumulation of dyes by biomass may involve a combination of active, metabolism-dependent and passive transport mechanisms starting with the diffusion of the adsorbed solute to the surface of the microbial cell [236,228,237]. Once the dye has diffused to the surface, it will bind to sites on the cell surface. The precise binding mechanisms may range from physical (i.e., electrostatic or van der Waal forces) to chemical binding (i.e., ionic and covalent). However, it is now recognized that the efficiency and the selectivity of adsorption by biomass are due to ion-exchange mechanisms. Biosorption processes are particularly suitable for the treatment of solutions containing dilute (toxic) dye concentration. They are a potential promising alternative to conventional processes for the removal of dyes. However, these technologies are still being developed and much more work is required.

10.11   Starch-Based Derivatives

There is an increasing interest in the production of novel materials with particular functionalities from renewable resources. Natural polymers are replacing synthetic polymers in many applications partly because of their low cost, biodegradability, numerous properties, and the relative ease with which they can be modified chemically [258,9]. Numerous studies propose the use of starch-based materials as useful adsorbent for removal of pollutants [259–261]. Starch, one of the most abundant polysaccharides on earth, represents an interesting alternative as a complexing material because of its intrinsic characteristics (renewable and biodegradable raw resource) and chemical properties (high reactivity, flocculation and adsorption properties) resulting from the presence of reactive hydroxyl groups in the macromolecular chains. However, for non-food uses, it is necessary to modify starch in order to obtain derivatives with properties suitable for environmental applications. In the literature, there have been numerous studies on the preparation, characterization, properties and applications of starch-based derivatives. Among them, crosslinking materials have attracted great attention due to their facile synthesis, high adsorption properties, and particular selectivity. Among all the reactions proposed, chemical crosslinking using epichlorohydrin as crosslinker agent is the most straightforward method to produce derivatives for environmental applications. The chemical modification of these crosslinked materials is also an interesting step to introduce specific properties in order to enlarge the field of their potential applications. Starch-based materials have demonstrated interesting removal abilities for certain dyes, as summarized in Table 10.10. The state-of-the-art in the field of biosorption of dyes by starch-based derivatives was reviewed by Crini and Badot [262]. Their literature survey showed that starch was a promising tool for the purification of dye-containing wastewater because of its interesting adsorption capacity.

Crini [9] reported that crosslinked starch-based material showed a higher capacity for adsorption of dyes than commercial activated carbons. Cationic crosslinked starch was a rather better adsorbent than crosslinked starch for acidic dyestuffs and its production is not costly [260]. The interaction between cationic starches and anionic dyes has also been intensively investigated by Riauka [263]. He prepared both raw starches (CS) and crosslinked starches (CCS) containing quaternary ammonium groups. His investigations clearly indicated that CCS materials were more suitable than CS, and they had a high selectivity for dye molecules. He concluded that crosslinked cationic starches were very useful for the treatment of wastewater from the textile industry. Wang's group [264–266] also reported that crosslinked amphoteric starch may be useful adsorbents for dye removal because of their high adsorption capacities. These adsorbents were efficient for the removal of both acid and basic dyestuffs from aqueous solution without changing the pH of the solution due to the particular electrical character of the materials. They found that the maximum adsorption

Table 10.10 Reported adsorption capacities (q_{max} in mg g^{-1}) for dye removal onto starch-based materials (selected papers).

Crosslinked material	Dye	q_{max}	Reference
Amphoteric starch	Methyl Violet	333.33	[265]
Amphoteric starch	Acid Light Yellow 2G	227.27	[265]
Amphoteric starch	Acid Red G	217.39	[265]
Amphoteric starch	Acid Yellow G	149.6	[266]
Amphoteric starch	Methyl Green	133.33	[265]
Amphoteric starch	Basic Green 4	104.75	[264]
Cationic starch	Monobasic dye	1462	[263]
Cationic starch	Acid Red 151	613	[263]
Cationic starch	Acid Blue 25	322	[259]
Cationic starch	Acid Blue 25	322	[260]
Cationic starch	Acid Blue 25	249	[267]
Cationic starch	Acid Green 25	151	[8]
Crosslinked starch	Methyl Violet	99.3	[268]
Crosslinked starch	Basic Blue 9	9.46	[269]
Grafted starch	Safranine T	204	[270]
Grafted starch	Basic Violet 7	8.3	[271]
Starch-based material	Brilliant Blue X-BR	122	[272]

capacities of starch for Basic Green 4 and Acid Yellow were 141.9 mg/g and 149.6 mg/g, respectively. However, it is important to note that the adsorption mechanisms are not fully understood because numerous interactions are possible including ion-exchange, complexation, electrostatic interactions, acid-base interactions, physisorption, hydrogen bonding and hydrophobic interactions [8,9,262]. It is possible that more than one of these interactions can occur simultaneously depending on the composition of the material, the dye structure and the solution conditions. Also, all the adsorbents proposed in the literature must now be thoroughly tested on an industrial scale.

10.12   Miscellaneous Adsorbents

Other materials have been studied as non-conventional adsorbents such as cotton waste, cellulose, alginates, cyclodextrins, calixarenes and cucurbituril [273–27]. Selected adsorption capacities are reported in Table 10.11. Cotton is the most abundant of all naturally occurring organic substrates, being planted on a large scale in some countries such as the United States, India, Brazil, Australia and China, and is widely used. This material characteristically exhibits excellent physical and chemical properties in terms of stability, water absorbency and dye ability [276]. Cotton plant wastes are composed primarily of cellulose (30–50%), hemicelluloses (20–30%) and lignin (20–30%). Tunç *et al.* [277] studied the potential use of cotton plant wastes as adsorbents for the removal of Reactive Black 5. Adsorption was strongly pH-dependent but slightly temperature-dependent. The authors showed that both external mass transfer and

Table 10.11 Reported adsorption capacities q_{max} (mg g^{-1}) for miscellaneous adsorbents (selected papers).

Adsorbent	Dye	q_{max}	Reference
Calcium alginate beads	Basic Black	57.7	[282]
Crosslinked cyclodextrin	Basic Blue 9	105	[287]
Crosslinked cyclodextrin	Basic Blue 9	56.5	[288]
Crosslinked cyclodextrin	Basic Green 4	91.9	[290]
Crosslinked cyclodextrin	Acid Blue 25	88	[289]
Crosslinked cyclodextrin	Basic Violet 10	53.2	[274]
Crosslinked cyclodextrin	Basic Blue 3	42.4	[274]
Crosslinked cyclodextrin	Congo Red	36.2	[275]
Crosslinked cyclodextrin	Basic Violet 3	35.8	[274]
Cyclodextrin/chitosan material	Acid Blue 25	77.4	[291]
Cyclodextrin/chitosan material	Basic Blue 9	50.12	[292]
Cotton waste	Basic Red 2	875	[27]
Cotton waste	Basic Blue 9	277	[27]
Cotton waste	Reactive Black 5	50.9	[277]
Modified cellulose	Basic Fuchsine	1155.7	[293]
Modified cellulose	Basic Fuchsine	31.92	[294]
Modified cellulose	Basic Blue 9	194.6	[294]
Modified cellulose	Reactive Red	78	[295]
Treated cotton	Acid Blue 25	589	[276]
Treated cotton	Acid Yellow 99	448	[276]
Treated cotton	Reactive Yellow 23	302	[276]

intraparticle diffusion played an important role in the adsorption mechanism of dye. Cellulose is also an inexpensive and promising polysaccharide for environmental purposes because of its abundance and renewability [278]. This biopolymer is the main constituent of plants and the most abundant biomass in the world. However, raw cellulose has low adsorption capacity as well as poor physical stability. Chemical modifications such as graft copolymerization, crosslinking or amination have been used to overcome these drawbacks. An excellent detailed discussion of cellulose-based materials for dye removal may be found in the comprehensive review by Hubbe *et al.* [279]. In particular, the authors discussed factors affecting adsorption processes. Alginate is another one of the most extensively investigated polysaccharides for removal of pollutants from aqueous solution, as it is inexpensive, nontoxic and efficient [280–282]. Alginate is a naturally linear carbohydrate polymer extracted from brown seaweeds. It has a capacity to remove toxic pollutants due to its high carboxyl group content as showed by Aravindhan *et al.* [282], and more recently by Khari *et al.* [283]. Innovative magnetic alginate beads for dye removal were also proposed by Rocher *et al.* [280].

Cyclodextrins, calix[n]arenes and cucurbituril have also been studied as non-conventional adsorbents for pollutant removal. Cyclodextrins (CDs) are torus-shaped cyclic oligosaccharides containing six to twelve glucose units. The most characteristic feature of CDs is the ability to form inclusion compound with various aromatic and phenolic molecules, and polymers. However, in spite of varied interesting characteristics and properties, a limited number of dye adsorption studies have been carried out on CD-based derivatives [9,8,284]. Morin-Crini and Crini [285] recently summarized the developments in the use of CD-based materials for environmental purposes. This comprehensive review provided a summary of information on color removal obtained using batch methods. In general, from a point of view of their high adsorption capacity to adsorb dyes, the CD materials are interesting, although they are nonporous and possess low surface area. Indeed, adsorption processes mainly occur by inclusion complex formation. Calix[n]arenes are the third major class of supramolecular host systems along with cyclodextrins and crown ethers. Their ease of synthesis and relative simplicity of chemical modification have produced increased interest in the host-guest chemistry over the last few years, although their cost can be a disadvantage for environmental purposes. Calix[n]arenes are cyclic oligomers composed of phenol units very well known as ionophores and for the fact that they provide a unique three-dimensional structure. The complexation properties of these molecules are highly dependent upon the nature, number of donor groups and the conformation of the calix[n]arene moiety. Calix[n]arenes have been widely employed as building blocks in the design of novel host molecules. These macrocyclic compounds are also attractive as a platform to develop novel host material for entrapping pollutants. The preparation and characterization of a calix[4] arene-based resin and its use for dye removal were reported by Kamboh *et al.* [286]. Cucurbituril, a cyclic polymer of glycoluril and formaldehyde, is another material also known to form host-guest complexes with aromatic compounds. This adsorbent presents interesting adsorption capacities for various types of textile dye [87]. However, because of its solubility, cucurbituril is not feasible as an adsorbent in aqueous solutions unless it is incorporated into fixed bed adsorption filters or covalently fixed onto a suitable support material. Cost is also another disadvantage. Adsorption mechanisms are explained by the formation of host-guest complexes and the presence of hydrophobic interactions and the formation of insoluble cucurbituril-dye-cation aggregates.

10.13   Concluding Remarks

This chapter has attempted to cover a wide range of non-conventional adsorbents so that the reader can get an idea about the various types of materials used for the removal of dye molecules from aqueous solutions. It is now accepted that inexpensive, locally available and effective non-conventional adsorbents could be used in place of commercial activated carbons and synthetic organic resins for dye removal [296]. It is important to note that the adsorption capacity exhibited by each material relates primarily to its textural and chemical properties. Other factors, however, such as operation difficulty, practicability, regeneration potential and environmental impact, need to be taken into consideration when selecting one adsorbent over another. It is also important to point out that a particular adsorbent is only applicable to a particular class of dye molecules. Thus, using only one type of non-conventional adsorbent is difficult for the treatment of the complicated composition of dye wastewaters. However, undoubtedly non-conventional materials offer a lot of promising benefits for commercial purposes in the future.

Although much has been accomplished from the use of new adsorbents and their adsorption process studies, much work is, however, necessary to predict the performance of dye removal from real industrial effluents under a range of operating conditions and to demonstrate the applicability of these materials on an industrial scale. Also, much work is still necessary to clearly identify the adsorption mechanism. Another aspect concerns the fact that little effort has been made to carry out a cost comparison between conventional and non-conventional adsorbents. This aspect needs to be investigated further to promote large-scale use of non-conventional adsorbents. Perhaps one reason why non-conventional adsorbents have not been widely used in industry in Europe is the lack of knowledge about the engineering of such materials. I think that, for novel adsorbents to be accepted by industry, it will be necessary to adopt a multidisciplinary approach in which chemists, biologists, microbiologists, polymerists and engineers work together [6].

References

1. A. Imran, 2014. Water treatment by adsorption columns: Evaluation at ground level. *Sep. Purif. Rev.* 43, 175–205.
2. M. T. Yagub, T. K. Sen, S. Afroze, H. M. Ang, 2014. Dye and its removal from aqueous solution by adsorption: A review. *Adv. Colloid. Int. Sci.* 209, 172–184.
3. S. T. Ong, P. S. Keng, S. L. Lee, Y. T. Hung, 2014. Low cost adsorbents for sustainable dye containing-wastewater treatment. *Asian J. Chem.* 26, 1873–1881.
4. G. Z. Kyzas, J. Fu, K. A. Matis, 2013. The changes from past to future for adsorbent materials in treatment of dyeing wastewaters. *Materials* 6, 5131–5158.
5. R. Sanghi, P. Verma, 2013. Decolorisation of aqueous dye solutions by low-cost adsorbents: A review. *Coloration Technol.* 129, 85–108.
6. G. Crini, P. M. Badot, 2010. *Adsorption Processes and Pollution.* PUFC, ed., Besançon, France.
7. Z. Aksu, 2005. Application of biosorption for the removal of organic pollutants: A review. *Proc. Biochem.* 40, 997–1026.
8. G. Crini, 2006. Non-conventional low-cost sorbents for dye removal: A review. *Bioresour. Technol.* 97, 1061–1085.
9. G. Crini, 2005. Recents developments in polysaccharide-based materials used as sorbents in wastewater treatment. *Prog. Polym. Sci.* 30, 38–70.
10. K. K. H. Choy, G. McKay, J. F. Porter, 1999. Sorption of acid dyes from effluents using activated carbon. *Resources Conservation Recycling* 27, 57–71.
11. Y. S. Ho, J. F. Porter, G. McKay, 2002. Equilibrium isotherm studies for the sorption of divalent metal ions onto peat: Copper, nickel and lead single component systems. *Water Air Soil Poll.* 141, 1–33.
12. Y. S. Ho, 2003. Removal of copper ions from aqueous solution by tree fern. *Water Res.* 37, 2323–2330.
13. G. Crini, P. M. Badot, 2007. *Traitement et Epuration des Eaux Industrielles Polluées (in French).* PUFC, ed., Besançon, France. p. 353.
14. S. Babel, T. A. Kurniawan, 2003. Low-cost adsorbents for heavy metals uptake from contaminated water: A review. *J. Hazard. Mat.* B97, 219–243.

15. F. Derbyshire, M. Jagtoyen, R. Andrews, A. Rao, I. Martin-Gullon, E. Grulke, 2001. Carbon materials in environmental applications. In: L. R. Radovic, editor. *Chemistry and Physics of Carbon*, Vol. 27. Marcel Dekker: New-York, pp. 1–66.

16. K. R. Ramakrishna, T. Viraraghavan, 1997. Dye removal using low cost adsorbents. *Water Sci. Tech.* 36, 189–196.

17. M. Streat, J. W. Patrick, M. J. Pérez, 1995. Sorption of phenol and para-chlorophenol from water using conventional and novel activated carbons. *Water Res.* 29, 467–472.

18. L. S. Oliveira, A. S. Franca, 2008. Low-cost adsorbents from agri-food wastes. In: *Food Science and Technology: New Research*. L. V. Greco and M. N. Bruno, eds., Nova Science Publishers, Inc. pp. 1–39.

19. R. S. Juang, F. C. Wu, R. L. Tseng, 2002. Characterization and use of activated carbons prepared from bagasses for liquid-phase adsorption. *Colloid Surf. A: Physicochem. Eng. Aspects* 201, 191–199.

20. M. Valix, W. H. Cheung, G. McKay, 2004. Preparation of activated carbon using low temperature carbonisation and physical activation of high ash raw bagasse for acid dye adsorption. *Chemosphere* 56, 493–501.

21. Z. Zhang, I. M. O'Hara, G. A. Kent, W. O. S. Doherty, 2013. Comparative study on adsorption of two cationic dyes by milled sugarcane bagasse. *Ind. Crops Products* 42, 41–49.

22. S. Venkata Mohan, P. Sailaja, M. Srimurali, J. Karthikeyan, 1999. Colour removal of monoazo acid dye from aqueous solution by adsorption and chemical coagulation. *Environ. Eng. Policy* 1, 149–154.

23. R. S. Juang, R. L. Tseng, F. C. Wu, 2001. Role of microporosity of activated carbons on their adsorption abilities for phenols and dyes. *Adsorption* 7, 65–72.

24. O. Hamdaoui, 2006. Batch study of liquid-phase adsorption of methylene blue using cedar sawdust and crushed brick. *J. Hazard. Mat.* B135, 264–273.

25. R. Han, Y. Wang, P. Han, J. Shi, J. Yang, Y. Lu, 2006. Removal of methylene blue from aqueous solution by chaff in batch mode. *J. Hazard. Mat.* B137, 550–557.

26. F. Ferrero, 2007. Dye removal by low cost adsorbents: Hazelnut shells in comparison with wood sawdust. *J. Hazard. Mat.* 142, 144–152.

27. G. McKay, J. F. Porter, G. R. Prasad, 1999. The removal of dye colours from aqueous solutions by adsorption on low-cost materials. *Water Air Soil Pollut.* 114, 423–438.

28. C. Namasivayam, M. Dinesh Kumar, K. Selvi, R. Begum Ashruffunissa, T. Vanathi, R. T. Yamuna, 2001. Waste coir pith – A potential biomass for the treatment of dyeing wastewaters. *Biomass and Bioenergy* 21, 477–483.

29. C. Namasivayam, D. Kavitha, 2002. Removal of Congo red from water by adsorption onto activated carbon prepared from coir pith, an agricultural solid waste. *Dyes and Pigments* 54, 47–58.

30. D. Kavitha, C. Namasivayam, 2007. Recycling coir pith, an agricultural solid waste, for the removal of procion orange from wastewater. *Dyes and Pigments* 74, 237–248.
31. F. Banat, S. Al-Asheh, L. Al-Makhadmeh, 2003. Evaluation of the use of raw and activated date pits as potential adsorbents for dye containing waters. *Process Biochem.* 39, 193–202.
32. A. M. M. Vargas, A. L. Cazetta, A. C. Martins, J. C. G. Moraes, E. E. Garcia, G. F. Gauze, W. F. Costa, V. C. Almeida, 2012. Kinetic and equilibrium studies: Adsorption of food dyes Acid Yellow 6, Acid Yellow 23, and Acid Red 18 on activated carbon from flamboyant pods. *Chem. Eng. J.* 181–182, 243–250.
33. B. H. Hameed, 2009. Grass waste: A novel sorbent for the removal of basic dye from aqueous solution. *J. Hazard. Mat.* 166, 233–238.
34. A. Aygün, S. Yenisoy-Karakas, I. Duman, 2003. Production of granular activated carbon from fruit stones and nutshells and evaluation of their physical, chemical and adsorption properties. *Microporous Mesoporous Mat.* 66, 189–195.
35. S. Karaca, A. Gürses, R. Bayrak, 2004. Effect of some pre-treatments on the adsorption of methylene blue by Balkaya lignite. *Energy Conversion Manag.* 45, 1693–1704.
36. P. K. Malik, 2003. Use of activated carbons prepared from sawdust and rice-husk for adsorption of acid dyes: A case study of acid yellow 36. *Dyes and Pigments* 56, 239–249.
37. B. H. Hameed, 2009a. Evaluation of papaya seeds as a novel non-conventional low-cost adsorbent for removal of methylene blue. *J. Hazard. Mat.* 162, 939–944.
38. R. L. Tseng, F. C. Wu, R. S. Juang, 2003. Liquid-phase adsorption of dyes and phenols using pinewood-based activated carbons. *Carbon* 41, 487–495.
39. T. Calvete, E. C. Lima, N. F. Cardoso, S. L. P. Dias, F. A. Pavan, 2009. Application of carbon adsorbents prepared from Brazilian pine-fruit-shell for the removal of Procion Red MX 3B from aqueous solution – Kinetic, equilibrium, and thermodynamic studies. *Chem. Eng. J.* 155, 627–636.
40. Y. Guo, S. Yang, W. Fu, J. Qi, R. Li, Z. Wang, H. Xu, 2003. Adsorption of malachite green on micro- and mesoporous rice husk-based active carbon. *Dyes and Pigments* 56, 219–229.
41. M. M. Mohamed, 2004. Acid dye removal: Comparison of surfactant-modified mesoporous FSM-16 with activated carbon derived from rice husk. *J. Colloid Int. Sci.* 271, 28–34.
42. V. Vadivelan, K. V. Kumar, 2005. Equilibrium, kinetics, mechanism, and process design for the sorption of methylene blue onto rice husk. *J. Colloid. Int. Sci.* 286, 90–100.
43. N. Kannan, M. M. Sundaram, 2001. Kinetics and mechanism of removal of methylene blue by adsorption on various carbons – A comparative study. *Dyes and Pigments* 51, 25–40.

44. Y. Safa, H. N. Bhatti, 2011. Kinetic and thermodynamic modeling for the removal of Direct Red 31 and Direct Orange 26 dyes from aqueous solutions by rice husk. *Desalination* 272, 313–322.

45. L. Li, S. Liu, T. Zhu, 2010. Application of activated carbon derived from scrap tires for adsorption of Rhodamine B. *J. Environ. Sci.* 22, 1273–1280.

46. M. J. Martin, A. Artola, M. Dolors Balaguer, M. Rigola, 2003. Activated carbons developed from surplus sewage sludge for the removal of dyes from dilute aqueous solutions. *Chem. Eng. J.* 94, 231–239.

47. M. Otero, F. Rozada, L. F. Calvo, A. I. Garcia, A. Moran, 2003. Kinetic and equilibrium modelling of the methylene blue removal from solution by adsorbent materials produced from sewage sludges. *Biochem. Eng. J.* 15, 59–68.

48. W. T. Tsai, C. Y. Chang, M. C. Lin, S. F. Chien, H. F. Sun, M. F. Hsieh, 2001. Adsorption of acid dye onto activated carbon prepared from agricultural waste bagasse by $ZnCl_2$ activation. *Chemosphere* 45, 51–58.

49. A. K. Jain, V. K. Gupta, A. Bhatnagar, Suhas, 2003. Utilization of industrial waste products as adsorbents for the removal of dyes. *J. Hazard. Mat.* B101, 31–42.

50. K. Okada, N. Yamamoto, Y. Kameshima, A. Yasumori, 2003. Adsorption properties of activated carbon from waste newspaper prepared by chemical and physical activation. *J. Colloid Int. Sci.* 262, 194–199.

51. T. Shichi, K. Takagi, 2000. Clay minerals as photochemical reaction fields. *J. Photochem. Photobiol. C: Photochem. Rev.* 1, 113–130.

52. M. S. Ranđelović, M. M. Purenović, B. Z. Matović, A. R. Zarubica, M. Z. Momčilović, J. M. Purenović, 2014. Structural, textural and adsorption characteristics of bentonite-based composite. *Microporous Mesoporous Mat.* 195, 67–74.

53. M. Bagane, S. Guiza, 2000. Removal of a dye from textile effluents by adsorption. *Ann. Chim. Sci. Mat.* 25, 615–626.

54. R. G. Harris, J. D. Wells, B. B. Johnson, 2001. Selective adsorption of dyes and other organic molecules to kaolinite and oxide surfaces. *Colloids Surf. A: Physicochem. Eng. Aspects* 180, 131–140.

55. D. Ghosh, Bhattacharyya K. G., 2002. Adsorption of methylene blue on kaolinite. *Appl. Clay Sci.* 20, 295–300.

56. A. G. Espantaleon, J. A. Nieto, M. Fernandez, A. Marsal, 2003. Use of activated clays in the removal of dyes and surfactants from tannery waste waters. *App. Clay Sci.* 24, 105–110.

57. M. A. Al-Ghouti, M. A. M. Khraisheh, S. J. Allen, M. N. Ahmad, 2003. The removal of dyes from textile wastewater: A study of the physical characteristics and adsorption mechanisms of diatomaceous earth. *J. Environ. Management* 69, 229–238.

58. R. A. Shawabkeh, M. F. Tutunji, 2003. Experimental study and modelling of basic dye sorption by diatomaceous clay. *App. Clay Sci.* 24, 111–120.

59. G. Atun, G. Hisarli, 2003. Adsorption of carminic acid, a dye onto glass powder. *Chem. Eng. J.* 95, 241–249.
60. O. Ozdemir, B. Armagan, M. Turan, M. S. Celik, 2004. Comparison of the adsorption characteristics of azo-reactive dyes on mezoporous minerals. *Dyes and Pigments* 62, 49–60.
61. A. S. Özcan, B. Erdem, A. Özcan, 2004. Adsorption of acid blue 193 from aqueous solutions onto Na-bentonite and DTMA-bentonite. *J. Colloid Int. Sci.* 280, 44–54.
62. W. T. Tsai, H. C. Hsu, T. Y. Su, K. Y. Lin, C. M. Lin, 2008. Removal of basic dye (methylene blue) from wastewaters utilizing beer brewery waste. *J. Hazard. Mat.* 154, 73–78.
63. O. Demirbas, M. Alkan, 2013. Removal of an anionic dye from aqueous solution by sepiolite using a full factorial experimental design. *Fresenius Environ. Bull.* 22, 3501–3510.
64. I. Feddal, A. Ramdani, S. Taleb, E. M. Gaigneaux, N. Batis, N. Ghaffour, 2014. Adsorption capacity of methylene blue, an organic pollutant, by montmorillonite clay. *Desalination and Water Treatment* 52, 2654–2661.
65. A. Vanaamudan, N. Pathan, P. Pamidimukkala, 2014. Adsorption of Reactive Blue 21 from aqueous solutions onto clay, activated clay, and modified clay. *Desalination and Water Treatment* 52, 1589–1599.
66. K. Bellir, I. Sadok Bouziane, Z. Boutamine, M. Bencheikh Lehocine, A. H. Meniai, 2012. Sorption study of a basic dye Gentian Violet from aqueous solutions using activated bentonite. *Energy Procedia* 18, 924–933.
67. Y. El Mouzdahir, A. Elmchaouri, R. Mahboub, A. Gil, S. A. Korili, 2010. Equilibrium modeling for the adsorption of methylene blue from aqueous solutions on activated clay minerals. *Desalination* 250, 335–338.
68. Y. S. Ho, C. C. Chiang, Y. C. Hsu, 2001. Sorption kinetics for dye removal from aqueous solution using activated clay. *Sep. Sci. Technol.* 36, 2473–2488.
69. S. Qiao, Q. Hu, F. Haghseresht, X. Hu, G. Q. Lu, 2009. An investigation on the adsorption of acid dyes on bentonite based composite adsorbent. *Sep. Purif. Technol.* 67, 218–225.
70. G. M. Walker, L. Hansen, J. A. Hanna, S. J. Allen, 2003. Kinetics of a reactive dye adsorption onto dolomitic sorbents. *Water Res.* 37, 2081–2089.
71. A. Gürses, Ç. Dogar, M. Yalçin, M. Açikyildiz, R. Bayrak, S. Karaca, 2006. The adsorption kinetics of the cationic dye, methylene blue, onto clay. *J. Hazard. Mat.* 131, 217–228.
72. A. B. Karim, B. Mounir, M. Hackar, M. Bakasse, A. Yaacoubi, 2009. Removal of Basic Red 46 dye from aqueous solution by adsorption onto Moroccan clay. *J. Hazard. Mat.* 168, 304–309.
73. D. S. Duc, N. V. Noi, D. Q. Trung, V. T. Quyen, V. Y. Ninh, 2012. Adsorption Basic Red 46 onto activated carbon. *Res. J. Chem. Environ.* 16, 169–173.
74. T. B. Iyim, G. Güçlü, 2009. Removal of basic dyes from aqueous solutions using natural clay. *Desalination* 249, 1377–1379.

75. A. Gürses, S. Karaca, C. Dogar, R. Bayrak, M. Acikyildiz, M. Yalcin, 2004. Determination of adsorptive properties of clay/water system: Methylene blue sorption. *J. Colloid Int. Sci.* 269, 310–314.

76. Y. S. Ho, C. C. Chiang, 2001. Sorption studies of acid dye by mixed sorbents. *Adsorption* 7, 139–147.

77. N. K. Lazaridis, T. D. Karapantsios, D. Geogantas, 2003. Kinetic analysis for the removal of a reactive dye from aqueous solution onto hydrotalcite by adsorption. *Water Res.* 37, 3023–3033.

78. B. K. Nandi, A. Goswami, M. K. Purkait, 2009. Adsorption characteristics of brilliant green dye on kaolin. *J. Hazard. Mat.* 161, 387–395.

79. K. Rida, S. Bouraoui, S. Hadnine, 2013. Adsorption of methylene blue from aqueous solution by kaolin. *Appl. Clay Sci.* 83–84, 99–105.

80. V. Vimonses, S. Lei, B. Jin, C. W. K. Chow, C. Saint, 2009. Adsorption of Congo Red by three Australian kaolins. *Appl. Clay Sci.* 43, 465–472.

81. T. A. Khan, S. Dahiya, I. Ali, 2012. Use of kaolinite as adsorbent: Equilibrium, dynamics and thermodynamic strudies on the adsorption of Rhodamine B from aqueous solution. *Appl. Clay Sci.* 69, 58–66.

82. C. A. P. Almeida, N. A. Debacher, A. J. Downs, L. Cottet, C. Mello, 2009. Removal of methylene blue from colored effluents by adsorption on montmorillonite clay. *J. Colloid Int. Sci.* 332, 46–53.

83. S. C. R. Santos, R. A. R. Boaventura, 2008. Adsorption modeling of textile dyes by sepiolite. *App. Clay Sci.* 42, 137–145.

84. I. Küncek, S. Sener, 2010. Adsorption of methylene blue onto sonicated sepiolite from aqueous solutions. *Ultrasonics Sonochem.* 17, 250–257.

85. Z. Al-Qodah, W. K. Lafi, Z. Al-Anber, M. Al-Shannag, A. Harahsheh, 2007. Adsorption of methylene blue by acid and heat treated diatomaceous silica. *Desalination* 217, 212–224.

86. M. Greluk, 2013. Sorption of acid and reactive dyes from aqueous solutions on minerals. *Przemysl Chemiczny* 92, 646–654.

87. S. J. Allen, B. Koumanova, 2005. Decolourisation of water/wastewater using adsorption (review). *J. Univ. Chem. Technol. Metallurgy* 40, 175–192.

88. M. Dogan, M. Alkan, A. Türkyilmaz, Y. Özdemir, 2004. Kinetics and mechanism of removal of methylene blue by adsorption onto perlite. *J. Hazard. Mat.* B109, 141–148.

89. M. Dogan, M. Alkan, Y. Onager, 2000. Adsorption of methylene blue from aqueous solution onto perlite. *Water Air Soil Pollut.* 120, 229–248.

90. M. Dogan, M. Alkan, 2003. Removal of methyl violet from aqueous solution by perlite. *J. Colloid Int. Sci.* 267, 32–41.

91. M. Dogan, M. Alkan, 2003. Adsorption kinetics of methyl violet onto perlite. *Chemosphere* 50, 517–528.

92. O. Demirbas, M. Alkan, M. Dogan, 2002. The removal of Victoria blue from aqueous solution by adsorption on a low-cost material. *Adsorption* 8, 341–349.

93. M. Özacar, A. I. Sengil, 2002. Adsorption of acid dyes from aqueous solutions by calcined alunite and granular activated carbon. *Adsorption* 8, 301–308.

94. M. Özacar, A. I. Sengil, 2003. Adsorption of reactive dyes on calcined alunite from aqueous solutions. *J. Hazard. Mat.* B98, 211–224.

95. I. H. Abd El Maksod, H. M. AlBishri, A. S. Al-bogami, 2014. Saudi Arabian white silica as a good adsorbent for pollutants. *Clean Soil Water* 42, 480–486.

96. A. Krysztafkiewicz, S. Binkowski, T. Jesionowski, 2002. Adsorption of dyes on a silica surface. *Appl. Surf. Sci.* 199, 31–39.

97. G. Crini, M. Morcellet, 2002. Synthesis and applications of adsorbents containing cyclodextrins. *J. Sep. Sci.* 25, 1–25.

98. C. D. Woolard, J. Strong, C. R. Erasmus, 2002. Evaluation of the use of modified coal ash as a potential sorbent for organic waste streams. *Appl. Geochem.* 17, 1159–1164.

99. T. N. T. Phan, M. Bacquet, M. Morcellet, 2000. Synthesis and characterization of silica gels functionalized with monochlorotriazinyl beta-cyclodextrin and their sorption capacities towards organic compounds. *J. Inclusion Phenom. Macrocyclic Chem.* 38, 345–359.

100. M. N. Ahmed, R. N. Ram, 1992. Removal of basic dye from wastewater using silica as adsorbent. *Environ. Pollut.* 77, 79–86.

101. D. Rodrigues, T. A. P. Rocha-Santos, A. C. Freitas, A. M. P. Gomes, A. C. Duarte, 2013. Strategies based on silica monoliths for removing pollutants from wastewater effluents: A Review. *Sci. Total Environ.* 461–462, 126–138.

102. S. T. Akar, T. Alp, D. Yilmazer, 2013. Enhanced adsorption of Acid Red 88 by an excellent adsorbent prepared from alunite. *J. Chem. Technol. Biotechnol.* 88, 293–304.

103. H. Ghobarkar, O. Schäf, U. Guth, 1999. Zeolites – From kitchen to space. *Prog. Solid St. Chem.* 27, 29–73.

104. S. Wang, Y. Peng, 2010. Natural zeolites as effective adsorbents in water and wastewater treatment. *Chem. Eng. J.* 156, 11–24.

105. G. Calzaferri, D. Brühwiler, S. Megelski, M. Pfenniger, M. Pauchard, B. Hennessy, H. Maas, A. Devaux, A. Graf, 2000. Playing with dye molecules at the inner and outer surface of zeolite L. *Solid States Sci.* 2, 421–447.

106. O. Altin, H. O. Ozbelge, T. Dogu, 1998. Use of general purpose adsorption isotherms for heavy metal-clay mineral interactions. *J. Colloid Int. Sci.* 198, 130–140.

107. S. G. Liu, Y. Q. Ding, P. F. Li, K. S. Diao, X. C. Tan, F. H. Lei, Y. H. Zhan, Q. M. Lib, B. Huang, Z. Y. Huang, 2014. Adsorption of the anionic dye Congo red from aqueous solution onto natural zeolites modified with N,N-dimethyl dehydroabietylamine oxide. *Chem. Eng. J.* 248, 135–144.

108. M. Canli, Y. Abali, S. U. Bayca, 2013. Removal of methylene blue by natural and Ca and K-exchanged zeolite treated with hydrogen peroxide. *Physicochem. Problems Mineral Proc.* 49, 481–496.

109. X. Jin, M. Jiang, X. Shan, Z. Pei, Z. Chen, 2008. Adsorption of methylene blue and orange II onto unmodified and surfactant-modified zeolite. *J. Colloid Int. Sci.* 328, 243–247.

110. M. Korkmaz, C. Ozmetin, B. A. Fil, E. Ozmetin, Y. Yasar, 2013. Methyl violet dye adsorption onto clinoptilolite (natural zeolite): Isotherm and kinetic study. *Fresenius Environ. Bull.* 22, 1524–1533.

111. V. Meshko, L. Markovska, M. Mincheva, A. E. Rodrigues, 2001. Adsorption of basic dyes on granular activated carbon and natural zeolite. *Water Res.* 35, 3357–3366.

112. S. Rangabhashiyam, N. Anu, N. Selvaraju, 2013. Sequestration of dye textile industry wastewater using agricultural waste products as adsorbents. *J. Environ. Chem. Eng.* 1, 629–641.

113. A. Shukla, Y. H. Zhang, P. Dubey, J. L. Margrave, S. S. Shukla, 2002. The role of sawdust in the removal of unwanted materials from water. *J. Hazard. Mat.* B95, 137–152.

114. M. H. V. Baouab, R. Gauthier, H. Gauthier, M. E. B. Rammah, 2001. Cationized sawdust as ion exchanger for anionic residual dyes. *J. Appl. Polym. Sci.* 82, 31–37.

115. V. K. Garg, M. Amita, R. Kumar, R. Gupta, 2004. Basic dye (methylene blue) removal from simulated wastewater by adsorption using Indian Rosewood sawdust: A timber industry waste. *Dyes and Pigments* 63, 243–250.

116. V. K. Garg, R. Kumar, R. Gupta, 2004. Removal of malachite green dye from aqueous solution by adsorption using agro-industry waste: A case study of Prosopis cineraria. *Dyes and Pigments* 62, 1–10.

117. M. Özacar, A. I. Sengil, 2005. Adsorption of metal complex dyes from aqueous solutions by pine sawdust. *Bioresour. Technol.* 96, 791–795.

118. V. K. Garg, R. Gupta, A. B. Yadav, R. Kumar, 2003. Dye removal from aqueous solution by adsorption on treated sawdust. *Bioresour. Technol.* 89, 121–124.

119. F. A. Batzias, D. K. Sidiras, 2004. Dye adsorption by calcium chloride treated beech sawdust in batch and fixed-bed systems. *J. Hazard. Mat.* B114, 167–174.

120. G. Palma, J. Freer, J. Baeza, 2003. Removal of metal ions by modified Pinus radiate bark and tannins from water solutions. *Water Res.* 37, 4974–4980.

121. S. E. Bailey, T. J. Olin, M. Bricka, D. D. Adrian, 1999. A review of potentially low-cost sorbents for heavy metals. *Water Res.* 33, 2469–2479.

122. L. C. Morais, O. M. Freitas, E. P. Gonçalves, L. T. Vasconcelos, C. G. Gonzalez Beça, 1999. Reactive dyes removal from wastewaters by adsorption on eucalyptus bark: Variables that define the process. *Water Res.* 33, 979–988.

123. Y. S. Ho, T. H. Chiang, Y. M. Hsueh, 2005. Removal of basic dye from aqueous solutions using tree fern as a biosorbent. *Process Biochem.* 40, 119–124.

124. G. Annadurai, R. S. Juang, D. J. Lee, 2002. Use of cellulose-based wastes for adsorption of dyes from aqueous solutions. *J. Hazard. Mat.* B92, 263–274.

125. C. Namasivayam, D. Prabha, M. Kumutha, 1998. Removal of direct red and acid brilliant blue by adsorption on to banana pith. *Bioresour. Technol.* 64, 77–79.

126. B. H. Hameed, M. I. El-Khaiary, 2008. Removal of basic dye from aqueous medium using a novel agricultural waste material: Pumpkin seed hull. *J. Hazard. Mat.* 155, 601–609.

127. B. H. Hameed, D. K. Mahmoud, A. L. Ahmad, 2008. Equilibrium modeling and kinetic studies on the adsorption of basic dye by a low-cost adsorbent: Coconut (Cocos nucifera) bunch waste. *J. Hazard. Mat.* 158, 65–72.

128. M. M. R. Khan, M. Ray, A. K. Guha, 2011. Mechanistic studies on the binding of Acid Yellow 99 on coir pith. *Bioresour. Technol.* 102, 2394–2399.

129. C. Namasivayam, R. Radhika, S. Suba, 2001. Uptake of dyes by a promising locally available agricultural solid waste: Coir pith. *Waste Management* 21, 381–387.

130. M. V. Sureshkumar, C. Namasivayam, 2008. Adsorption behavior of Direct Red 12B and Rhodamine B from water onto surfactant-modified coconut coir pith. *Colloids Surf. A: Physicochem. Eng. Aspects* 317, 277–283.

131. Y. S. Ho, G. McKay, 2003. Sorption of dyes and copper ions onto biosorbents. *Process Biochem.* 38, 1047–1061.

132. B. Chen, C. W. Hui, G. McKay, 2001. Film-pore diffusion modelling and contact time optimisation for the adsorption of dyestuffs on pith. *Chem. Eng. J.* 84, 77–94.

133. S. D. Khattri, M. K. Singh, 2000. Colour removal from synthetic dye wastewater using a bioadsorbent. *Water Air Soil Pollut.* 120, 283–294.

134. B. H. Hameed, M. I. El-Khaiary, 2008. Sorption kinetics and isotherm studies of a cationic dye using agricultural waste: Broad bean peels. *J. Hazard. Mat.* 154, 639–648.

135. S. Rajeshwari, C. Namasivayam, K. Kadirvelu, 2001. Orange peel as an adsorbent in the removal of acid violet 17 (acid dye) from aqueous solutions. *Waste Manag.* 21, 105–110.

136. M. M. Nassar, Y. H. Magdy, 1997. Removal of different basic dyes from aqueous solutions by adsorption on palm-fruit bunch particles. *Chem. Eng. J.* 66, 223–226.

137. F. Deniz, S. Karaman, 2011. Removal of Basic Red 46 dye from aqueous solution by pine tree leaves. *Chem. Eng. J.* 170, 67–74.

138. B. H. Hameed, M. I. El-Khaiary, 2008. Batch removal of malachite green from aqueous solutions by adsorption on oil palm trunk fibre: Equilibrium isotherms and kinetic studies. *J. Hazard. Mat.* 154, 237–244.

139. K. S. Chou, J. C. Tsai, C. T. Lo, 2001. The adsorption of Congo Red and vacuum pump oil by rice hull ash. *Bioresour. Technol.* 78, 217–219.

140. Z. Aksu, I. A. Isoglu, 2006. Use of agricultural waste sugar beet pulp for the removal of Gemazol turquoise blue-G reactive dye from aqueous solution. *J. Hazard. Mat.* 137, 418–430.

141. S. D. Khattri, M. K. Singh, 1999. Colour removal from dye wastewater using sugar cane dust as an adsorbent. *Adsorption Sci. Technol.* 17, 269–282.

142. Y. H. Magdy, A. A. M. Daifullah, 1998. Adsorption of a basic dye from aqueous solutions onto sugar-industry mud. *Waste Management* 18, 219–226.

143. H. Lata, S. Mor, V. K. Garg, R. K. Gupta, 2008. Removal of a dye from simu-
     lated wastewater by adsorption using treated parthenium biomass. *J. Hazard.
     Mat.* 153, 213–220.

144. S. J. Allen, Q. Gan, R. Matthews, P. A. Johnson, 2003. Comparison of opti-
     mised isotherm models for basic dye adsorption by kudzu. *Bioresour. Technol.*
     88, 143–152.

145. Y. S. Ho, G. McKay, 1998. Kinetic models for the sorption of dye from aque-
     ous solution by wood. *Trans. Institution Chem. Eng.* 76, 183–191.

146. S. Netpradit, P. Thiravetyan, S. Towprayoon, 2004. Adsorption of three azo
     reactive dyes by metal hydroxide sludge: Effect of temperature, pH and elec-
     trolytes. *J. Colloid Int. Sci.* 270, 255–261.

147. S. Netpradit, P. Thiravetyan, S. Towprayoon, 2004. Evaluation of metal
     hydroxide sludge for reactive dye adsorption in a fixed-bed column system.
     Water Res. 38, 71–78.

148. S. Netpradit, P. Thiravetyan, S. Towprayoon, 2003. Application of "waste"
     metal hydroxide sludge for adsorption of azo reactive dyes. *Water Res.* 37,
     763–772.

149. P. Janos, H. Buchtova, M. Ryznarova, 2003. Sorption of dyes from aqueous
     solutions onto fly ash. *Water Res.* 37, 4938–4944.

150. V. K. Gupta, D. Mohan, S. Sharma, M. Sharma, 2000. Removal of basic dye
     (Rhodamine B and Methylene blue) from aqueous solutions using bagasse
     fly ash. *Sep. Sci. Technol.* 35, 2097–2113.

151. S. Wang, Y. Boyjoo, A. A. Choueib, 2005. Comparative study of dye removal
     using fly ash treated by different methods. *Chemosphere* 60, 1401–1407.

152. B. Acemioglu, 2004. Adsorption of Congo red from aqueous solution onto
     calcium-rich fly ash. *J. Colloid Int. Sci.* 274, 371–379.

153. S. K. Khare, K. K. Panday, R. M. Srivastava, V. N. Singh, 1987. Removal of
     victoria blue from aqueous solution by fly ash. *J. Chem. Technol. Biotechnol.*
     38, 99–104.

154. Y. S. Ho, G. McKay, 1999. Comparative sorption kinetic studies of
     dye and aromatic compounds onto fly ash. *J. Environ. Sci. Health* A34,
     1179–1204.

155. S. Wang, Y. Boyjoo, A. Choueib, Z. H. Zhu, 2005. Removal of dyes from
     aqueous solution using fly ash and red mud. *Water Res.* 39, 129–138.

156. C. Namasivayam, D. J. S. E. Arasi, 1997. Removal of Congo red from waste-
     water by adsorption onto red mud. *Chemosphere* 34, 401–471.

157. C. Namasivayam, B. Chandrasekaran, 1991. Treatment of dyeing industry
     wastewaters using waste Fe (II)/Cr (III) sludge and red mud. *J. Indian Assoc.
     Environ. Manag.* 18, 93–99.

158. L. C. A. Oliveira, M. Gonçalves, D. Q. L. Oliveira, M. C. Guerreiro, L. R. G.
     Guilherme, R. M. Dallago, 2007. Solid waste from leather industry as sorbent
     of organic dyes in aqueous-medium. *J. Hazard. Mat.* 141, 344–347.

159. Z. Mi-Na, L. Xue-Pin, S. Bi, 2005. Adsorption of surfactants on chromium
     leather waste. *J. Soc. Leather Technol. Chemists* 90, 1–6.

160. K. C. de Souza, M. L. P. Antunes, F. T. da Conceicao, 2013. Adsorption of Reactive Blue 19 dye in aqueous solution by red mud chemically treated with hydrogen peroxide. *Quimica Nova* 36, 651–656.

161. M. Shirzad-Siboni, S. J. Jafari, O. Giahi, I. Kim, S. M. Lee, J. K. Yang, 2014. Removal of acid blue 113 and reactive black 5 dye from aqueous solutions by activated red mud. *J. Ind. Eng. Chem.* 20, 1432–1437.

162. A. Olgun, N. Atar, 2009. Equilibrium kinetic adsorption study of Basic Yellow 28 and Basic Red 46 by a boron industry waste. *J. Hazard. Mat.* 161, 148–156.

163. C. Lee, K. Low, S. Chow, 1996. Chrome sludge as an adsorbent for color removal. *Environ. Technol.* 54, 183–189.

164. C. Namasivayam, S. Sumithra, 2005. Removal of direct red 12B and methylene blue from water by adsorption onto Fe (III)/Cr (III) hydroxide, an industrial solid waste. *J. Environ. Manag.* 74, 207–215.

165. S. H. Chang, K. S. Wang, H. C. Li, M. Y. Wey, J. D. Chou, 2009. Enhancement of Rhodamine B removal by low-cost fly ash sorption with Fenton pre-oxidation. *J. Hazard. Mat.* 172, 1131–1136.

166. K. V. Kumar, V. Ramamurthi, S. Sivanesan, 2005. Modeling the mechanism involved during the sorption of methylene blue onto fly ash. *J. Colloid. Int. Sci.* 284, 14–21.

167. T. Viraraghavan, K. R. Ramakrishna, 1999. Fly ash for colour removal from synthetic dye solutions. *Water Qual. Res. J. Can.* 34, 505–517.

168. S. C. R. Santos, V. J. P. Vilar, R. A. R. Boaventura, 2008. Waste metal hydroxide sludge as sorbent for a reactive dye. *J. Hazard. Mat.* 153, 999–1008.

169. M. A. A. Zaini, T. Y. Cher, M. Zakaria, M. J. Kamaruddin, S. H. M. Setapar, M. A. C. Yunus, 2014. Palm oil mill effluent sludge ash as adsorbent for methylene blue dye removal. *Desalination and Water Treatment* 52, 3654–3662.

170. A. E. Yilmaz, R. Boncukcuoglu, M. Kocakerim, I. H. Karakas, 2011. Waste utilization: The removal of textile dye (Bomaplex Red CR-L) from aqueous solution on sludge waste from electrocoagulation as adsorbent. *Desalination* 277, 156–163.

171. N. N. Bambalov, 2012. Use of peat as an organic raw material for chemical processing. *Solid Fuel Chem.* 46, 282–288.

172. P. A. Brown, S. A. Gill, S. J. Allen, 2000. Metal removal from wastewater using peat. *Water Res.* 34, 3907–3916.

173. A. N. Fernandes, C. A. P. Almeida, N. A. Debacher, M. M. de Souza Sierra, 2010. Isotherm and thermodynamic data of adsorption of methylene blue from aqueous solution onto peat. *J. Molecular Structure* 982, 62–65.

174. S. J. Allen, G. McKay, J. F. Porter, 2004. Adsorption isotherm models for basic dye adsorption by peat in single and binary component systems. *J. Colloid Int. Sci.* 280, 322–333.

175. S. J. Allen, M. Murray, P. Brown, O. Flynn, 1994. Peat as an adsorbent for dye-stuffs and metals in wastewater. *Resources Conservation Recycling* 11, 25–39.

176. S. J. Allen, G. McKay, K. Y. H. Khader, 1988. The adsorption of acid dye onto peat from aqueous solution-solid diffusion model. *J. Colloid Int. Sci.* 126, 517–524.

177. S. J. Allen, G. McKay, K. Y. H. Khader, 1988. Multi-component sorption isotherms of basic dyes onto peat. *Environ. Poll.* 52, 39–53.

178. Y. S. Ho, G. McKay, 1998. Kinetic models for the sorption of dye from aqueous solution by wood. *Trans. Institution Chem. Eng.* 76, 183–191.

179. Q. Sun, L. Yang, 2003. The adsorption of basic dyes from aqueous solution on modified peat-resin particle. *Water Res.* 37, 1535–1544.

180. D. Couillard, 1994. The use of peat in wastewater treatment. *Water Res.* 28, 1261–1274.

181. E. F. Smith, P. McCarthy, T. C. Yu, H. B. Mark Jr, 1977. Sulfuric acid treatment of peat for cation exchange. *Chem. Eng. J.* 49, 633–638.

182. X. Vecino, R. Devesa-Rey, J. M. Cruz, A. B. Moldes, 2013. Entrapped peat in alginate beads as green adsorbent for the elimination of dye compounds from vinasses. *Water Air Soil Pollut.* 224, 1448–1456.

183. L. A. Sepulveda, C. C. Santana, 2013. Effect of solution temperature, pH and ionic strength on dye adsorption onto peat. *Environ. Technol.* 34, 967–977.

184. Y. S. Ho, G. McKay, 1998. Sorption of dye from aqueous solution by peat. *Chem. Eng. J.* 70, 115–124.

185. L. Rusu, M. Harja, A. I. Simion, D. Suteu, G. Ciobanu, L. Favier, 2014. Removal of Astrazone Blue from aqueous solutions onto brown peat. Equilibrium and kinetics studies. *Korean J. Chem. Eng.* 31, 1008–1015.

186. A. W. M. Ip, J. P. Barford, G. McKay, 2009. Reactive Black dye sorption/desorption onto different sorbents: Effect of salt, surface chemistry, pore size and surface area. *J. Colloid Int. Sci.* 337, 32–38.

187. G. Z. Kyzas, M. Kostoglou, 2014. Green adsorbents for wastewaters: A critical review. *Materials* 7, 333–364.

188. B. J. Liu, D. F. Wang, G. L. Yu, X. H. Meng, 2013. Adsorption of heavy metal ions, dyes and proteins by chitosan composites and derivatives – A review. *J. Ocean Univ. China* 12, 500–508.

189. M. Borgogna, B. Bellich, A. Cesaro, 2011. Marine polysaccharides in micro-encapsulation and application to aquaculture: From sea to sea. *Marine Drugs* 9, 2572–2604.

190. G. Crini, P. M. Badot, 2008. Application of chitosan, a natural aminopolysaccharide, for dye removal from aqueous solutions by sorption processes using batch studies: A review of recent literature. *Prog. Polym. Sci.* 33, 399–447.

191. E. Guibal, 2004. Interactions of metal ions with chitosan-based sorbents: A review. *Sep. Purif. Technol.* 38, 43–74.

192. A. J. Varma, S. V. Deshpande, J. F. Kennedy, 2004. Metal complexation by chitosan and its derivative: A review. *Carbohydr. Polym.* 55, 77–93.

193. M. N. V. Ravi Kumar, 2000. A review of chitin and chitosan applications. *React. Funct. Polym.* 46, 1–27.

194. J. I. Houghton, J. Quarmby, 1999. Biopolymers in wastewater treatment. *Current Opinion Biotechnol.* 10, 259–262.

195. F. C. Wu, R. L. Tseng, R. S. Juang, 2000. Comparative adsorption of metal and dye on flake- and bead-types of chitosan prepared from fishery wastes. *J. Hazard. Mat.* B73, 63–75.

196. M. S. Chiou, H. Y. Li, 2003. Adsorption behaviour of reactive dye in aqueous solution on chemical cross-linked chitosan beads. *Chemosphere* 50, 1095–1105.

197. M. S. Chiou, P. Y. Ho, H. Y. Li, 2004. Adsorption of anionic dyes in acid solutions using chemically cross-linked chitosan beads. *Dyes and Pigments* 60, 69–84.

198. W. H. Cheung, Y. S. Szeto, G. McKay, 2007. Intraparticle diffusion processes during acid dye adsorption onto chitosan. *Bioresour. Technol.* 98, 2897–2904.

199. Y. C. Wong, Y. S. Szeto, W. H. Cheung, G. McKay, 2004. Adsorption of acid dyes on chitosan-equilibrium isotherm analyses. *Proc. Biochem.* 39, 693–702.

200. G. Annadurai, L. Y. Ling, J. F. Lee, 2008. Adsorption of reactive dye from an aqueous solution by chitosan: Isotherm, kinetic and thermodynamic analysis. *J. Hazard. Mat.* 152, 337–346.

201. L. Wang, A. Wang, 2008. Adsorption properties of Congo red from aqueous solution onto N, O-carboxymethyl-chitosan. *Bioresour. Technol.* 99, 1403–1408.

202. M. S. Chiou, H. Y. Li, 2002. Equilibrium and kinetic modelling of adsorption of reactive dye on cross-linked chitosan beads. *J. Hazard. Mat.* B93, 233–248.

203. K. Azlan, W. N. Wan Saime, L. Lai Ken, 2009. Chitosan and chemically modified chitosan beads for acid dyes sorption. *J. Environ. Sci.* 21, 296–302.

204. Z. Bekçi, C. Özveri, Y. Seki, K. Yurdakoç, 2008. Sorption of malachite green on chitosan bead. *J. Hazard. Mat.* 154, 254–261.

205. S. Chatterjee, S. Chatterjee, B. P. Chaterjee, A. K. Guha, 2007. Adsorptive removal of Congo Red, a carcinogenic textile dye by chitosan hydrobeads: Binding mechanism, equilibrium and kinetics. *Colloids Surf. A: Physicochem. Eng. Aspects* 299, 146–152.

206. J. S. Piccin, M. L. G. Vieira, J. O. Goncalves, G. L. Dotto, L. A. A. Pinto, 2009. Adsorption of FD&C red N° 40 by chitosan: Isotherm analysis. *J. Food Eng.* 95, 16–20.

207. G. Dotto, L. A. A. Pinto, 2011. Adsorption of food dyes onto chitosan: Optimization process and kinetic. *Carbohydr. Polym.* 84, 231–238.

208. M. S. Chiou, G. S. Chuang, 2006. Competitive adsorption of dye metanil yellow and RB15 in acid solutions on chemically cross-linked chitosan beads. *Chemosphere* 62 731–740.

209. C. A. Demarchi, M. Campos, C. A. Rodrigues, 2013. Adsorption of textile dye Reactive Red 120 by the chitosan-Fe (III)-crosslinked batch and fixed-bed studies. *J. Environ. Chem. Eng.* 1, 1350–1358.

210. M. Hasan, A. L. Ahmad, B. H. Hameed, 2008. Adsorption of reactive dye onto cross-linked chitosan/oil palm ash composite beads. *Chem. Eng. J.* 136, 164–172.

211. S. Rosa, M. C. M. Laranjeira, C. M. Mauro, H. G. Riela, V. T. Favere, 2008. Cross-linked quaternary chitosan as an adsorbent for the removal of the reactive dye from aqueous solutions. *J. Hazard. Mat.* 155, 253–260.

212. G. Z. Kyzas, N. K. Lazaridis, 2009. Reactive and basic dyes removal by sorption onto chitosan derivatives. *J. Colloid Int. Sci.* 331, 32–39.

213. Z. Zhengkun, L. Shiqi, Y. Tianlu, L. Tung-CHing, 2014. Adsorption of food dyes from aqueous solution by glutaraldehyde crosslinked magnetic chitosan nanoparticles. *J. Food Eng.* 126, 133–141.
214. S. Chatterjee, D. S. Lee, M. W. Lee, S. H. Woo, 2009. Enhanced adsorption of Congo Red from aqueous solutions by chitosan hydrogel beads impregnated with cetyl trimethyl ammonium bromide. *Bioresour. Technol.* 100, 2803–2809.
215. G. M. Gadd, 2009. Biosorption: Critical review of scientific rationale, environmentale importance and significance for pollution treatment. *J. Chem. Technol. Biotechnol.* 84, 13–28.
216. I. Michalak, K. Chojnacka, A. Witek-Krowiak, 2013. State of the art for the biosorption Process – A review. *Appl. Biochem. Biotechnol.* 6, 1389–1416.
217. R. Khan, P. Bhawana, M. H. Fulekar, 2013. Microbial decolorization and degradation of synthetic dyes: A review. *Rev. Environ. Sci. Biotechnol.* 12, 75–97.
218. M. D. Chengalroyen, E. R. Dabbs, 2013. The microbial degradation of azo dyes: Minireview. *World J. Microbiol. Biotechnol.* 29, 389–399.
219. M. Solis, A. Solis, H. I. Pérez, N. Manjarrez, M. Flores, 2012. Microbial decolourization of azo dyes: A review. *Proc. Biochem.* 47, 1723–1748.
220. T. Robinson, G. McMullan, R. Marchant, P. Nigam, 2001. Remediation of dyes in textile effluent: A critical review on current treatment technologies with a proposed alternative. *Bioresour. Technol.* 77, 247–255.
221. Y. Fu, T. Viraraghavan, 2001. Fungal decolorization of dye wastewaters: A review. *Bioresour. Technol.* 79, 251–262.
222. M. Bustard, G. McMullan, A. P. McHale, 1998. Biosorption of textile dyes by biomass derived from Klyveromyces marxianus IMB3. *Bioprocess. Eng.* 19, 427–430.
223. P. Nigam, I. M. Banat, D. Singh, R. Marchant, 1996. Microbial process for the decolorization of textile effluent containing azo, diazo and reactive dyes. *Process. Biochem.* 31, 435–442.
224. Y. Fu, T. Viraraghavan, 2002. Removal of Congo red from an aqueous solution by fungus Aspergillus niger. *Adv. Environ. Res.* 7, 239–247.
225. Y. Fu, T. Viraraghavan, 2002. Dye biosorption sites in Aspergillus niger. *Bioresour. Technol.* 82, 139–145.
226. Y. Fu, T. Viraraghavan, 2001. Removal of C. I. acid blue 29 from an aqueous solution by Aspergillus niger. *Am. Assoc. Textile Chem. Colorists Rev.* 1, 36–40.
227. Y. Fu, T. Viraraghavan, 2000. Removal of a dye from a aqueous solution by fungus Aspergillus niger. *Water Qual. Res. J. Canada* 35, 95–111.
228. Z. Aksu, S. Tezer, 2000. Equilibrium and kinetic modeling of biosorption of Remazol Black B by Rhizopus arrhizus in a batch system: Effect of temperature. *Proc. Biochem.* 36, 431–439.
229. P. Waranusantigul, P. Pokethitiyook, M. Kruatrachue, E. S. Upatham, 2003. Kinetic of basic dye (methylene blue) biosorption by giant duckweed (Spirodela polyrrhiza). *Environ. Poll.* 125, 385–392.
230. K. H. Chu, K. M. Chen, 2002. Reuse of activated sludge biomass: I. Removal of basic dyes from wastewater by biomass. *Process Biochem.* 37, 595–600.

231. K. H. Chu, K. M. Chen, 2002. Reuse of activated sludge biomass: II. The rate processes for the adsorption of basic dyes on biomass. *Process Biochem.* 37, 1129–1134.

232. Z. Aksu, G. Dönmez, 2003. A comparative study on the biosorption characteristics of some yeasts for remazol blue reactive dye. *Chemosphere* 50, 1075–1083.

233. Z. Aksu, 2003. Reactive dye bioaccumulation by Saccharomyces cerevisiae. *Proc. Biochem.* 38, 1437–1444.

234. A. K. Mittal, S. K. Gupta, 1996. Biosorption of cationic dyes by dead macro fungus Fomitopsis carnea: Batch studies. *Water Sci. Technol.* 34, 81–87.

235. F. Kargi, S. Ozmihci, 2004. Biosorption performance of powdered activated sludge for removal of different dyestuffs. *Enzyme Microbial Technol.* 35, 267–271.

236. T. O'Mahony, E. Guibal, J. M. Tobin, 2002. Reactive dye biosorption by Rhizopus arrhizus biomass. *Enzyme Microbial Technol.* 31, 456–463.

237. F. Veglio, F. Beolchini, 1997. Removal of heavy metal ions by biosorption: A review. *Hydrometallurgy* 44, 301–316.

238. Z. Aksu, 2001. Biosorption of reactive dyes by dried activated sludge: Equilibrium and kinetic modeling. *Biochem. Eng. J.* 7, 79–84.

239. O. Gulnaz, A. Kaya, F. Matyar, B. Arikan, 2004. Sorption of basic dyes from aqueous solution by activated sludge. *J. Hazard. Mat.* B108, 183–188.

240. O. Gulnaz, A. Kaya, S. Dincer, 2006. The reuse of dried activated sludge for adsorption of reactive dye. *J. Hazard. Mat.* 134, 190–196.

241. V. J. P. Vilar, C. M. S. Bothelho, R. A. R. Boaventura, 2007. Methylene blue adsorption by algal biomass based materials: Biosorbents characterization and process behavior. *J. Hazard. Mat.* 147, 120–132.

242. E. Rubin, P. Rodriguez, R. Herrero, J. Cremades, I. Barbare, M. E. S. Vicente, 2005. Removal of methylene blue from aqueous solutions using as biosorbent Sargassum muticum: an invasive macroalga in Europe. *J. Chem. Tech. Biotechnol.* 80, 291–298.

243. J. Y. Farah, N. S. El-Gendy, L. A. Farahat, 2007. Biosorption of Astrazone Blue basic dye from an aqueous solution using dried biomass of Baker's yeast. *J. Hazard. Mat.* 148, 402–408.

244. M. Kousha, E. Daneshvar, M. S. Sohrabi, M. Jokar, A. Bhatnagar, 2012. Adsorption of acid orange II dye by raw and chemically modified brown macroalga Stoechospermum marginatum. *Chem. Eng. J.* 192, 67–76.

245. Z. Aksu, S. Tezer, 2005. Biosorption of reactive dyes on the green alga Chlorella vulgaris. *Proc. Biochem.* 40, 1347–1361.

246. M. Iqbal, A. Saeed, 2007. Biosorption of reactive dye by loofa sponge-immobilized fungal biomass of Phanerochaete chrysosporium. *Process Biochemistry* 42, 1160–1164.

247. B. Wang, Y. Y. Hu, 2007. Comparison of four supports for adsorption of reactive dyes by immobilized Aspergillus fumigates beads. *J. Environ. Sci.* 19, 451–457.

248. M. Yakup Arica, G. Bayramoglu, 2007. Biosorption of Reactive Red-120 dye from aqueous solution by native and modified fungus biomass preparations of Lentinus sajor-caju. *J. Hazard. Mat.* 149 499–507.

249. P. Punjongharn, K. Meevasana, P. Pavasant, 2008. Influence of particle size and salinity on adsorption of basic dyes by agricultural waste: Dried Seagrape (Caulerpa lentillifera). *J. Environ. Sci.* 20, 760–768.

250. K. S. Low, C. Lee, K. K. Tan, 1995. Biosorption of basic dyes by water hyacinth roots. *Bioresour. Technol.* 52, 79–83.

251. F. Çolak, N. Atar, A. Olgun, 2009. Biosorption of acidic dyes from aqueous solution by Paenibacillus macerans: Kinetic, thermodynamic and equilibrium studies. *Chem. Eng. J.* 150, 122–130.

252. O. Anjaneya, M. Santoshkumar, S. N. Anand, T. B. Karegoudar, 2009. Biosorption of acid violet dye from aqueous solutions using native biomass of a new isolate of Penicillium sp. *International Biodeterioration* 63, 782–787.

253. G. Bayramoglu, G. Çelik, M. Y. Arica, 2006. Biosorption of Reactive Blue 4 dye by native and treated fungus Phanerocheate chrysosporium: Batch and continuous flow system studies. *J. Hazard. Mat.* 137, 1689–1697.

254. S. Cengiz, F. Tanrikulu, S. Aksu, 2012. An alternative source of adsorbent for the removal of dyes from textile waters: Posidonia oceanic (L.). *Chem. Eng. J.* 189–190, 32–40.

255. M. C. Ncibi, B. Mahjoub, A. M. B. Hamissa, R. B. Mansour, M. Seffen, 2009. Biosorption of textile metal-complexed dye from aqueous medium using Posidonia oceanica (L.) leaf sheaths: Mathematical modelling. *Desalination* 243, 109–121.

256. K. Kumari, T. E. Abraham, 2007. Biosorption of anionic textile dyes by nonviable biomass of fungi and yeast. *Bioresour. Technol.* 98, 1704–1710.

257. G. Bayramoglu, M. Y. Arica, 2007. Biosorption of benzidine based textile dyes "Direct Blue 1 and Direct Red 128" using native and heat-treated biomass of Trametes versicolor. *J. Hazard. Mat.* 143, 135–143.

258. R. S. Blackburn, 2004. Natural polysaccharides and their interaction with dye molecules: Applications in effluent treatment. *Environ. Sci. Technol.* 38, 4905–4909.

259. F. Renault, N. Morin-Crini, F. Gimbert, P. M. Badot, G. Crini, 2008. Cationized starch-based material as a new ion-exchanger adsorbent for the removal of C. I. Acid Blue 25 from aqueous solutions. *Bioresour. Technol.* 99, 7573–7586.

260. F. Gimbert, N. Morin-Crini, F. Renault, P. M. Badot, G. Crini, 2008. Adsorption isotherm models for dye removal by cationized starch-based material in a single component system: Error analysis. *J. Hazard. Mat.* 157, 34–46.

261. E. Yilmaz Ozmen, M. Sezgin, A. Yilmaz, M. Yilmaz, 2008. Synthesis of β-cyclodextrin and starch based polymers for sorption of azo dyes from aqueous solutions. *Bioresour. Technol.* 99, 526–531.

262. G. Crini, P. M. Badot, 2010. Starch-based biosorbents for dyes in textile wastewater treatment. *Int. J. Environ. Technol. Manage.* 12, 129–150.

263. A. Riauka, 2006. Binding of anionic dyes by cationic starch. Doctoral Dissertation, Technological Sciences, Kaunas University of Technology, Kaunas, Lithuania.

264. S. Xu, J. Wang, R. Wu, J. Wang, 2006. Effect of degree of substitution on adsorption behavior of Basic Green 4 by highly crosslinked amphoteric starch with quaternary ammonium and carboxyl groups. *Carbohydr. Polym.* 66, 55–59.
265. S. Xu, J. Wang, R. Wu, J. Wang, H. Li, 2006. Adsorption behaviors of acid and basic dyes on crosslinked amphoteric starch. *Chem. Eng. J.* 117, 161–167.
266. J. L. Wang, S. Xu, R. Wu, J. Wang, J. Wei, X. Li, H. Li, 2006. Adsorption behavior of Acid Yellow G by highly-crosslinked amphoteric starch. *J. Polym. Res.* 13, 91–95.
267. F. Delval, G. Crini, J. Vebrel, M. Knorr, G. Sauvin, E. Conte, 2003. Starch-based modified filters used for the removal of dyes from waste water. *Macromolecular Symposia* 203, 165–171.
268. L. Qintie, P. Jianxin, L. Qinlu, L. Qianjun, 2013. Microwave synthesis and adsorption performance of a novel crosslinked starch microsphere. *J. Hazard. Mat.* 263, 517–524.
269. G. Lei, L. Guiying, L. Junshen, M. Yanfeng, T. Yanfeng, 2013. Adsorptive decolorization of methylene blue by crosslinked porous starch. *Carbohydr. Polym.* 93, 374–379.
270. G. Güçlü, S. Keles, 2007. Removal of basic dyes from aqueous solutions using starch-graft-acrylic acid copolymers. *J. Appl. Polym. Sci.* 106, 2422–2426.
271. S. E. Abdel-Aal, Y. H. Gad, A. M. Dessouki, 2006. Use of rice straw and radiation-modified maize starch/acrylonitrile in the treatment of wastewater. *J. Hazard. Mat.* B129, 204–215.
272. G. X. Xing, S. L. Liu, Q. Xu, Q. W. Liu, 2012. Preparation and adsorption behavior for Brilliant Blue X-BR of the cost effective cationic starch intercalated clay composite matrix. *Carbohydr. Polym.* 87, 1447–1452.
273. V. V. Panić, S. I. Šešlija, A. R. Nešić, S. J. Veličković, 2013. Adsorption of azo dyes on polymer materials. *Hemijska Industrija* 67, 881–900.
274. G. Crini, 2008. Kinetic and equilibrium studies on the removal of cationic dyes from aqueous solution by adsorption onto a cyclodextrin polymer. *Dyes and Pigments* 77, 415–426.
275. E. Yilmaz Ozmen, M. Yilmaz, 2007. Use of β-cyclodextrin and starch based polymers for sorption of Congo red from aqueous solutions. *J. Hazard. Mat.* 148, 303–310.
276. I. Bouzaida, M. B. Rammah, 2002. Adsorption of acid dyes on treated cotton in a continuous system. *Mat. Sci. Eng. C* 21, 151–155.
277. Ö. Tunç, H. Tanaci, Z. Aksu, 2009. Potential use of cotton plant wastes for the removal of Remazol Black B reactive dye. *J. Hazard. Mat.* 163, 187–198.
278. D. W. O'Connell, C. Birkinshaw, T. F. O'Dwyer, 2008. Heavy metal adsorbents prepared from the modification of cellulose: A review. *Bioresour. Technol.* 99, 6709–6724.
279. M. A. Hubbe, K. R. Beck, G. O'Neal, Y. C. Sharma, 2012. Cellulosic substrates for removal of pollutants from aqueous systems: A review. 2. *Dyes. BioResources* 7, 2592–2687.

280. V. Rocher, A. Bee, J. M. Siaugue, V. Cabuil, 2010. Dye removal from aqueous solution by magnetic alginate beads crosslinked with epichlorohydrin. *J. Hazard. Mat.* 178, 434–439.

281. V. Rocher, J. M. Siaugue, V. Cabuil, A. Bee, 2008. Removal of organic dyes by magnetic alginate beads. *Water Res.* 42, 1290–1298.

282. R. Aravindhan, N. N. Fathima, J. R. Rao, B. U. Nair, 2007. Equilibrium and thermodynamic studies on the removal of basic black dye using calcium alginate beads. *Colloids Surf. A: Physicochem. Eng. Aspects* 299, 232–238.

283. F. A. Khari, M. Khatibzadeh, N. M. Mahmoodi, K. Gharanjig, 2013. Removal of anionic dyes from aqueous solution by modified alginate. *Desalination and Water Treatment* 51, 2253–2260.

284. B. Gidwani, A. Vyas, 2014. Synthesis, characterization and application of epichlorohydrin-β-cyclodextrin polymer. *Colloids Surf. B: Biointerfaces* 114, 130–137.

285. N. Morin-Crini, G. Crini, 2013. Environmental applications of water-insoluble beta-cyclodextrin-epichlorohydrin polymers. *Prog. Polym. Sci.* 38, 344–368.

286. M. A. Kamboh, I. B. Solangi, S. T. H. Sherazi, S. Memon, 2009. Synthesis and application of calix[4]arene based resin for the removal of azo dyes. *J. Hazard. Mat.* 172, 234–239.

287. D. Zhao, L. Zhao, C. S. Zhu, W. Q. Huang, 2009. Water-insoluble β-cyclodextrin polymer crosslinked by citric acid: Synthesis and sorption properties toward phenol and methylene blue. *J. Incl. Phenom. Macrocyl. Chem.* 63, 195–201.

288. G. Crini, H. N. Peindy, 2006. Sorption of C. I. Basic Blue 9 on cyclodextrin-based material containing carboxylic groups. *Dyes Pigments* 70, 204–211.

289. G. Crini, 2003. Studies of adsorption of dyes on beta-cyclodextrin polymer. *Bioresour. Technol.* 90, 193–198.

290. G. Crini, H. N. Peindy, F. Gimbert, C. Robert, 2007. Removal of C. I. Basic Green 4 (Malachite Green) from aqueous solutions by adsorption using cyclodextrin-based adsorbent: Kinetic and equilibrium studies. *Sep. Purif. Technol.* 53, 97–110.

291. B. Martel, M. Devassine, G. Crini, M. Weltrowski, M. Bourdonneau, M. Morcellet, 2001. Preparation and sorption properties of a beta-cyclodextrin-linked chitosan derivative. *J. Polym. Sci. Part A: Polym. Chem.* 39, 169–176.

292. L. Fan, C. Luo, M. Sun, H. Qiu, W. Li, 2013. Synthesis of magnetic β-cyclodextrin-chitosan/graphene oxide as nanoadsorbent and its application in dye adsorption and removal. *Colloids Surf. B: Biointerfaces* 103, 601–607.

293. Y. M. Zhou, M. Zhang, X. Y. Hu, X. H. Wang, J. Y. Niu, T. S. Ma, 2013. Adsorption of cationic dyes on a cellulose-based multicarboxyl adsorbent. *J. Chem. Eng. Data* 58, 413–421.

294. Y. Zhou, Q. Jin, X. Hu, Q. Zhang, T. Ma, 2012. Heavy metal ions and organic dyes removal from water by cellulose modified with maleic anhydride. *J. Mater. Sci.* 47, 5019–5029.

295. L. S. Silva, L. C. B. Lima, F. C. Silva, J. M. E. Matos, M. R. M. C. Santos, L. S. S. Junior, K. S. Sousa, E. C. da Silva Filho, 2013. Dye anionic sorption in aqueous solution onto a cellulose surface chemically modified with aminoethanethiol. *Chem. Eng. J.* 218, 89–98.

296. N. Morin-Crini, G. Crini, 2012. Review of current physical and chemical technologies for the decontamination of industrial wastewater. *Int. J. Chem. Eng.* 5, 143–186.

11

Hen Feather: A Remarkable Adsorbent for Dye Removal

Alok Mittal and Jyoti Mittal

Department of Chemistry, Maulana Azad National Institute of Technology, Bhopal, India

Abstract

This chapter outlines the role of Hen Feather as a potential adsorbent for the eradication of the hazardous azo class of dyes from wastewater. It first describes the
imperative need for clean water for mankind and the importance of the adsorption technique over other physicochemical methods in the removal of hazardous
pollutants from water.

In the next section of the chapter, an elaborative discussion of Azo dyes, their
uses, importance and toxicity, are described by exhibiting the experimental
results of the adsorption of two important azo dyes, Tartrazine and Amaranth,
over the adsorbent Hen Feather. This section also includes a detailed literature
survey on the physical characteristics of Hen Feather and its use by authors and
various workers as an adsorbent to remove metals, dyes and other pollutants from
wastewater.

The results of the adsorption of the undertaken dyes over Hen Feather include
preliminary studies by monitoring the effects of pH of solution, concentration
of dyes, amount of adsorbent, temperature, etc. The ongoing adsorption was
monitored by Langmuir, Freundlich, Tempkin and D-R adsorption isotherms,
and their detailed results are presented in this section along with various evaluated thermodynamic parameters. Kinetics of the dye–Hen Feather adsorption is
also presented on the basis of rate constant studies and elucidation of reaction
mechanism.

Keywords: Azo dye, hen feather, tartrazine, amaranth, adsorption, isotherm,
kinetics

*Corresponding author: aljymittal@gmail.com

Sanjay K. Sharma (ed.) Green Chemistry for Dyes Removal from Wastewater, (409–457)
© 2015 Scrivener Publishing LLC

11.1   Introduction

Water is the most fathomless and abundant resource on Earth, which has been most lavishly used by human beings. This unusual compound with unique physical properties is also known as "the compound of life." It covers three-fourth of the earth's surface with an average depth of 3000 meters. Despite so much of an available amount, only 1% of it is usable to us because 97% is salty seawater and 2% is frozen in glaciers and polar ice caps. Thus, approximately 1% of the World's water supply is a precious commodity necessary for our survival [1–3]. Despite so many intellectual, scientific and economic advancements, we have done practically nothing about the conservation and purity of water. We have literally polluted the air, land and water as if there was no tomorrow. Then eventually it started affecting our health and we started realizing that the carrying capacity of the natural environment certainly has its limits. In the twenty-first century, as humans have evaluated their relationship with water and the rest of the biosphere, pollution and resource depletion has become an increasing problem [3]. Thus, the importance of water has increased manifold.

Safe, healthy and disease-free drinking water is an essential prerequisite for the existence of humanity. It is therefore not surprising that we have conscientiously guarded the sources of water and over the centuries many skirmishes have taken place over water rights. Experts have already warned of the possibility of a water crisis and with the rapid increase in population this possibility can be foreseen. Hence, careful and apposite management and conservation of natural water is required globally. Although Mother Nature has a great ability to diminish environmental damage, the growing demand for water sources still necessitates exploiting the skills of mankind to maintain their quality as well as quantity. Hence, whatever our needs for industry, agriculture, energy, irrigation, etc., for the growth of civilization, we have to handle water with great care.

Today water pollution has become one of the major environmental problems and the control of water pollution is one of the prime concerns of society. It is a proven fact that about half of the known chemicals are found dissolved in natural water and even a sparkling clear running stream of water may contain complex mixtures of organics and inorganics [4]. The actual cause of water pollution is the recklessness of human activities. For many of us it is easy to dispose of waste by dumping it into a river or lake. The waste disposed of in large or small amounts, dumped intentionally or accidentally, is carried away by the water current, but will never disappear [5–7]. It will emerge downstream, sometimes in a changed form, or just diluted. Mother Nature has a great ability to break down waste materials, but not in the quantities discarded by today's society. The overload that results, called pollution, eventually puts the ecosystem out of balance. Most often our waterways are being polluted by municipal, agricultural and industrial wastes, including many toxic synthetic chemicals, which cannot be broken down at all by natural processes. Thus, the environmental threat posed by contamination in water has to be lessened for a healthy and disease-free life, and for this to happen united efforts have to be made for effectual wastewater treatment and recycling technologies [8,9].

Water pollution is even a greater problem in the Third World, where millions of people obtain water for drinking and sanitation from unprotected streams and ponds that are contaminated with human waste. This type of contamination has been estimated to cause more than three million deaths annually from diarrhea in Third World countries, most of them children. Clearly, the problems associated with water pollution have the capability to disrupt life on our planet to a great extent [3].

It can definitely be said that one of the major causes of water pollution is rapid industrialization. Chief sources contributing to this contamination are mining and smelting, disposal of municipal industrial wastes, use of fertilizers, pesticides, chemicals and automobiles [10]. In view of the volume discharged and effluent composition, the wastewater generated by the textile industries is rated as the most polluting among all industrial sectors. In 1980, Clark and Anliker [11] reported that world production of dyes in 1978 were estimated at 640,000 tons. By now, surely this figure must have grown up to 100 million tons per year. The majority of this quantity is used in the textile and dying industries, which includes many types of different compounds, and their environmental behavior is largely unknown. Industries discharge massive amounts of the most toxic pollutants into the water system. There are various industries like nuclear power projects, petroleum refineries, chemical fertilizers, pesticides, pharmaceuticals, synthetic rubber, etc., that are put in the red category, and the dye industry is also one of them.

Dyes, the most impending material used in the industrial sector as coloring material, are synthetic organic aromatic compounds that are molecularly dispersed and bound to the substrates by intermolecular forces. The textile industry ranks first in the consumption of the dyes and effluents released from textile dyeing, which is intensely colored and poses serious problems to various segments of the environment. The global production of these dyes is about 7×10^5 tons per annum. There are no reliable published statistics on the financial size of the color market, however, on the universal scale a reasonable guesstimate would be 940 million dollars [12]. Zollinger [13] revised the synthesis, properties and applications of organic dyes and pigments and found that about fifteen percent of the total world production of dyes is lost during the dyeing process and is released in the textile effluent. The persisting color and the non-biodegradable, toxic and inhibitory nature of spent dye baths have considerable deleterious effects on the water and soil environment. The presence of coloring stuff in extensive quantities of receiving water not only reduces light access and photosynthetic activity, but also renders its appearance unaesthetic [14]. Therefore, it becomes imperative that color must be removed from dye effluents before disposal of the effluent. Use of colossal quantities of coloring matter (around 1.3 million tons annually), having more than 9000 chemically distinct types of dyestuffs, need increased dogmatic requirements and public awareness. Thus, a comprehensive approach is indisputably needed along with strenuous efforts to solve the problems of color pollution control. Keeping the requirements in mind, in recent years the removal of color from effluents by various types of treatment methodologies has been developed.

Different physical methods like nanofiltration, osmosis, ion exchange, air stripping, etc., have also been employed for treatment, and it has been found that these simply transfer the pollutants to another phase instead of completely eradicating them. Some of the techniques mentioned above have been proven quite effective, although they have various shortcomings. The main drawback of the above technologies is that they generally lack the broad scope of treatment efficiency required to reduce all types of pollutants present in textile wastewater. However, when one approach does look promising, its capital costs or operating costs often become prohibitive when applied to the large water needs common to any industry. It has become a major problem to treat wastewater by an economically and environmentally acceptable approach. Thus, in the field of water treatment, the adsorption process has revealed itself to be a highly competent technique for the removal of toxic pollutants.

The adsorption technique has been rapidly gaining importance as a wastewater treatment process to treat textile effluents. The adsorption process has been found more advantageous for water pollution control as it needs less investment in terms of both initial cost and land. Secondly, the treatment equipment is simply designed and easy to operate. The adsorption process imparts no side effects or toxicity to the water, and this accounts for the superior removal of organic waste constituents as compared to conventional treatments. The adsorption process provides an attractive alternative treatment, especially if the adsorbent is inexpensive and readily available. Textile dye color removal by adsorption onto activated carbon has proven to be highly efficient and reliable [15].

In recent years, adsorption phenomenon has emerged as a most powerful tool for purification and separation. The beauty of the phenomenon is that it is controlled by various parameters, viz. initial adsorbate concentration, pH, adsorbent dose, contact time, particle size of adsorbent, temperature, etc., and this process is operative in most natural, physical, biological and chemical systems [16]. Adsorption operations employing solids such as activated carbon and synthetic resins as adsorbents are used widely in industrial applications and for purification of water and wastewater. In the past few years, several types of adsorbents have been lucratively employed for the removal of a wide range of pollutants including metals, dyes, phenols, insecticides, poly-hydrocarbons, etc., from aqueous solutions. Voluminous literature has piled up in the last two decades on this subject and efforts have been made to keep pace with the latest developments reported in this direction.

In order to reduce the operation cost there has been considerable interest in using low-cost adsorbents for decolorization of wastewater. These materials include chitosan, zeolite, fly ash, coal and oxides, agricultural wastes, lignocellusic wastes, etc. Very limited work has been so far reported, where animal waste material such as hairs, bones, etc., have been used as adsorbent. This laboratory developed Hen Feathers as potential adsorbent and received exceptionally good results, particularly for the removal of dyes from wastewater [17–26]. This chapter is an attempt to exhibit the efficacy of hen feathers to eradicate azo dyes from wastewater. In order to demonstrate the adsorption capabilities of hen feathers, azo dyes are deliberately chosen due to the wide applicability and hazardous nature of azo dyes. This chapter will present the results on the adsorption of two azo dyes, Tartrazine and Amaranth, by Hen Feathers.

low operating costs

11.2 Adsorbate Materials – Azo Dyes

Azo dyes are the largest (more than 50% of all dyes) and most important class of synthetic organic dyes due to their bright colors, excellent color fastness, easy application, chemical stability and versatility. Azo dyes are the class of dyes which are characterized by the presence of one or more azo (–N=N–) groups, which is attached to aromatic ring(s). These dyes possess different chemical structures, which are primarily based on substituted aromatic and heterocyclic groups. The presence of aromatic groups in the structure of the dye accounts for the stable nature of the organic azo molecules. Presence of the steady azo-group and complex aromatic structure makes these dyes resistant to light, acids, bases, oxygen, biological or even chemical degradation [27,28]. This renders azo dyes highly nonde-gradable compounds [29]. Azo dyes, their precursors and biotransformation products such as aromatic amines are toxic [30], carcinogenic [31,32] and mutagenic in nature [33]. Thus, the removal of azo dyes from wastewater becomes highly important.

In the present case, two azo dyes, Tartrazine and Amaranth, have been considered for the removal process. Their important physicochemical properties like molecular weight, physical state, color, melting point, odor, solubility in water, specific gravity and absorption maximum, etc., are presented in Tables 11.1 and 11.2.

11.2.1   Tartrazine

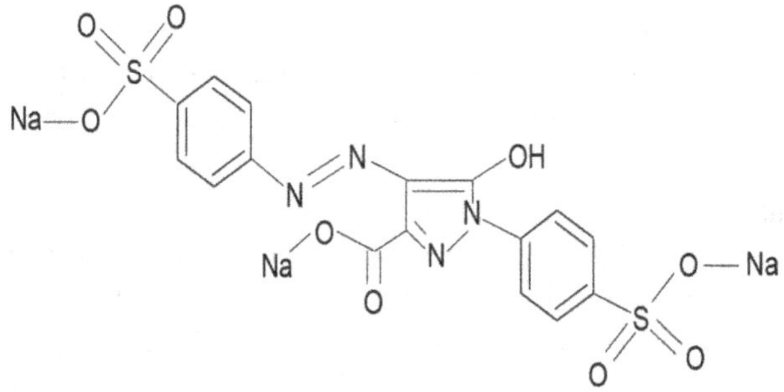

Scheme 11.1 Chemical structure of Azo Dye–Tartrazine.

Tartrazine ($C_{16}H_9N_4Na_3O_9S_2$, molecular weight 534.4; Scheme 11.1), IUPAC name Trisodium-5-hydroxy–1-(4-sulfonatophenyl)-4-(4-sulfona-tophenylazo)-H- pyrazole-3-carboxylate is a coal tar dye belonging to the azo class of dyes (Table 11.1). It has a light orange color and being that it is the least expensive synthetic color it is widely used for the coloring of wool, silk and other textile materials, cosmetics [34] and pharmaceuticals [35,36]. The presence of polar groupings renders it highly water soluble. It is used as food additive in a variety of food materials [37–40]. It is also used extensively in laboratories as either biological stains or pH indicators. The wide applicability of the dye, especially in foodstuffs, has initiated global interest regarding its toxicological impact on living systems and the environment.

Various review articles have appeared in the literature describing Tartrazine as an initiator of allergies, asthma [41,42], and dermal diseases in humans [43,44], and corneal staining of the eyes [45]. Many publications have appeared highlighting the effects of Tartrazine on the lungs, leading

Table 11.1 Important facts about Azo Dye–Tartrazine.

Synonyms	Acid Yellow 23, Filter Yellow, Food Yellow, Food Yellow # 4
Chemical Formula	$C_{16}H_9N_4Na_3O_9S_2$
IUPAC Name	Trisodium-5-hydroxy-1-(4-sulfonatophenyl)-4-(4-sulfonato phenylazo)-H- pyrazole-3-carboxylate
Color Index	19140
CAS Number	1934–21–0
Molecular Weight	534.4
Physical State	Solid
Color	Solid Granular, Light Orange
Melting Point	> 300°C (Decomposes at 250°C)
Odor	Odorless
Solubility in Water	Soluble (140 g/l)
Absorption Maxima	426 nm
Specific Gravity	1.85–1.95

Table 11.2 Important facts about Azo Dye–Amaranth.

Synonyms	Acid Red 27, Food Red 9, FD & C Red 2, Azorubin S
Chemical Formula	$C_{20}H_{11}N_2Na_3O_{10}S_3$
IUPAC Name	Trisodium 3-hydroxy-4-(4-sulfonato-1-naphthylazo)-2,7-naphthalene disulfonate
Color Index	16185
CAS Number	915–67–3
Molecular Weight	604.48
Physical State	Solid
Color	Reddish – Brown Powder
Melting Point	Decomposes at 120°C (Without Melting)
Odor	Odorless
Solubility in Water	Soluble (50 g/L)
Absorption Maxima	520 nm
Specific Gravity	1.5

to tuberculosis [46,47]. The dye is also reported to cause hypersensitivity reactions in humans [48].

Tartrazine is also well known for its carcinogenicity and mutagenic characteristics [49]. Chung [50] established that the use of Tartrazine increases the chances of intestinal cancer, which is due to reduction of dyes to aromatic amines by intestinal microflora. The carcinogenicity of the dye has been reported in the publication of Stefanidou *et al.* [51], where it is established that Tartrazine causes a significant increase in DNA contents which stimulates the mitotic division of cells in the living system.

Sobotka and coworker [52] found that Tartrazine affects the postnatal development of the central nervous system of female offspring of mice, marked by depressed body weight, an apparent reduction in thymus weight and a slight elevation of red blood cells and hemoglobin. The effects of a Tartrazine diet on the growth and survival of rats have been studied by Ershoff [53], and it was found that the dye causes noticeable retardation in growth and death of 50% or more of the rats within an experimental period of 14 days. Tartrazine also shows a significant increase in chromosomal aberrations at higher concentrations [54].

11.2.2   Amaranth

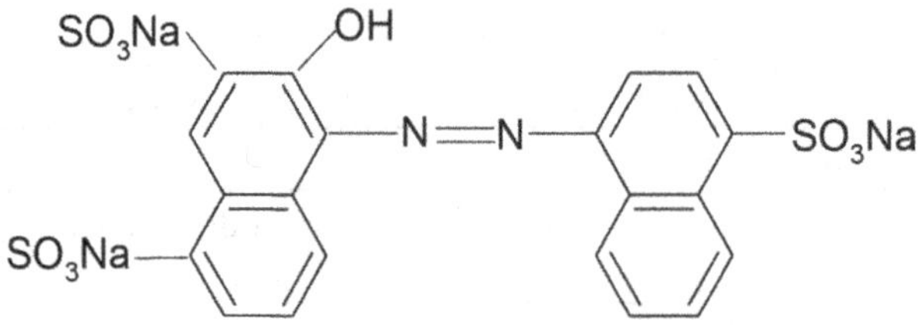

Scheme 11.2 Chemical structure of Azo Dye–Amaranth.

Amaranth (Scheme 11.2) has the chemical name Trisodium 3-hydroxy-4-(4-sulfonato-1-naphthylazo)-2,7-naphthalene disulfonate (Table 11.2). It is a dark red to purple colored anionic azo dye. The presence of trisulphonic and a hydroxyl group provides it with exceptionally good water solubility (50g/L). Amaranth is widely used as colorant in industries like textiles, paper, cosmetics, pharmaceuticals and leather, etc. This sulphonic acid-based napthyl azo dye was also used as food additive in ice creams, jams, canned fruit pie fillings, soups, jellies and ketchup until 1976, when it was banned by the US Food and Drug Administration agency for use in food and drugs [55].

Critical investigations on the toxicity of the dye Amaranth to living systems establish that amongst humans the dye causes allergic and/or intolerance reactions. Patients with problems of aspirin intolerance and asthma are more prone to the harmful effects of this dye [56,57]. It is also associated with behavioral problems like hyperactivity. Being an azo dye, if it enters into the body through ingestion, it metabolizes to aromatic amines by intestinal microorganisms after reduction. These degradable products of Amaranth are toxic and can cause tumors [58]. Its reductive enzymes in the liver can also catalyze the reductive cleavage of the azo linkage. Chung *et al.* [59] reported that a wide variety of anaerobic bacteria isolated from caecal or fecal contents from experimental animals and humans have the ability to cleave the azo linkage(s) to produce aromatic amines. Chung [59] also reduced Amaranth to aromatic amines by intestinal microflora and reviewed the mutagenicity of the dye. It is now well established that reductive ring fission of the azo linkage results in the formation and accumulation of colorless aromatic amines [60,61] and these reduction products are toxic, mutagenic, and carcinogenic to animals and humans [62].

The importance of wastewater treatment in the present scenario has been the initiating factor in research dedicated to the removal of dyes from wastewater in an effective and efficient manner. The application of dyes in various fields, their toxicological impact on the environment and living systems, their separation and identification, along with the removal techniques exploited so far, are thoroughly discussed in the present chapter.

11.2.3 Dye Procurement

Both of the dyes employed in the present investigation were procured from M/s Merck and have been used as obtained without any further purification. All the test solutions employed in the experiments have been prepared using double-distilled water. All other reagents were of A.R. grade.

The pH of each test solution was measured by using a microprocessor-based pH meter model no. HI 8424 (M/s Henna Instruments, Italy). A UV/ visible spectrophotometer model no. 117 (M/s Systronics, Ahmedabad, India) was employed to carry out absorbance measurements over the wavelength range of 200–700 nm for both dyes being investigated for their absorption ability on Hen Feathers.

11.3   Adsorbent Material – Hen Feather

Amongst all living organisms, it is only birds which have been gifted with highly decorative, beautiful and soft feathers. In other words, we may also classify birds as those living beings which carry feathers at some point in their life span. Feathers possess marvelous and proficient structures. They are flexible but at the same time strong [63]. They help keep birds warm and dry in winter, yet help them to keep cool and hold and transport water during scorching summer weather. Apart from assisting birds to fly, feathers keep them protected from injury, and allow them to send signals to their friends and warnings to their enemies.

Feathers are composed of keratin. Chemically keratin is similar to the substance that makes up the fur of most mammals, scales of reptiles, horns of animals like rhinoceros and fingernails of humans [64]. Though feathers come in several forms, they are all made up of the same basic parts. These parts may be absent or rearranged a bit, depending on the main function of the feather. Every feather possesses a main shaft, which is also known as rachis. This shaft or rachis supports the entire structure of the feather. The shaft has blood vessels within it. During the growth of the feather, these blood vessels help in carrying nutrients to the growing parts of the feather. On maturity, these blood vessels die out and the rachis is sealed at the base, leaving the feather shaft hollow. Because of this phenomenon the feather is very light in weight [65].

Branching off the rachis are barbs; barbs are branched into barbules, and the barbules have branches called barbicels. Thus, these three parts—barbs, barbules and barbicels—give the feather its "feather-like" shape. Barbicels are very tiny, and a good magnifying glass or microscope is needed to observe them. They are generally hook-shaped, and interweave with each other [65]. They hold the vane of the feather together in a similar manner as Velcro strips. If you've rubbed a feather the "wrong way" and then smoothed it back to its original shape, what you've done is un-hook and re-hook the barbicels. The barbicels can hold the feather vanes together so tightly that water cannot go through.

A survey of literature reveals that the use of hen feathers as a candidate for potential adsorbent for the removal of hazardous dyes was an innovative initiative first created in a laboratory in the year 2006 [17–26]. Before the year 2006, the use of hen feathers as adsorbent was limited to the removal of metal ions only and Al-Asheh and coworkers [66–68] were the major contributors in this area. In a similar type of study [68] binary systems of copper, zinc and nickel ions have been removed through batch adsorption processes using chicken feathers as an adsorbent. Teixeira et al. [69] described a biological route for direct sorption of aqueous As (III) species over a waste biomass with a high fibrous protein content obtained from chicken feathers. In recent years attempts have been made to use chemically modified feathers for the removal of heavy metal (Zn^{2+}) from polluted water [70]. Feathers are modified by tannic acid and kinetics of adsorption of Zn^{2+} by treated and untreated feather has been studied as a function of the weight gain of tannic acid, pH of metal solution, etc.

11.3.1   Development of Adsorbent Material   Cleaning the materials

As described earlier, hen feathers usually contain a soft barb part and a hard rachis. The feathers procured from poultry were dirty and about 1 cm in length. To remove the dirt, blood stains and odor from the feathers they were first agitated in a pool of distilled water bath and then rinsed several times by doubly distilled water. The washed feathers were then dried and their soft barbs were cut into small pieces, each of about 0.1 mm length, with the help of a sharp blade, and middle rachis were removed and discarded. The barbs thus obtained were then treated with 30% v/v hydrogen peroxide for about 24 hours to oxidize the adhering organic material. Next, for the removal of moisture, the material obtained was kept in an oven at 100°C for 12 hours and activated adsorbent thus obtained was stored in a vacuum desiccator until used.

11.3.2   Characterization of Adsorbents

11.3.2.1   Chemical Analysis

The chemical analysis of activated Hen Feather has also been carried out by standard methods [71] and reveals that Hen Feather contains protein contents of around 84 percent along with several inorganic constituents like calcium, magnesium, selenium, zinc, etc. The raw feather is relatively insoluble and possesses a very low digestibility of five percent due to the high keratin contents and the strong disulphide bonding of the amino acids. The chemical constituents of feather have been portrayed elsewhere [17–26].

11.3.2.2   Analysis of Physical Properties

The porosity and density of activated Hen Feathers have been ascertained by standard procedures using mercury porosimeter and specific gravity bottles respectively. Amongst various physical parameters of activated Hen Feathers, surface area was found to be 1170.6 $(cm^2.g^{-1})$, density 0.3834 $(g.mL^{-1})$, porosity 74% and loss on ignition was obtained at 2.63%.

The nature of Hen Feather has been determined by dipping a weighed quantity of each adsorbent in 25 mL of distill water (pH = 7.0) in a 100 mL measuring flask, separately. The test solutions are stirred and left undisturbed for 24 hours in airtight measuring flasks. The solutions are then filtered and pH of each solution is measured. The nature of adsorbents is confirmed as acidic, as pH values of all three solutions are found to decrease.

11.3.2.3   Scanning Electron Microscopic Analysis

To analyze the shape and nature of the surface of both the materials, scanning electron microscopic (SEM) studies are carried out. The photographs with different magnifications reveal the porosity of the Hen Feathers. A careful study of these pictures clearly indicates that hen feathers are a needle-shaped fibrous material and cannot be considered spherical. The surface morphologies of the Hen Feathers were analyzed on the basis of SEM photographs, and it is clear from the photographs that the feathers have a shaft/stick-like shape. They possess smooth and compact surface with a smooth stripe along the parallel section, and a branch knot on the vertical section. In addition, some barbs/barbules can also be seen as adhering to the surface of the feathers.

11.3.2.4   Infrared Spectrophotometric Analysis

The infrared spectrum of activated Hen Feathers exhibits two primary bands of IR spectra of proteins which are obtained at 1600–1700 cm^{-1} for the amide I and 1500–1560 cm^{-1} for amide II, which arise from specific stretching and bending vibrations of the protein backbone. The presence of protein constituents can be ascertained by the amide I and amide II bands appearing at 1641 and 1531 cm^{-1}, respectively. The band at 2966–2877 cm^{-1} attributes to the stretching vibration of C-H in methylene, and at 1110 cm^{-1} corresponds to the CO stretching vibration of ether bond.

11.4   Preliminary Investigations

The foremost requirement of the developed dye removal process is to optimize the working conditions by carrying out preliminary experiments on various involved parameters. During the preliminary investigations, parameters like pH of the solution, concentrations of the dye, amount of adsorbent, temperature, contact time, etc., were altered and their effect on the dye uptake were measured. Thus, for the dye-adsorbent system, complete efficiency of the ongoing adsorption process depends upon the optimum conditions, which are provided by the preliminary studies.

It is a well-known fact that the adsorption of any chemical substance depends upon its protonation ability in the working solution, hence the adsorption is always dependent upon the pH of the dye solution. The effect of the concentration of the dye solution and contact time of adsorbate and adsorbent alter the rate of transference of the adsorbate to the exterior of the adsorbent; therefore concentration and contact time studies are helpful in observing the dynamics of the adsorbate material. Moreover, study of the amount of adsorbent is also of utmost importance as adsorption depends upon the amount of the adsorbent material in the solution. During the present study, adsorption of the dyes over Hen Feather is carried out under different conditions and a comparative analysis supported by figures and tables has been made. The entire test performed during the course of the studies is helpful in evaluating and designing isothermal and kinetic aspects of the adsorption process and in understanding the basic adsorption characteristics.

11.4.1   Experimental Methodology

During the preliminary investigations, batch technique, i.e., finite bath systems, has been employed. The batch technique involves a continuous interaction between the dye particles with adsorbent particles at predetermined conditions in the reaction vessel [72]. Experimentation has been carried out by taking 25 mL of the dye solution of a particular concentration in a series of 100 mL volumetric flasks. An appropriate amount of the adsorbent was then added to these flasks and intermittent uniform shaking was carried out by a mechanical shaker for a homogeneous mixing. Throughout the studies, the temperature was kept constant at 30, 40 or 50°C using a water bath. After a fixed period of time these solutions were filtered using Whatman filter paper (No. 41) and analyzed spectrophotometerically for measuring the dye uptake. All the analytical grade reagents and chemicals (hydrogen peroxide, sodium hydroxide, hydrochloric acid, etc.) used in these studies were procured from M/s Merck.

11.4.1.1   Effect of pH

The pH of the dye solution plays an important role in the adsorption of any adsorbate. Batch studies have been carried out for the adsorption of all the eight dyes over the adsorbents by varying the pH over a wide pH range from acidic to alkaline region. The pH of each test solution was adjusted using NaOH and HCl solutions. Each solution was homogeneously mixed for about 24 hours and then the uptake of the dye was measured. All pH measurements have been carried out at constant experimental conditions by taking a fixed initial dye concentration, amount of Hen Feather and temperature.

11.4.1.2   *Effect of Adsorbate Concentration*

The rates of absorptive reactions are directly proportional to the concentration of the solute. Thus, examining the effect of adsorbate concentration becomes essential [73]. The adsorption behaviors of both the dyes on the Hen Feather have been investigated at concentrations ranging from 1×10^{-5} mol L^{-1} to 10×10^{-5} mol L^{-1}, at fixed pH and different temperatures (30, 40 and 50°C). For each adsorption system a series of 100 mL graduated volumetric flasks, each containing 25 mL of adsorbate solutions at a definite range of concentrations, has been employed at optimum pH, followed by the procedure mentioned above.

11.4.1.3   *Effect of Temperature*

Temperature is an important parameter in the adsorption studies which provides valuable information about several thermodynamic parameters like Gibb's free energy, enthalpy and entropy, etc., along with feasibility and endothermic or exothermic nature of the ongoing adsorption process. The effect of temperature on the adsorption of both the azo dyes was investigated at a fixed pH, definite concentration of the dye solution and with a fixed amount of Hen Feather. All the experiments of the present studies have been carried out at temperatures of 30, 40 and 50°C. These temperatures were deliberately chosen, as temperature 30°C is treated as room temperature and further two increments of 10°C each provide a wider range applicable for the adsorption studies.

11.4.1.4   *Effect of Adsorbent Dose*

In order to study the effect of the amount of adsorbent for both the dyes, various amounts of Hen Feathers were used and batch experiments were conducted to determine the adsorption capacities at 30, 40 and 50°C temperatures for 24 hours, at a particular concentration of the adsorbate and pH of the solution. The amount of Hen Feathers added varied from 0.005 to 0.025 g and 0.005 to 0.01 g for 25 mL solutions of Tartrazine and Amaranth, respectively.

11.4.1.5   *Effect of Contact Time*

In order to innovate effective modeling of the adsorption process, the effect of contact time on the adsorption equilibrium was investigated. The contact time study is helpful in determining the duration of equilibrium attainment, i.e., the time beyond which no further significant adsorption takes place over the surface of adsorbent. Thus, a contact time study is very much helpful in calculating the kinetic data for the rate-determining step. Under the contact time studies, uptake capacities of the dyes are determined by taking a fixed concentration of the dye and varying the amount of the Hen Feather at different temperatures. When the equilibrium is thought to have been established, the solutions are filtered after definite time intervals and noted for the change in the absorbance. Thus, contact time studies are helpful in determining the uptake capacities of Tartrazine–Hen Feather and Amaranth–Hen Feather systems at 30, 40 and 50°C temperatures.

11.4.2   Results and Discussions

11.4.2.1   Effect of pH

Figure 11.1 presents variations of the adsorption of undertaken Azo dyes by Hen Feather over a wide range of pH in terms of percentage of the dye adsorbed. The results indicate that the extent of adsorption of each dye is

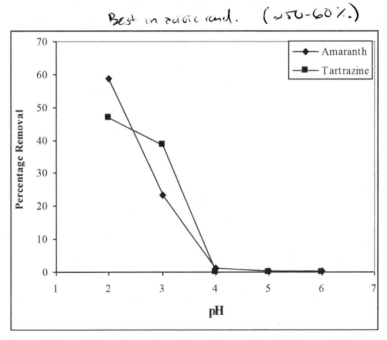

Figure 11.1 Effect of pH on adsorption of Azo dyes over Hen Feathers at 30°C. (Amaranth: Dye Concentration = 10×10^{-5} M, Adsorbent Dose = 0.01 g/25mL; Tartrazine: Dye Concentration = 9×10^{-5} M, Adsorbent Dose = 0.01 g/25mL).

affected significantly by pH due to its dependence on the surface binding site of the adsorbent and the ionization process of the dye molecule. In both cases the adsorption was found to decrease almost linearly with the increase in pH. It may be noted that for both dyes the maximum uptake of the dye takes place at around pH 2.0, and beyond pH 4.0 the adsorption of the dye was almost negligible. Thus, pH 2.0 was selected for all subsequent studies.

The higher adsorption of the dye at low pH may be due to increased protonation by neutralization of the negative charge at the surface of the adsorbents. This phenomenon helps in the preference of the dye for active sites and facilitates the diffusion process in the working solution. While, with an increase in alkaline conditions or pH, protonation is reduced and electrostatic repulsive force becomes operative, which thereby retards diffusion and adsorption.

11.4.2.2   Effect of Adsorbate Concentration

Different concentrations of the Azo dyes have been investigated with fixed amount of adsorbents by increasing the temperature from 30 to 50°C. The concentration study reveals that for both the Azo dyes an increase in concentration of the dye solution increases the extent of adsorption almost linearly over Hen Feathers (Figures 11.2 and 11.3). The amount of dye

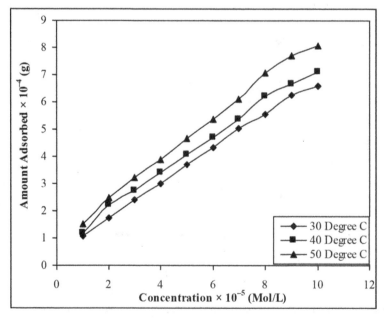

Figure 11.2 Effect of concentration of Amaranth (pH = 2.0) on adsorption over 0.01 g/25 mL Hen Feathers at different temperatures.

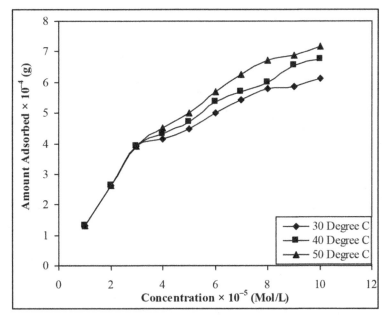

Figure 11.3 Effect of concentration of Tartrazine (pH = 2.0) on adsorption over 0.01 g/25mL Hen Feathers at different temperatures.

adsorbed obtained for each adsorption system indicates that an increase in concentration of the dye decreases the resistance towards the dye uptake, which increases the mass driving force between adsorbent and adsorbate, thereby enhancing the percentage adsorption of the dye.

11.4.2.3   Effect of Temperature

The effect of temperature on the adsorption of the Azo dyes can be observed in the Figures 11.2 (Amaranth–Hen Feather system) and 11.3 (Tartrazine–Hen Feather system). It can be observed that the Amaranth and Tartrazine dyes exhibit an increase in the adsorption with increasing temperature, thereby indicating the purely endothermic nature of the ongoing adsorption.

11.4.2.4   Effect of Adsorbent Dosage

It can be observed that in both cases the adsorption of the dye increases with an increase in the amount of Hen Feather. The graphical representation of this behavior can be observed for the Amaranth–Hen Feather (Figure 11.4) and Tartrazine–Hen Feather (Figure 11.5) systems. The increase in adsorption with increasing amounts of the adsorbent is mainly

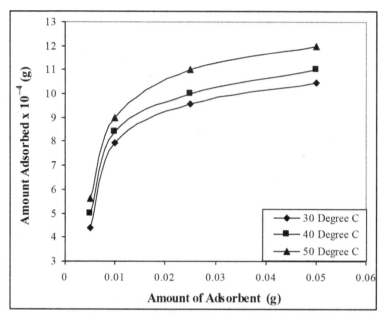

Figure 11.4 Effect of amount of adsorbent on the uptake of Amaranth (10×10^{-5} M, pH = 2.0) over Hen Feathers at different temperatures.

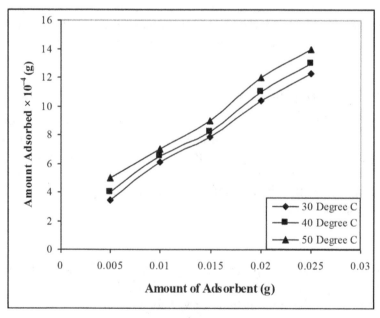

Figure 11.5 Effect of amount of adsorbent on the uptake of Tartrazine (9×10^{-5} M, pH = 2.0) by Hen Feathers at different temperatures.

due to the increase in surface area of Hen Feather and availability of more adsorption sites. However, after an optimum dose there is no significant change in the adsorption rate for both the adsorption systems, which may be due to saturation of all the available active sites of the adsorbents.

11.4.2.5 Effect of Contact Time

For both the Azo dyes, contact time studies reveal that in general, as time passes, adsorption increases and finally attains saturation. It is found that Hen Feather, Tartrazine takes about 4 hours (Figure 11.6) and Amaranth takes about 2 hours (Figure 11.7) to attain saturation.

11.5 Adsorption Isotherm Models

Preliminary investigations help in determining various working conditions to achieve optimum adsorption of the dye. The results obtained during preliminary investigations are now used to verify various isothermal models. This would be another step to develop efficient dye removal methodology because equilibrium sorption isotherms are fundamentally very significant

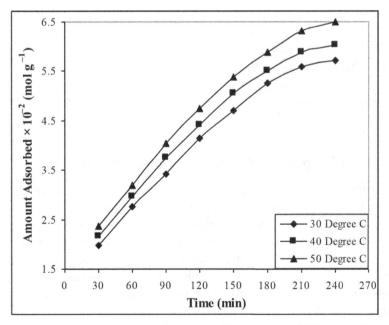

Figure 11.6 Effect of contact time on the uptake of Tartrazine (9×10^{-5} M, pH 3.0) by Hen Feathers at different temperatures.

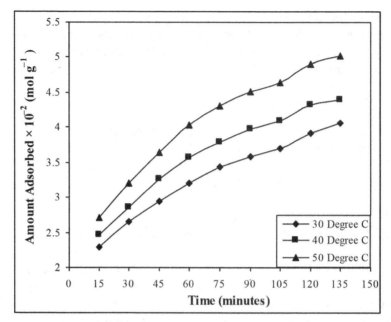

Figure 11.7 Effect of contact time for the uptake of Amaranth (9×10^{-5} M, pH 2.0) by Hen Feathers at different temperatures.

for the designing of any sorption system. The information gathered under various isothermal conditions during the course of investigations is utilized to evaluate useful thermodynamic parameters, which suggest the feasibility, favorability and spontaneity of the ongoing adsorption process in each dye-adsorbent system. Moreover, the chemical or physical character of the adsorption has also been investigated in the present studies for both the dyes.

11.5.1   Adsorption and Adsorption Isotherm Models

Adsorption is a surface phenomenon which involves the accumulation or concentration of substances at a surface or interface. This process can occur at an interface of any two phases, such as liquid-liquid, gas-liquid, gas-solid or liquid-solid. Therefore, adsorption is a process in which the molecules or atoms of one phase interpenetrate nearly uniformly among those of another phase to form a solution with this phase. As such the solute remaining in the solution is in dynamic equilibrium with that at the surface of the adsorbed phase.

Adsorption is related to the phenomenon of surface energy. In a bulk material, all the bonding requirements, whether ionic, covalent or metallic, of the constituent atoms of the material are satisfied. But atoms on the adsorbent surface experience a bond deficiency and the preferential concentration of molecules in the proximity of adsorbent surface arises. The adsorption of various substances is due to increased free surface energy of the solids because of their extensive surface. According to the second law of thermodynamics, this energy has to be reduced. This is achieved by reducing the surface tension via the capture of extrinsic substances. The exact nature of the bonding depends on the details of the species involved and is generally classified as physisorption or chemisorption.

Thus, phenomenon can be graphically understood by plotting a graph with concentration (C) term on the abscissa and amount of adsorbate adsorbed (q_e) on the ordinate. The concentration (C) represents the concentration of the migrating substance in the liquid phase in contact with the solid phase at a liquid-solid interphase, or in other words, the concentration of the solute remaining in solution at equilibrium. However, the term q_e depicts the amount of adsorbate adsorbed, i.e., the substance that has moved across the interphase. A concentration versus q_e graph, when plotted under constant temperature conditions, is also termed as adsorption isotherm.

Adsorption capacity of an adsorbent is the amount of adsorbate taken up by the per unit mass of it. It depends upon fluid phase concentration, temperature and other conditions like surface area of the adsorbent, amount of adsorbent, contact time, etc. Adsorption capacity data are gathered at a fixed temperature and different adsorbate concentrations and the results obtained are plotted as isotherm, which is termed as adsorption isotherm plots.

In the present investigations, equilibrium adsorption data is described using well-known Freundlich, Langmuir, Tempkin and Dubinin-Radushkevich (D-R) isotherm models.

11.5.1.1 Langmuir Isotherm Model

In 1916, Irving Langmuir developed an empirical isotherm which depends upon the monolayer adsorption capacity of an adsorbent by assuming unimolecular layer formation of the adsorbate at the adsorption saturation. This sorption model is based on two hypotheses [74,75]: (a) that uptake of adsorbate occurs on a homogenous surface by monolayer adsorption and (b) there is no interaction between the ions of adsorbate. The model also considers that there is no transmigration of the adsorbate in the plane of the surface of the adsorbent. The model is quite versatile and the mechanism involved for all the adsorbate-adsorbent systems is treated similarly. The simplified mathematical form of the Langmuir equation can be expressed as:

$$\frac{1}{q_e} = \frac{1}{q_o} + \frac{1}{bq_oC_e} \qquad (11.1)$$

where q_e is the amount of adsorbate adsorbed (mol.g^{-1}), C_e is the equilibrium molar concentration of the dye (mol. L^{-1}), q_o is the maximum adsorption capacity over unit mass of the adsorbent (mol.g^{-1}) and b is the energy of adsorption (L.mol^{-1}). The Langmuir constants q_o and b are calculated from the slopes and intercept of $1/q_e$ against $1/C_e$ that gives a straight line at each temperature, thereby confirming that the Langmuir isotherm is followed in the adsorption process. The Langmuir constant 'b' obtained by the above expression is helpful in determining various thermodynamic parameters and also the favorability of the adsorption process.

11.5.1.2   Freundlich Isotherm Model

In the Freundlich model, it is considered that the binding affinities on the adsorbent surface vary with the interactions between the adsorbed molecules. Consequently, the sites with stronger affinity are occupied first. The equation describing the Freundlich model represents adsorption of solutes from a liquid to a solid surface. This experimental model can be applied to non-ideal adsorption on heterogeneous surfaces as well as multilayer adsorption. The Freundlich model is chosen to estimate the adsorption intensity of the adsorbate on the adsorbent surface [74, 5]. The mathematical expression for the Freundlich adsorption isotherm model is:

$$\log q_e = \log K_F + (1/n) \log C_e \qquad (11.2)$$

Here C_e denotes the equilibrium concentration (M) of the adsorbate and q_e, the amount adsorbed (g.mol^{-1}) and K_F and n are the Freundlich constants related to the adsorption capacity and adsorption intensity of the adsorbate-adsorbent system, respectively. Thus, the Freundlich adsorption isotherm plot is sketched by taking $\log C_e$ on abscissa and $\log q_e$ on the ordinate. When the experimental data obtained in terms of $\log C_e$ and $\log q_e$, fit into straight lines with regression coefficients close to unity, the Freundlich adsorption isotherm is assumed to be verified. Values of K_F and n are derived from the intercepts and slope of these straight lines, respectively.

11.5.1.3   Tempkin Isotherm Model

The Tempkin isotherm model [76] describes the effect of some indirect interactions amongst adsorbate particles. The model assumes that the heat of adsorption of all the molecules in the layer decreases linearly with surface coverage due to adsorbent-adsorbate interactions. Moreover, the adsorption process is characterized by a uniform distribution of the binding energies, up to maximum binding energy. The linear form of the Tempkin isothermal model is expressed as:

$$q_e = k_1 \ln k_2 + k_1 \ln C_e \qquad (11.3)$$

where q_e is the amount of adsorbate adsorbed per unit mass of adsorbent at equilibrium ($mol.g^{-1}$), C_e is the final concentration at equilibrium ($mol. L^{-1}$), k_1 is the Tempkin isotherm energy constant ($L.mol^{-1}$) related to the heat of adsorption [77] and k_2 is the Tempkin isotherm constant. Thus, a plot of $\ln C_e$ as function of amount adsorbed at equilibrium gives straight lines suggesting the uniform distribution of binding energy arising due to interaction of the adsorbate molecules. The straight lines obtained from the graphs are also helpful in determining the Tempkin isotherm constants.

11.5.1.4   Dubinin-Radushkevich (D-R) Isotherm Model

Dubinin and Radushkevich have made a great contribution to Surface Science by, developing an isothermal model based on the heterogeneous characteristics of the adsorbents which is helpful in understanding the interaction between adsorbate and adsorbent [78]. The isotherm model suggested by Dubinin and Radushkevich [79] has been used to describe liquid-phase adsorption and on the basis of the Dubinin-Radushkevich equation adsorption energy can be estimated. The model, which is commonly known as the D-R Adsorption Model, is also chosen to estimate the characteristic porosity and the apparent free energy of adsorption. It suggests that adsorption data can be analyzed to distinguish between chemical and physical adsorption by employing the following equation:

$$\ln C_{ads} = \ln X_m - \beta \in^2 \qquad (11.4)$$

where C_{ads} is the amount of dye adsorbed per unit weight of adsorbent ($mol.g^{-1}$), X_m is the maximum adsorption capacity ($mol.g^{-1}$), β is the activity coefficient ($mol^2. J^{-2}$) related to mean adsorption energy and $\in$ is the Polanyi potential, which is given as:

$$\in = RT \ln(1 + \frac{1}{C_e}) \qquad (11.5)$$

where R is the universal gas constant ($J.mol^{-1}. K^{-1}$), T is the temperature (K) and C_e is the concentration at equilibrium ($mol. L^{-1}$). Polanyi sorption potential ($\in$) is the work required to remove a molecule to infinity from its location in the sorption space, independent of temperature. This model assumes the heterogeneity of sorption energies within this space. The slopes of straight-lines of graphs between $\ln C_{ads}$ against $\in^2$ give activity coefficient and intercept yields adsorption capacity. The applicability of the isotherm is related with determination of the nature of the adsorption process and mean sorption energy is the decisive factor for distinguishing between chemical and physical adsorption. It is given by the following form:

$$E = \frac{1}{\sqrt{-2\beta}} \qquad (11.6)$$

It has been postulated that in any adsorbate-adsorbent system, when mean sorption energy 'E' estimated by the above expression is less than 8 kJ.mol^{-1}, physisorption dominates the sorption mechanism, whereas if 'E' is between 8 to 16 kJ.mol^{-1}, chemisorption is the governing factor of the process [80]. Thus, the calculated values of E play a significant role in deciding the operative nature of the ongoing adsorption.

11.5.1.5   Thermodynamic Parameters

An isothermal model like the Langmuir model has been proven useful in the determination of thermodynamics of the adsorption process. Using Langmuir adsorption data, particularly the Langmuir constant, energy of adsorption (b), various thermodynamic parameters of the adsorption systems such as change in Gibb's free energy ($\Delta G°$), change in enthalpy ($\Delta H°$) and change in entropy ($\Delta S°$) are calculated from the following well-known relations [81]:

$$\Delta G° = - R T \ln b \qquad (11.7)$$

$$\Delta H° = -R\left(\frac{T_2 T_1}{(T_2 - T_1)}\right) \times \ln\left(\frac{b_2}{b_1}\right) \qquad (11.8)$$

$$\Delta S° = \frac{\Delta H° - \Delta G°}{T} \qquad (11.9)$$

where b, b_1 and b_2 are Langmuir constants (L.mol^{-1}) at different temperatures (K) and R is universal gas constant (J.mol^{-1}. K^{-1}).

Thus, Gibb's free energy ($\Delta G°$) is calculated at different temperatures by putting the corresponding values of Langmuir constant 'b' in the

Equation 11.7. The negative values of Gibb's free energy ($\Delta G°$) indicate feasibility of the processes, while positive values suggest unfeasibility of the adsorption process.

Similarly the idea about the exothermic and endothermic nature of the ongoing adsorption reaction can be derived by determining the sign of ΔH in Equation 11.8. If it is positive, the reaction is endothermic, and the heat is absorbed by the system due to greater enthalpy than the reactants. On the other hand, if $\Delta H°$ is negative, the reaction is exothermic, that is the overall decrease in enthalpy is achieved by the generation of heat. Similarly with the help of Equation 11.9, an increase or decrease in randomness of the process can be postulated by getting positive and negative values of change in entropy ($\Delta S°$), respectively. Positive value of entropy change ($\Delta S°$) shows increased degree of randomness at the solution-solid interface with some structural changes in the adsorbent. It also corresponds to an increased degree of freedom of the adsorbed species. The opposite trends of these parameters occur when negative value of entropy change ($\Delta S°$) is obtained during the adsorption.

11.5.1.6   Favorability Determination

To define favorability of a chemical or physical process a dimensionless constant called separation factor (r) has been introduced by Weber and Chakravorti [82]. For the separation factor the calculation can done by employing the following relation:

$$r = \frac{1}{1 + bC_o} \tag{11.10}$$

where b denotes the Langmuir constant (L.mol^{-1}) derived from Equation 11.2 and C_o is the initial concentration (mol. L^{-1}). Thus, Langmuir constant 'b' is helpful in determining the favorability of the process. The significance of separation factor lies in ascertaining the favorability of adsorption and the shape of the adsorption isotherms [82]. The nature of favorability is determined on the basis of 'r' values obtained. The parameters indicate the shape of the isotherm accordingly. If r is greater than unity and equal to one, the process is said to be unfavorable and linear, respectively. Irreversible condition arises when 'r' value is obtained to be equal to zero. The only favorable condition is when 0 < r < 1. Thus, isothermal models expand their applicability by determining the favorability through separation factor.

11.5.2   Experimental Methodology

For the verification of isotherm models batch adsorption has been performed over Hen Feather for both the Azo dyes by taking 25 mL of the dye solution of known concentration and definite pH, in a 100 mL volumetric flask at 30, 40 and 50°C. A fixed amount of the Hen Feather is now added into the solution and adsorption is monitored after gradually mixing the dye-adsorbent mixture in a mechanical shaker for about 24 hours. After 24 hours these solutions are filtered with Whatman filter paper (No. 41) and the amount of the dye uptake is analyzed spectrophotometerically at definite wavelengths, λ_{max} corresponding to the dye. The complete experimental procedure involved in the study has already been explained in the previous chapter. The data obtained is then applied to various isothermal models and the best-fit straight lines are ascertained in each adsorption model.

11.5.3   Results and Discussions

11.5.3.1   Langmuir Isotherm Model

For the verification of the Langmuir isotherm model $1/C_e$ versus $1/q_e$ graphs have been plotted for the data obtained during adsorption of both the Azo dyes at temperatures 30, 40 and 50°C. The Langmuir plots for the adsorption systems Amaranth–Hen Feather and Tartrazine–Hen Feather are presented in Figures 11.8 and 11.9, respectively. In all the cases, straight lines with appreciable R^2 values are obtained at all the temperatures, thereby indicating verification of the Langmuir adsorption model and involvement of monolayer adsorption in each case. These straight lines are helpful in calculating the Langmuir constant 'b' and the number of moles of the dye adsorbed per unit weight of the adsorbent (q_0) through their slopes and intercepts, respectively (Table 11.3).

11.5.3.2   Freundlich Isotherm Model

The linear plots of log C_e versus log q_e at temperatures 30, 40 and 50°C give straight lines with regression coefficient value close to unity. This clearly reveals that adsorption of Azo dyes on Hen Feathers (Figures 11.10 and 11.11) follows the Freundlich model at all the temperatures. With the help of these Freundlich linear adsorption isotherm plots, Freundlich constants are evaluated. The values of K_F and n derived from the intercepts and slopes of straight lines obtained are collectively listed in Table 11.3.

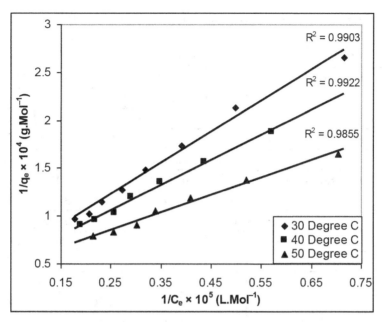

Figure 11.8 Langmuir adsorption isotherm for Amaranth (pH = 2.0) – Hen Feather (0.01 g/25mL) system at different temperatures.

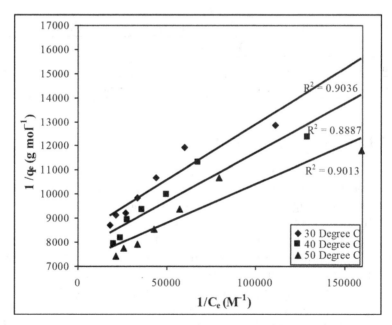

Figure 11.9 Langmuir adsorption isotherm for Tartrazine (pH = 2.0) – Hen Feather (0.01g/25mL) system at different temperatures.

Table 11.3 Various isotherm constants for the removal of Azo Dyes at different temperatures.

Langmuir Constants						
Adsorption System	$q_o \times 10^{-4}$ (mol g^{-1})			$b \times 10^3$ (L mol^{-1})		
	30°C	40°C	50°C	30°C	40°C	50°C
Amaranth – Hen Feather	2.44	2.49	2.50	1.25	1.73	2.19
Tartrazine – Hen Feather	1.20	1.32	1.45	179.14	188.98	220.24
Freundlich Constants						
Adsorption System	n			K_F		
	30°C	40°C	30°C	40°C	30°C	40°C
Amaranth – Hen Feather	0.13	0.15	0.15	0.999	0.999	0.999
Tartrazine – Hen Feather	4.32	3.99	4.03	0.001	0.002	0.002
Tempkin Constants						
Adsorption System	$k_1 \times 10^{-5}$ (L.mol^{-1})			$k_2 \times 10^4$		
	30°C	40°C	50°C	30°C	40°C	50°C
Amaranth – Hen Feather	5.00	5.00	6.00	16.275	16.275	11.662
Tartrazine – Hen Feather	10	20	20	17.848	9.873	16.272
D-R Constants						
Adsorbent	$\beta \times 10^{-9}$ (mol^2 J^{-2})			E(kJ mol^{-1})		
	30°C	40°C	50°C	30°C	40°C	50°C
Amaranth – Hen Feather	6	5	4	9.13	10.00	11.18
Tartrazine – Hen Feather	2	2	2	15.81	15.81	15.81

11.5.3.3   Tempkin Isotherm Model

The graphical presentations for the Tempkin adsorption model of both the adsorption systems at 30, 40 and 50°C are presented in Figures 11.12 to 11.13. The straight lines obtained clearly indicate that in both of the cases the regression coefficient values close to unity are obtained. This indicates that by and large adsorption of Azo dyes follows Tempkin adsorption isotherm and uniform distribution of binding energy takes place due to interaction of the dye molecules. The straight lines obtained from the graphs are also helpful in determining the Tempkin isotherm constants and Tempkin isotherm energy constants for both the adsorption systems. The values of these constants are presented in Table 11.3.

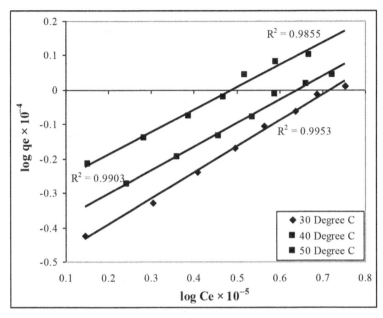

Figure 11.10 Freundlich adsorption isotherm for Amaranth (pH = 2.0) – Hen Feather (0.01 g/25mL) system at different temperatures.

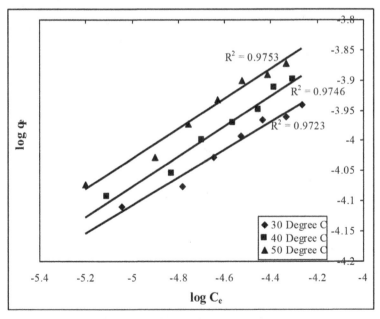

Figure 11.11 Freundlich adsorption isotherm for Tartrazine (pH = 2.0) – Hen Feather (0.01 g/25mL) system at different temperatures.

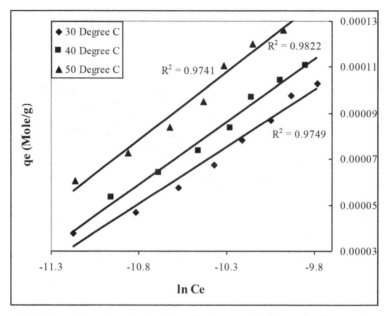

Figure 11.12 Tempkin adsorption isotherm for Amaranth (pH = 2.0) – Hen Feather (0.01 g/25mL) system at different temperatures.

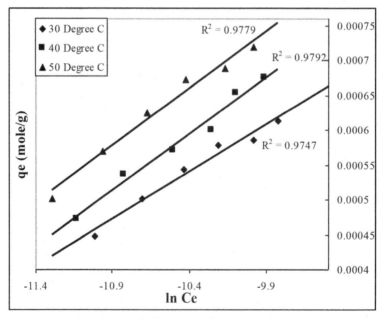

Figure 11.13 Tempkin adsorption isotherm for Tartrazine (pH = 2.0) – Hen Feather (0.01 g/25mL) system at different temperatures.

11.5.3.4 Dubinin-Radushkevich (D-R) Isotherm Model

To verify the applicability of the D-R isotherm model graphs between $\ln C_{ads}$ against ϵ^2 are plotted for the adsorption of Azo dyes over Hen Feathers (Figures 11.14 and 11.15) at 30, 40 and 50°C. As evident from the graphs, the values of regression coefficients of the obtained straight line are found close to unity in all cases, thereby indicating the applicability of D-R adsorption isotherm at all the temperatures.

The slopes of straight-lines of the graphs between $\ln C_{ads}$ against ϵ^2 give activity coefficient (β) and intercept yields adsorption capacity (X_m) of the adsorption systems. The values of activity coefficient (β) are presented in Table 11.3. It is clear from the data that for both the dye-adsorbent systems the value of activity coefficient is found to the tune of 10^{-9} mol^2 J^{-2} and remains almost constant at all the temperatures.

Table 11.3 also presents the values of mean sorption energy (E) for the adsorption of Azo dyes over Hen Feathers at 30, 40 and 50°C. It is interesting to note that for both the systems the values of mean sorption energy (E) are found between 8 to 16 kJ/mole at all the temperatures, which clearly indicates that chemisorption operates in the case of Azo dyes adsorption.

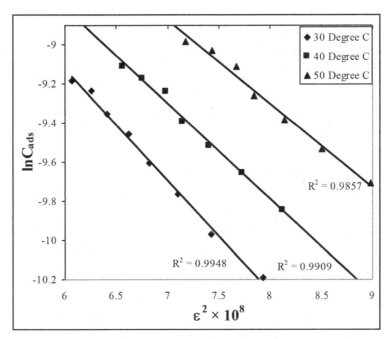

Figure 11.14 D-R adsorption isotherm for Amaranth (pH = 2.0) – Hen Feather (0.01 g/25mL) system at different temperatures.

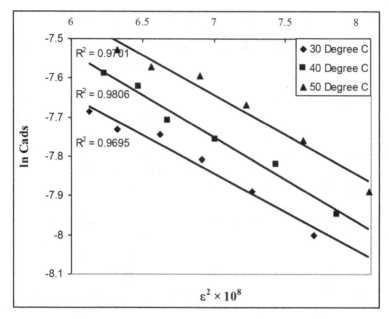

Figure 11.15 D-R adsorption isotherm for Tartrazine (pH = 2.0) – Hen Feather (0.01 g/25mL) system at different temperatures.

Table 11.4 Thermodynamic parameters for the uptake of Azo Dyes.

Adsorbent	$-\Delta G°$ (kJ mol^{-1})			$\Delta H°$ (kJ mol^{-1})	$\Delta S°$ (JK^{-1} mol^{-1})
	30°C	40°C	50°C		
Amaranth – Hen Feather	17.96	19.40	20.66	22.75	11.00
Tartrazine – Hen Feather	30.47	31.62	33.04	8.50	74.16

11.5.3.5   Thermodynamic Parameters

The theory presented above indicates that thermodynamic dependency and various thermodynamic parameters of the adsorption of Azo dyes over Hen Feather can be evaluated by using Langmuir adsorption constant 'b'. The values of free energy ($\Delta G°$), enthalpy change ($\Delta H°$) and change in entropy ($\Delta S°$) are calculated and presented in Table 11.4.

For all the Azo dyes $\Delta G°$ values are negative, indicating that the sorption process brings a decrease in Gibbs free energy and that the adsorption process is feasible and spontaneous at all the temperatures. Equilibrium experiments performed for the adsorption of Amaranth and Tartrazine over Hen Feathers at different temperatures (30, 40 and 50°C) show an increase in the amount of dye adsorbed, implying an endothermic nature of the sorption process for all four adsorption systems. The endothermic nature of the adsorption process is further confirmed by the positive values of $\Delta H°$ obtained in each case. The positive values of entropy change ($\Delta S°$) for these adsorption systems show decreased randomness at the interface and no considerable change in the internal structure of the adsorbents. Thus, values of $\Delta G°$, $\Delta H°$ and $\Delta S°$ confirm the spontaneous and endothermic nature of the adsorption process for both dye-adsorbent systems.

11.5.3.6   Favorability of the Processes

By applying the Weber-Chakravorti equation (Equation 11.10) to both the adsorption systems, the separation factor 'r' is calculated for each system. It is evident from Table 11.3 that the value of the Langmuir constant 'b' is positive and so is the dye initial concentration (C_o). Therefore the denominator of Equation 11.10 will always be greater than unity and the value of separation factor 'r' will stand between zero to unity. This indicates favorability of the adsorption process for both the Azo dye-adsorbent systems at all temperatures.

11.6   Kinetics Measurements

The main focus of the present work is to couple kinetic studies with the adsorption process so that on the basis of the results obtained, a deeper understanding of the ongoing adsorption process can be developed, thereby unfolding many hidden intricacies of the adsorption of Azo dyes over Hen Feathers. The principal application of teaming adsorption technique with kinetic studies includes deriving necessary information about the order and mechanism of the process through which the accumulation of the dye takes place over the surface of the adsorbent material, and also determining various other kinetic parameters and sorption characteristics of the adsorbents. The purpose of carrying out chemical kinetic study of any adsorption process is to observe the influence of different experimental conditions over rate of adsorption and adsorption mechanism. In order to determine the rate-controlling steps, the kinetic behavior of the overall adsorption and mechanism of the present ongoing adsorption processes finite bath method has been chosen. The models describing the order of the reaction have also been analyzed along with detailed quantitative investigations of the mechanisms involved in the present adsorption processes.

11.6.1   Theory of Kinetic Measurements

11.6.1.1   Effect of Contact Time

Contact time studies are used to monitor the influence of system parameters on the rate of adsorption of the dye on the adsorbent. They also help in determining the time of equilibrium attainment, i.e., the time beyond which no further significant adsorption takes place over the surface of adsorbent, rate-determining step of the process. Thus, equilibrium tests are carried out to ascertain the time at which complete saturation of adsorbent occurs and steady rate is reached.

Under the contact time study, uptake capacity of the adsorbent is determined at different temperatures, fixed amount of adsorbent, definite concentration of the dye and constant pH. When the equilibrium is reached, the solution is filtered and dye uptake (q_e) is calculated using the following equation:

$$q_e = \frac{(C_o - C_e) \cdot V}{W} \tag{11.11}$$

where C_e is the concentration of the dye at equilibrium in mol/L, C_o is the initial concentration in mol/L, V is volume in liters and W is the mass (g) of the adsorbent taken in the reaction flask.

11.6.1.2   Rate Constant Study

Kinetics of any adsorption process is helpful in determining the efficacy of the adsorbent material for a particular dye and efficiency of the ongoing adsorption process. The monitoring of kinetics of the adsorption process is useful in ascertaining the rate-determining step, order of the process and value of rate constant. These parameters mainly provide information about the rapidity of the adsorption. A literature survey confirms that there are several models and methods which can be used for calculating the rate and rate-determining step of any adsorption process. However, in the present cases, the determination of specific rate constant of adsorption of dyes on each adsorbent has been made by using Lagergren's pseudo-first-order rate expression [83]. The verification of the model has been done by calculating the regression coefficients of the best-fit straight lines.

The adsorption of the dye molecules from the aqueous phase to a solid phase can be considered a reversible process in which equilibrium is established between the dye in the solution and dye at the surface of adsorbent at equilibrium. The kinetics of such type of adsorption is described through Lagergren's first order rate equation [84], with an understanding that initially there is no adsorbate present on the surface of the adsorbent. Thus, by assuming the dye molecule is a non-dissociating substance the sorption phenomenon is treated as the diffusion controlled process [85]. Therefore for the adsorption of small amount of the adsorbate (dq) over a small period of time (dt), the rate of reaction can be calculated using expressions:

$$\frac{dq}{dt} \propto (q_e - q_t) \tag{11.12}$$

or
$$\frac{dq}{dt} = k_{ad}(q_e - q_t) \tag{11.13}$$

where q_e and q_t (mol.g^{-1}) denote the amount adsorbed at equilibrium and at any time t respectively and k_{ad} (s^{-1}) is the first order rate constant of adsorption. By integrating the Equation 11.13 with the boundary conditions $q_t = 0$ at t = 0 to $q_t = q_t$ at t = t, we get the following linear form of the Lagergren's pseudo-first-order kinetics [86]:

$$\log (q_e - q_t) = \log q_e - \frac{k_{ad}}{2.303} \times t \tag{11.14}$$

Thus, by plotting a graph between log (q_e-q_t) versus time (t) a straight line is obtained with the intercept on y-axes providing the value of log q_e and slop giving value of the rate constant (k_{ad}).

11.6.1.3   Elucidation of Reaction Mechanism

A survey of literature reveals that for the proper interpretation of the experimental data, it is essential to identify the steps in the adsorption process which govern the overall removal rate in each case. To identify whether the ongoing process is particle diffusion or film diffusion, the kinetic data obtained by finite batch method has been treated by an ingenious mathematical treatment suggested by Boyd et al. [87] and Reichenberg [88].

It is considered that the adsorption of an organic/inorganic compound over a porous adsorbent normally involves the following three steps:

i. Film diffusion mechanism, in which transport of the ingoing ions (adsorbate) to the external surface of the adsorbent takes place.

ii. Particle diffusion mechanism, in which transport of adsorbate within the pores of the adsorbent takes place except for a small amount of adsorption, which occurs on the external surface.
iii. Adsorption of adsorbate on the interior surface of the adsorbent.

Out of the above three processes the third one is very rapid and does not represent the rate-limiting step in the uptake of organic compounds [89]. While the other two steps give rise to the following three distinct cases:

Case I :   External Transport > Internal Transport
            (Here rate is governed by particle diffusion)
Case II :   External Transport < Internal Transport
            (Here rate is governed by film diffusion)
Case III: External Transport ≈ Internal Transport

In Case III the transport of ions to the boundary may not be possible at a considerable rate, thereby leading to the formation of a liquid film with a concentration gradient surrounding the sorbent particles.

In accordance to Fick's first law the flux J (in moles per unit time and unit cross-section, normal to the path of flow) of the diffusing ion [89] is given by equation:

$$J = \frac{D_i \delta C}{R} \tag{11.15}$$

where C is concentration (in moles per unit volume), 'D_i' is effective diffusion coefficient and 'r' gives the direction along which transport is taking place.

Using Fick's second law, the governing differential equation used for a spherical exchanger bead of radius r_o in a solution can be written as:

$$\frac{\delta C}{\delta t} = D_i \left(\frac{\delta^2 C}{\delta r^2} + 2\frac{\delta C}{r \delta C} \right) \tag{11.16}$$

Using the proper boundary conditions, the solution of this equation has been obtained as an infinite series, i.e.,

$$F = \frac{Q_t}{Q_\infty} \tag{11.17}$$

where Q_t and Q_∞ are amounts adsorbed after time t and after infinite time respectively

$$F = 1 - \frac{6}{\pi^2} \sum_1^\infty (1/n^2) \exp(-n^2 B_t) \qquad (11.18)$$

where F is the fractional attainment of equilibrium at time 't' and is obtained by using Equation 11.18 and 'n' is the Freundlich constant of the adsorbate.

$$B_t = \frac{\pi^2 D_i}{(r_o^2)} = \text{Time Constant} \qquad (11.19)$$

where B_t = time constant, D_i = effective diffusion coefficient of ion in the adsorbent phase $(cm^2.s^{-1})$ and r_o = radius of the adsorbent particle assumed to be spherical.

Although the equation is a mathematical infinite series, only a few terms can give a good approximation coefficient D_i for an adsorption system. For every observed value of F, a corresponding value of B_t is derived from Reichenberg's table [88]. The plot of B_t versus time distinguishes between film diffusion and particle diffusion controlled rates of adsorption. The plot of 'B_t' versus time (t) giving a straight line passing through the origin, shows a particle diffusion mechanism [87].

It is to be noted that the treatment of diagnosing the particle diffusion process as a rate-controlling process is the same as that for the isotopic exchange. The only difference in the two processes is that in the developed mathematical equation the self-diffusion coefficient is replaced by the effective diffusion coefficient of exchange ions [88–90].

For the system undergoing particle diffusion mechanism, the plot of log D_i versus 1/T is linear and permits the use of the Arrhenius equation [91]. Activation energy E_a and D_o values for such processes can be calculated from the slope and intercept of this plot.

$$D_i = D_o \exp\left[-\frac{E_a}{RT}\right] \qquad (11.20)$$

where D_i is the diffusion constant, D_o is the maximum diffusion constant, E_a is the activation energy $(KJ.mol^{-1})$, R is universal gas constant $(J.mol^{-1}K^{-1})$ and T is temperature (K).

By increasing the temperature the increase in value of D_i ascertains increased mobility of the diffusing ions due to decreased retarding force. The plot between 1/Temperature and log D_i defines the diffusibility of ions and helps in determining entropy of activation using the following expression:

The entropy of activation ΔS can be calculated by using the following equation:

$$D_o = (2.72 \ d^2 \ kT/h) \ exp\left(\frac{\Delta S}{R}\right) \qquad (11.21)$$

where d is the average distance between two successive sites of the adsorbent, k is the Boltzmann constant ($J. \ K^{-1}$), h is the Plank's constant ($J. \ K^{-1}$), T is the temperature (K) and R is the Universal Gas Constant ($J. \ K^{-1}.mol^{-1}$). Usually the distance between two successive sites of the adsorbent is taken as 5×10^{-8} cm [92]. The value of change in entropy (ΔS) plays an important role in diagnosing the change in the internal structure of the adsorbent after the adsorption of the dye. The negative value of ΔS indicates no change in the internal structure of the adsorbents, while positive value suggests some change in the internal structure of the adsorbent, which may affect the adsorbing ability of the adsorbent.

11.6.2   Experimental Methodology

In all the adsorption experiments 25 mL dye solution of known concentration is taken in a 100 mL graduated volumetric flask at a fixed pH and amount of adsorbent. The mixture is then shaken on a mechanical agitator and after a fixed interval of time, the solution is withdrawn and uptake of the dye is analyzed spectrophotometrically.

11.6.3   Results and Discussions

11.6.3.1   Effect of Contact Time

The uptake of the dye molecules by the Hen Feather and time of establishment of equilibrium suggest the effectiveness of this material for wastewater treatment. In order to determine the equilibrium time for maximum dye uptake, a contact time study is carried out. For the contact time studies of the adsorption over Hen Feathers, it is observed that at temperatures 30, 40 and 50°C the adsorption of the Amaranth is faster in the case of low concentration of the dye solution and equilibrium is established in about 2 hours, while at higher concentration the equilibration time is around 3 hours. In both the concentration ranges the amount of adsorbate increases with increasing temperatures. A careful inspection suggests that at 30, 40 and 50°C Hen Feathers adsorb around 47.5, 52 and 65% of the Amaranth, respectively, within the first one hour of adsorption from 5×10^{-5} M solution of the dye, while for the 9×10^{-5} M solution the amount adsorbed is 42, 47 and 60% at the same temperatures (Table 11.5). A further confirmation of the endothermic nature of the ongoing process is made by calculating the half-life of the process at each concentration and it is found to decrease with increase in temperature. Almost similar observations have been made

Table 11.5 Effect of contact time on the adsorption of Azo Dyes over Hen Feathers at different concentrations of the dye and different temperatures (Amaranth & Tartrazine: Amount of Hen Feather = 0.01 g/25mL, pH = 2.0).

Amaranth (0.09 mM)				Amaranth (0.05 mM)			
Time (Seconds)	Amount Adsorbed × 10^{-5} (mole/g)			Time (Seconds)	Amount Adsorbed × 10^{-5} (mole/g)		
	30°C	40°C	50°C		30°C	40°C	50°C
900	2.29	2.47	2.71	900	0.94	1.25	1.45
1800	2.65	2.85	3.20	1800	2.07	2.27	2.54
2700	2.95	3.26	3.64	2700	2.93	3.14	3.47
3600	3.20	3.57	4.03	3600	3.58	3.82	4.23
4500	3.44	3.79	4.31	4500	4.05	4.29	4.73
5400	3.59	3.97	4.51	5400	4.71	4.93	5.30
6300	3.70	4.09	4.65	6300	4.95	5.15	5.41
7200	3.91	4.33	4.90	7200	4.99	5.20	5.46
8100	4.06	4.40	5.02	8100	5.00	5.20	5.50
Tartrazine (0.09 mM)				Tartrazine (0.05 mM)			
Time (Seconds)	Amount Adsorbed × 10^{-4} (mole/g)			Time (Seconds)	Amount Adsorbed × 10^{-4} (mole/g)		
	30°C	40°C	50°C		30°C	40°C	50°C
1800	1.97	2.16	2.36	1800	1.42	1.56	1.77
3000	2.76	2.97	3.19	3000	1.98	2.15	2.45
5400	3.42	3.75	4.05	5400	2.44	2.68	3.05
7200	4.15	4.42	4.75	7200	3.01	3.31	3.79
9000	4.71	5.05	5.38	9000	3.56	3.82	4.34
10800	5.26	5.50	5.88	10800	3.71	4.03	4.45
12600	5.59	5.89	6.32	12600	3.88	4.15	4.55
14400	5.71	6.03	6.51	14400	4.00	4.25	4.68

for the adsorption of Tartrazine and the half-life of each process is found to decrease with increase in temperature (Table 11.6).

In all the above-mentioned cases the rate of removal of dye decreases with time, which may be due to aggregation of dye molecules around the adsorbent. This aggregation disallows the migration of adsorbate, as the adsorption sites get filled up, and also because resistance to diffusion of dye molecules in the adsorbent increases.

11.6.3.2   Rate Constant Study

In order to identify the order of the reaction for the adsorption of the Azo dyes, Lagergren's pseudo-first-order rate equations have been applied and order of the ongoing adsorption process is ascertained on the basis of comparison of regression constant values of the best-fit straight lines.

For the Hen Feather adsorption of Azo dyes–Amaranth (Figure 11.16) and Tartrazine (Figure 11.17), the R^2 values are found closer to unity in the case of Lagergren's pseudo-first-order. It is therefore concluded that dyes Amaranth and Tartrazine exhibit pseudo-first-order kinetics for adsorption over Hen Feather. A higher molecular weight of Amaranth (Molecular Weight = 604.48) and Tartrazine (Molecular Weight = 534.4) may be the possible reason for lower order in these cases.

In the case of systems following first order kinetics the values of rate constants presented in Table 11.7 follow the order Amaranth–Hen Feather > Tartrazine–Hen Feather, which may be due to more numbers of attachment sites ($-SO_3^-$) with Amaranth. Moreover, no significant

Table 11.6 Effect of adsorbent amount on rate of adsorption of Azo Dyes over Hen Feathers and half life of the process at different temperatures.

Dye	Amount of Adsorbent (g)	Rate Constant (per hour)			Half Life (Hour)		
		30°C	40°C	50°C	30°C	40°C	50°C
Amaranth	0.005	0.014	0.017	0.019	48.28	41.35	35.88
	0.01	0.031	0.034	0.038	22.33	20.46	18.35
	0.025	0.042	0.045	0.054	16.46	15.32	12.76
	0.05	0.049	0.054	0.066	14.12	12.76	10.50
Tartrazine	0.005	0.019	0.021	0.027	37.41	32.62	25.86
	0.01	0.034	0.037	0.041	20.31	18.83	17.00
	0.015	0.048	0.052	0.062	14.32	13.29	11.19
	0.02	0.087	0.107	0.263	7.92	6.47	2.63

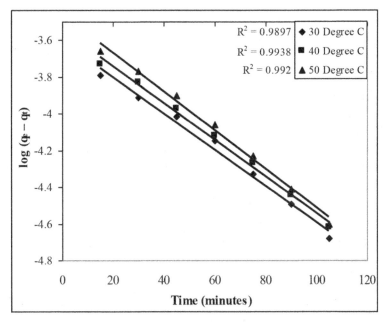

Figure 11.16 Lagergren's plot of time versus log (q_e –q_t) for Amaranth adsorption on Hen Feathers at different temperatures.

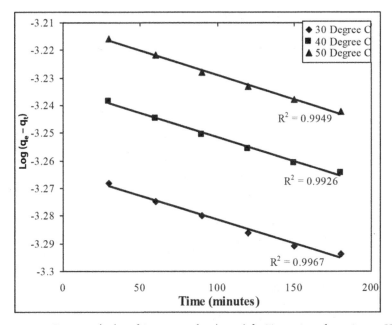

Figure 11.17 Lagergren's plot of time versus log (q_e –q_t) for Tartrazine adsorption on Hen Feathers at different temperatures.

variation is observed in the values of the rate constants with the change in temperature.

11.6.3.3   Elucidation of Reaction Mechanism

In order to ascertain the operative mechanisms of the undertaken systems as particle diffusion or film diffusion, B_t versus time graphs are plotted. In the cases of adsorption of Azo dyes Amaranth and Tartrazine, linearity of the line is deviated at all the temperatures, thereby suggesting involvement of a film diffusion mechanism as the rate-determining step. Figure 11.18 represents typical time versus B_t plot in the lower as well as higher concentration ranges for the Amaranth–Hen Feather system, while

Table 11.7 Rate constant values for the removal of Azo Dyes at different temperatures.

Adsorbate	Adsorbent	Rate Constant			Unit of Rate Constant
		30°C	**40°C**	**50°C**	
Amaranth	Hen Feather	2.2×10^{-2}	2.3×10^{-2}	2.4×10^{-2}	min^{-1}
Tartrazine	Hen Feather	4.1×10^{-4}	2.4×10^{-4}	4.1×10^{-4}	sec^{-1}

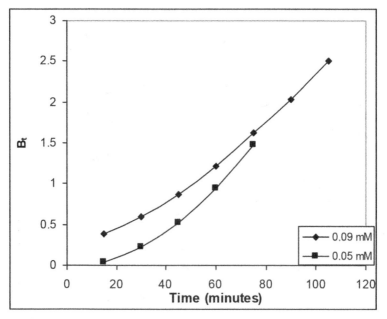

Figure 11.18 Correlation of time versus B_t for Amaranth–Hen Feather system at different concentrations at 30°C.

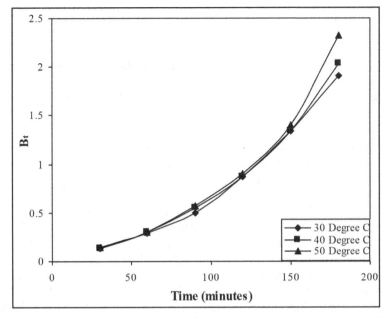

Figure 11.19 Correlation of time versus B_t for Tartrazine–Hen Feather system at different temperatures.

Figure 11.19 present a graph of Tartrazine–Hen Feather adsorption at all the temperatures.

11.7   Conclusions

The main focus of this research was to develop an economic, fast and versatile method for the removal of hazardous Azo dyes from wastewater and to utillize waste material Hen Feather as potential adsorbent. The research presented in this chapter clearly establishes that all Hen Feather can be successfully employed as adsorbent for the removal of toxic Azo dyes from wastewaters. Since this waste material is available in plenty and its disposal has always been a problem, its utilization for such a noble cause is indeed a thriving attempt for the benefit of mankind. Results presented in this chapter evidently indicate that detailed and systematic studies have been made throughout the development of the dye eradication process.

Thus, it is very safe to conclude that Hen Feather act as an effective and eco-friendly adsorbent for the removal of highly hazardous organic pollutants–Azo dyes from wastewater and the adsorption processes developed during the course of present investigations are efficient, eco-friendly and economic.

References

1. B. M. Linde, *Water on Earth*, Bench Education Co. New York (2005).
2. L. Hubbell, *The Earth*, National Science Teachers Assoc., National Aeronautics and Space Administration, USA (1964).
3. D. McNab and N. D. W. Jig, *Earth, Water, Air and Fire: Studies in Canadian Ethno History* (1998).
4. S. D. Kenneth and J. A. Day, *Water: The Mirror of Science*, Garden City, New York. Double Day, Anchor Book, p. 133 (1961).
5. T. A. Ternes, Occurrence of rugs in German sewage treatment plants and rivers, *Water Research* 32 (1998) 3245–3252.
6. B. Halling-Sorensen, S. N. Nielsen, P. F. Lanzky, F. Ingerslev, H. C. Holten Outshoot, and S. E. Jorgensen, Occurrence, fate and effects of pharmaceutical substances in the environment – A review. *Chemosphere* 36 (1998) 357–393.
7. D. W. Kolpin, E. T. Furlong, M. T. Meyer, E. M. Thurman, S. D. Zaugg, L. B. Barber, and H. T. Buxton, Pharmaceuticals, hormones, and other organic wastewater contaminants in U. S. streams, 1999–2000: A national reconnaissance. *Environmental Science and Technology* 36 (2002) 1202–1211.
8. R. L. Droste, Theory and *Practice of Water and Wastewater*, John Wiley and Sons, Inc., New York (1977).
9. L. B. Franklin, *Wastewater Engineering: Treatment, Disposal and Reuse*, McGraw Hill, Inc., New York (1991).
10. U. N. Joshi and Y. P. Luthra, An overview of heavy metals: Impact and remediation. *Current Science* 78 (2000) 773–775.
11. A. E. Clarke and R. Anliker, Organic dyes and pigments, in: *Handbook of Environmental Chemistry*, Springer-Verlag, New York (1980).
12. A. Downham and P. Collins, Colouring our foods in the last and next millennium. *International Journal of Food Science and Technology*, 35 (2000) 5–22.
13. H. Zollinger, *Color Chemistry: Synthesis, Properties and Applications of Organic Dyes and Pigments*, 2nd revised ed., VCH, New York (1991).
14. S. V. Mohan, P. Sailaja, M. Srimurali, and J. Karthikeyan, Color removal of monoazo acid dye from aqueous solution by adsorption and chemical coagulation. *Environmental Engineering and Policy* 1 (1999) 149–154.
15. S. H. Lin, Adsorption of disperse dye by various adsorbents. *Journal Chemical Technology and Biotechnology* 58 (1993) 159–163.
16. J. E. Kilduff and C. J. King, Effect of carbon adsorbent surface properties on the uptake and solvent regeneration of phenol. *Industrial and Engineering Chemistry Research* 36 (1997) 1603–1613.
17. A. Mittal, L. Kurup, and J. Mittal, Freundlich and Langmuir adsorption isotherms and kinetics for the removal of Tartrazine from aqueous solutions using hen feathers. *Journal of Hazardous Materials* 146(1–2) (2007) 243–248.
18. A. Mittal, J. Mittal, and L. Kurup, Utilization of hen feathers for the adsorption of a hazardous dye, Indigo Carmine from its simulated effluent. *Journal Environmental Protection Science* 1 (2007) 92–100.

19. V. K. Gupta, A. Mittal, L. Kurup, and J. Mittal, Adsorption of a hazardous dye, Erythrosine, over hen feathers. *Journal of Colloid & Interface Science* 304 (2006) 52–57.

20. A. Mittal, Adsorption kinetics of removal of a toxic dye, Malachite Green, from wastewater by using hen feathers. *Journal of Hazardous Materials* 133(1–3) (2006) 196–202.

21. A. Mittal, Use of hen feathers as potential adsorbent for the removal of a hazardous dye, Brilliant Blue FCF, from waste water. *Journal of Hazardous Materials* 128(2–3) (2006) 233–239.

22. A. Mittal, Removal of the dye Amaranth from waste water using hen feathers as potential adsorbent. *Electronic Journal of Environmental, Agricultural and Food Chemistry* 5(2) (2006) 1296–1305.

23. J. Mittal, V. Thakur, H. Vardhan, and A. Mittal. Batch removal of hazardous azo dye Bismark Brown R using waste material hen feather. *Ecological Engineering* 60 (2013)

24. A. Mittal, V. Thakur, J. Mittal, and H. Vardhan, Process development for the removal of hazardous anionic azo dye Congo-red from wastewater by using hen feather as potential adsorbent. *Desalination and Water Treatment* 10.1080/19443994.2013.785030.

25. A. Mittal, V. Thakur, and V. Gajbe, Adsorptive removal of toxic azo dye Amido Black 10B by hen feather. *Environmental Science and Pollution Research* 20 (2013) 260–269.

26. A. Mittal, V. Thakur, and V. Gajbe, Evaluation of adsorption characteristics of an anionic azo dye Brilliant Yellow onto hen feathers in aqueous solutions. *Environmental Science & Pollution Res.* 19 (2012) 2438–2447.

27. W. Jiangning and W. Tingwei, Ozonation of aqueous azo dye in a semi-batch reactor. *Water Research* 35 (2000) 1093–1099.

28. U. Pagga and D. Brown, The degradation of dyestuffs. Part II. Behavior of dyestuffs in aerobic biodegradation tests. *Chemosphere* 15 (1986) 479–491.

29. K. R. Ramakrishna and T. Viraraghavan, Dye removal using low cost adsorbents. *Water Science and Technology* 36 (1997) 189–196.

30. M. F. Boeniger, *Carcinogenity of Azo Dyes Derived from Benzidine*, Department of Health and Human Services (NIOSH), Cincinnati, USA (1980).

31. *Kirk-Othomer Encyclopedia, Chemical Technology*, 8th ed. (1994) 547–672.

32. S. Padmavathy, S. Sardhya, K. Swaminathan, and V. V. Subrahmanyam, Aerobic decolourisation of reactive dyes in presence of various cosubstratres. *Chemical and Biochemical Engineering* 17 (2003) 147–151.

33. K. T. Chung and C. E. Cerniglia, Mutagenicity of azo dyes: Structure-activity relationships. *Mutation Research* 277 (1992) 201–220.

34. C. Desiderio, C. Marra and S. Fanali, Quantitative analysis of synthetic dyes in lipstick by micellar electrokinetic capillary chromatography. *Electrophoresis* 19 (1998) 1478–1483.

35. V. L. Bagirova and L. I. Mitkina, Determination of Tartrazine in drugs. *Pharmaceutical Chemistry Journal* 37 (2003) 558–559.

36. S. D. Lockey Sr., Hypersensitivity to Tartrazine (FD & C Yellow No. 5) and other dyes and additives present in foods and pharmaceutical products. *Annals of Allergy* 38 (1977) 206–210.

37. F. Ishikawa, M. Oishi, K. Kimura, A. Yasui, and K. Saito, Determination of synthetic food dyes in food by capillary electrophoresis. *Journal of the Food Hygienic Society of Japan* 45 (2004) 150–155.

38. P. Rao, R. V. Bhat, R. V. Sudershan, T. P. Krishna, and N. Naidu, Exposure assessment to synthetic food colours of a selected population in Hyderabad, India. *Food Additives and Contaminants* 21 (2004) 415–421.

39. S. S. Chou, Y. H. Lin, C. C. Cheng, and D. F. Hwang, Determination of synthetic colors in soft drinks and confectioneries by micellar electrokinetic capillary chromatography. *Journal of Food Science* 67 (2002) 1314–1318.

40. M. Perez-Urquiza and J. L. Beltran, Determination of dyes in foodstuffs by capillary zone electrophoresis. *Journal of Chromatography A* 898 (2000) 271–275.

41. D. D. Stevenson, R. A. Simon, W. R. Lumry, and D. A. Mathison, Adverse reactions to Tartrazine. *Journal of Allergy and Clinical Immunology* 78 (1986) 182–191.

42. T. Itoh, Y. Yasuda, and Y. Totani, Tartrazine-induced asthma without aspirin intolerance. *Japanese Journal of Allergology* 32 (1983) 1005–1009.

43. V. M. Verallo, Dermatologic uses and adverse reactions to new drugs. *Philippine Journal of Internal Medicine* 20 (1982) 243–246.

44. A. P. Kaplan, Drug-induced skin disease. *Journal of Allergy and Clinical Immunology* 74 (1984) 573–579.

45. S. D. Gettings, D. L. Blaszcak, M. T. Roddy, A. S. Curry, and G. N. McEwan Jr., Evaluation of the cumulative (repeated application) eye irritation and corneal staining potential of FD and C Yellow No. 5, FD and C Blue No. 1 Aluminium Lake. *Food and Chemical Toxicology* 30 (1992) 1051–1055.

46. J. Meadway, Lung disease due to inhaled Tartrazine. *British Journal of Diseases of the Chest* 73 (1979) 420–421.

47. S. J. Pearce, Adverse effects of drugs on the lung. *Adverse Drug Reaction Bulletin* 94 (1982) 344–347.

48. A. Ellsworth and D. Gross, Desipramine not Tartrazine induced drug eruption. *Drug Intelligence and Clinical Pharmacy* 21 (1987) 510–512.

49. N. Sankaranarayanan and M. S. S. Murthy, Testing of some permitted food colours for the induction of gene conversion in diploid yeast. *Mutation Research* 67 (1979) 309–314.

50. K. T. Chung, The significance of azo-reduction in the mutagenesis and carcinogenesis of azo dyes. *Mutation Research* 114 (1983) 269–281.

51. M. Stefanidou, G. Alevisopoulos, A. Chatziioannou, and A. Koutselinis, Assessing food additive toxicity using a cell model. *Veterinary and Human Toxicology* 45 (2003) 103–105.

52. T. J. Sobotka, R. E. Brodie, and S. L. Spaid, Tartrazine and the developing nervous system of rats. *Journal of Toxicology and Environmental Health* 2 (1977) 1211–1220.

53. B. H. Ershoff, Effects of diet on growth and survival of rats fed toxic levels of Tartrazine (FD&C Yellow No. 5) and Sunset Yellow FCF (FD&C Yellow No. 6). *Journal of Nutrition* 107 (1977) 822–828.

54. A. K. Giri, S. K. Das, G. Talukder, and A. Sharma, Sister chromatid exchange and chromosome aberrations induced by Curcumin and Tartrazine on mammalian cells in-vivo. *Cytobios* 62 (1990) 111–117.

55. R. A. Simon, Adverse reactions to food and drug additives. *Immunology and Allergy Clinics of North America* 16 (1996) 137–176.

56. C. C. Blanco, A. M. G. Campana, and F. A. Barrero, Derivative spectrophotometric resolution of mixtures of the food colourants Tartrazine, Amaranth and Curcumin in a micellar medium. *Talanta* 43 (1996) 1019–1027.

57. D. Talmage, *Biologic Markers in Immunotoxicology*. The National Academies Press, Washington D. C., USA (1992) p. 37.

58. E. Rindle and W. J. Troll, Analysis of sulphonated and sulphonated-azo dyes in water samples by capillary electrophoresis. *National Cancer Institute* 55 (1975) 181–187.

59. K. T. Chung, S. E. Stevens Jr., and C. E. Cerniglia, The reduction of azo dyes by the intestinal microflora. *Critical Reviews in Microbiology* 18 (1992) 175–190.

60. K. T. Chung and S. E. Stevens Jr., Degradation of azo dyes by environmental microorganisms and helminthes. *Environmental Toxicology and Chemistry* 12 (1993) 2121–2132.

61. H. Zollinger, *Colour Chemistry: Synthesis, Properties and Applications for Organic Dyes and Pigments*. VCH Publishers, Inc., New York, USA (1987).

62. W. G. Levine, Metabolism of azo dyes: Implication for detoxification and activation. *Drug Metabolism Reviews* 23 (1991) 253–309.

63. P. P. Purslow and J. F. V. Vincent, Mechanical properties of primary feathers from the pigeon. *Journal of Experimental Biology* 72 (1978) 251–260.

64. N. Reddy and Y. Yang, Structure and properties of chicken feather barbs as natural protein fibers. *Journal of Polymers and the Environment* 15 (2007) 81–87.

65. S. Leeson and T. Walsh, Feathering in commercial poultry II: Factors influencing feather growth and feather loss. *World's Poultry Science Journal* 60 (2004) 52–63.

66. S. Al-Asheh, F. Banat, and D. Al-Rousan, Beneficial reuse of chicken feathers in removal of heavy metals from wastewater. *Journal of Cleaner Production* 11 (2003) 321–326.

67. F. Banat, S. Al-Asheh, and D. Al-Rousan, Comparison between different keratin-composed biosorbents for the removal of heavy metal ions from aqueous solutions. *Adsorption Science and Technology* 20 (2002) 393–416.

68. S. Al-Asheh, F. Banat, and D. Al-Rousan, Adsorption of copper, zinc and nickel ions from single and binary metal ion mixtures on to chicken feathers. *Adsorption Science and Technology* 20 (2002) 849–864

69. M. C. Teixeira and V. S. T. Ciminelli, Development of a biosorbent for arsenite: Structural modeling based on x-ray spectroscopy. *Environmental Science and Technology* 39(3) (2005) 895–900.
70. C. Yang, L. Guan, Y. Zhao, Z. Su, and T. Cai, Adsorption of Zn (II) on TA-modified feather. *Lizi Jiaohuan Yu Xifu/Ion Exchange and Adsorption* 23(3) (2007) 259–266.
71. A. I. Vogel, *A Text Book of Quantitative Inorganic Analysis.* London Longmans, Green & Co. Ltd., 1939.
72. F. A. DiGiano, W. J. Weber Jr., Sorption kinetics in finite bath experiments. *Journal of Water Pollution Control Federation* 45 (1973) 713–725.
73. F. Helferrrich, *Ion Exchange.* Mc Graw-Hill, New York, 1962.
74. M. F. Carvalho, A. F. Duque, I. C. Goncalves, and P. M. L. Castro, Adsorption of fluorobenzene onto granular activated carbon: Isotherm and bioavailability studies. *Bioresource Technology* 98 (2007) 3424–3430.
75. V. K. Gupta, R. Jain, and S. Varshney, Removal of Reactofix Golden Yellow 3 RFN from aqueous solution using wheat husk – An agricultural waste. *Journal of Hazardous Materials* 142 (2007) 443–448.
76. D. Kavitha and C. Namasivayam, Experimental and kinetic studies on Methylene Blue adsorption by coir pith carbon. *Bioresource Technology* 98 (2007) 14–21.
77. S. J. Allen, G. McKay, and J. F. Porter, Adsorption isotherm models for basic dye adsorption by peat in single and binary component systems. *Journal of Colloid Interface Science* 280 (2004) 322–333.
78. D. Kavitha and C. Namasivayam, Recycling coir pith, an agricultural solid waste, for the removal of Procion Orange from wastewater. *Dyes and Pigments* 74 (2007) 237–248.
79. M. M. Saeed, M. S. Hasany, and M. Ahmed, Adsorption and thermodynamic characteristics of Hg (II)-SCN complex onto polyurethane. *Talanta* 50 (1999) 625–634.
80. S. Q. Memon, N. Memon, A. R. Solangi, and J. Memon, Sawdust: A green and economical sorbent for thallium removal. *Chemical Engineering Journal* 140 (2008) 235–240.
81. V. Gopal and K. P. Elango, Equilibrium, kinetic and thermodynamic studies of adsorption of fluoride onto plaster of Paris. *Journal of Hazardous Materials* 141 (2007) 98–105.
82. T. W. Weber and R. K. Chakravorti, Pore and solid diffusion models for fixed bed adsorbers. *Journal of American Institute of Chemical Engineering* 20 (1974) 228–238.
83. I. A. Sengil, M. Ozacar, and H. Turkmenler, Kinetic and isotherm studies of Cu (II) biosorption onto valonia tannin resin. *Journal of Hazardous Materials* 162 (2009) 1046–1052.
84. M. A. Abdullah, L. Chiang, and M. Nadeem, Comparative evaluation of adsorption kinetics and isotherms of a natural product removal by amberlite polymeric adsorbents. *Chemical Engineering Journal* 146 (2009) 370–376.

85. V. C. Srivastava, M. M. Swamy, I. D. Mall, B. Prasad, and I. M. Mishra, Adsorptive removal of phenol by bagasse fly ash and activated carbon: Equilibrium, kinetics and thermodynamics. *Colloid Surface A: Physicochemical and Engineering Aspects* 272 (2006) 89–104.

86. Y. S. Ho and G. McKay, The sorption of lead (II) ions on peat. *Water Research* 33 (1999) 578–584.

87. G. E. Boyd, A. W. Adamson, and L. S. Meyers, The exchange adsorption of ions from aqueous solution by organic zeolites II. Kinetics. *Journal of the American Chemical Society* 69 (1947) 2836–2848.

88. D. Reichenberg, Properties of ion exchange resins in relation to their structure. III. Kinetics of exchange. *Journal of the American Chemical Society* 75 (1953) 589–597.

89. J. Crank, *The Mathematics of Diffusion*. Clarenden Press, Oxford (1956).

90. E. Glueckauf, *Ion Exchange and Its Application*, edited by Society of Chemical Industry, London (1955).

91. I. P. Saraswat, S. K. Srivastava, and A. K. Sharma, Kinetics of ion exchange of some complex cations on chromium ferro cyanide gel. *Canadian Journal of Chemistry* 54 (1979) 1214–1217.

92. H. J. Fornwalt and R. A. Hutchins, Purifying liquids with activated carbon. *Chemical Engineering Journal* 73 (1966) 179–184.

_ What happens to activated carbon
after it is all used up.
- Would answer the question of
what happens to adsorbents.

Index

459

Also of Interest

Check out these other related titles from Scrivener Publishing

Biogas Production, Edited by Ackmez Mudhoo, ISBN 9781118062852. This volume covers the most cutting-edge pretreatment processes being used and studied today for the production of biogas during anaerobic digestion processes using different feedstocks, in the most efficient and economical methods possible. *NOW AVAILABLE!*

Bioremediation and Sustainability: Research and Applications, Edited by Romeela Mohee and Ackmez Mudhoo, ISBN 9781118062845. Bioremediation and Sustainability is an up-to-date and comprehensive treatment of research and applications for some of the most important low-cost, "green," emerging technologies in chemical and environmental engineering. *NOW AVAILABLE!*

Green Chemistry and Environmental Remediation, Edited by Rashmi Sanghi and Vandana Singh, ISBN 9780470943083. Presents high quality research papers as well as in depth review articles on the new emerging green face of multidimensional environmental chemistry. *NOW AVAILABLE!*

Bioremediation of Petroleum and Petroleum Products, by James Speight and Karuna Arjoon, ISBN 9780470938492. With petroleum-related spills, explosions, and health issues in the headlines almost every day, the issue of remediation of petroleum and petroleum products is taking on increasing importance, for the survival of our environment, our planet, and our future. This book is the first of its kind to explore this difficult issue from an engineering and scientific point of view and offer solutions and reasonable courses of action. *NOW AVAILABLE!*

A Guide to the Economic Removal of Metal from Aqueous Solutions, by Yogesh Sharma, ISBN 9781118137154. Presents data and practical solutions as to how to extract toxic metals from water yielding higher purity levels at signifcant lower costs. *NOW AVAILABLE!*